Cellular Calcium

The Practical Approach Series

SERIES EDITORS

D. RICKWOOD
Department of Biology, University of Essex
Wivenhoe Park, Colchester, Essex CO4 3SQ, UK

B. D. HAMES
Department of Biochemistry and Molecular Biology, University of Leeds
Leeds LS2 9JT, UK

Affinity Chromatography
Anaerobic Microbiology
Animal Cell Culture
Animal Virus Pathogenesis
Antibodies I and II
Biochemical Toxicology
Biological Membranes
Biosensors
Carbohydrate Analysis
Cell Growth and Division
Cellular Calcium
Cellular Neurobiology
Cytokines
Centrifugation (2nd edition)
Clinical Immunology
Computers in Microbiology
Crystallization of Proteins and Nucleic Acids
Cytokines
Directed Mutagenesis
DNA Cloning I, II, and III
Drosophila
Electron Microscopy in Biology
Electron Microscopy in Molecular Biology
Essential Molecular Biology I and II
Fermentation
Flow Cytometry
Gel Electrophoresis of Nucleic Acids (2nd edition)
Gel Electrophoresis of Proteins (2nd edition)
Genome Analysis
HPLC of Small Molecules
HPLC of Macromolecules
Human Cytogenetics
Human Genetic Diseases
Immobilised Cells and Enzymes
Iodinated Density Gradient Media
Light Microscopy in Biology
Liposomes
Lymphocytes
Lymphokines and Interferons
Mammalian Cell Biotechnology
Mammalian Development
Medical Bacteriology
Medical Mycology

Microcomputers in Biology
Microcomputers in Physiology
Mitochondria
Molecular Neurobiology
Molecular Plant Pathology I and II
Mutagenicity Testing
Neurochemistry
Nucleic Acid and Protein Sequence Analysis
Nucleic Acids Hybridisation
Nucleic Acids Sequencing
Oligonucleotide Synthesis
Oligonucleotides and Analogues
PCR
Peptide Hormone Action
Peptide Hormone Secretion
Photosynthesis: Energy Transduction
Plant Cell Culture
Plant Molecular Biology
Plasmids
Postimplantation Mammalian Embryos
Prostaglandins and Related Substances
Protein Architecture
Protein Function
Protein Purification Applications
Protein Purification Methods
Protein Sequencing
Protein Structure
Proteolytic Enzymes
Radioisotopes in Biology
Receptor Biochemistry
Receptor–Effector Coupling
Ribosomes and Protein Synthesis
Solid Phase Peptide Synthesis
Spectrophotometry and Spectrofluorimetry
Steroid Hormones
Teratocarcinomas and Embryonic Stem Cells
Transcription and Translation
Virology
Yeast

Cellular Calcium

A Practical Approach

Edited by
JAMES G. McCORMACK

Department of Pharmacology,
Syntex Research Centre, Research Avenue South,
Heriot-Watt University Research Park
Riccarton, Edinburgh, UK

and

P. H. COBBOLD

Department of Human Anatomy and Cell Biology,
University of Liverpool, UK

IRL PRESS
at
OXFORD UNIVERSITY PRESS
Oxford New York Tokyo

Oxford University Press, Walton Street, Oxford OX2 6DP
Oxford New York Toronto
Delhi Bombay Calcutta Madras Karachi
Petaling Jaya Singapore Hong Kong Tokyo
Nairobi Dar es Salaam Cape Town
Melbourne Auckland
and associated companies in
Berlin Ibadan

Oxford is a trade mark of Oxford University Press

Published in the United States
by Oxford University Press, New York

A catalogue record for this book is available from the British Library

Library of Congress Cataloging in Publication Data
Cellular calcium: a practical approach/edited by James G. McCormack and P. H. Cobbold.
(The Practical approach series)
Includes bibliographical references and index.
1. Calcium—Physiological effect—Research—Methodology.
2. Calcium—Analysis. 3. Second messengers (Biochemistry)—Research—Methodology. 4. Cellular signal transduction—Research—Methodology. I. Series.
QP535.C2C46 1991 599'.087—dc20 91–10924
ISBN 0–19–963131–X (hbk)
ISBN 0–19–963130–1 (pbk)

Typeset by Cambrian Typesetters, Frimley, Surrey
Printed in Great Britain by
Information Press Ltd, Eynsham, Oxford

Preface

THE key second-messenger role that the intracellular concentration of calcium ions plays in signal transduction processes in many eukaryotic cell systems is now widely known and appreciated. For instance, one can easily think of vital life events such as muscle and non-muscle motility and both exocrine and endocrine secretion. Or alternatively, one could think of the enormous impact which drugs such as the Ca^{2+}-antagonists have had on health care. The study of this important area thus concerns very many life scientists from many different disciplines who are investigating many different aspects of normal cellular physiology and also pathophysiology. The range of techniques available to study this key parameter has been considerably expanded and refined over the last few years, as has the knowledge of the molecular and functional details of the transport proteins responsible for controlling Ca^{2+} concentrations, and of the other intracellular Ca^{2+}-binding proteins which sense and respond to changes in Ca^{2+} concentration and hence lead to the biological effects which are the end-points of the signal-transduction processes. The methods associated with the study of these major topics thus form the subject matter of this book, and now would appear to be a timely point at which to try and gather together this information concerning the methods used to investigate the signal transduction processes revolving around cellular calcium.

The book is initially concerned with the various methods used to measure the concentration of ionic Ca^{2+} in biological systems and within cells and thus begins with chapters which describe, sequentially, the use of fluorescent Ca^{2+}-indicators, Ca^{2+}-sensitive photoproteins, Ca^{2+}-selective electrodes, metallochromic Ca^{2+}-indicators, and finally the use of an NMR probe. These chapters also deal with the various means of accessing the inside of cells with the various indicators, and also the limitations, pitfalls, and problems that can arise. X-ray probe microanalysis is described next, which rather measures total calcium content but can be used to do this in very small regions of the cell in snap-frozen tissue or cell sections derived directly from *in vivo* circumstances and which can then be correlated with likely free Ca^{2+} concentrations.

There then follows a chapter which is chiefly focused on methods for controlling and calculating free Ca^{2+} and the use of buffer systems, and which includes descriptions of the newly developed photo-activatable compounds. A use of these agents in studying the kinetics of Ca^{2+}-troponin C interactions is described in part of the next chapter which is the first of three on cytosolic protein Ca^{2+}-ligands, the other two being one on calmodulin which details some of the molecular biological approaches now being used in this area, and one on the annexins, an intriguing family of proteins which can constitute very significant proportions of the total protein content of many cells, but whose functions remain largely unknown.

The next part of the book contains chapters describing the methods associated with studies on the major membrane-bound Ca^{2+}-transporting proteins in mammalian cells, and begins at the plasma membrane with chapters on Ca^{2+}-channels, the Na^{+}–Ca^{2+} exchanger, and the ATP-driven Ca^{2+} pump. There then follows a chapter on reticulum membrane Ca^{2+}-transport proteins, and then one on Ca^{2+}-transport across the mitochondrial inner membrane. The final chapter deals with the rather special case of cellular calcium in plants and with the particular problems encountered in studying this topic.

We have tried to be as comprehensive and up to date as possible in covering this important topic, and all of our contributors are actively engaged in using the methods that they have described, thus giving the best possible insight into the techniques, and any problems likely to be encountered, from a practical point of view. Inevitably in such a wide topic area there are a few omissions; for instance, we decided not to have specific chapters on ^{45}Ca fluxes or on atomic absorption as we felt that these were more established techniques, and also appeared to some extent throughout the rest of the chapters. Similarly, the relationship between Ca^{2+} and phosphoinositide metabolism is alluded to throughout, and has already been dealt with in some detail elsewhere in this series.

We are very grateful to our contributors for helping to accomplish this project and in particular for adhering to the laboratory-based spirit of this excellent series. Special thanks are also due to the staff at OUP for their help, persistence and encouragement throughout. We hope that this book may help and inspire others to carry out experiments which will further advance our knowledge of this fascinating and important subject.

Edinburgh
Liverpool
July 1991

Jim McCormack
Peter Cobbold

Contents

11. Quantification of the affinity of drugs acting at the calcium channel 267

Barry A. Kenny, Andrew T. Kilpatrick, and Michael Spedding

12. Sodium–calcium exchange activity in plasma membrane vesicles 283

John P. Reeves

Appendix

Index

Contributors

C.C. ASHLEY
University Laboratory of Physiology, Parks Road, Oxford OX1 3PT, UK.

INDRANI C. BAGCHI
Department of Cell Biology, Baylor College of Medicine, 1 Baylor Plaza, Texas Medical Center, Houston, Texas 77030, USA.

CATHERINE M. BOUSTEAD
Department of Biochemistrry and Molecular Biology, University of Leeds, Leeds LS2 9JT, UK.

M. R. BOYETT
Department of Physiology, University of Leeds, Leeds LS2 9JT, UK.

DALE A. CALLAHAM
Department of Botany, University of Massachusetts, Amherst, MA 01003, USA.

ERNESTO CARAFOLI
Laboratory of Biochemistry, Swiss Federal Institute of Technology (ETH), 8092 Zurich, Switzerland.

P. H. COBBOLD
Department of Human Anatomy and Cell Biology, University of Liverpool, PO Box 147, Liverpool L69 3BX, UK.

MARTIN CROMPTON
Department of Biochemistry, University College London, Gower Street, London WC1E 6BT, UK.

FERNANDO DELAVILLE
Department of Pathology and Cell Biology, Thomas Jefferson University, Philadelphia, PA 19107, USA.

ALEXANDRE FABIATO
Department of Physiology, Box 551, Medical College of Virginia, Richmond, VA 23298–0551, USA.

C. H. FRY
Sherrington School of Physiology, St Thomas's Hospital Medical School, Lambeth Palace Road, London SE1 7EH, UK.

P. J. GRIFFITHS
University Laboratory of Physiology, Parks Road, Oxford OX1 3PT, UK.

SUSAN L. HAMILTON
Department of Molecular Physiology and Biophysics, Baylor College of Medicine, Houston, Texas 77030, USA.

PETER K. HEPLER
Department of Botany, University of Massachusetts, Amherst, MA 01003, USA.

M. HUNTER
Department of Physiology, University of Leeds, Leeds LS2 9JT, UK.

GERRIT ISENBERG
Department of Physiology, University of Cologne, Robert-Koch-Str. 39, 5000 Köln 41, Germany.

BARRY A. KENNY
Department of Pharmacology, Syntex Research Centre, Research Park, Riccarton, Edinburgh EH14 4AP, UK.

ANDREW T. KILPATRICK
Department of Pharmacology, Syntex Research Centre, Research Park, Riccarton, Edinburgh EH14 4AP, UK.

T. J. LEA
University Laboratory of Physiology, Parks Road, Oxford OX1 3PT, UK.

J. A. C. LEE
Department of Human Anatomy and Cell Biology, University of Liverpool, PO Box 147, Liverpool L69 3BX, UK.

JAMES G. McCORMACK
Department of Pharmacology, Syntex Research Centre, Research Avenue South, Heriot-Watt University Research Park, Riccarton, Edinburgh EH14 4AP, UK.

ANTHONY R. MEANS
Department of Cell Biology, Baylor College of Medicine, 1 Baylor Plaza, Texas Medical Centre, Houston, Texas 77030, USA.

JAMES C. METCALFE
Department of Biochemistry, University of Cambridge, Tennis Court Road, Cambridge CB2 1QW, UK.

I. P. MULLIGAN
University Laboratory of Physiology, Parks Road, Oxford OX1 3PT, UK.

C. H. ORCHARD
Department of Physiology, University of Leeds, Leeds LS2 9JT, UK.

R. E. PALMER
University Laboratory of Physiology, Parks Road, Oxford OX1 3PT, UK.

COLIN D. RASMUSSEN
Department of Cell Biology, Baylor College of Medicine, 1 Baylor Plaza, Texas Medical Centre, Houston, Texas 77030, USA.

JOHN P. REEVES
Roche Institute of Molecular Biology, Roche Research Center, Nutley, NJ 07110, USA.

SARAH J. SIMNETT
University Laboratory of Physiology, Parks Road, Oxford OX1 3PT, UK.

GERRY A. SMITH
Department of Biochemistry, University of Cambridge, Tennis Court Road, Cambridge CB2 1QW, UK.

MICHAEL SPEDDING
Department of Pharmacology, Syntex Research Centre, Research Park, Riccarton, Edinburgh EH14 4AP, UK.

CHARLOTTE A. TATE
Department of Pharmacology, Institute for Cardiovascular Studies, University of Houston, Houston, Texas 77204, USA.

ANDREW P. THOMAS
Department of Pathology and Cell Biology, Thomas Jefferson University, Philadelphia, PA 19107, USA.

M. V. THOMAS
Cairn Research Ltd, Cairn House, Newnham, Sittingbourne, Kent ME9 0LQ, UK.

MARK F. A. VANBERKUM
Department of Cell Biology, Baylor College of Medicine, 1 Baylor Plaza, Texas Medical Center, Houston, Texas 77030, USA.

JOHN H. WALKER
Department of Biochemistry and Molecular Biology, University of Leeds, Leeds LS2 9JT, UK.

MARIA FIORA WENDT-GELLITELLI
Institute of Physiology II, University of Tübingen, Gmelinstr. 5, 7400 Tübingen, Germany.

Abbreviations

AM	acetoxymethyl
AME	acetoxymethyl ester
BAPTA	1,2 bis (*o*-amino phenoxy)ethane-*N*,*N*,*N*′,*N*′-tetraacetic acid
BSA	bovine serum albumin
CCD	charge coupled device
CTnC	cardiac troponin C
DEAE	diethylaminoethyl
DHP	dihydropyridine
DIC	differential interference contrast
DMSO	dimethyl sulfoxide
DTPA	diethylenetriamine pentaacetic acid
DTT	dithiothreitol
EDTA	ethylene diaminetetraacetic acid
EGTA	ethyleneglycol-bis-(β-aminoethyl ether) *N*,*N*,*N*′,*N*′-tetraacetic acid
EPMA	electron probe microanalysis
FID	free induction decay
FITC	fluorescein isothiocyanate
HPLC	high performance liquid chromatography
ICDH	isocitrate dehydrogenase
IOPP	iso-octylphenyl phosphate
IP_3	inositol 1,4,5-trisphosphate
MCP	microchannel plate
MLCK	myosin light chain kinase
OD	optical density
OGDH	2-oxoglutarate dehydrogenase
PDE	phosphodiesterase
PDH	pyruvate dehydrogenase
PEG	polyethylene glycol
PEI	polyethylenimine
PEP	phosphoenolpyruvate
PK	pyruvate kinase
PMSF	phenylmethylsulfonyl fluoride
POPC	palmitoyl-oleoyl-phosphatidylcholine
POPE	palmitoyl-oleoyl-phosphatidylethanolamine
POPS	palmitoyl-oleoyl-phosphatidylserine
PVC	polyvinyl chloride
PVP	polyvinyl pyrrolidone
PMT	photomultiplier tube
RF	radiofrequency

SDS	sodium dodecyl sulphate
SEM	standard error of the mean
SL	sacrolemma
SIT	silicon intensified target
SNR	signal-to-noise ratio
SR	sarcoplasmic reticulum
STnC	skeletal troponin C
THF	tetrahydrofuran
TnC	troponin C
TPEN	*N,N,N′,N′*-tetrakis (2-pyridylmethyl)-ethylenediamine
VSCC	voltage-sensitive Ca^{2+} channels

1

The use of fluorescent indicators for measurements of cytosolic-free calcium concentration in cell populations and single cells

ANDREW P. THOMAS and FERNANDO DELAVILLE

1. Introduction

The development of fluorescent indicators of free Ca^{2+} ion concentration which could be loaded into cells in a non-disruptive manner initiated a revolution in studies of cellular Ca^{2+} homeostasis. This revolution has been led primarily by R. Y. Tsien and co-workers, who first popularized quin-2 for studies of cytosolic-free Ca^{2+} ($[Ca^{2+}]_i$) and have since contributed a series of additional Ca^{2+}-sensitive dyes with a number of improved properties (1–5). Indicators for measurements of other ions, including Mg^{2+}, Na^+, K^+, and H^+ have also been developed, and new dyes continue to be introduced. The major use for which these dyes have been employed is in studies of the regulation of $[Ca^{2+}]_i$ by hormones and other agonists. Indeed, quin-2 and the subsequently developed Ca^{2+} indicators have facilitated the rapid and very substantial advances which have been made in elucidating the mechanisms involved in agonist-induced Ca^{2+} mobilization. Fluorescent Ca^{2+} indicators have also proved useful for studies of Ca^{2+} regulation in permeabilized cells and subcellular fractions. These dyes have found less acceptance for studies of electrically-triggered events in neuronal and certain muscle systems, due to limitations in temporal resolution. Some advances in this direction have, however, been made (see below). Overall, in terms of ease of use and calibration, fluorescent Ca^{2+} indicators offer the most straightforward and accessible of the available techniques for measuring $[Ca^{2+}]_i$. The objective of this chapter is to discuss principals of use and required instrumentation for studies using this approach and to describe, in general terms, optimal conditions for carrying out experiments and appropriate calibrations.

2. General concepts in the use of fluorescent Ca^{2+} indicators

2.1 Principles of operation

2.1.1 The use of fluorescent probes for biological measurements

Fluorescence is a molecular process in which emission of electromagnetic radiation results from the absorption of radiation from an extra-molecular source. The excited state of the molecule generally has a life-time in the order of 10^{-8} sec, and as such fluorescence does not persist once exposure to the excitation source is terminated. Fluorescent radiation has a longer wavelength (lower energy) than that of the absorbed radiation. For a more complete discussion of the nature of fluorescence, see ref. 6. For most biological applications, the useful range of wavelengths of electromagnetic radiation falls in the near ultraviolet and through the range of visible light. Each fluorescent molecular species has one or more optimal wavelengths for excitation, which usually coincide with absorption maxima, and a related optimum wavelength of emission. Under conditions of constant illumination and in a fixed chemical and physical environment, the intensity of fluorescence from a molecular species is proportional to its concentration (providing that the concentration is not so high that self-quenching occurs). Thus, fluorescence intensity can be used as a quantitative tool. However, unlike absorbance, which is dependent only on the path length of the sample, the relationship between fluorescence and concentration is very much dependent on the instrumentation used, necessitating the use of standards for absolute measurement of concentrations.

Fluorescent probes have been used in two major ways in cell biology. They have been used as indicators of the location and quantity of particular cellular structures (e.g. fluorescent antibodies against particular proteins). Alternatively, fluorescent molecules which are sensitive to specific aspects of their environment have been used to report the parameter of interest within cellular compartments and to allow continuous monitoring of how this parameter changes with various manipulations. Some of the first fluorescent probes of this type were designed to monitor pH, and the fluorescent Ca^{2+} indicators to be discussed here also fall into this classification. A third, intermediate, group is represented by probes which have been attached to specific protein sites to allow direct monitoring of conformational changes.

In order for a fluorescent probe to provide useful information about its environment, it is necessary that its fluorescence properties be altered in a suitable manner by the parameter to be measured. For most biological applications, any one or combination of the following three property changes are appropriate:

- a change in fluorescence yield (fluorescence output as a function of excitation intensity)
- a shift in the excitation spectrum
- a shift in the emission spectrum

Other properties such as the fluorescence lifetime of the probe, fluorescence depolarization or self-association can also be used, but these generally require more sophisticated instrumentation. Measurements of fluorescence intensity changes resulting from alterations in fluorescence yield at a single wavelength are the most straightforward to carry out. However, calibration of these signals requires some measure of the absolute amount of probe present, and in spatially resolved signals it is generally not possible to distinguish between fluorescence differences due to real variations in the parameter of interest and differences resulting from non-homogeneous dye distribution. For dyes which give shifts of excitation or emission spectra in response to the parameter to be measured, it is possible to overcome the problems mentioned above by carrying out measurements at two wavelengths. Wavelengths are selected to give signals which either change inversely in response to the variable being measured, or one of the two wavelengths is selected to be an 'iso-fluorescent' point, where fluorescence intensity does not change in response to changes in the parameter to be measured. Using the dual wavelength approach (either dual emission or dual excitation), it is possible to calculate the proportion of the dye in one of two forms (e.g. Ca^{2+}-bound vs. Ca^{2+}-free) from the ratio of fluorescence measured at the two wavelengths (see Section 6). This then becomes independent of dye distribution or concentration.

2.1.2 General characteristics of fluorescent Ca^{2+} indicators

The available fluorescent indicators for measuring Ca^{2+} have been modelled on the widely used Ca^{2+}-selective chelator, EGTA. EGTA has a selectivity for Ca^{2+} over Mg^{2+} of 5 to 6 orders of magnitude, and is well suited to buffering Ca^{2+} in the physiological range of 10^{-8}–10^{-6} M in the presence of physiological Mg^{2+} concentrations of 10^{-4}–10^{-3} M. *Figure 1* shows the structures of EGTA, quin-2 and fura-2 for comparison. While the tetracarboxylic acid Ca^{2+} binding site is largely unchanged in the fluorescent Ca^{2+} indicators, additional groups have been added which act as fluorescent reporters. Binding of Ca^{2+} withdraws the nitrogen lone pair electrons from the aromatic rings resulting in altered fluorescence properties (4). These fluorescence changes are used to calculate the proportion of dye in the Ca^{2+}-bound and Ca^{2+}-free form, and hence the concentration of free Ca^{2+} with which the indicator is in equilibrium. By modifying the aromatic nucleus to make it more or less electron withdrawing, the affinity of the binding site for Ca^{2+} can be decreased or increased, respectively. For optimal measurements of Ca^{2+}, the binding affinity of the indicator should ideally be in the same range as the

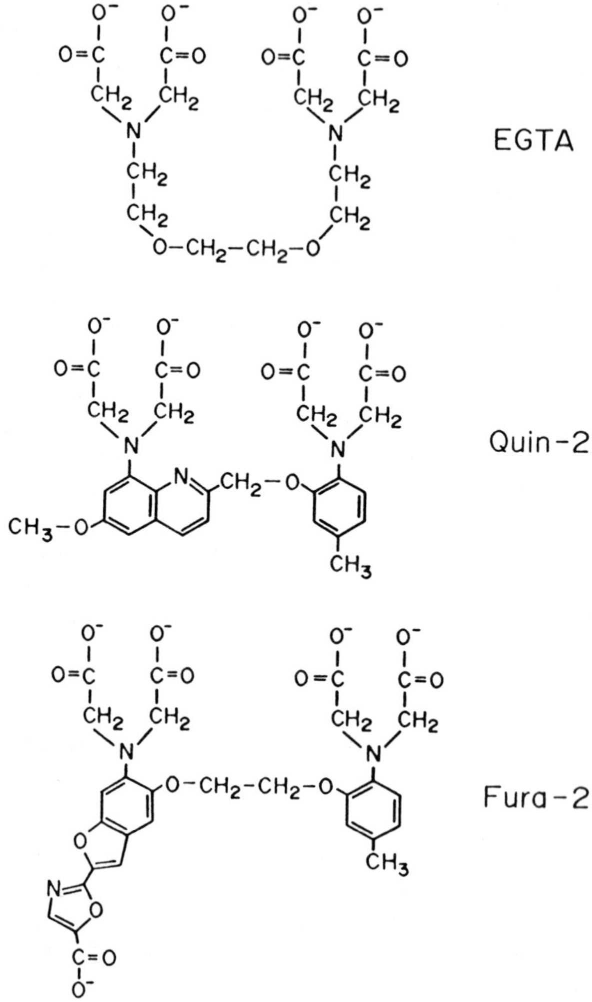

Figure 1. Comparison of the structures of EGTA, quin-2 and fura-2.

free Ca^{2+} concentration to be measured. This ensures that the largest possible signal changes are obtained for a given change in [Ca^{2+}]. At the extremes of the [Ca^{2+}] response curve, the proportion of Ca^{2+}-bound to Ca^{2+}-free dye becomes more difficult to determine.

As noted above, an important property of the Ca^{2+} indicators is their selectivity for Ca^{2+} over Mg^{2+}. However, a number of other ions can bind to EGTA, including Mn^{2+}, Zn^{2+}, Fe^{2+}, and other divalent transition metal ions which can be found at low concentrations in cells. These metal ions also bind to the fluorescent Ca^{2+} indicators and can interfere with calibration of the dyes (see below). Another ion which can affect Ca^{2+} binding sites is H^+, and EGTA is very sensitive to pH in the physiological range (pH 6.5–8). By contrast, the fluorescent Ca^{2+} indicators are relatively insensitive to pH in this range because the highest pK_a's are less than 6.5, in comparison with a highest pK_a of 9.5 for EGTA.

2.1.3 Non-disruptive loading of Ca^{2+} indicators into cells

For studies of cytosolic free Ca^{2+}, an important consideration is the method by which the indicators are introduced into the cell. Techniques involving microinjection, that have been used for Ca^{2+}-sensitive photoproteins and absorbance dyes, can also be utilized to introduce fluorescent Ca^{2+} indicators. However, the latter indicators also lend themselves to loading as membrane permeant esters, without any requirement for disruption of the cell membrane. This is usually achieved by incubating intact cells with an esterified form of the indicator (acetoxymethyl esters have proved to be the most useful). Since the esterified esters are uncharged and hydrophobic, they readily cross cell membranes. Once in the cytosol, endogenous esterases 'activate' the Ca^{2+} indicators releasing the free acid forms which do not permeate the cell membrane. This results in the accumulation of trapped indicator in the cell cytosol. Once extracellular dye has been removed, fluorescence signals from the cytosolic dye can be recorded and used to calculate $[Ca^{2+}]_i$.

2.2 Comparison of fluorescent indicator methods with other approaches to cytosolic Ca^{2+} measurement

2.2.1 Advantages

Other major approaches to measuring $[Ca^{2+}]_i$ in living cells (reviewed elsewhere in this volume) include Ca^{2+}-sensitive photoproteins, Ca^{2+}-selective microelectrodes, metallochromic absorbance indicators, fluorescent Ca^{2+} binding proteins, NMR of fluorine-containing Ca^{2+} chelators and indirect measurements through other Ca^{2+}-sensitive processes (e.g. Ca^{2+}-activated Cl^- channels). A major element in the popularity of fluorescent Ca^{2+} indicator approaches to measuring $[Ca^{2+}]_i$ is the ease of use. No special skills are required for loading cells with the dyes and instrumentation suitable for monitoring the fluorescence changes is readily available in most research settings. The ester loading technique permits all cells in a population to be included in $[Ca^{2+}]_i$ measurements, permitting direct correlation of $[Ca^{2+}]_i$ with metabolites or secretory products which have to be measured on a population basis. Since the cell membrane remains intact throughout, concerns relating to elevated Ca^{2+} leakage resulting from damage to the plasma membrane are eliminated. In addition, many small or fragile cells which are not amenable to microinjection can readily be loaded with esterified fluorescent Ca^{2+} indicators (blood platelets, diameter <3 µm, are an example of the usefulness of this technique). Other advantages of the fluorescent Ca^{2+} indicators include the improved signal-to-noise compared to photoproteins and the better temporal resolution when compared with ion-selective electrodes and NMR approaches. At the level of spatially resolved measurements utilizing imaging techniques, the fluorescent Ca^{2+} indicators have far surpassed any other method available for measuring $[Ca^{2+}]_i$ in living cells. Fluorescent Ca^{2+} indicators also have advantages in terms of calibration for calculation of absolute levels of free Ca^{2+}, although as with all methods a number of assumptions and controls are required. At resting levels of $[Ca^{2+}]_i$ (<200 nM), most other techniques cannot compete with the fluorescence approach. Finally, the specificity for Ca^{2+} and lack of sensitivity to most other physiological ions of the fluorescent Ca^{2+} indicators is remarkably good compared to many other Ca^{2+} indicators.

2.2.2 Disadvantages

The advantages of using fluorescent Ca^{2+} indicators should be balanced against a number of disadvantages associated primarily with the loading of a 'foreign' molecule into the cell. Since the dyes must bind Ca^{2+} in order to report its concentration, they can perturb the true level of $[Ca^{2+}]_i$. This problem is directly related to the level of dye loading. With quin-2, which has a relatively low fluorescence yield requiring cytosolic concentrations in the order of several hundred micromolar for optimal signals, Ca^{2+} buffering can be substantial. Indicators introduced more recently, such as fura-2 and indo-1, offer much higher efficiency of fluorescence, reducing the required loading levels by at least an order of magnitude. It should also be noted that because of the basic cellular mechanisms regulating $[Ca^{2+}]_i$, Ca^{2+} buffering by the indicators tends not to have much effect on basal or steady state elevated levels of $[Ca^{2+}]_i$. Ca^{2+} buffering by the indicators can, however, alter the kinetics of Ca^{2+} changes and buffer transient $[Ca^{2+}]_i$ responses. Other problems associated with the use of fluorescence Ca^{2+} indicators include compartmentation of the dyes and residual partially esterified dye forms and interference by heavy-metal ions. Most of these problems can be resolved using suitable controls and calibrations. These are discussed in detail in Sections 6 and 8.

2.3 Relative merits of cell population measurements and single-cell studies

With the advent of fluorescent Ca^{2+} indicators with high fluorescence yields, it has become possible to use this approach to measure $[Ca^{2+}]_i$ at the single-cell level and at the subcellular level, using imaging methods. There are a number of advantages in examining $[Ca^{2+}]_i$ at the single cell level. Clearly, this is the only approach that can yield information about the spatial organization of $[Ca^{2+}]_i$ changes within the cell. Furthermore, due to the somewhat asynchronous nature of most $[Ca^{2+}]_i$ responses, complex patterns may be averaged out in population measurements (e.g. 7–10). The kinetics of $[Ca^{2+}]_i$ changes may also be distorted. Even a relatively small cell-to-cell variation in the time of onset of a simple Ca^{2+} rise can lead to a substantial underestimate of the rate of $[Ca^{2+}]_i$ increase at the cellular level. Single-cell Ca^{2+} measurements also circumvent problems associated with heterogeneous cell populations, allowing data to be obtained from an identifiable cell type and eliminating contributions from damaged or foreign cells. For some types of study, however, it may be desirable to obtain a population average. This would be the case when $[Ca^{2+}]_i$ must be correlated with measurements of metabolites or secretory products which can only be measured using relatively large numbers of cells. The main advantage of cell population measurements is that these are simpler to carry out, calibration is more straightforward and specialized instrumentation is not required. In addition, with population measurements, the signal-to-noise is often better, especially with high temporal resolution.

3. Instrumentation

3.1 Fluorometers for cell population measurements

3.1.1 Basic equipment for single-wavelength measurements

The essential elements of a fluorometer are a light source, a monochromator or filter for selecting the excitation wavelength, a sample compartment, an emission light collector and wavelength selector, a detector (usually a photomultiplier tube, PMT), and an amplification and recording system. Almost any instrument that meets these basic requirements can be used to record Ca^{2+} changes using fluorescent Ca^{2+} indicators in a single-wavelength mode. Quin-2 has always been used in this way, and dual-wavelength indicators, such as fura-2 and indo-1, can also be used to measure $[Ca^{2+}]_i$ in the same manner. For the majority of cell types that sediment rapidly under gravity, a mixing device is required, such as an under-cuvette magnetic drive. The sample compartment should also be temperature-regulated and it may be important to allow gassing of the cell suspension. Although near-UV excitation is used for several of the Ca^{2+} indicators, quartz cuvettes are not required. Glass cuvettes are suitable, as are some types of plastic cuvettes with extended transmission characteristics. Monochromators are preferred over filters since the former offer greater flexibility and allow spectra to be taken, which can be very useful in diagnosis of problems relating to incomplete dye hydrolysis or simply to follow the process of intracellular de-esterification. For dyes such as quin-2 and fura-2, the excitation light needs to provide sufficient intensity in the near UV. Xenon arc lamps are the most popular, since these have a more continuous spectrum than the mercury arc lamp (see below) and offer greater intensity and wavelength range than tungsten, halogen, or deuterium lamps. Stability of the light source is important, particularly for single wavelength measurements; recording times of tens of minutes are often used for studies of $[Ca^{2+}]_i$. Many fluorometers have a ratio mode, in which a reference detector is used to correct for fluctuations in the excitation beam. A chart recorder is adequate to record the fluorescence signals. However, some form of data storage, such as a personal computer fitted with an analogue digital converter board and appropriate data acquisition software, allows plotting of data on appropriate scales and on a linear scale of Ca^{2+}, following calibration.

3.1.2 Equipment for dual-wavelength Ca^{2+} indicators

As noted above, dual-wavelength indicators, such as fura-2 and indo-1, can be used for Ca^{2+} measurements in a single wavelength fluorometer. However, a dual-wavelength fluorometer is required to make optimal use of these indicators and to utilize the ratio calibration technique (see below). Different instrument configurations are required for different dyes. Indo-1 shifts its emission wavelength on binding Ca^{2+} and therefore requires a dual emission system. This is most readily accomplished by using a 'T-format', in which two monochromators (or appropriate filters), each with its own PMT, are set on opposite sides of the sample compartment at right angles to the excitation beam. Other configurations using light guides or chopping systems are practical alternatives. Where two detectors are used, the two wavelengths can be recorded continuously and simultaneously, yielding optimal temporal resolution and maximizing signal-to-noise ratios. For dual excitation (e.g. with fura-2), it is necessary to alternate the excitation beam wavelength, since the detector would be unable to discriminate fluorescence resulting from two distinct excitation wavelengths applied simultaneously. Several commercial designs exist for this purpose. Most depend on a single light source with either a rapid filter changer or a chopping device which alternates the excitation light through two separate static monochromators or filters (see *Figure 2*). In both cases, some form of computer or electronic monitoring of the alternating excitation light is used to discriminate and separate the fluorescence signals at a single emission detector. The chopped designs of dual excitation fluorometers have an advantage in the flexibility and quality of light provided by the monochromators, but filter changer systems have simpler light paths and can often accommodate more than two excitation wavelengths. One other approach to multi-wavelength fluorometry, which is offered on some standard research fluorometers, is rapid slewing of monochromators between wavelengths. The time-resolution in such systems is currently inferior to dedicated dual-wavelength systems, and there must be some concern for wear on the monochromator drives. However, instruments of this type may be appropriate where a less specialized fluorometer is required.

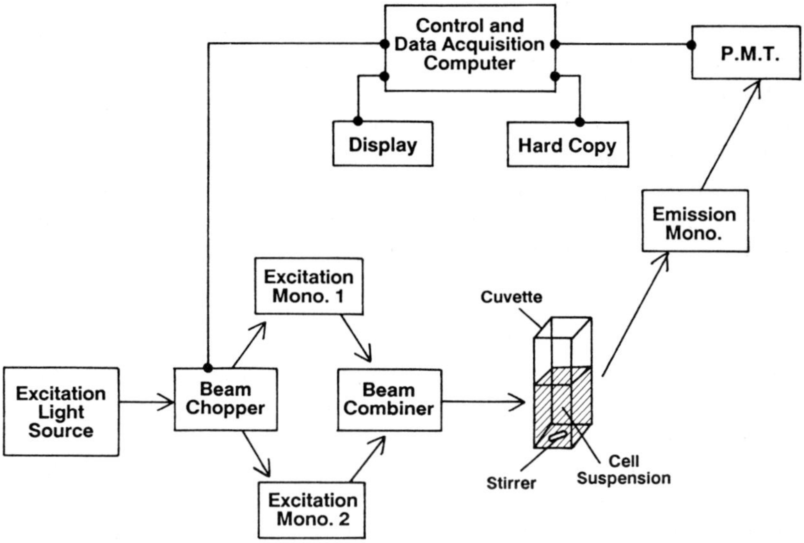

Figure 2. Schematic diagram of a dual excitation wavelength fluorometer.

3.1.3 Population studies of adherent cells

The majority of cell population measurements of $[Ca^{2+}]_i$ have utilized cells in free suspension. However, a number of cells either do not function in suspension or function differently when attached to a surface. For the most part such cells are best studied at the single-cell level. However, it is possible to measure $[Ca^{2+}]_i$ of cells in attached monolayers using fluorometers of the type described above. Cells can be grown on suitably coated glass or plastic coverslips of an appropriate size to fit into a standard cuvette (9, 11). The coverslip is positioned diagonally within the cuvette, so that its front face is at 45 degrees to the excitation beam and 45 degrees to the emission detector. ACLAR plastic sheets have been found to be suitable for this purpose, since this material can be cut to required shapes and sizes and has sufficient flexibility to be 'spring-loaded' into the cuvette (9). Some additional precautions need to be taken for cells in monolayers. This configuration is more prone to photobleaching, because cells do not pass in and out of the excitation beam. Controls must be carried out to ensure cells do not detach, which will give false Ca^{2+} signals. Calibration must be carried out without cell lysis (see Section 6). Because of these concerns, the monolayer approach is most appropriate for dual wavelength fluorometer systems.

3.2 Photometer approaches to single-cell measurements

Studies of $[Ca^{2+}]_i$ at the single-cell level can be subdivided into imaging approaches, which give spatially resolved information, and non-spatially resolved approaches, which measure the total fluorescence signal from a defined region. Both approaches rely on similar microscopes and illumination systems and these elements will be discussed in Section 3.3. For non-spatially resolved fluorescence measurements at the single-cell level, a microscope configured for epifluorescence is used (see *Figure 3*). Emission light is diverted to the fluorescence detector through a diaphragm or pin-hole in an image plane of the microscope. This allows measurements of fluorescence from a region selected specifically by the operator. A system of this type can be viewed as a microfluorometer that allows fluorescence measurements from a single cell or even a subregion of a cell. Photometer systems can be either single or dual wavelength, but reliable calibration requires a dual-wavelength approach. Electronics similar to those described above for dual-wavelength fluorometers are used for subsequent separation and discrimination of the

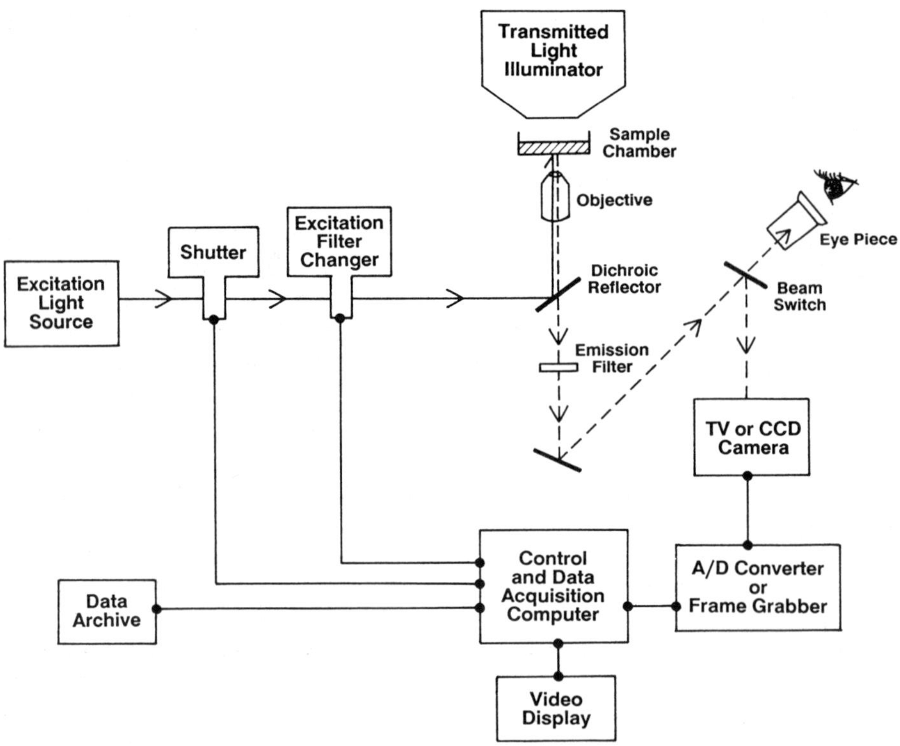

Figure 3. Schematic diagram of an epifluorescence microscope digital imaging system.

individual fluorescence signals. Examples of data obtained using dual-wavelength photometer systems can be found in refs 7 and 12.

The detector most commonly used for non-spatially resolved measurements is a PMT, which has a rapid response and a low noise count. Recently, high quality photodiodes for fluorescence detection have also been used. Both detector types can be used in two modes: photon counting or analogue amplifier mode. A PMT converts photons to electrons with a photocathode at the detector surface, these electrons are then accelerated and amplified by a cascade of dynodes (by secondary emission). The biggest source of noise for the PMT is thermal. In the case of a PMT in photon-counting mode, the size of the pulse can be used to discriminate some of the thermal noise from the signal. The quantum efficiency (QE = arriving photons/detected photons × 100) of a PMT is dependent on the photocathode surface, and is usually low in the visible region of the spectrum (20% at 500 nm, green emission). Photodiodes use silicon wells for the photoconversion process. These have a higher quantum efficiency (60%, at 500 nm). However, with photodiodes electrons resulting from thermal and other noise are indistinguishable from photon-generated electrons.

3.3 Imaging cytosolic Ca^{2+} at the single-cell and subcellular level

3.3.1 Microscopes

i. Choice of microscope

The heart of a cellular imaging system is the microscope. An epifluorescence microscope is required for imaging fluorescence emission from cells. The microscope contains the optical components necessary to deliver the excitation beam to the sample, isolate the fluorescence emission signal, and focus it as an image at the detector. The microscope will also provide transmission optics for direct observation and positioning of cells, and this will also be accessible to the imaging device for correlation of $[Ca^{2+}]_i$ images with morphological features of the cells. To obtain a clear transmitted light image of living cells, some form of optical contrast enhancement is desirable. The simplest of these is phase contrast, which is available on most microscopes and does not complicate the fluorescence light path unduly (although the phase rings will reduce light throughput somewhat). Better quality transmitted light images are obtained with differential interference contrast (DIC), but this requires that polarizers be introduced into the light path and these must be removed for optimal fluorescence imaging. Both upright and inverted microscopes can be used for single-cell Ca^{2+} measurements, the choice will depend primarily on the desired sample configuration. For most applications, an inverted microscope is preferred, since this allows attached cells to be superfused with buffer and imaged from below with high numerical aperture, short working distance objectives (see below). With a long working distance condenser, access from above the cell incubation chamber is relatively unobstructed, facilitating the use of micropipettes, patch electrodes, etc. *Figure 3* is a schematic of a typical imaging system showing the major components in the light path.

ii. Objectives

The objective is the single most important optical element of the microscope and the characteristics of this element will make a major contribution to the performance of the system as a whole. In an epifluorescence microscope the objective both delivers the excitation beam and collects the fluorescent emission from the sample. In choosing an objective for fluorescence studies, four parameters should be considered:

- magnification
- numerical aperture (NA)
- transmission at the wavelengths to be used
- chromatic and spatial aberrations.

Clearly, the objective should be as free as possible from chromatic and spatial aberration, at least within the field which will fall on the detector. Standard microscope objectives do not transmit light in the 340 nm range required for several of the fluorescent Ca^{2+} indicator dyes. Optimal light transmission is obtained with quartz objectives. However, these are very expensive and most microscope manufacturers offer more economically priced Fluor-objectives which give adequate transmission at wavelengths down to (but not below!) 340 nm. It should be noted that there are a number of other optical elements in the excitation path of most microscopes, and these also should be optimized for the wavelengths to be used.

The available light at the detector decreases with the square of the magnification factor, so the lowest magnification consistent with the required resolution should be used. *NA* is an index of the light-collecting capacity of the objective; the higher the number the greater the efficiency. The *NA* is particularly critical in epifluorescence, because light passes through the objective in both directions. For a given illumination intensity, the fluorescence intensity increases with the fourth power of the *NA* (14). Thus, a lens of higher *NA* will tend to overcome the loss of signal associated with a higher magnification. The use of immersion fluids, with higher refractive indices than air, between the objective and coverslip increases the *NA*; oil, glycerol, or water are frequently used for this purpose (note that different lenses are designed for use with a specific immersion fluid). The *NA* (rather than magnification) also determines the spatial resolving power of the objective. For a self-luminous body the minimum distance by which two points can be separated and still resolved (r) is given by:

$$r = \frac{1.22 \times \lambda_0}{2 \times NA}$$

Where λ_0 is the wavelength of the light in air. For a more detailed description of how to optimize the resolving power and optical efficiency of a microscope, see ref. 14.

3.3.2 Fluorescence illumination system

The delivery of excitation light to the specimen and subsequent isolation of the emitted light requires the following elements:

- light source and focusing optics
- excitation wavelength selector
- dichroic beam splitter for wavelength discrimination
- emission wavelength selector.

All of these components are provided in a standard epifluorescence microscope (see *Figure 3*). For dual excitation measurements an additional control unit is necessary to allow alternation of wavelengths. The following discussion outlines key features in the selection of these components.

i. Light source and related optics

The light-source requirements for fluorescence imaging are similar to those needed for fluorometric measurements of Ca^{2+} (Section 3.1). Since fura-2 is currently the indicator of choice for ratio imaging, a good UV light source is essential. *Figure 4* compares the output of xenon and mercury arc lamps, the two most commonly used fluorescence microscope excitation sources. As can be seen, the xenon lamp provides a much more even spectrum of illumination, allowing any required wavelength in the near UV and throughout the visible range to be selected using appropriate filters or monochromators. By contrast, mercury lamps offer only a limited number of discrete wavelengths with suitable intensity for fluorescence measurements.

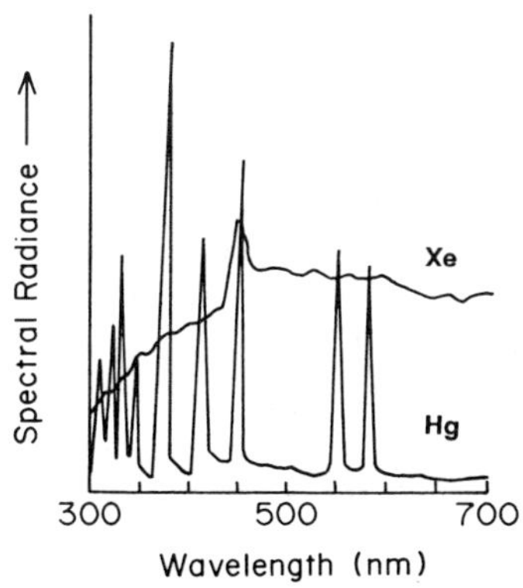

Figure 4. Comparison of output of xenon (Xe) and mercury (Hg) arc lamps. (Redrawn from data in ref. 15.)

With a mercury arc lamp the high intensity lines do not fall at optimal wavelengths for fura-2 excitation and there is a substantial amount of light at around 360 nm, where fura-2 fluorescence is not changed by Ca^{2+} binding. Thus, the xenon arc lamp is recommended for imaging with fluorescent Ca^{2+} indicators. Although xenon arc lamps are available with a range of power outputs, a 75-W lamp (the bottom of the range) is sufficient for most applications. In fact, it is often necessary to attenuate the excitation light to minimize photobleaching. One other important consideration is the heat output of the excitation lamp. Xenon arc lamps produce a large amount of energy in the infra-red region. It is good practice to include a heat filter early in the light path for all illuminators, including the transmitted light source.

Lasers offer a more coherent light source but these only provide acceptable power at discrete wavelengths, depending on the type of laser (see ref. 15). Argon and nitrogen lasers, alone, or in conjunction with dye modules can be used to excite at some appropriate wavelengths. However, lasers are expensive to purchase and maintain compared to arc lamp systems and are usually a less flexible alternative for illumination if a variety of fluorescent dyes are to be used. Pulsed lasers or pulsed strobe lights can be used to 'freeze' images without motion blur. The exposure can be timed to external events or can be used to obtain fast time-sequences. However, as the exposure time is reduced, the excitation light intensity must be increased to retain the same photon count. Care must be taken under these conditions, because the higher power output may saturate the fluorescent dye or damage the cell tissue.

The optical elements in the illumination path should all be capable of transmission in the relevant UV wavelengths. This is particularly important for complex illuminators, since the transmission of the system as a whole is the product of the transmission of each individual element. Thus, if the transmission of each element was 50% at the desired wavelength, with n elements the total transmission would be 0.5^n. For a 6-lens system this would give only 1.5% transmission.

ii. Excitation wavelength selector

In a standard epifluorescence system, the excitation wavelength is selected using a filter. Single-layer glass (or gelatine) filters generally pass a relatively wide band of light, while more selective, narrower wavelength ranges can be obtained using multilayer interference filters. Interference filters usually have lower peak transmission than pass filters. For fura-2 studies, interference filters with about 10 nm bandwidth should be used. These are usually centred around 340 nm and 380 nm, but when the UV transmission of the system is limiting it may be preferable to use a 350 nm filter for monitoring Ca^{2+}-bound fura-2. In cases where a single excitation wavelength is to be used, a wider bandwidth may be preferable to obtain more light throughput. For dual excitation measurements it is necessary to have an automated mechanism for switching between filters. There are a number of configurations for achieving this, including mechanical reflective choppers, rotating filter wheels, and motorized sliders. The filter changer is generally located in the excitation path between the arc bulb housing and the microscope. Several commercial filter wheel assemblies are available that provide computerized control modules to allow synchronization of filter changes with image frame acquisition. Some systems are also available in which the excitation wavelength is selected by monochromators, utilizing similar illumination systems to those used in dual excitation fluorometers (Section 3.1.2). These systems offer greater flexibility than filter changers, but are much more expensive. With appropriate mechanical design and electronic control systems, the switching time for both filter and monochromator systems can be reduced to the tens of milliseconds range (video rates or less).

One other feature which is generally built into the excitation light control system is a shutter to prevent illumination when data is not being collected. Ideally, this is a fast action electronic shutter which can be controlled by the image acquisition computer. This allows the excitation light to be interrupted between sessions of image collection, and at slow imaging rates (>300 msec) it may even be advantageous to close the shutter between each frame. This helps to minimize photobleaching and cellular photodamage.

iii. Dichroic beam splitter

A dichroic beam splitter is used to reflect the excitation light into the objective, but to pass the longer wavelength emission light to the eye-piece or imaging detector (*Figure 3*). The choice of the correct dichroic reflector is critical. Dichroic reflectors are constructed in the same way as interference filters (which also reflect wavelengths which they do not pass), but fewer layers are used and the aim is to pass most light and reflect just a limited wavelength range. The dichroic reflector should be chosen to give peak reflectance at the required excitation wavelength and to pass all longer wavelengths in the region of the desired emission wavelength. In dual excitation systems, some compromise may have to be made, but suitable reflectors are available for fura-2 which give good reflectance (70–90%) at both 340 and 380 nm, while passing most light above 400 nm.

iv. Emission wavelength selector

In most systems a fixed filter is used to select the required emission wavelengths. Since emitted light is generally at a premium, a wide bandpass filter is preferable over a narrow band interference filter. In the absence of other fluorophores or interfering cellular fluorescence, a filter which passes light in the range 460–600 nm can be used for fura-2.

3.3.3 Cell incubation chambers

For imaging of Ca^{2+} in living cells a suitable incubation chamber is required. With an inverted microscope configuration, the base of the chamber must be made of a material that transmits the required wavelengths of light and gives a good quality image. With short working distance and immersion objectives, a glass or quartz coverslip of <0.2-mm thickness is generally necessary. Adherent cells maintained in culture can be grown in the chamber or plated on suitable coverslips in Petri dishes and then inserted into the chamber. Cell incubation chambers may have open or closed tops, depending on whether access is required. Closed chambers are designed with a buffer perfusion system to allow the introduction of test agents, etc. Open chambers can also have a perfusion flow system, with some form of continual removal of excess buffer. If the buffer is equilibrated with an atmosphere other than air, either a closed chamber can be used or the required (humidified) gas phase can be gently blown across the liquid surface. A number of commercial designs for microscope chambers are available. A basic design can be fabricated simply by cutting a hole in the base of a Petri dish and gluing an appropriate coverslip across the hole. Many cells will attach directly to glass, but coatings such as polylysine, collagen laminin, etc., are frequently used.

For most biological measurements it is desirable to work at a physiological temperature (around 37 °C for mammalian cells). Temperature regulation may be achieved in a number of ways, from situating the entire microscope in a warm room or temperature-controlled chamber, to providing direct temperature control using a heating element within the chamber. Other options include directing warm air or infra-red light on to the chamber and internally-heated microscope stages. Whatever the method of heating chosen, it is important to avoid significant heating and cooling cycles, such as those generated by simple thermostats. Changes of as little as 0.1 °C can alter the focus and/or position of the specimen. Ideally, a thermocouple temperature probe is located as close as possible to the cells under examination and this is used to control the heating device via a proportional output control circuit. With immersion objectives, the lens itself represents a significant heat sink. The lens temperature can be warmed to the specimen temperature by circulating water through coils of copper or plastic tubing around the objective. However, excessive temperatures should be avoided as damage to the objective may result.

3.3.4 Imaging devices

To record spatial information from fluorescently labelled cells, a camera capable of subdividing the photon flux according to the points of origin in the sample is required. Once digitized, these individual fluorescence intensity measurements are used to make up the elements of the image, known as pixels. Thus, each pixel is the equivalent of a separate fluorometric detector, analogous to the single PMT in a photometer system. Since the emission light from a single cell is distributed between many pixels, the photon flux available for each measurement is very small. As smaller and smaller regions of a cell are quantized (higher spatial resolution), the photon flux per pixel is decreased. In addition to the low levels of fluorescent light, Ca^{2+} measurements in living cells tend to require a fairly rapid time-resolution in the order of seconds or less. The combined effect of the various constraints on imaging of fluorescent Ca^{2+} indicators is that the imaging camera needs to be very sensitive. The most common two-dimensional detectors in use are:

i. Video camera coupled to an image intensifier assembly
In a standard video camera tube the incoming image is focused on to a photoconductive target layer which is scanned by an electron beam. Incident photons lower the resistance and generate electron–hole pairs at the target, increasing the current of the electron beam. This current is used to generate a video signal voltage which is proportional to the photon flux arriving between electron beam scans at each point on the target. Relatively high light levels are required to generate a detectable signal. However, standard video cameras, including vidicons, plumbicons, newvicons, etc. as well as solid state video cameras relying on CCDs (see below), can be used to detect faint fluorescent images if they are coupled to an image intensifier amplification stage. In generation I intensifiers, incoming photons eject photoelectrons from a photocathode and these are accelerated by a high-voltage electric potential on to a phosphor, which emits many photons for each incident electron. When several stages of this type are cascaded together, the brightness of the initial signal can be increased by up to 5 or 6 orders of magnitude. However, noise increases with each amplification stage and there is a progressive loss of spatial resolution. The most common image intensifiers in use today are the generation II and generation III intensifiers, which multiply the initial signal by cascading secondary electrons. These intensifiers consist of a photocathode, an electron accelerator/focusing mechanism, an electron multiplier, and an output phosphor. Electrons generated at the photocathode are accelerated at high voltage and focused into a microchannel plate (MCP) wafer, with several million parallel hollow glass cylinders coated so as to release secondary electrons. Multiple collisions of the electrons with the coating causes each cylinder to act as an individual electron multiplier. The electrons are then guided to a phosphor surface where the electron to photon conversion takes place. The output phosphor of the image intensifier can be coupled with lenses or a fibre optic faceplate to the video camera. Because the output of the camera is a standard analogue TV signal (RS-170 or CCIR standard), the output can be viewed directly on a TV monitor or stored using commercial videotape machines.

ii. Silicon intensified target (SIT) cameras
A SIT camera has an intensifier built into the video camera tube. The incoming image arriving at a photocathode, generates photoelectrons which are accelerated (increased energy) by a high-voltage electrical potential and focused onto a target consisting of a silicon semiconductor layer. Each

incident photoelectron causes a localized accumulation of a large number of electron–hole pairs. An electron beam is scanned over the surface of this detector and a video image is generated in a similar manner to a normal video tube. The output of SIT cameras conforms to standard analogue video conventions (RS-170 or CCIR). Image intensifiers can also be coupled to SIT cameras increasing their sensitivity by an order of magnitude. These are called ISITs.

iii. Cooled charge coupled device (CCD) cameras (digital output)
A CCD is a solid-state imaging device which differs fundamentally from video tube technology (although CCDs are widely used to generate video-type signals and are now the preferred device in home video cameras). No vacuum tube is required and there is no high-voltage stage, unless an intensifier is added. The CCD sensor consists of a semiconductor layer which is divided into discrete, electrically isolated electronic traps or 'wells' defined by gates in the surface. Arriving photons create electron–hole pairs and the signal accumulates as a potential which is proportional to the incident photon flux at each pixel. The signals are read out by shifting the electron charge in the potential wells in a sequential manner. The electron charge is then amplified, sampled and digitized so that it can be fed to an imaging computer for storage and display. Thus, unlike a video tube, there is a separate detector for each pixel and signals are integrated continuously at all points simultaneously until read out. In addition to improving signal-to-noise, this offers the flexibility that exposure time can be varied as required. The major source of noise in CCD cameras is from thermal generation of free electrons. This is generally minimized by cooling the detector with liquid nitrogen or a thermoelectric cooling device. Thinned, back illuminated CCD cameras are inherently linear in their response to light, have quantum efficiencies of between 50% and 80% and are largely free from geometric distortion.

iv. Confocal microscopes
Confocal microscopes suitable for imaging fluorescence in living cells are a relatively recent innovation. The optics of these instruments (discussed in detail in ref. 16) allow improved resolution, particularly in the *Z*-axis, where out of focus information can be rejected almost completely. This is achieved by illuminating the sample with the image of a pin-hole and viewing the fluorescence through a second pin-hole at an identical image plane. Fluorescence light from out of focus regions of the sample is also out of focus when it reaches the observation pin-hole, and as such is almost totally excluded from the detector. An image is built up by scanning the confocal pin-holes in two dimensions over the sample. Since the spatial information is built up on a point-by-point basis, a single PMT can be utilized as the flourescence detector. The main disadvantages of currently available confocal microscopes are the relatively low temporal resolution and the high level of

excitation light required. Lasers are generally used as the excitation source in these instruments and dual excitation is not offered commercially. Although dual emission detectors can be fitted on confocal microscopes, lasers with sufficient power in the near UV suitable for indo-1 measurements are very expensive. Visible light lasers suitable for use with either fluo-3 or rhod-2 are more readily available, and a number of workers have had success with these dyes in confocal fluorescence microscopy.

3.3.5 Selection of an appropriate imaging detector

There are a number of parameters which determine the suitability of a detector for fluorescence imaging. The characteristics which should be considered when choosing a camera or other imaging device are:

i. Linearity

For quantitative microscopy the output signal from the camera must increase linearly with light input. Linearity is dependent on the gamma factor of the camera. It is the exponent of the power function that relates output response to light input:

$$\text{responsivity} = \text{constant} \times (\text{light input})^{\text{Gamma}}.$$

Thus, a gamma factor of one is required to use a camera as a photometric device. Cameras with a gamma value of less than unity can be used to increase the contrast and dynamic range in transmission microscopy. Some cameras provide variable gamma settings. Most solid state detectors have a gamma factor of 1. In general, image intensifiers also have a unitary gamma factor.

ii. Quantum efficiency

The ratio of photons arriving (P_a) over photons detected (P_d) for any detector is called quantum efficiency (QE). QE becomes a key factor when imaging low photon count events, either due to low light-levels or very short exposure times. The statistics of photon emission by fluorophores follow a Poisson curve; that is, the variance is equal to the mean value. Statistical noise limits the theoretical signal-to-noise ratio (SNR) of any measurement to the square root of the detected photons. The QE of the first detector surface will set this limit at:

$$\text{SNR} <= \surd\, P_d = \surd\, (QE \times P_a).$$

The signal increases linearly with QE and the SNR will increase as the square root of QE. Later steps of amplification and other electronic manipulations will further deteriorate the data, until the image is encoded in a digital form.

iii. Spatial resolution

The spatial resolution of a detector is limited by the number of TV scan lines in the case of tube cameras or the number of pixels for solid-state detectors. It should be mentioned that the 'spot size' function on tube-type cameras is usually wider and smoother than in solid-state cameras. This means that the response to closely located objects will tend to fuse or 'blur'. The function that describes this process is called the Point Spread Function (17). When video images are subsequently digitized, the number of resolvable points will be the lower of the acquisition board resolution and detector resolution. Spatial resolution decreases at low light-levels, because the SNR decreases. According to the Raleigh criteria (14) spatial resolution is defined as the minimum distance between two identical point sources which can be resolved with a 26.5% 'dip' in intensity between them.

iv. Temporal resolution

For most studies of $[Ca^{2+}]_i$, spatial resolution is less critical than temporal response. Tube and image intensified cameras have a degree of image persistence, also called 'decay lag'. This can be expressed as a percentage of the signal left on the phosphor output after a complete video frame (33 msec). It varies from 30% to 60% in tube-type cameras. However, some image intensifiers with faster output phosphors have lower lag values. Persistence tends to average out some noise but decreases the time-resolution of any measurement so that it is lower than the video rate (30 frames per second). Solid-state detectors have negligible persistence if operated below the saturation level, but at high light levels these can demonstrate bloom and streaking due, respectively, to charge spill over between adjacent pixels and incomplete charge transfer during read-out. Image intensifiers also demonstrate persistence, which can be substantial at very low light-levels when high gain is applied. Persistence can pose a serious problem if the same camera is imaging different sets of data in rapid succession. For example, with dual-wavelength data obtained for ratiometric measurements, a frame taken at one wavelength would be contaminated by the preceding frame.

In the case of cooled CCD cameras the read-out rate is kept low in order to reduce the system noise. Electronic white noise and environmental noise contributions are reduced by limiting the bandwidth of the overall signal acquisition and amplification process. The cooled cameras are usually read at a rate of between 50 kHz to 500 kHz (a 460 × 460 pixel detector is read out in 200 to 2000 msec). The factor limiting the speed of cooled CCD cameras then, is not the shifting of the charge itself but the read-out and digitization process. Several approaches can be used to improve the temporal resolution of these detectors: Some cameras offer the capability to combine (bin) several adjacent pixels into a single super pixel prior to digitization. This increases the readout rate but reduces spatial resolution. Alternatively, subregions of the detector can be digitized without loss of spatial resolution but with a loss of part of the image field. In another technique (18), the detector itself is used to store a short time series of subimages collected at very high speeds (e.g. 5 msec) and then the entire time-series is read-out as a single image.

v. Sensitivity and noise characteristics
The effective sensitivity of a detector can be considered to be the lowest light-level that can be detected with an acceptable SNR. Different manufacturers and authors utilize different SNR criteria (anywhere from 3 to 10). The noise in low light-level imaging can be divided into photon-dependent (statistical or 'shot' noise), which is equal to the square root of the detected signal, and noise sources which are independent of photon flux. The photon-independent noise includes thermal noise (dark current of the detector) and the read-out noise. With increasing light input, the noise due to photon-independent processes remains constant and therefore contributes a decreasing proportion to the total signal, while at low light levels the dark current and the read-out noise tend to dominate. Cooled CCD cameras have a very low dark current and read-out noise and since they are not limited by video frame rates, they can integrate light for any period of time before digitization. In most video format cameras one frame leaves the camera every 30th of a second and any integration to reduce noise must be carried out by averaging images after they have been digitized. The disadvantage of this approach is that the read-out component of the noise is added on every image as opposed to the cooled CCD where it is included only once.

Table 1 displays a qualitative comparison of some types of commercial low light-level cameras. The current generation of cooled CCD cameras offer the best combination of sensitivity, image quality, dynamic range and overall flexibility. The main disadvantage of these detectors is the slow read-out rate. If video speed and/or video storage is required, at the expense of SNR, intensified video cameras are a cost-effective alternative.

Table 1. Comparison of performance of frequently used imaging devices

	SIT (video)	Intensified CCD	Cooled CCD
Sensitivity	+ +	+ +	+ + +
Spatial resolution	+ +	+ +	+ +
Lack of persistence	+	+ +	+ + +
Read-out speed	+ + +	+ + +	+

3.3.6 Computer requirements for image acquisition and processing

The objective of this section is to discuss the basic computing requirements appropriate for a digital imaging fluorescence microscopy system suitable to carry out measurements of $[Ca^{2+}]_i$. It is expected that most users will purchase commercial hardware and software packages. Building such a system from components is time-consuming and requires considerable electronic and computer programming skills. It is recommended, however, that any commercial software program selected should offer the user the ability to modify acquisition and processing features.

The computer used should have sufficient processing speed to allow basic image manipulations such as subtraction and division to occur reasonably rapidly. As a minimum, IBM-type PCs with at least an 80286 processor is required. The speed and capacity of the host computer is a major determinant of the power of the system. Since more time is often spent analysing image data than acquiring it, the capability for speedy image processing will quickly be appreciated. In addition to the basic computer, the following hardware requirements should be met:

- frame grabber or other image acquisition hardware
- graphic display monitor (preferably colour) with appropriate graphics display board (this can often be the frame grabber)
- an interactive graphic device (mouse or similar)
- controller boards to interface with filter changers, etc.
- random access memory with sufficient capacity to hold at least four images during image calculation: processing of image time series is greatly facilitated if an entire series is held in memory (a minimum of 8–16 megabytes is recommended for this)
- image processing will be facilitated further if some form of image arithmetic capabilities are included in the frame grabber board or an additional board
- an appropriate form of mass storage medium is essential (see below).

In most systems, fluorescent data is stored at acquisition time for later processing, $[Ca^{2+}]_i$ calculation, and display. The storage requirements are determined by the number of images to be stored, the image size (number of pixels) and the data precision. The precision is given by the number of distinct levels allowed for the intensity at each pixel. It is usually expressed in terms of the number of bits used for the digitization. The number of levels is equal to 2^n, where n is the number of bits. Therefore an 8-bit digitizer will provide 256 signal levels of precision for every sample point. This is the minimum recommended for $[Ca^{2+}]_i$ measurements, 6 or 7 bit digitization does not provide sufficient precision for calculation of $[Ca^{2+}]_i$. For a standard 512 × 512 image at 8-bit resolution, more than two million bits of information must be stored, requiring 256 kbytes of memory. Clearly a time series of dual wavelength images, usually composed of several hundred images, can rapidly fill the memory of any computer and most internal storage media. Thus, some form of mass storage medium is required for temporary storage and, where required, for archiving image data. The following describes some suitable media.

i. Hard disc drives

Hard drives are typically found in sizes ranging from 20 Mbytes to 300 Mbytes. They are read-write random access devices and can be easily interfaced with any computer. New models offer the option of removable disc, which may be an advantage for multiple user systems. They probably offer the best temporary storage medium, but are not an effective alternative for long-term storage.

ii. Magnetic tape cartridges

Magnetic tape offers a large storage capacity per removable cartridge. This is an inexpensive option for long-term storage, and the cartridges can be re-used. The sequential storage pattern and slow access time makes digital tape unsuitable for real time data storage or image processing. Its primary use is as an archiving medium.

iii. CD rom and digital optical disc drives

These are based on similar technology to that found in audio compact discs. They provide a medium to high-speed random access storage medium. Most commercial models have a write once read mostly (WORM) configuration. Their large capacity and speed are an advantage when compared to the magnetic tape cartridges used for long term storage.

iv. Analogue optical disc and video tape

Images can also be stored in the video format using optical disc or video tape recorders. The advantage of these is that they represent about the only mass storage media which can store images at video rates. However, since these are analogue devices, there is inevitable signal degradation at each read or write operation (this is much more pronounced with video tape). Furthermore, most devices have a line resolution which is approximately half that of the original video camera. A video frame digitizer is needed to store and retrieve the images. Optical video discs are WORM devices, but give reasonably rapid random access. Video tape suffers from all of the access problems of digital magnetic tape, in addition to the substantial image degradation. However, many users have found that commercial video tape machines can provide inexpensive data storage of useable quality.

3.3.7 Image processing

Most imaging software provides some form of image processing facilities. This can range from simple image arithmetic and intensity measurements, to sophisticated image enhancement, feature extraction and morphometric measurements. Many image analysis programs are attuned primarily to structural features and these are often not well-suited to fluorescence Ca^{2+} imaging, which relies mostly on intensity quantitation. Some requirements for $[Ca^{2+}]_i$ calculations are the ability to carry out arithmetic operations on images (subtraction, multiplication and division, both by constants and by other images on a point-by-point basis); facilities for extracting intensity values from user selected regions of interest, and obtaining the intensity information from the same region for an entire time series; methods for locating the edges of objects in fluorescence images; spatial filtering routines to improve sharpness or reduce noise (these should be used with care prior to generating a ratio image). Pseudocolour display and on screen graphic representations of data are additional features which facilitate analysis.

In addition to calculating the final $[Ca^{2+}]_i$ of single cells and subcellular regions, the image processing package will also be used to carry out necessary corrections to the images. The detected signal does not only include the fluorescent emission from the cell, but is also offset by other signal sources and corrupted by statistical and electronic noise. Among the most common factors that call for correction are:

i. Dark current of the detector
This can be estimated by obtaining an image from the detector without any light input. If the detector is kept at a constant temperature this image can be subtracted from all subsequent images.

ii. Other sources of light
Any signal due to ambient light and fluorescence from the cell and media that are independent of the Ca^{2+} dye should also be subtracted from the fluorescence image data. These values can be estimated using calibration procedures outlined in Section 6.

iii. Spatial differences in camera gain or the illumination field
This effect is usually called 'shading' and it poses a problem mostly for single-wavelength dyes. An image of homogeneous background illumination can be used to correct for shading if it is taken under identical conditions. Every data image is divided pixel by pixel by the shade correction image, and then multiplied by the average mean background intensity.

v. Changes in cell shape (the use of dual-wavelength or volume markers)
With ratio imaging the calculated Ca^{2+} concentration is independent of dye distribution and cell shape. If a single wavelength is used in conjunction with a volume marker the calcium dye images can be normalized for concentration by using the volume marker images. If the cells are actively changing shape, the speed of image acquisition at both wavelengths must be enough to eliminate motion artefacts.

4. Characteristics of fluorescent Ca^{2+} indicators

In this section the properties and use of currently available dyes is discussed. A summary of spectral properties can be found in *Table 2* and excitation and emission spectra of the free acid forms of the dyes at various levels of $[Ca^{2+}]$ are shown in *Figure 5*.

Table 2. Fluorescence properties of Ca^{2+} indicators

	Fluorescence maxima (nm)		Extinction coef. (× 10^{-3})	Optimal wavelengths for measurement (nm)		K_d for Ca^{2+} (nM)
	Ex	Em		Ex	Em	
Quin-2	352	492	5	339	492	
Quin-2-Ca	332	492	5	339	492	126
Quin-2/AM	339	430	5			
Fura-2	362	512	30	380	510	
Fura-2-Ca	335	505	35	340	510	224
Fura-2/AM	369	478	30			
Indo-1	349	485	34	355	485	
Indo-1-Ca	331	410	34	355	405	250
Indo-1/AM	361	479	34			
Fluo-3	506	526	78	490	525	
Fluo-3-Ca	506	526	83	490	525	400
Fluo-3/AM	464	NF[a]	23			
Rhod-2	556	576	80	550	575	
Rhod-2-Ca	553	576	80	550	575	1000
Rhod-2/AM	550	574	80			

[a] NF = Not fluorescent (Data from refs. 3, 5, 13)

4.1 Quin-2

This was the first fluorescent Ca^{2+} indicator to be used widely. It does not show a usable wavelength shift on Ca^{2+} binding and has a lower extinction coefficient and fluorescence yield than dyes developed later, necessitating the use of much higher loading levels to obtain an equivalent signal. These factors make quin-2 unsuitable for single-cell studies. Other disadvantages of quin-2 compared to other dyes are the low K_d (limiting the maximum measurable Ca^{2+} level to about 1 μM) and its relatively poor discrimination of Ca^{2+} over Mg^{2+} (Mg^{2+} K_d = 1 mM). For these reasons, quin-2 has been largely superseded by other dyes, even for single wavelength measurements. However, quin-2 has some advantages. It is more water-soluble in the acetoxymethyl ester form, has generally given less problems in loading, and is the least problematic of the dyes in terms of unwanted intracellular compartmentation. It is also the dye of choice when experiments are carried out in which buffering of $[Ca^{2+}]_i$ is an objective. In fact, in all studies carried out with quin-2, Ca^{2+} buffering is likely to be a significant factor. When excited in the range 300–350 nm, quin-2 shows a fluorescence increase of about sevenfold. The preferred wavelengths for measuring Ca^{2+} with quin-2 are 340 nm excitation and 395 nm emission. Higher excitation wavelengths should be avoided, because Mg^{2+} binding has a proportionally increasing effect on fluorescence as the excitation wavelength is increased. There is an iso-fluorescence point with respect to $[Ca^{2+}]$ at about 360 nm (i.e. wavelength at which fluorescence is independent of Ca^{2+} binding), which can be used to monitor fluorescence of other species.

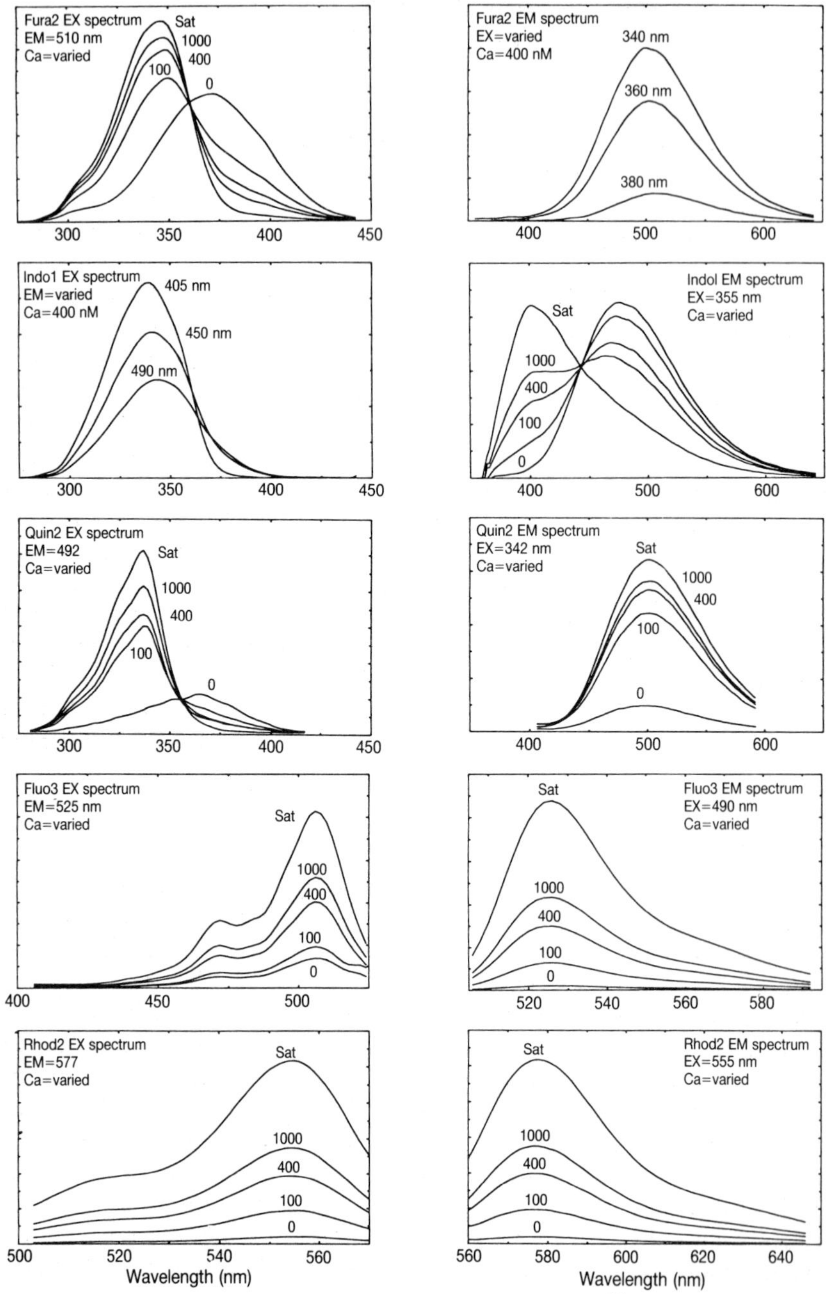

Figure 5. Excitation and emission spectra of fluorescent Ca^{2+} indicators. Spectra of commercial preparations of free acid forms of the Ca^{2+} indicators were obtained in buffer composed of 120 mM KCl, 20 mM Hepes, 10 mM NaCl, 0.5 mM $MgCl_2$ with Ca^{2+} buffered to the indicated concentrations with 1 mM EGTA, at pH 7.2. For spectra where $[Ca^{2+}]$ was not varied, the concentration was set to 400 nM. Conditions with saturating Ca^{2+} (Sat.) were obtained with 1 mM excess $CaCl_2$.

4.2 Fura-2

Fura-2 is currently the most widely used of the fluorescent Ca^{2+} indicator dyes and is the indicator of choice for imaging studies. Fura-2 is about 30 times more fluorescent than quin-2, has a higher K_d for Ca^{2+}, less Mg^{2+} sensitivity, is less susceptible to photobleaching and can be used in the ratio mode. Typically, fura-2 is used at intracellular concentrations in the tens of micromolar as opposed to the millimolar range used for quin-2, and as a result has much less of a Ca^{2+} buffering effect. With a K_d for Ca^{2+} of about 225 nM, fura-2 is suitable for $[Ca^{2+}]_i$ measurements up to several micromolar. The main problems which have been encountered with fura-2 relate to the incomplete hydrolysis of the acetoxymethyl ester (AME) form and the propensity of this dye to become compartmentalized in subcellular organelles during loading. Several studies have also suggested that fura-2 may have slightly different fluorescence properties in the cytosolic milieu as compared to standard buffer solutions. The fluorescence excitation maximum of fura-2 shifts to a lower wavelength on Ca^{2+} binding with negligible shift in the emission maximum. This allows fura-2 to be utilized as a dual excitation indicator. For optimal separation of fluorescence due to the two forms of the indicator, Ca^{2+}-free fura-2 is usually monitored at 380 nm and Ca^{2+}-bound dye at 340 nm. Fura-2 has an iso-fluorescence point at 360 nm which is useful for monitoring Ca^{2+}-independent fluorescence or for following quenching by heavy-metal ions.

4.3 Indo-1

Indo-1 is the only currently available Ca^{2+} indicator other than fura-2 that can be used for ratio measurements of $[Ca^{2+}]_i$. In contrast to fura-2, indo-1 shows a shift in its emission maximum rather than an excitation shift on Ca^{2+} binding. Because accurate co-registration of images obtained at different wavelengths is difficult, indo-1 is less suitable than fura-2 for imaging $[Ca^{2+}]_i$ by fluorescence microscopy (although in principle a filter changer could be placed directly in front of the detector if this did not cause too much image distortion). However, the dual emission character of indo-1 allows simultaneous measurements of the fluorescence of both Ca^{2+}-bound and Ca^{2+}-free dye, which enhances the potential for temporal resolution and generally improves signal-to-noise. Indo-1 is ideally suited to single-cell photometer measurements, using a secondary dichroic reflector to split the two required emission wavelengths between two detectors. A similar approach could be used in laser scanning confocal microscopy, where spatial information is determined from the position of the scanned beam on the sample, rather than at the level of the detector. Indo-1 is also an excellent indicator for use in fluorometer population studies (especially if a 'T-format' instrument is available) and for measurements using a flow cytometer. Apart from differences in the method of use in the ratio mode, indo-1 shares most of the advantages (and drawbacks) of fura-2. It is highly fluorescent and has a K_d for Ca^{2+} of about 250 nM. It is somewhat more susceptible to photobleaching than fura-2 but seems to give less problems in terms of incomplete hydrolysis and compartmentation. In monochromator systems the preferred excitation wavelength for indo-1 is 355 nm, which is also optimal for microscope systems with glass elements in the excitation path, while if a mercury lamp is to be used the 365 nm line is appropriate. Emission wavelengths of 485 nm (Ca^{2+}-free dye) and 405 nm (Ca^{2+}-bound dye) give optimal separation of fluorescence signals due to the two forms of the Ca^{2+} indicator. There is an iso-fluorescence point for indo-1 at an emission wavelength of about 445 nm.

4.4 Fluo-3

The fluorophore component of fluo-3 is based on fluorescein and as a result it is excited in the visible range of light rather than the UV required for the indicators discussed above. Fluo-3 does not undergo a wavelength shift on Ca^{2+} binding and so must be used in a single-wavelength mode similar to quin-2. However, the fluorescence enhancement on Ca^{2+} binding is about 40-fold, which is the largest change of any of the available indicators. The fluorescence intensity of Ca^{2+}-bound fluo-3 is similar to that of fura-2, but the Ca^{2+}-free dye is almost non-fluorescent. This property allows significant signal changes for even relatively small Ca^{2+} changes at the low end of the range of $[Ca^{2+}]_i$. The K_d of fluo-3 for Ca^{2+} (400 nM) is higher than for any of the preceding indicators and this extends the range of $[Ca^{2+}]$ over which it can be used. Fluo-3 is similar to fura-2 in its selectivity for Ca^{2+} over Mg^{2+} and is essentially insensitive to pH above 7.0 (but is somewhat more pH sensitive than fura-2 in the range pH 6.5–7.0). As with the other Ca^{2+} indicators, heavy-metal ions such as Mn^{2+} and Zn^{2+} bind with high affinity. An important distinction, however, is that Mn^{2+} does not quench fluo-3 fluorescence in the way it does the UV-excited dyes, instead Mn^{2+} binding gives a fluorescence which is intermediate between that of Ca^{2+}-bound and Ca^{2+}-free dye (5). Because fluo-3 cannot be used in the ratio mode it is less useful than fura-2 for single-cell measurements of $[Ca^{2+}]_i$. However, its higher wavelength of excitation has some advantages. These include:

(a) It can be used with UV-excited fluorescence indicators for other ions or pH and can be used with caged compounds which are photolysed using UV light.
(b) It is less demanding on optics and excitation light sources for effective illumination.
(c) The excitation wavelength does not overlap with pyridine nucleotide autofluorescence.
(d) It can be used with relatively low power visible light lasers such as the argon ion laser 488 nm line used in scanning confocal microscopy.

Apart from the lack of ratio capability, the only drawback of fluo-3 is the rather small separation between excitation and emission wavelengths, which necessitates the use of relatively narrow band filters or monochromators. Fluo-3 can be excited at 490 nm and fluorescence monitored at 525 nm, a standard fluorescein filter set is suitable.

4.5 Rhod-2

Rhod-2 has a number of features in common with fluo-3. It is excited in the visible light range (rhod-2 is based on the rhodamine fluorophore rather than fluorescein), it shows a fluorescence enhancement on Ca^{2+} binding but no wavelength shift which can be used for ratio measurements, and it has a relatively high K_d for Ca^{2+}. In fact, the Ca^{2+} K_d is about 1000 nM, which allows rhod-2 to be sensitive to $[Ca^{2+}]$ at least up to 10 μM. There are, however, some disadvantages of rhod-2 as compared to fluo-3. The fluorescence enhancement on Ca^{2+} binding is only 3.4-fold and the Ca^{2+}-free dye has significant fluorescence. It has also been reported that rhod-2 is more prone to intracellular compartmentation (5). Excitation of rhod-2 is at 550 nm with emission at 575 nm, rhodamine filter sets for fluorescence microscopes are appropriate.

4.6 Selecting an appropriate Ca^{2+} indicator dye

For Ca^{2+} measurements using imaging the indicator of choice is fura-2, since this can be used to obtain ratio images by use of a dual wavelength excitation system. Either fura-2 or indo-1 are appropriate for photometer single-cell Ca^{2+} measurements, with the dual emission indicator indo-1 being preferable if dual detectors are available to make simultaneous fluorescence emission measurements. In the case of cell population measurements, a ratio approach is also preferred. Thus if dual excitation or dual emission features are available on a fluorometer, then fura-2 or indo-1 should be used, respectively. Even when the fluorometer to be used is not equipped for continuous dual wavelength measurements, it is generally better to utilize one of the ratio indicators because calibrations can be verified using the ratio approach by manual adjustment of the wavelengths. In cases where UV excitation is a problem (see Section 4.4 for examples), fluo-3 is currently the preferred Ca^{2+} indicator. Notwithstanding the theoretical considerations, in many cases the choice of Ca^{2+} indicator may be influenced by unpredictable problems of obtaining efficient loading and avoiding compartmentation of the dye into organelles. In this respect, trial and error of loading conditions and different indicator dyes is the only way to select the best Ca^{2+} indicator. Generally quin-2 has been the least problematic of the available fluorescent Ca^{2+} indicators. One other consideration is the Ca^{2+} buffering effects of the indicator. In most instances it is preferable to minimize the intracellular Ca^{2+} indicator concentration to avoid altering the kinetics of Ca^{2+} changes, but if

Ca^{2+} buffering is an objective, quin-2 is by far the easiest indicator to load at high concentrations.

4.7 Sources of Ca^{2+} indicator dyes

The fluorescent Ca^{2+} indicators described above are available as both the active free acid (or potassium salt) and cell permeant AME forms from a number of suppliers of research chemicals and biochemicals world-wide, including (for addresses see Appendix):

Molecular Probes Inc. (*The Handbook of Fluorescent Probes and Research Chemicals* by R. P. Haugland, ref. 13 which comprises the catalogue of this company is a useful information resource).

Calbiochem Corp. (The Calbiochem catalogue also provides technical information on some of the Ca^{2+} indicators.)

Sigma Chemical Co.

5. Loading Ca^{2+} indicators into cells

5.1 Loading cells with Ca^{2+} indicators as the AME

The loading of fluorescent Ca^{2+} indicators as the AME is the most commonly used approach (although it may also be the most prone to artefacts in Ca^{2+} measurements). The main advantages of ester loading are that no special equipment is required and that this method is applicable to cell population measurements. A variety of esters of the Ca^{2+} indicator dyes have been found to penetrate the cell membrane, but the AME is the most readily activated by cellular esterases.

5.1.1 Ester loading protocols

Optimal *intracellular* indicator concentrations are in the range 10–50 μM for most of the dyes (>100 μM for quin-2). Loading is less than 50% efficient, but a sufficiently high proportion of added dye is often taken up by the cells that cell density can affect the level of loading. Thus, with high cell densities a higher concentration of Ca^{2+} indicator (AME form) is required. An amount of Ca^{2+} indicator in the range of 0.2–4 nmol/mg cell protein is appropriate. With very low cell densities (e.g. cells in monolayer culture) the proportion of dye to cells may need to be increased somewhat. Since cell types vary greatly in their ability to accumulate the Ca^{2+} indicator dyes, a trial and error approach is inevitable with previously untested cells. The AM-esters are not soluble in water and are generally dissolved in dimethyl sulfoxide (DMSO). Since the ester forms of the dyes are rather susceptible to hydrolysis, it is recommended that stock solutions be stored in aliquots sufficient for a single experiment. In addition, the use of anhydrous DMSO may prolong the storage life of the dyes.

Dye loading is carried out in an appropriate physiological cell incubation medium. Loading proceeds most rapidly at 37 °C, but lower temperatures (25–30 °C) can reduce compartmentation. To improve solubilization of the ester forms of the dyes, serum albumin can be used in the medium. As an additional dispersion aid, low levels of Pluronic F-127 (a non-ionic surfactant polyol) can be premixed with the AM-esters. Continuous mixing of cells in suspension or gentle shaking of cells in monolayer culture facilitates even and efficient loading. Cells can be incubated at relatively high densities to improve the economy of loading, but because the hydrolysis products of the de-esterification may be toxic, cells are often diluted after sufficient incubation time to allow completion of uptake. In fact, it may be advantageous to incubate the cells for a period of time after washing away excess dye to facilitate completion of the de-esterification reactions. Typical cell loading buffers and a suggested loading procedure are given in *Protocol 1*. Incubation times and dye concentrations will vary depending on cell type.

Protocol 1. Loading cells with AME of Ca^{2+} indicator dyes

Materials

- cell incubation medium: any appropriate physiological medium is acceptable, e.g. 120 mM NaCl, 20 mM Hepes, 4.7 mM KCl, 1.2 mM KH_2PO_4, 1.2 mM $MgSO_4$, 1.25 mM $CaCl_2$, 10 mM glucose, pH 7.4
- pluronic F-127 solution: 10% (w/v) in DMSO
- acetoxymethyl ester of Ca^{2+} indicator at 1–10 mM in dry DMSO
- bovine serum albumin (BSA) solution of 2% (w/v) in incubation buffer

Protocol

1. Pre-incubate cells for 10 min at the required temp. (25–37 °C) to allow stabilization and equilibration of ion gradients.
2. Prepare a dispersion of the AME by vigorously mixing 4 μl/ml indicator solution, 10 μl/ml Pluronic F-127 solution in the BSA-containing medium.
3. Add 0.25 ml of this dispersion per 0.75 ml cell incubation and mix well. Depending on the concentration of the initial stock solution, this will give a final indicator–AM concentration of 1–10 μM (see Section 5.1 for selecting optimal loading). Pluronic F-127 is 0.025% final.
4. Continue incubation for 15–60 min as required (Section 5.1). If a relatively high cell density is used the cell suspension can be diluted after about one-third of the time.
5. Wash the cells three times in the required final incubation medium.

5.1.2 Assessing hydrolysis of added ester and determining cytosolic dye content

When establishing loading conditions for a new cell type, and at any time when problems are encountered, it is important to determine the time-course and extent of de-esterification and to measure the cytosolic content of active dye. Several of the available Ca^{2+} indicators have sufficiently different emission or excitation spectra in the ester form for the hydrolysis to be followed during the loading incubation by repeated wavelength scans. This approach does not, however, distinguish between dye hydrolysed inside the cell and that hydrolysed in the extracellular medium. In addition, since the de-esterification is a 4- or 5-step process, the spectra may not reveal the presence of partially hydrolysed forms. The most direct approach to assessing hydrolysis of the ester groups is to release the indicator from the cells and to test its Ca^{2+} sensitivity. This is best done after the cells have been washed free of extracellular indicator. Ideally, cells are lysed using a method which is specific for the plasma membrane (e.g. digitonin). Cells are initially lysed in a Ca^{2+}-free medium containing EGTA, and then a series of spectra are taken (similar to those in *Figure 5*) with increasing levels of added Ca^{2+}, buffered to known concentrations. If significant extracellular dye is present (see Section 6) it may be necessary to correct for this by carrying out a parallel Ca^{2+} titration without cell lysis. In cases where a spectrofluorometer is not available, it is sufficient to determine the Ca^{2+} sensitivity of the released dye at a single wavelength. Details on the preparation and use of EGTA–Ca buffers can be found elsewhere in this volume (Chapter 7).

To determine intracellular dye content, a similar cell lysis approach to that described above can be used. However, in this case the magnitude of fluorescence change on transition from the Ca^{2+} bound to Ca^{2+}-free form at a single wavelength is compared with the equivalent fluorescence change for an added standard (free acid form) of the Ca^{2+} indicator. An example of this approach is included at the end of the calibration procedure depicted in *Figure 6*. If the intracellular volume of the cells is known the cytosolic dye concentration can be calculated. For example, hepatocytes have a cytosolic volume of approximately 2 µl/mg cell protein. If the dye content of a cell suspension containing 3 mg protein/ml was found to be equivalent to 0.2 µM added indicator, this would give a cellular dye content of: $0.2 \times 10^{-6}/3 \times 10^{3} = 67$ pmol/mg protein and an intracellular concentration of: $67 \times 10^{-12}/2 \times 10^{-6} = 33$ µM.

5.2 Other techniques for loading cells with fluorescent Ca^{2+} indicators

Although the ester loading approach is the most widely used, other approaches may be preferable under certain circumstances. Microinjection

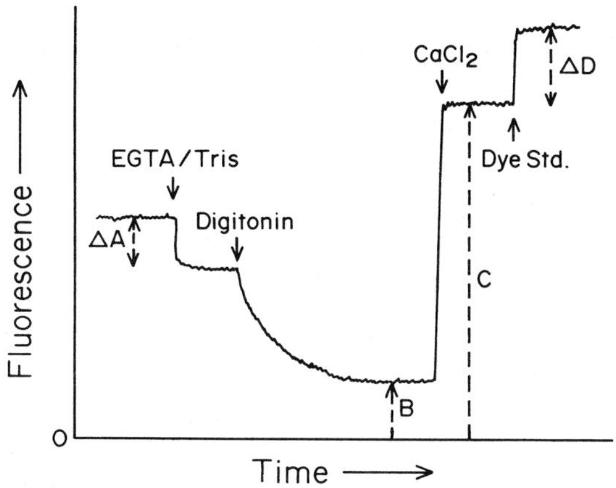

Figure 6. Example of calibration of single-wavelength data by cell lysis.

allows direct introduction of free acid forms of the indicator and reduces the chances of intracellular compartmentation. Appropriate injection techniques are also described in (19–21). These methods are used to introduce a number of other cytosolic Ca^{2+} indicators and are described in detail in other chapters of this book. Intracellular dye concentrations in the order of 30–100 μM are required (higher for quin-2). Thus, if a pressure injection of 1% of cell volume is to be used, the micropipette filling solution should contain about 3–10 mM indicator. On the other hand, when cells are loaded by intracellular perfusion from a patch-pipette, the filling solution will usually contain the required final Ca^{2+} indicator concentration. One other technique for loading fluorescent Ca^{2+} indicators which has achieved some success in plant cells, which are difficult to load with acetoxymethyl ester forms, is the use of acidic extracellular pH to increase the protonation of the indicators making them more membrane permeant (22).

6. Calibration

6.1 Basic concepts underlying calibration

Ca^{2+} measurements using the fluorescent indicators rely on the change in fluorescent properties following Ca^{2+} binding to report the degree of Ca^{2+} saturation of the indicator. At equilibrium, this is related to the free Ca^{2+} concentration according to the mass action equation:

$$K_d = \frac{[X] \times [Ca^{2+}]}{[CaX]} \text{ giving: } Ca^{2+} = K_d \times \frac{[CaX]}{[X]},$$

where X is the fluorescent indicator which binds Ca^{2+} with dissociation constant K_d. In a single wavelength mode of measurement, where fluorescence increases with Ca^{2+} binding, the amount of CaX for any given fluorescence signal is proportional to the difference between the measured fluorescence and the minimum fluorescence at zero [Ca^{2+}]. Likewise, the amount of free X is proportional to the difference between maximal fluorescence at saturating Ca^{2+} and the measured fluorescence. Thus, the term [CaX]/[X] can be replace by the measured fluorescence parameters to yield the following equation (2):

$$[Ca^{2+}] = K_d \times \frac{(F - F_{min})}{(F_{max} - F)}$$

where F_{min} is the fluorescence in the absence of Ca^{2+}, F_{max} is the fluorescence at saturating Ca^{2+}, and F is the fluorescence at any unknown Ca^{2+} concentration. F_{min} and F_{max} are determined by exposing the indicator to Ca^{2+}-free and Ca^{2+}-saturating conditions using the calibration procedures described below. The single-wavelength equation can be used for any Ca^{2+} indicator, but is most commonly used for indicators which are not suitable for ratio measurements.

With dyes such as fura-2 or indo-1, which give shifts in fluorescence spectra, measurements can be carried out at two wavelengths to obtain signals which are proportional to Ca^{2+}-bound and Ca^{2+}-free indicator. The ratio of fluorescence at the two wavelengths is directly related to the ratio of the two forms of the dye and therefore can be used to calculate $[Ca^{2+}]_i$. The advantage of the ratio approach is that indicator concentration terms, path length (important for single-cell work) and instrument sensitivity parameters cancel out at the level of the real-time measurement. There is no requirement that these parameters remain constant throughout the experiment and during calibration, as there is in the single-wavelength mode. The equation which relates the measured Ca^{2+}-bound/Ca^{2+}-free fluorescence ratio (R) to $[Ca^{2+}]_i$ has a similar form to the single-wavelength equation (see ref. 3 for derivation):

$$[Ca^{2+}] = K_d \times \frac{(R - R_{min})}{(R_{max} - R)} \times \frac{S_{f2}}{S_{b2}}$$

where R_{max} and R_{min} are the fluorescence ratio values under saturating and Ca^{2+}-free conditions, respectively, and S_{f2}/S_{b2} is the ratio of fluorescence values for Ca^{2+}-bound/Ca^{2+}-free indicator measured at the wavelength used to monitor the Ca^{2+}-free indicator (the denominator wavelength of R). Importantly, these are all unitless values which depend only on the chemical properties of the indicator and the relative sensitivity of the instrument as a whole for the two wavelengths of measurement. As with the single-wavelength method, calibration measurements at zero and saturating levels of Ca^{2+} are required to determine the constants R_{max}, R_{min}, and S_{f2}/S_{b2}.

Overall, the objective of calibration (for either single-wavelength or dual-wavelength measurements) is to determine the fluorescence levels under conditions of maximum and minimum Ca^{2+} binding. This can be achieved by releasing the Ca^{2+} indicator to the incubation medium and manipulating Ca^{2+} there, or by manipulating intracellular Ca^{2+} without releasing the indicator. The latter approach gives values under conditions that most closely mimic the conditions of $[Ca^{2+}]_i$ measurement, but it is frequently difficult to manipulate $[Ca^{2+}]_i$ over the required range. With indicators used in the ratio mode, where the absolute amount of indicator is not relevant, calibration can also be carried out using solutions of the free acid form of the Ca^{2+} indicator. More details of the underlying assumptions and derivation of the calibration procedures can be found in refs. 4, 23, 24.

6.2 Calibration of single-wavelength measurements

6.2.1 Cell lysis calibration at a single wavelength

This is the most straightforward calibration, in which the Ca^{2+} indicator is released from the cells and then medium Ca^{2+} is adjusted to give F_{min} and F_{max}. This approach works well for cells maintained in suspension in a cuvette, but is not suitable for cell preparations in which the released indicator redistributes out of the light path (e.g. single-cell studies or monolayers). *Figure 6* shows an example of cell lysis calibration and the procedure is given in *Protocol 2*. In most preparations there is some extracellular dye which has not been removed by washing. To correct for this, the fluorescence due to extracellular Ca^{2+}-bound indicator is determined by addition of excess EGTA (in the presence of Tris-base to elevate the final pH >8, increasing the affinity of EGTA for Ca^{2+}). The rapid fluorescence signal change following EGTA–Tris addition (ΔA) is subtracted from all fluorescence measurements carried out in the presence of extracellular Ca^{2+} prior to calibration. An alternative approach to determine extracellular dye is to use the ability of Mn^{2+} to quench the fluorescence as described below for *in situ* calibration (note this will not work with fluo-3). Immediately after the rapid change induced by EGTA–Tris, the cells are lysed with detergent. To help minimize errors due to compartmentalized dye, the use of a detergent which primarily affects the plasma membrane, such as digitonin, is recommended. This should be titrated to determine the optimum concentration (using Trypan Blue exclusion or lactate dehydrogenase release as a marker for plasma membrane permeabilization). Alternatively, any detergent which causes a minimum light scattering change such as low concentrations of Triton X-100 can be used. Cell lysis may take several minutes. When a stable minimum fluorescence is reached the value (B) is taken as F_{min}. An excess of $CaCl_2$ is next added to saturate all of the Ca^{2+} indicator. F_{max} for intracellular dye is then given by the maximum fluorescence at saturating Ca^{2+} after subtracting the signal due to what was originally the extracellular dye component ($F_{max} = C - \Delta A$).

Protocol 2. Calibration of single-wavelength fluorescence data using cell lysis

1. Use this calibration with cells in a well-mixed suspension.
2. Start with cells in physiological extracellular $[Ca^{2+}]$.
3. Estimate the contribution of extracellular indicator by adding EGTA equivalent to 3–4 times the medium Ca^{2+} content plus Tris-base to elevate the pH above 8. For cell incubation medium containing 1 mM $CaCl_2$ add 10 μl/ml of stock solution: 400 mM EGTA, 3 M Tris-base, pH 8.7.
4. Release cytosolic Ca^{2+} indicator using digitonin, 1–10 μl/ml of a 10 mg/ml solution (amount determined by titration).
5. Add excess $CaCl_2$ to give at least 100 μM free $[Ca^{2+}]$. With 4 mM final [EGTA] and 1 mM $CaCl_2$ in the initial incubation medium (step 2), use 8 μl/ml of 500 mM $CaCl_2$.
6. Determine cellular dye content by adding a suitable standard. For quin-2 use 10 μl/ml of 100 μM quin-2 (free acid dissolved in water) to give 1 μM final. For more highly fluorescent indicators use 10 μl/ml of 10 μM stock solutions to give a final concentration of 0.1 μM.
7. Determine whether significant autofluorescence changes occur during calibration by repeating steps 2–5 with cells not loaded with Ca^{2+} indicator.
8. Calculate values for F_{min} and F_{max} using the fluorescence measurements indicated in *Figure 6* as described in Section 6.2.1.

Figure 6 also shows the addition of a dye standard of the free acid form of the Ca^{2+} indicator. The fluorescence increase resulting from this addition (ΔD) can be used to calculate the intracellular concentration of the indicator as described in Section 5.1.2. To relate the fluorescence change following addition of the standard to the cellular dye content, it is necessary to know what percentage of the signal is Ca^{2+}-dependent. This is best determined in a separate aliquot of cell suspension buffer. The Ca^{2+} indicator is added to the buffer and the fluorescence change noted. EGTA–Tris is then added and the resulting fluorescence decrease as a percentage of the total dye fluorescence gives a correction factor for converting between Ca^{2+}-dependent and total fluorescence. For quin-2 this is about 83%. Thus, for the experiment depicted in *Figure 6*, the intracellular dye content is given by: $(C - \Delta A - B)/(\Delta D \times 0.83) \times$ Std. Where Std. is the amount (pmol) of added Ca^{2+} indicator.

The contributions of autofluorescence cancel out in the two expressions ($F - F_{min}$) and ($F_{max} - F$) so long as these do not change. Changes of autofluorescence are most likely to occur during cell lysis but may also result from agonist addition (e.g. Ca^{2+}-mobilizing hormones elevate NADH levels in hepatocytes with a resulting increase in autofluorescence; ref. 25). If changes of autofluorescence occur, these are most easily corrected by carrying out a parallel experiment with an *equal concentration* of unloaded cells and subtracting the autofluorescence values at each point from the fluorescence in dye-loaded cells. This should be done prior to determining the calibration parameters.

6.2.2 *In situ* calibration at a single wavelength using ionophore

F_{min} and F_{max} would best be determined with the dye retained inside the cell. This can be achieved by making the cell membrane specifically permeable to Ca^{2+}, using an ionophore, and then manipulating extracellular [Ca^{2+}] as required. Since A23187 is fluorescent, the ionophores most commonly used are ionomycin or Br-A23187. In common with A23187, both of these ionophores exchange Ca^{2+} or other divalent metal ions, including Mg^{2+} and Mn^{2+}, for H^+. The main problem with using ionophores is that these agents do not easily equilibrate $[Ca^{2+}]_i$ with the very low levels of extracellular Ca^{2+} required to give F_{min}, due to the small transmembrane gradient of Ca^{2+}. An alternative approach to the measurement of F_{min} is to use the ability of Mn^{2+} to alter the fluorescence of the Ca^{2+} indicators. Since Mn^{2+} can enter cells in the presence of ionophore and binds to the indicators with higher affinity than Ca^{2+}, the fluorescence under conditions of Mn^{2+} saturation can be used to calculate the fluorescence of the Ca^{2+}-free form of the indicator *in situ*. For dyes such as quin-2 and fura-2, Mn^{2+} quenches all fluorescence leaving only the autofluorescence, while with fluo-3 the Mn^{2+}-bound dye has an intermediate fluorescence, making this approach more complicated. However, in all cases the fluorescence of the Ca^{2+}-free indicator has a fixed relationship to the difference in fluorescence between the Mn^{2+}-bound and Ca^{2+}-bound forms and consequently this difference can be used to calculate F_{min}. In some cells, it is also difficult to fully saturate cytosolic Ca^{2+} indicators during treatment with Ca^{2+} ionophores in the presence of extracellular Ca^{2+}. This may be due to the activity of the Ca^{2+} extrusion pumps in these cells. In some cases this can be overcome by increasing extracellular Ca^{2+} to 10–20 mM or by using inhibitors to reduce Ca^{2+} efflux. The *in situ* calibration is the only alternative which is suitable for use with single wavelength dyes used in single-cell studies.

Figure 7 shows a typical calibration using ionomycin to determine F_{min} and F_{max} for a single-wavelength Ca^{2+} indicator *in situ*. Details of the sequence of additions is given in *Protocol 3*. First, a low Mn^{2+} concentration is used to quench extracellular dye. The resulting fluorescence change (ΔA) can be subtracted from all measurements in the presence of extracellular Ca^{2+}, as

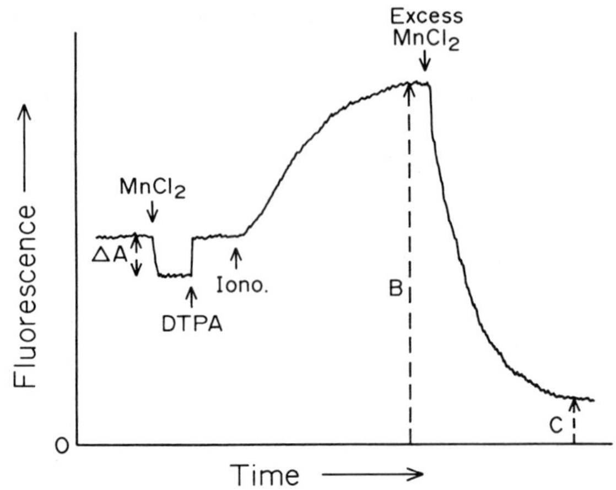

Figure 7. Example of calibration of single-wavelength data using a Ca^{2+} ionophore.

described above for the initial EGTA addition. As soon as the rapid Mn^{2+}-induced fluorescence change is complete, the Mn^{2+} is chelated using diethylenetriaminepentaacetic acid (DTPA, ref. 26) to prevent quenching of intracellular dye. Ionomycin is then added and the fluorescence at the plateau of maximal $[Ca^{2+}]_i$ (B) is corrected for extracellular dye and the difference $(B - \Delta A)$ is taken as F_{max}. Subsequently, excess Mn^{2+} is added and this now penetrates the cytosol using the ionophore and quenches both intracellular and extracellular dye fluorescence. The relationship between the fluorescence of Ca^{2+}-bound, Mn^{2+}-bound and free indicator needs to be known or determined in a separate cell-free experiment. For example, with quin-2 the Ca^{2+}-free dye has about 17% the fluorescence of the Ca^{2+}-bound form and Mn^{2+} quenches almost completely. Thus, F_{min} for quin-2 is given by: $(B - \Delta A - C) \times 0.17 + C$. Clearly, with Ca^{2+} indicators other than quin-2 the value 0.17 is different, and in fact this parameter is also dependent on the precise wavelength used for the measurements. Note that ΔA must be subtracted from all values of F used to calculate $[Ca^{2+}]_i$. In addition, since for most indicators C is effectively the autofluorescence, this parameter can be subtracted from all calibration and measurement values prior to calibration, in which case it is removed from the above equation. With single-cell or monolayer measurements, where extracellular indicator can be washed away, autofluorescence can be determined by releasing cytosolic dye with digitonin. In this approach the equation: $F_{min} = (B - \Delta A - C) \times \text{Factor} + C$ is still valid; with C = autofluorescence and Factor = the fluorescence of the Ca^{2+} free indicator as a percentage of that given by the Ca^{2+}-bound indicator. An example of this calibration for cells loaded with fluo-3 is given in ref. 27. For systems where significant autofluorescence changes occur, a separate experiment with unloaded cells must still be carried out as described in Section 6.2.1.

Protocol 3. Calibration of single-wavelength fluorescence data *in situ* using calcium ionophore

1. Use this calibration for single-wavelength measurements with cells in suspension, monolayers maintained in a fluorometer or single-cell studies carried out with a fluorescence microscope.
2. Estimate the contribution of extracellular indicator by adding 0.1 mM $MnCl_2$. After 10–20 sec add 0.25 mM DTPA. (Correction for extracellular dye is generally not necessary in single-cell measurements).
3. Add sufficient ionophore to give a maximal signal increase. With ionomycin, a final concentration of 2–40 μM from a stock solution in DMSO is usually sufficient. (*Note*: serum albumins bind ionomycin and should be eliminated from the buffer if possible to maximize the efficacy of the ionophore).
4. Add 2 mM $MnCl_2$ to be in excess of the [DTPA] added at step 2. The Mn^{2+} utilizes the ionophore to obtain access to the intracellular Ca^{2+} indicator.
5. Test for autofluorescence changes by repeating steps 2–4 with cells not loaded with Ca^{2+} indicator.
6. Calculate values for F_{min} and F_{max} using the fluorescence measurements indicated in *Figure 7* as described in Section 6.2.2.

6.3 Calibration using ratio approach

The values R_{min} and R_{max} can be obtained by manipulating medium Ca^{2+} as described above for the single-wavelength measurements. However, because autofluorescence does not cancel out in the ratio equation, this parameter must be measured separately at both wavelengths and subtracted from all data and calibration values *prior to calculating the ratios*. Extracellular Ca^{2+} indicator can be corrected for by using the brief Mn^{2+} treatment followed by chelation with DTPA, as outlined above, since this quenches both measurement wavelengths for fura-2 and indo-1.

6.3.1 Cell lysis calibration of ratio

When the cell lysis approach is used (see above), R_{min} is given by the ratio at steady state following addition of digitonin in the presence of EGTA–Tris, and R_{max} is the ratio after addition of excess $CaCl_2$. R_{min} and R_{max} may be obtained in any order after cell lysis. In the calibration depicted in *Figure 8* digitonin is added in the presence of normal (high) extracellular Ca^{2+} and

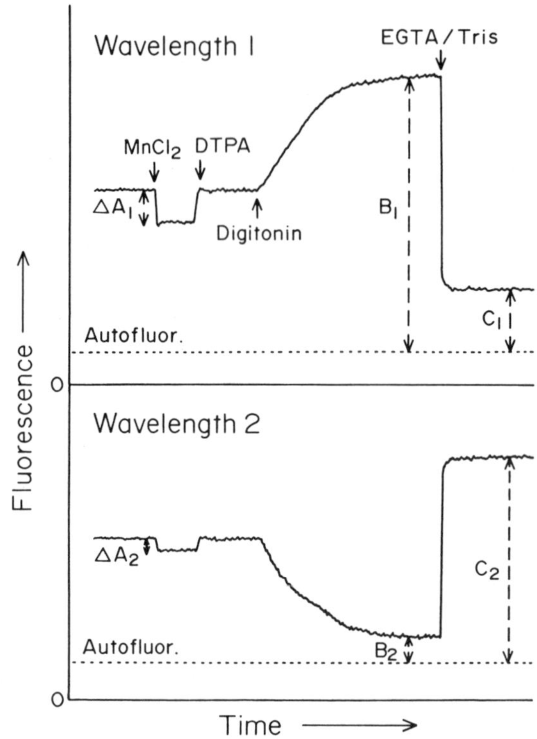

Figure 8. Example of ratio calibration following cell lysis.

EGTA–Tris is added subsequently. *Protocol 4* describes the additions for this procedure. Note that in the ratio calibration the fluorescence values do not need to be corrected for extracellular dye, since the ratio is independent of dye concentration. Thus ΔA_1 and ΔA_2 are only subtracted from the pre-calibration experimental values. R_{max} is given by B_1/B_2; R_{min} is given by C_1/C_2; and $S/_{f2}/S_{b2}$ is given by C_2/B_2. These parameters are calculated *after* subtraction of autofluorescence.

Protocol 4. Calibration of dual-wavelength fluorescence data using cell lysis

1. Use this calibration only for fluorometer measurements from cells in suspension, where released dye remains in the light path.
2. Estimate the contribution of extracellular indicator by adding 0.1 mM $MnCl_2$, followed a short time later by 0.25 mM DTPA.
3. Release cytosolic Ca^{2+} indicator by adding 10–100 μg/ml digitonin.

Protocol 4. *Continued*

4. Add EGTA equivalent to 3–4 times the medium Ca^{2+} content plus Tris-base to elevate the pH above 8. For medium containing 1 mM $CaCl_2$ add 10 µl/ml of 400 mM EGTA, 3 M Tris-base, pH 8.7.
5. Measure autofluorescence, repeating steps 2–4 with cells not loaded with Ca^{2+} indicator.
6. Subtract appropriate autofluorescence values from all experimental measurements and calibration data at both wavelengths.
7. Calculate values for R_{min}, R_{max}, and S_{f2}/S_{b2} using the fluorescence measurements indicated in *Figure 8* as described in Section 6.3.1.

6.3.2 *In situ* calibration of ratio using ionophores

Quenching by Mn^{2+} cannot readily be used as a substitute for low $[Ca^{2+}]_i$ in ratio calibration methods carried out *in situ* using ionophores, since a meaningful value of R_{min} is not obtained. Correction as described above using the difference between Ca^{2+}-saturated and Mn^{2+}-quenched dye at each wavelength is complex and subject to relatively large errors. Thus, it is necessary to rely on direct lowering of $[Ca^{2+}]_i$ using low extracellular Ca^{2+}. *Figure 9* shows an example of such a calibration. This pair of traces represents a single-cell experiment in which extracellular indicator is negligible. After subtraction of autofluorescence, the final ratio in the presence of EGTA–Tris plus ionomycin (A_1/A_2) is R_{min} and the maximal ratio with saturating $CaCl_2$ (B_1/B_2) is R_{max}. The constant S_{f1}/S_{b2} is given by A_2/B_2. In a single-cell experiment the extracellular medium can easily be exchanged and the excess EGTA washed out. Under this condition it is possible to add excess $MnCl_2$ in the continued presence of ionophore to quench intracellular dye, yielding a value for autofluorescence. This is shown in *Figure 9*. The procedure for this type of calibration is detailed in *Protocol 5*.

Protocol 5. Calibration of dual-wavelength fluorescence data *in situ* using calcium ionophore

1. Use this calibration for any measurement configuration with dual-wavelength indicators in which the cells respond sufficiently well to ionophore to allow the necessary manipulation of $[Ca^{2+}]_i$.
2. If necessary, estimate the contribution of extracellular indicator using $MnCl_2$/DTPA (*Protocol 3*, step 2).
3. Add EGTA–Tris (*Protocol 2*, step 3) to chelate extracellular Ca^{2+}.
4. Add 2–40 µM ionomycin as required to yield maximal signal changes at the steady-state plateau levels (*Protocol 3*, step 3).

Protocol 5. *Continued*

5. After a minimum $[Ca^{2+}]_i$ has been achieved, add 10 mM $CaCl_2$ to allow saturation of intracellular Ca^{2+} indicator.
6. Add 5 mM $MnCl_2$ which quenches intracellular dye fluorescence (by virtue of the ionophore), yielding an estimate of autofluorescence at each wavelength.
7. Check for autofluorescence changes by repeating steps 2–6 with cells not loaded with Ca^{2+} indicator.
8. Subtract appropriate autofluorescence values from all experimental measurements and calibration data at both wavelengths.
9. Calculate values for R_{min}, R_{max}, and S_{f2}/S_{b2} using the fluorescence measurements indicated in *Figure 9* as described in Section 6.3.2.

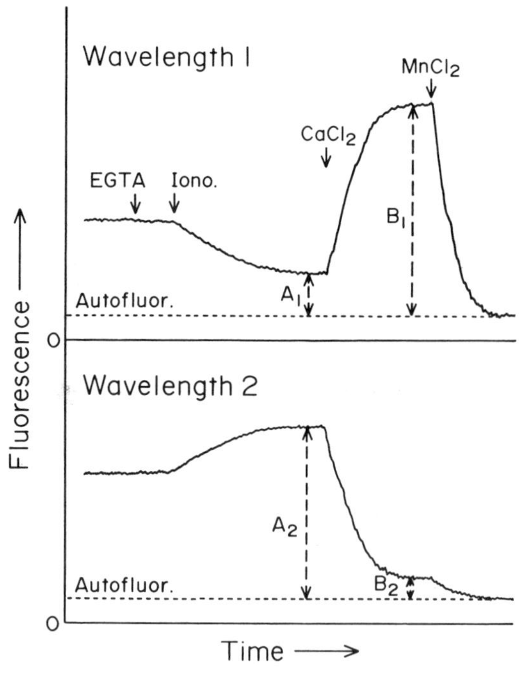

Figure 9. Example of ratio calibration using a Ca^{2+} ionophore.

6.3.3 Cell-free calibration of ratio measurements

Because the ratio is independent of dye concentration or path length, it is possible to obtain values for R_{min} and R_{max} in the absence of cells for the particular instrument configuration and wavelengths to be used with dye-loaded cells. This is achieved by taking a solution of the free acid form of the

Ca^{2+} indicator in a buffer designed to mimic, as far as possible, the intracellular milieu. Measurements are taken with Ca^{2+} chelated using EGTA (can be at physiological pH if sufficient excess is used) followed by addition of excess $CaCl_2$. *Protocol 6* gives a suggested procedure. This *in vitro* calibration can be used for both single-cell and cuvette measurements.

Protocol 6. Calibration of dual-wavelength fluorescence data using values obtained in free solution

1. Use this protocol to calculate R_{min}, R_{max}, and S_{f2}/S_{b2} for ratio calibration in any instrument where fluorescence measurements can be made from the indicator in solution, including fluorometer cuvettes and microscope cell incubation chambers.
2. Prepare a solution composed of 120 mM KCl, 20 mM Hepes, 2 mM KH_2PO_4, 0.5 mM $MgCl_2$, 0.2 mM EGTA, pH 7.2.
3. Obtain initial fluorescence measurements at the two required wavelengths. These background values are subtracted from the subsequent measurements.
4. Add 0.5 μM fura-2 or indo-1. The ratio of fluorescence of this solution at the two wavelengths (after background subtraction) gives R_{min}.
5. Add 1 mM $CaCl_2$. The fluorescence ratio of this solution gives R_{max}.
6. The calibration parameter S_{f2}/S_{b2} is obtained from the ratio of fluorescence values measured before/after $CaCl_2$ addition at wavelength 2 (the wavelength optimized for the Ca^{2+}-free indicator form).

6.4 Additional considerations in the calibration of Ca^{2+} images

Images of the fluorescence of Ca^{2+} indicators in single cells present a number of special problems. The first group of these relate to the correction of the images for non-homogeneity of illumination, camera gain, or spatial distortions (see Section 3.3.7). Because of variations in cell thickness and dye distribution, it is essential that some form of ratio measurement be carried out. With dual wavelength indicators accurate co-registration of the images obtained at the two wavelengths is an important prerequisite for calculation of the image ratio (even a small offset can introduce misleading 'edge effects'). If cells move slowly, the dual-wavelength approach can be expected to provide reliable images of the fluorescence ratio. However, if the frequency of data collection is relatively slow, rapidly moving cells can also generate edge effects. Another problem is to obtain a satisfactory correction for autofluorescence. This is most often dealt with by using sufficient dye loading to minimize the contribution of autofluorescence. Providing that autofluorescence remains constant throughout the experiment, it is possible to correct for it using the Mn^{2+} quench approach described above (*Figure 9*).

7. Examples of cytosolic Ca^{2+} measurements

In this section, Ca^{2+} measurements with a number of cell types and different instrument configurations carried out in our laboratory are described as examples of the types of data which may be obtained using fluorescent Ca^{2+} indicators.

7.1 Ca^{2+} measurements in cell populations

Figure 10 shows studies carried out using quin-2-loaded hepatocytes. The cells were loaded at 10 mg cell protein/ml in a Hepes-buffered medium with 80 μM quin-2/AM for 20 min. Hepatocytes have a high level of autofluorescence (due primarily to pyridine nucleotides) which changes as a result of the Ca^{2+} mobilization induced by hormone treatment. To distinguish autofluorescence changes resulting from increases in pyridine nucleotide redox state from quin-2 fluorescence changes, measurements were carried out using a T-format fluorometer with emission filters selected to allow discrimination of the fluorescence from the two species. Excitation was at 335 nm (mercury lamp) and emission measurements were made at 420–450 nm, reflecting primarily pyridine nucleotide fluorescence and at 490–570 nm, where quin-2 is the major fluorescent species (see ref. 25 for details). The upper trace of *Figure 10* shows the increase in $[Ca^{2+}]_i$ induced by the α_1-adrenergic agonist phenylephrine at a submaximal dose, followed by a saturating dose of vasopressin (10 nM), which also acts by increasing $[Ca^{2+}]_i$ in these cells. In the lower trace the fluorescence signal resulting from pyridine nucleotides is

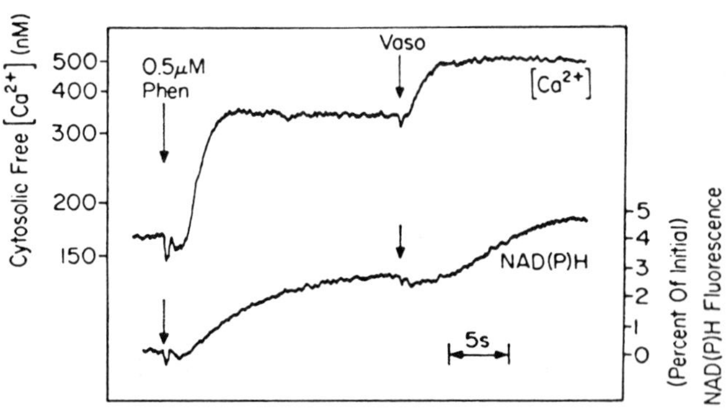

Figure 10. Hormone-induced Ca^{2+} increases and parallel changes of pyridine nucleotide fluorescence in quin-2-loaded hepatocytes maintained in suspension. See text for details.

plotted, showing that the Ca^{2+}-induced increase in redox state follows slower kinetics than the $[Ca^{2+}]$ increase.

An experiment utilizing fura-2 to follow $[Ca^{2+}]_i$ changes in isolated rat ventricular heart cells maintained in suspension is shown in *Figure 11*. In this case a dual excitation fluorometer was used with a rapid chopper capable of alternating excitation wavelengths at up to 500 Hz (similar to that shown schematically in *Figure 2*). The cardiomyocytes were incubated at 5 mg cell protein/ml with 4 μM fura-2/AM for 25 min. After washing, the cells were incubated in a bicarbonate buffer with gentle stirring and continuous gassing in the fluorometer cuvette fitted with Pt electrodes on either side of the light path. A series of short duration electrical pulses (bottom of figure) were used to stimulate the cells to beat synchronously and the $[Ca^{2+}]_i$ changes responsible for excitation–contraction coupling were monitored. The top traces show the transient spikes of $[Ca^{2+}]_i$ calculated from the individual fluorescence measurements at 340 nm (middle trace) and 380 nm (lower trace). The first set of recordings were obtained in the absence of agonist and the second set of recordings were obtained after addition of the *β*-adrenergic agonist isoproterenol (see ref. 28 for more details).

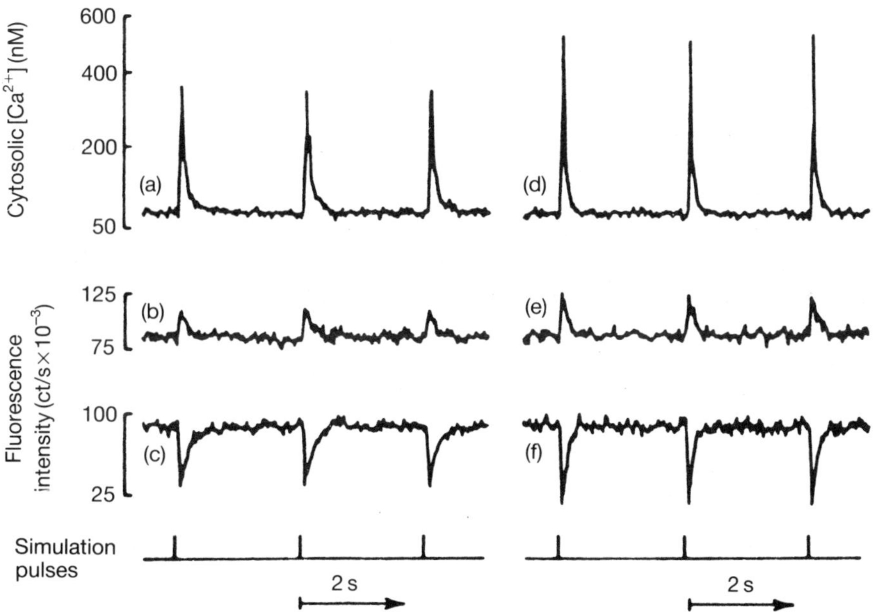

Figure 11. Population measurements of electrically-triggered Ca^{2+} transients in fura-2-loaded isolated ventricular heart cells. Upper traces (a, d) show cytosolic Ca^{2+} levels calculated from fluorescence measurements at 340 nm (b, e) and 380 nm (c, f). Electrical field stimulation pulses are shown in the bottom trace. For (d–f) cells were exposed to isoproterenol. See text for further details.

An experiment in which fura-2 was used to follow $[Ca^{2+}]_i$ changes in response to the growth factor PDGF in NIH-3T3 cells is shown in *Figure 12*. These cells were loaded with 3 μM fura-2/AM by incubation in Hepes buffer for 20 min. The dual excitation wavelength fluorometer used in the experiment of *Figure 11* was also used in this experiment. PDGF (20 ng/ml and 50 ng/ml) was added at zero time (arrow) and remained present throughout the rest of the traces. The $[Ca^{2+}]_i$ changes induced by PDGF were calculated from the ratio of 340 nm/380 nm. Note the relatively long delay between PDGF addition and the initiation of the $[Ca^{2+}]_i$ increase. As with many cell types, this latency period decreases with increasing agonist dose.

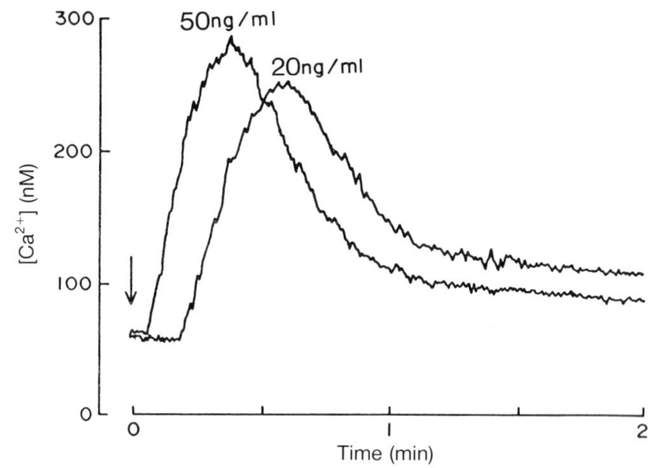

Figure 12. Response to PDGF of fura-2-loaded NIH-3T3 cells maintained in suspension. Calculated Ca^{2+} responses to 20 and 50 ng/ml PDGH added at arrow.

7.2 Ca^{2+} measurements at the single-cell level

Figure 13 shows traces of the fluorescence and calculated $[Ca^{2+}]_i$ changes for a single fura-2-loaded hepatocyte treated with phenylephrine (added at the arrow and present for the remainder of the experiment). These data were extracted from digital images of fluorescence obtained using a liquid-N_2-cooled CCD camera attached to an inverted epifluorescence microscope. A computer-controlled filter changer was used to alternate between 340 nm and 380 nm excitation wavelengths provided by a xenon arc lamp. In the top panel, the fluorescence values at 380 nm (upper trace) and 340 nm (lower trace) for the selected cell obtained from each pair of images is plotted and the $[Ca^{2+}]_i$ time-course calculated from these is plotted in the lower panel (data from ref. 29). Note the fluorescence at 340 nm is lower than that at 380 nm due to the optical limitations of the microscope. This experiment

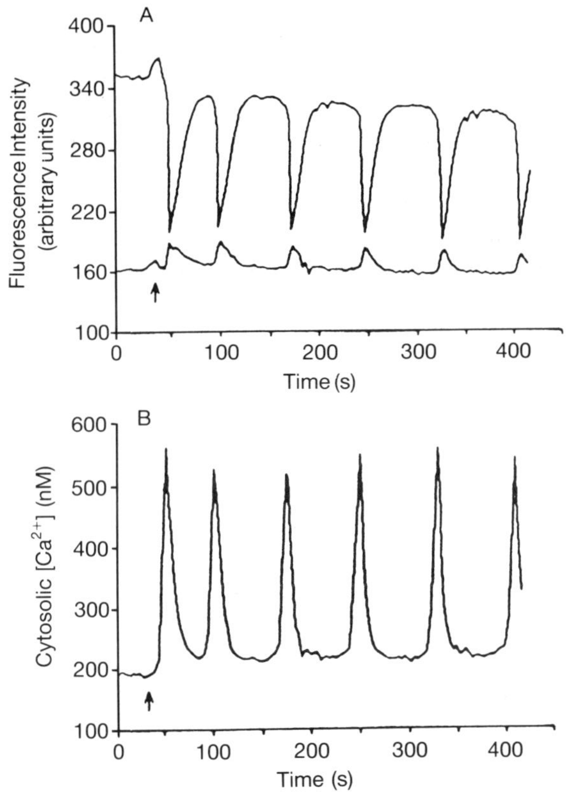

Figure 13. Fluorescence changes and calculated $[Ca^{2+}]_i$ in a single fura-2-loaded hepatocyte. Data were obtained using a digital imaging fluorescence microscopy system. Phenylephrine (10 μM) was present from the arrow to the end of the trace. (Data from ref. 29.)

illustrates some of the information that can be revealed by studies at the single-cell level that would not be apparent from population measurements. The most striking difference from the population experiment shown in *Figure 10* is that the individual cells do not respond to phenylephrine with a simple monophasic Ca^{2+} increase, but give a series of discrete $[Ca^{2+}]_i$ spikes. In addition, the rise time for these spikes is faster than observed for the population as a whole. This fine structure is lost in the population measurements due to the asynchronous nature of the individual cell responses to agonist. The discovery and characterization of these oscillatory $[Ca^{2+}]_i$ responses to hormones which act through the inositol lipid-dependent signal transduction system (29, 30) has provided one of the major spurs towards studies of $[Ca^{2+}]_i$ at the single-cell level.

Most of the information about the kinetics of $[Ca^{2+}]_i$ changes in single cells can be analysed by extracting data from images on a cell by cell basis, as outlined in the preceding paragraph. However, there is a further level of information available from such digital images, which relates to possible subcellular spatial heterogeneity of the $[Ca^{2+}]_i$ changes. In the case of the $[Ca^{2+}]_i$ oscillations induced by hormones in primary cultured hepatocytes, we have found that these are spatially organized. Each $[Ca^{2+}]_i$ spike originates from a specific subcellular locus, (apparently due to the polarized location of the receptors) and then spreads as a self-propagating wave through the rest of the cell. An example of this type of response is shown in *Figure 14*, where a series of images of $[Ca^{2+}]_i$ obtained at 0.9 sec intervals are shown, using a pseudocolour representation (spectral scale of blue = low $(Ca^{2+}]_i$ to red = high $[Ca^{2+}]_i$). Several cells appear in this image series and two of these show $[Ca^{2+}]_i$ waves during the time frame shown (for more details see ref. 31). This example demonstrates the utility of using colour to encode $[Ca^{2+}]_i$ images.

Another example of $[Ca^{2+}]_i$ images in a single cell is shown in *Figure 15*, where the $[Ca^{2+}]_i$ transient in response to electrical stimulation is followed. The upper and lower image series follow the time-course of single $[Ca^{2+}]_i$ transients in the absence (top) and presence (bottom) of isoproterenol. The stimulus was applied after the second cell image and can be seen to have caused a fairly homogeneous transient increase of $[Ca^{2+}]_i$ which was accompanied by contraction and subsequent relaxation of the rod-shaped heart cell. These images were obtained using a frame-transfer-cooled CCD camera with a time-base of 40 msec per image (rates up to 5 msec are feasible using this technique), although separate contractions had to be recorded to obtain fluorescence images at each wavelength. Details of the methodology and hardware are described in ref. 18. This experiment demonstrates another advantage of the imaging approach to $[Ca^{2+}]_i$ measurements, which is the ability to correlate Ca^{2+} changes with structural changes (in this case the Ca^{2+}/contraction relationship can be determined).

8. Problems encountered using fluorescent Ca^{2+} indicators

There are a number of potential problems with using the fluorescent Ca^{2+} indicators. Possible ways of identifying and getting round many of these have already been discussed in preceding sections. Further information relating to the most common concerns are given here.

8.1 Fluorescence contributions from non-cytosolic dye and Ca^{2+}-insensitive indicator

8.1.1 Extracellular dye

In almost all cell suspension measurements there is a finite amount of extracellular dye which resists removal by repeated washing of the cells. This

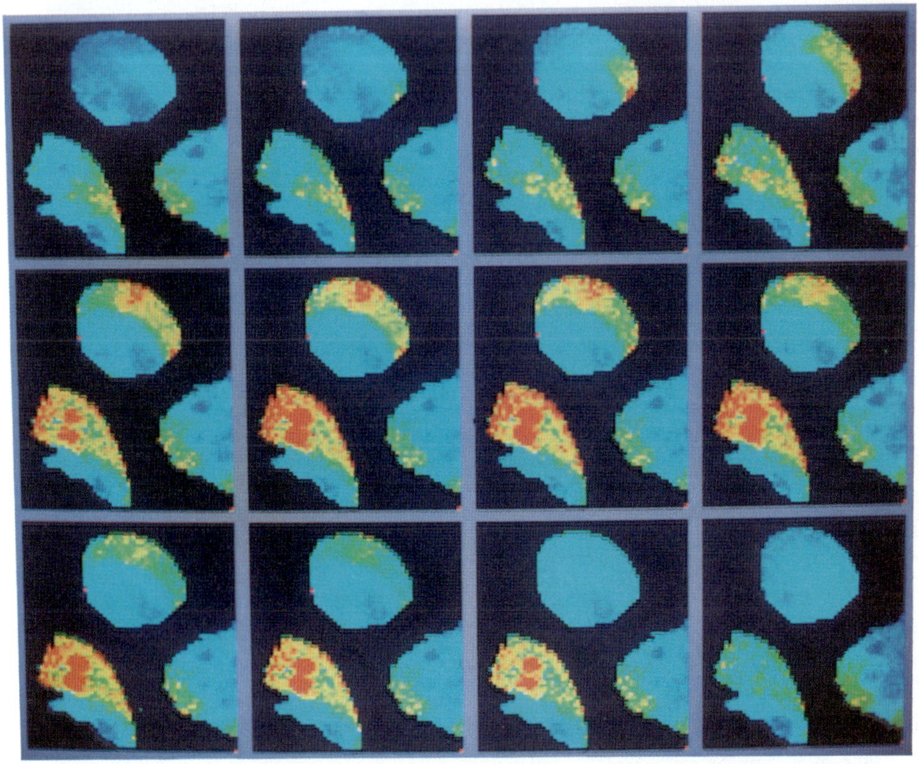

Figure 14. Images of $[Ca^{2+}]_i$ changes in fura-2-loaded hepatocytes maintained in primary culture. A series of 12 images at 0.9 sec intervals are shown during Ca^{2+} spikes of the type shown in *Figure 13*. Two cells (crescent in top group of cells and wedge-shaped cell in lower left group) show Ca^{2+} spikes which are composed of Ca^{2+} waves that propagate through the cells. See text and ref. 31 for further details.

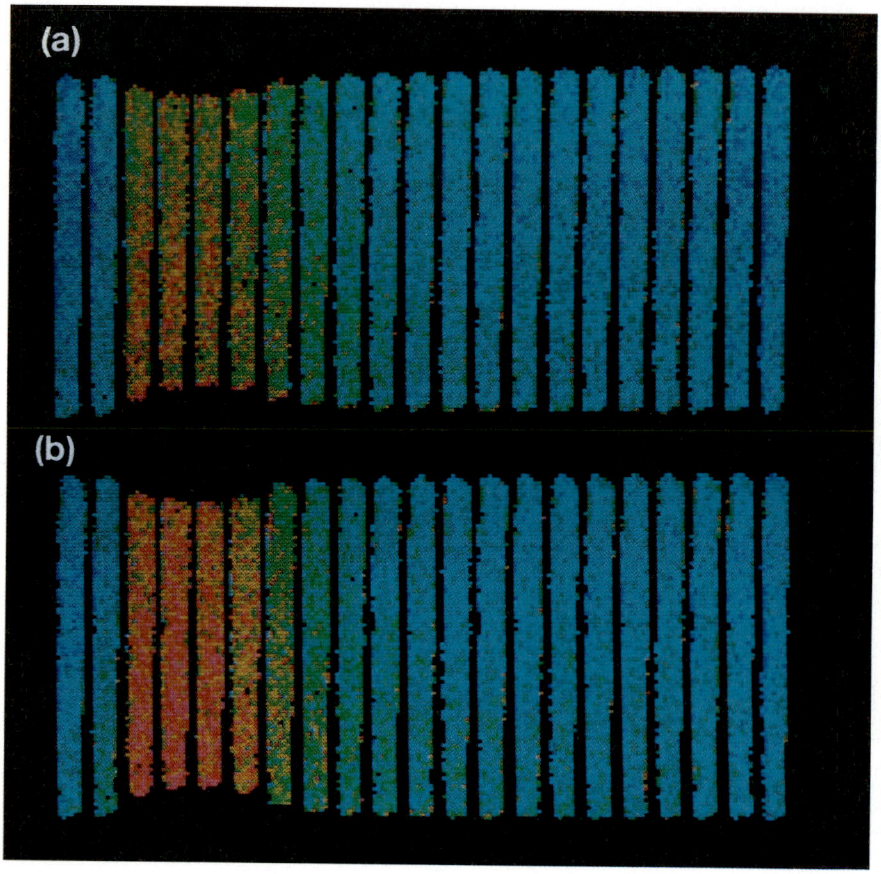

Figure 15. Time-series images of an isolated heart cell during electrically-induced contractions in the absence (a) or presence (b) of isoproterenol. Fast image acquisition using the frame transfer mode of a cooled CCD camera was used to obtain a series of images of fura-2 fluorescence at 40 msec intervals during a single contraction. Separate contractions were monitored for measurement of 340 nm and 380 nm fluorescence. Full details of the methodology can be found in ref. 18.

can be accounted for by the calibration procedures described in Section 6. However, accurate $[Ca^{2+}]_i$ measurements rely on the assumption that extracellular dye does not change significantly during the time-course of the experiment. If substantial leakage occurs it may be possible to correct the data prior to calibration. Since dye leakage is generally relatively linear with time, the leak rate can be determined in a separate experiment and this rate used to extrapolate to give a calculated extracellular dye signal for correction of the data at each time-point. It is, however, preferable to minimize leakage rather than correcting for it. One approach is to lower the working temperature to 25 ° or 30 °C. In many cells the indicators are actually extruded from the cells by an anion transporter. This can be blocked by inhibitors such as 200 μM sulfinpyrazone (dissolve at alkaline pH) or 2.5 mM probenecide (32).

8.1.2 Compartmentalized indicator and residual esterified dye

Compartmentalized indicators may be the most troublesome of the problems encountered with calibrating fluorescence signals from Ca^{2+} indicators. The amount of dye in non-cytosolic compartments can be reduced by lowering the temperature at which loading is carried out (lower incubation temperatures may also help). It can also be helpful to load for longer times with lower concentrations of the AM-esters. Where compartmentalized dye exists, its signal can often be subtracted from measurements by using an appropriate calibration. For example, the use of digitonin to lyse cells releases primarily cytosolic dye (27) and this feature can be used in a number of calibration procedures (Section 6). The problem of residual esterified dye can be dealt with in a similar manner, since this usually remains associated with cellular membranes. Many of the problems discussed in this section can, of course, be circumvented if the free acid form of the dye is injected directly into the cell cytosol.

8.1.3 Binding to cell constituents

Even the fully de-esterfieid indicator or injected free acid forms can interact with cellular constituents to alter their properties. This may involve actual binding (suggested from measurements of fluorescence depolarization) or effects of the highly viscous protein solution which constitutes the cytosol. This problem is most easily overcome by performing the calibration measurements *in situ* using a Ca^{2+} ionophore. However, the intracellular viscosity can be mimicked, using solutions of gelatin or sucrose for *in vitro* calibration (4).

8.1.4 Impurities in commercial preparations of Ca^{2+} indicators

Most researchers assume that any commercial reagent is essentially pure or contains only those contaminants and other foreign materials listed by the supplier. With complex organic compounds this is probably the exception rather than the rule. The fluorescent Ca^{2+} indicators are relatively unstable, light-sensitive, and are derived from complex syntheses. Thus, it is quite possible to obtain a poor quality preparation. In some cases even fluorescent contaminants may be relatively benign, contributing only to the total background signal or even not penetrating the cells. However, other contaminants may bind Ca^{2+} or other ions, give spurious fluorescence signals or even interfere with normal cell physiology (33, 34). Fluorescence spectra (before and after de-esterification) can be informative as to the purity of the preparation. If a problem is suspected, the AM-esters can be examined by TLC; for example, using silica gel plates developed in 1:1 (v/v) hexane–ethyl acetate (26). Kawanishi *et al.* (33) have also utilized HPLC to identify fura-2 contaminants.

8.2 Effects of other ions

For the most part the fluorescent Ca^{2+} indicators are relatively insensitive to pH, but if a slightly acidic environment is to be probed the Ca^{2+} sensitivity may be shifted. Monovalent cations are essentially without effect. All of the indicators have high selectivity for Ca^{2+} over Mg^{2+}, but the cytosolic $[Mg^{2+}]$ is sufficiently high (10^3 to 10^4 times basal $[Ca^{2+}]_i$) that this ion generally has some influence on the Ca^{2+} measurement. Under most conditions $[Mg^{2+}]$ does not change sufficiently to generate artefactual fluorescence signals, but it is advisable to use a K_d for Ca^{2+} determined at an appropriate level of Mg^{2+}. The most significant effects on Ca^{2+} indicator fluorescence derive from certain transition metal ions, including Mn^{2+}, Zn^{2+}, and Fe^{2+}. These ions bind with much higher affinity than Ca^{2+} and once bound cause important fluorescence changes (most often quenching). The ability of Mn^{2+} to quench dye fluorescence and to act as a Ca^{2+} surrogate in penetrating plasma membrane channels has been exploited by a number of workers for measuring the effects of agonists on stimulated Ca^{2+} entry into cells (see, for example, ref. 35). When already present in cells, the heavy-metal ions can interfere with the calibration of $[Ca^{2+}]_i$ changes because they can be carried by Ca^{2+} ionophores and will bind to other chelators such as EGTA. In some cultured cell-lines these metal ions may be present in sufficient amount to interfere directly with the Ca^{2+} measurements. One way in which the effects of these ions can be assessed is to use the membrane permeant chelator TPEN (*N*,*N*,*N'*,*N'*-tetrakis (2-pyridylmethyl)-ethylenediamine) (26, 36). Any increase in signal following TPEN addition gives an indication of the amount of heavy-metal quenching. Experiments can be carried out in the presence of 10–50 μM TPEN (from a 10 mM stock solution in DMSO). For the most part this does not appear to affect cellular responses, but obviously this should be examined for each cell type.

8.3 Changes in cellular dye-independent fluorescence

As discussed in the section on calibration, autofluorescence must be subtracted prior to calculating $[Ca^{2+}]_i$ by the ratio method from dual-wavelength fluorescence measurements, and must also be subtracted or otherwise corrected for if it changes significantly during the course of the experiment. In most cells the primary source of autofluorescence at the near-UV wavelengths used to excite quin-2, fura-2, and indo-1 is reduced pyridine nucleotides (NADH and NADPH). Lipofuscin, a lipid degradation product, contributes significant autofluorescence in the 500-nm excitation range. The experiment depicted in *Figure 10* shows how significant autofluorescence changes can occur during measurements. In this case the increased autofluorescence is probably due to activation of mitochondrial dehydrogenases by the increased Ca^{2+} which leads to enhanced NAD(P)H production (see Chapter 15, this volume). Apart from the simultaneous measurement approach utilized for the experiment of *Figure 10*, the most direct way to deal with large autofluorescence changes during a measurement is to carry out a parallel experiment with unloaded cells. Changes of autofluorescence in most cells loaded to a reasonable extent with the most highly fluorescent indicators is generally not of sufficient magnitude to interfere with the $[Ca^{2+}]_i$ measurement. One other comment that should be made in this section is that many drugs are fluorescent. Any spurious signal change should be checked by adding the drug to buffer or unloaded cells.

8.4 Photobleaching

Since the intensity of fluorescence is dependent on the intensity of the excitation light, relatively high illumination intensities are often used, particularly for single-cell studies. This can result in photobleaching, which gives the appearance of dye loss. For single-wavelength dyes this is difficult to correct for unless a constant bleach rate is assumed, similar to the approach for dye leakage discussed in Section 8.1.1. In single-cell studies, the ratio approach to Ca^{2+} measurement cancels out the affect of signal loss due to photobleaching. However, since photodegradation can give rise to other fluorescent species or molecules which interfere with the measurement of $[Ca^{2+}]_i$ (34), it is preferable to limit photobleaching by using the minimum combination of illumination intensity and exposure time consistent with good signal-to-noise. Obviously this will be facilitated by the selection of optimal optical elements and the most sensitive detector available. In single-cell measurements, the addition of a shutter mechanism to limit the duration of illumination to the times during which data is actually being accumulated also helps to reduce photobleaching.

8.5 Buffering of cytosolic Ca^{2+} changes

All of the Ca^{2+} indicators must bind Ca^{2+} in order to report its concentration. Even with the most fluorescent of the dyes, intracellular concentrations in the order of 30–100 μM are commonly used. When compared with calculated intracellular concentrations of Ca^{2+} binding proteins this can represent significant additional Ca^{2+} binding. However, the homeostatic processes responsible for regulating $[Ca^{2+}]_i$ are controlled by the free Ca^{2+} concentration and as such steady-state Ca^{2+} levels (both in basal and stimulated states) are relatively insensitive to the additional Ca^{2+} buffering. The effect of the presence of the Ca^{2+} indicators becomes more apparent during rapid changes of $[Ca^{2+}]_i$, when proportionally more Ca^{2+} must be moved to get the same change in free Ca^{2+}. This can result in a slowing of the absolute rate of Ca^{2+} change without affecting the final level or in the case of transient changes, such as those observed with electrically stimulated heart cells (*Figure 11*), the absolute magnitude of the Ca^{2+} signal may also be depressed (18, 37). For most experiments the loading of the Ca^{2+} indicator should be kept to the minimum which is consistent with a suitable signal. In some cases it may be useful to deliberately overload cells with Ca^{2+} buffer to examine the role of the changes in $[Ca^{2+}]_i$ in a given response (see, for example, ref. 38).

8.6 Temporal resolution of indicators

The kinetics of Ca^{2+} binding to fura-2 have been studied in depth utilizing temperature-jump relaxation measurements of fluorescence. The rate-constants for Ca^{2+} binding and dissociation were $6 \times 10^8\ M^{-1}\ sec^{-1}$ and $97\ sec^{-1}$, respectively, at 20 °C (39). In practical terms this means that fura-2 is capable of following Ca^{2+} changes which occur on the order of 5 msec or slower. This is adequate for almost all conceivable biological applications. In most cases, rapid $[Ca^{2+}]_i$ changes are likely to be limited by the buffering effects of the indicators before kinetic limitations come into play.

8.7 Toxicity of indicators

The loading of cells with Ca^{2+} indicators can be expected to affect normal cell responses in two ways; one, by the Ca^{2+} buffering effects discussed in Section 8.5, and second, by direct chemical effects. Surprisingly few problems of the latter type have been reported. The by-products of AME hydrolysis are the most likely cause of cell damage or abnormal function. These are easily washed out. Generally, when a new cell type is to be used, the effects of dye loading on relevant physiological processes of the cells should be assessed.

References

1. Tsien, R. Y. (1980). *Biochemistry*, **19**, 2396.
2. Tsien, R. Y., Pozzan, T., and Rink, T. J. (1982). *Journal of Cell Biology*, **94**, 325–34.
3. Grynkieicz, G., Poenie, M., and Tsien, R. Y. (1985). *Journal of Cell Biology*, **260**, 3440–50.
4. Tsien, R. Y. (1989). *Annual Review of Neuroscience*, **12**, 227–53.
5. Minta, A., Kao, J. P., and Tsien, R. Y. (1989). *Journal of Biological Chemistry*, **264**, 8171–8.
6. Lakowicz, J. R. (1983). *Principles of fluorescence spectroscopy*. Plenum Press, New York.
7. Prentki, M., Glennon, M. C., Thomas, A. P., Morris, R. L., Matschinsky, F. M., and Corkey, B. E. (1988). *Journal of Biological Chemistry*, **263**, 11044–7.
8. Millard, P. J., Gross, D., Webb, W. W., and Fewtrell, C. (1988). *Proceedings of the National Academy of Sciences, USA*, **85**, 1854–8.
9. Monck, J. R., Reynolds, E. E., Thomas, A. P., and Williamson, J. R. (1988). *Journal of Biological Chemistry*, **263**, 4569–75.
10. Rooney, T. A., Sass, E. J., and Thomas, A. P. (1989). *Journal of Biological Chemistry*, **264**, 17131–41.
11. Moolenaar, W. H., Tertoolen, L. G. J., and deLaat, S. W. (1984). *Journal of Biological Chemistry*, **259**, 8006.
12. Cleeman, L. and Morad, M. (1991). *Journal of Physiology*, **432**, 283–312.
13. Haugland, R. P. (1989). *Handbook of fluorescent probes and research chemicals*. Molecular Probes Inc., Eugene, Oregon, USA.
14. Inoue, S. (1986). *Video microscopy*. Plenum Press, New York.
15. *Electro-optics handbook*. (Manufacturer's handbook.)
16. White, J. G., Amos, W. B., and Fordham, M. (1987). *Journal of Biological Chemistry*, **105**, 41–8.
17. Fay, F. S., Fogarty, K. E. and Coggins, J. M. (1986). In *Optical methods in cell physiology* (ed. P. DeWeer and B. Salzburg), pp. 51–63. Wiley, New York.
18. O'Rourke, B., Reibel, D. K., and Thomas, A. P. (1990). *American Journal of Physiology*, **259**, H230–42.
19. Saez, J. C., Connor, J. A., Spray, D. C., and Bennett, M. V. L. (1989). *Proceedings of the National Academy of Sciences, USA*, **86**, 2708–12.
20. Cannel, M. B., Berlin, J. R., and Lederer, W. J. (1987). *Science*, **238**, 1419–23.
21. Penner, R., Mathews, G., and Neher, E. (1988). *Nature*, **334**, 499–504.
22. Bush, D. S. and Jones, R. L. (1987). *Cell Calcium*, **8**, 455–472.
23. Roe, M. W., Lemasters, J. J., and Herman, B. (1990). *Cell Calcium*, **11**, 63–73.
24. Hansen, C. A., Monck, J. R., and Williamson, J. R. (1991). *Methods in enzymology*. (In press.)
25. Thomas, A. P., Martin-Requero, A., and Williamson, J. R. (1985). *Journal of Biological Chemistry*, **260**, 5963–73.
26. Tsien, R. Y. and Pozzan, T. (1989). *Methods in enzymology*, Vol. 172, (ed. S. Fleischer and B. Fleischer), pp. 230–62. Academic Press, Orlando, Florida.
27. Kao, J. P. Y., Harootunian, A. T., and Tsien, R. Y. (1989). *Journal of Biological Chemistry*, **264**, 8179–84.

28. Thomas, A. P., Sass, E. J., Tun-Kirchmann, T. T., and Rubin, E. (1989). *Journal of Molecular and Cellular Cardiology*, **21**, 555–65.
29. Rooney, T. A., Sass, E. J., and Thomas, A. P. (1989). *Journal of Biological Chemistry*, **264**, 17131–41.
30. Woods, N. M., Cuthbertson, K. S. R., and Cobbold, P.H. (1986). *Nature*, **329**, 719–21.
31. Rooney, T. A., Sass, E. J., and Thomas, A. P. (1990). *Journal of Biological Chemistry*, **265**, 10792–6.
32. DiVirgilio, F., Steinberg, T. H., Swanson, J. A., and Silverstein, S. C. (1988). *Journal of Immunology*, **140**, 915–20.
33. Kawanishi, T., Blank, L. M., Harootunian, A. T., Smith, M., and Tsien, R. Y. (1989). *Journal of Biological Chemistry*, **264**, 12859–66.
34. Becker, P. L. and Fay, F. S. (1987). *American Journal of Physiology*, **253**, C613–18.
35. Merritt, J. E. and Hallam, T. J. (1988). *Journal of Biological Chemistry*, **263**, 6161–4.
36. Arslan, P., DiVirgilio, F., Beltrame, M., Tsien, R. Y., and Pozzan, T. (1985). *Journal of Biological Chemistry*, **260**, 2719–27.
37. Thomas, A. P., Selak, M., and Williamson, J. R. (1986). *Journal of Molecular and Cellular Cardiology*, **18**, 541–5.
38. Renard, D., Poggioli, J., Berthon, B., and Claret, M. (1987). *Biochemical Journal*, **243**, 391–8.
39. Kao, J. P. Y. and Tsien, R. Y. (1988). *Biochemical Journal*, **53**, 635–9.

2

Aequorin measurements of cytoplasmic free calcium

P. H. COBBOLD and J. A. C. LEE

1. Introduction

The photoproteins aequorin and obelin generate blue-green light upon binding Ca^{2+} ions. Aequorin has been used over the past twenty years to measure cytoplasmic free Ca^{2+} (Ca_f) in single cells by injecting the cell with purified protein and then simply measuring the light emitted from the cell with a photomultiplier. The technique has been used in giant cells [giant barnacle muscle fibre (1), amoeba (2)] down to single mammalian cells [oocytes (3), and cells from heart (4), liver (5), and adrenal gland (6)]. Populations of mammalian cells have been loaded with obelin by liposome fusion, or with aequorin by various reversible permeabilization procedures such as hypo-osmotic shock, ATP-EGTA, scrape-loading, or spin-loading (for review see ref. 7).

Photoproteins have several advantages over fluorescent indicators, including the need for much simpler recording instrumentation and freedom from factors such as autofluorescence, intracellular compartmentalization, rapid leakage of the probe from the cell and buffering of calcium by the probe. Relative disadvantages are the need (at present) to microinject single cells, and lack of sufficient signal to make meaningful Ca_f images in mammalian cells. Recently it has been possible to compare data obtained with aequorin and fura-2 in single hepatocytes (5, 8, 9), when similar cellular Ca_f responses were found during hormone-stimulation; a gratifying result in view of the complexities of the intracellular environment and possible sources of spurious probe behaviour.

2. Properties of photoproteins

(a) The long history of use of aequorin and obelin as intracellular indicators is reflected in the number of comprehensive reviews (7, 10, 11).

(b) *Luminescence mechanism.* Aequorin is a 21 kd protein containing three consensus Ca^{2+}-binding sites. A gene has been cloned and the sequence published (12–14). Apoprotein obtained by expression of recombinant DNA in *Escherichia coli*, or by treating spent native aequorin with EDTA, can be recharged by adding the luminophoric molecule coelenterazine. Coelenterazine can be chemically synthesized (15) and, upon peroxidative attack by the apoprotein, releases a photon. This is a one-off reaction; fresh coelenterazine would need to be supplied to regenerate the photoprotein, along with oxygen. When using aequorin as an intracellular probe coelenterazine is not normally supplied since native aequorin binds and retains coelenterazine over years of storage.

(c) *Intracellular localization.* Aequorin appears to remain in the cytosol of cells for many hours after microinjection. In the jellyfish it is a cytosolic protein, so it probably lacks signal sequence for transport into vesicular cellular compartments. Entry into the Ca-rich lumen of endoplasmic reticulum would result in immediate discharge of the protein.

(d) Native aequorin is in fact a mixture of several iso-aequorins whose Ca-sensitivities differ somewhat (16).

(e) Blue-green light of 470 nm is emitted. About 15% of aequorin molecules release a photon. The photon is generated by peroxidation of coelenterazine but oxygen is not required for the reaction, being tightly bound to the apoprotein.

(f) Calcium sensitivity. In zero calcium at about 20 °C a slow rate of discharge of the photoprotein occurs (approx. 10^{-7} sec^{-1}; the 'Ca-independent light'). As calcium is raised the rate of discharge rises rapidly because of co-operativity between the three Ca^{2+}-binding sites such that at 10^{-6} M Ca^{2+}, about 10^{-3} of the protein is being consumed per second. At 10^{-5} M Ca the rate is close to maximal, about 10^{-1} sec^{-1}. This steep relationship between luminescence and calcium concentration (see *Figure 8*) makes aequorin ideal for recording changes in Ca_f in the physiological range, although it lacks precision for measuring small changes in resting Ca_f. Calibration of signals relies upon normalizing the signal as a fraction of the total aequorin in the cell being discharged per second. Of course, during isolation and handling Ca levels must be kept low to avoid unintentional discharge of the protein.

(g) The kinetics of the luminous reaction are reasonably fast; rate-constant about 100 sec^{-1}.

(h) The only cellular ion to discharge aequorin is Ca^{2+}, but Sr^{2+} and La^{3+} (and probably other lanthanides) will discharge aequorin at comparable rates.

(i) Mg^{2+} at millimolar concentrations depresses the sensitivity of aequorin to Ca^{2+}. Precision in calibrating aequorin signals requires cytosolic $[Mg^{2+}]_{free}$ to be known.

(j) Ionic strength depresses Ca-sensitivity, but aequorin will function as a probe in the cytoplasm of freshwater amoebae (approx. 30 mM KCl) and in marine invertebrate cells (approx. 450 mM). Again, calibration needs to be carried out at the correct ionic strength.

(k) Calcium chelators such as EDTA and EGTA at physiological ionic strength (about 100 mM and above) have no direct effect on aequorin and can be used to prepare Ca buffers. However, at lower ionic strength and at concentrations of *c.*10 mM and over both EDTA and EGTA depress the sensitivity of aequorin to Ca (17). This is not a problem in mammalian cells. No information is yet available concerning the new generation of pH-insensitive Ca buffers such as BAPTA.

(l) Aequorin's sensitivity to Ca is little affected by pH in the physiological range (6.6–7.4).

3. Aequorin in single mammalian cells

The procedures described here have been developed in Liverpool over the past several years and allow Ca_f to be measured in single mammalian cells such as mouse oocytes (3), rat heart (4) and liver (5) cells, and single chromaffin cells (6). These cells have diameters of approx. 70 μm (oocyte), 25–30 μm (hepatocyte), and 12–14 μm (chromaffin). The technique allows detection of a resting signal (about 200 nM Ca_f) and above. In principle, the Liverpool technique is the same as other single-cell aequorin techniques, utilizing microinjection and photomultiplier recordings. We have, however, optimized every aspect of the technique, which is published here in detail for the first time. The novel features are: (i) microdialysis method for dialysing approx. 200 nl vol. of highly concentrated aequorin solution (approx. 150 mg/ml), (ii) a simple method for filling a micropipette through its tip, (iii) improvements for microinjecting a single cell, (iv) very-low-noise photon counting detection of the signal, (v) computerized off-line calibration and print-out.

3.1 Hints on the handling of aequorin

(a) Rigorous control of free Ca must be observed by using double glass-distilled water, EDTA and EGTA buffering, and plastic containers.

(b) Aequorin is catalytically destroyed by trace levels of silver and mercury. Any solution that is to come into contact with aequorin must not therefore have pH electrodes dipped into it; aliquots must be removed in order to adjust the pH of the solution. Likewise platinum wire must be used to replace Ag–AgCl electrodes in electrophysiological studies.

(c) *Stability*. Aequorin can be freeze-dried and is stable indefinitely at −70 °C, and sufficiently stable at room temp. to permit a few days' airmail transport. Aequorin precipitated in ammonium sulphate is similarly temperature-tolerant. In solution at −70 °C it is stable for months and tolerates freeze–thaw cycles. Freeze-dried aequorin can be redissolved in Ca-free buffer (e.g. 10 mM EDTA, 10 mM Pipes, pH 7.0) and dialysed at high concentration (*c*.10–150 mg ml^{-1}) on a micro-scale in about 3 h (see below). Dialysis of dilute aequorin solutions is inadvisable.

(d) *Source*. Aequorin of very high quality, in which spent aequorin has been separated from active aequorin is available from Prof. O. Shimomura (Marine Biological Laboratory, Woods Hole, MA 02543, USA) or Dr J. R. Blinks (PO Box 3050, Friday Harbor, WA98250, USA). It is supplied in 1 mg aliquots freeze-dried from 1 or 10 mg ml^{-1} solutions in isotonic KCl buffer.

3.2 Aequorin preparation

Outline. We buy vials containing 1 mg aequorin which has been freeze-dried from 100 µl of 10 mg ml^{-1} solution of aequorin in isotonic KCl buffer. This yellow-tinged powder is dissolved in 7 µl of buffer to give a stock solution of about 150 mg ml^{-1} aequorin and *c*.2.5 M KCl. Aequorin for injection is prepared by dialysing *c*.200 nl (nanolitres) of this stock solution, twice per week, providing enough solution to fill hundreds of micropipettes per week. Annual consumption of aequorin in a group of 10 workers is approx. 5 mg.

Protocol 1. Dissolving freeze-dried aequorin to make a stock solution

1. The 1 mg freeze-dried aequorin is tapped into a corner of the vial which is held with 'Blu-Tak' horizontally at a shallow angle under a stereomicroscope.

2. (a) Using an externally silanized (Repelcote) glass pipette with the end snapped off to give a 0.1 mm (approx.) bore, 7 µl of EDTA-Pipes buffer (1 mM each, pH 7.0) is added to the powder, collapsing it into a small pool of solution (*Figure 1*, a).

(b) Movement of fluid into and out of the pipette is controlled by a 4-cm-diameter (approx.) rubber bulb connected with approx. 35 cm of silicon rubber tubing to the pipette. The apex of the rubber-bulb is pierced with a small hole which can be occluded by the forefinger to control whether or not squeezing the bulb transmits pressure or suction to the pipette (*Figure 1*, b) (i.e. a giant version of the Drummond 'microcap' system).

3. (a) The aequorin solution is sucked back into the pipette, avoiding bubbles, and then aliquoted into three 60-mm plastic Petri dishes containing about 3 mm depth of liquid paraffin (BDH) (also called mineral oil).

Protocol 15. *Continued*

(b) It is advisable to pre-wet the patch of plastic in which the 2 μl droplet is to be placed, and to mark the position on the underside of the dish. (*Figure 1*, c.)

(c) These dishes form the aequorin 'stock solution' which is kept at −70 °C and which is stable for months. Whenever a dish is removed from the −70 °C freezer it is placed in a box containing dry silica gel to prevent condensation of water droplets on the oil surface.

4. We advise practising every manipulation in this procedure, using some freeze-dried BSA, before opening the aequorin vial.

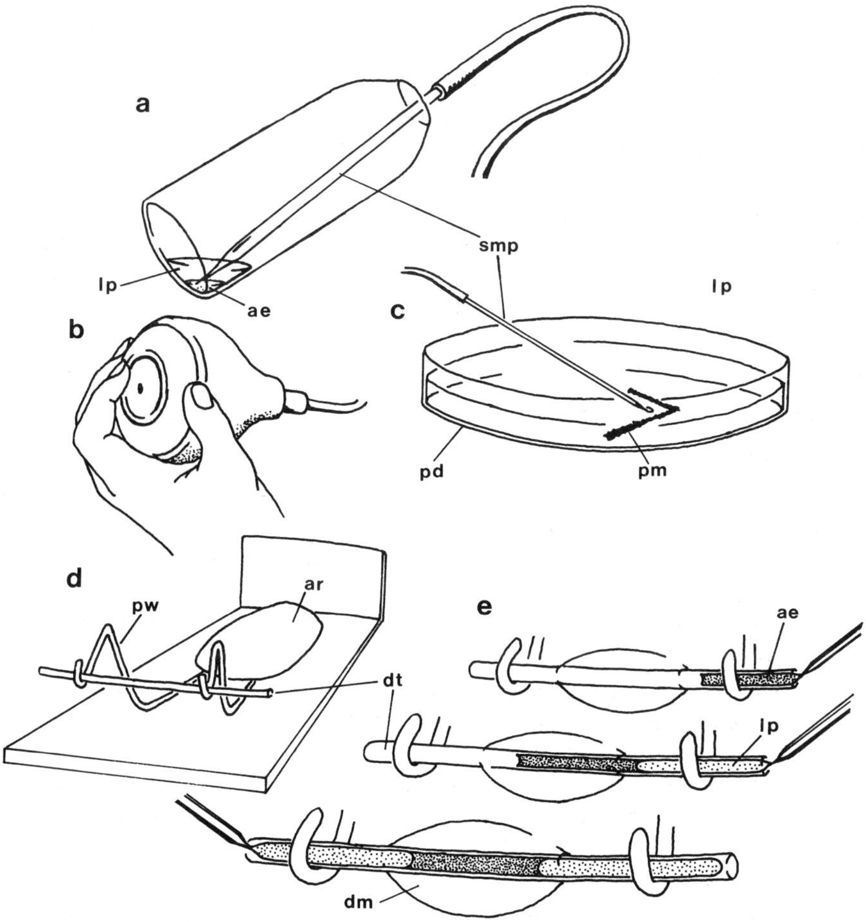

Figure 1. Storage and microdialysis of aequorin. ae = aequorin, ar = 'araldite' (epoxy resin), dm = dialysis medium, dt = dialysis tubing, lp = liquid paraffin, pd = Petri dish, pm = pen mark on base of Petri dish, pw = Teflon-coated platinum wire, smp = silanized micropipette.

3.2.1 Preparation of injection buffer

Injection buffer. The stock solution of aequorin contains too much EDTA and KCl to inject into a cell, so we dialyse 200 nl aliquots against injection buffer whose composition is; KCl 150 mM (BDH 'Specpure'), Pipes 1 mM, EDTA 100 μM, EGTA 25 μM, dithiothreitol 1 mM, pH 7.2 using KOH ('Specpure'), in double-distilled water. The injection buffer is kept in 500-ml plastic tissue-culture flasks, and pH measurements are carried out on 10-ml aliquots, which are then discarded to avoid silver contamination of the buffer. Dithiothreitol improves the flow of aequorin into and out of micropipettes presumably by reducing the rate of oxidative denaturation. For the use of PVP as an osmotic buffer, see Section 3.2.3.

3.2.2 Construction of microdialysis apparatus

(a) We use transparent microdialysis tubule of approx. 150 μm bore, supplied in the 1970s by Bio-Rad as Biofiber 50A. This tubule is no longer available but a suitable substitute is 'Microtubing' from Medicell International (London), which is transparent with a bore of approx. 180 μm and a cut-off of about 5 kd.

(b) A 4- to 5-mm length of tubule is held in two V-shaped clefts formed from Teflon-coated platinum wire (130 μm diam) (PT.5T Clark Electromedical Instruments) (see *Figure 1*, d). The wire is glued (Araldite) to a piece of plastic cut from the base of a plastic Petri dish so that the dialysis tubule is held about 2 mm above the plastic base. By using fine forceps to collapse the wire to lightly compress the tubule, the tubule can be held without occluding its bore.

(c) Checking the tubule for leaks. A micropipette with snapped-off tip is used to add a large drop of wash buffer (EDTA-Pipes 1 mM each, pH 7) to the outside of the dry tubule. Small water droplets will form inside the bore, but if an inward rush of buffer occurs the tubule is probably punctured and should be replaced.

(d) An intact tubule is first washed by filling with wash buffer and is then immersed for several minutes. This is carried out in a 60-mm plastic culture dish. (Transfer of the dialysis tubule holder is easy if its plastic base has been cut from the edge of a Petri dish, the vertical side-wall giving a convenient handle for the foreceps to grip.) (*Figure 1d.*)

Protocol 2. Microdialysis procedure

1. The washed dialysis tubule in its wire cradle is placed under the stereomicroscope (magnification *c.*20×) on a clean plastic dish lid. Wash buffer is sucked off the tubule, and out of its lumen. A droplet of injection buffer is then placed outside the tubule to minimize drying-out.

Protocol 2. *Continued*

2. A new micropipette with tip snapped off to about 50 μm is then loaded with a short column of liquid paraffin followed by approx. 200 nl of stock aequorin solution, as judged by the length of the column in the tip. This procedure is also carried out under stereomicroscope.

3. The aequorin solution is run into the dialysis tubule, which will then be about one-third full. A short column of the oil is run in after the aequorin to leave an equal-length column of air at the other end. This end is then also plugged with oil, trying not to trap too much air (see *Figure 1*, e).

4. (a) The microtubule in its wire cradle is then submerged in about 4-mm depth of injection buffer in a 60-mm plastic culture dish. Because of the high osmotic pressure of the solution there will be a tendency for the aequorin column to swell, sometimes rapidly, with the risk of displacing the oil plugs from the end of the tubule and loss of the aequorin. This is avoided by removing the tubule and evaporating the aequorin column back to its original length. After approx. 20 min rapid volume changes do not occur.

(b) However, a slow increase in volume can occur, presumably because of the colloid osmotic pressure of the aequorin. This can be offset by including 28 mg ml^{-1} of polyvinylpyrrolidone (PVP; about 40 000 mol. wt; Sigma) in the injection buffer. This trick allows aequorin to be dialysed without attention overnight, at 4 °C. Without PVP, inspection should be made every 20 min or so.

5. About 2 to 4 h dialysis is probably sufficient to reach completion. At the end of dialysis a pipette whose tip has been snapped off to about 50 μm and which has been washed in injection buffer is used to remove the aequorin to place it as a hemispherical droplet on to a pre-wetted patch under 3 mm approx. of liquid paraffin (BDH, light grade) in a 60-mm plastic tissue-culture Petri dish. This sample of aequorin can be frozen at −70 °C and thawed several times, but will slowly lose activity over approx. one week. It is possible to subdivide the sample into three dishes. We routinely perform two dialyses per week.

3.2.3 Filling micropipettes

(a) Micropipettes are pulled on a simple vertical puller from borosilicate stock (length 100 mm, bore 1 mm, wall 0.5 mm). Filament-lined glass is not used, nor are the capillaries washed beforehand, but steel forceps are used to handle the glass capillary while fitting it into the chucks of the puller to avoid contaminating the region that will form the tip. The puller is set to give a relatively long tapering tip (approx. 11 mm from tip to full diameter). Under SEM the bore of the pipette is about 0.3 μm, and they have a resistance of *c.*10 MΩ when filled with 3 M KCl. Pipettes are used straight out of the puller without bevelling or washing or other laborious procedures.

(b) By simply dipping the tip for a few seconds in the liquid paraffin covering the aequorin droplet, a short column of oil (approx. 50 μm) will be seen to enter the tip using a stereomicroscope (× 80) (*Figure 2*, c). When the tip is dipped into the aequorin micro-droplet a short column of this will also enter the tip; sometimes it needs to be assisted by gently scraping the tip across the plastic surface. This procedure is quick and simple.

3.2.4 Advantages and hints

(a) One pipette is made for each cell to be injected. If the tip blocks the pipette is discarded.

(b) There is no need to filter small volumes of aequorin solution. The microdroplet is sufficient to fill hundreds of pipettes.

(c) It is essential to include adequate pH buffer in the injection buffer (1 mM Pipes suffices, 10 μM does not) in order to avoid a rise in pH consequent upon ion-exchange of H^+ for Na^+ in the glass; this can lead to very rapid discharge of aequorin in the pipette tip and a toxic pH in the injected solution (possibly pH 9–10).

3.3 Preparing cells for microinjection

The techniques described in Sections 3.3.1 to 3.3.4 allow single cells to be held in an agarose gel coccoon inside an optically-flat capillary ('microslide') during microinjection. This capillary can then be transferred in a pair of forceps to a small perfusable chamber in order to record the aequorin signal. The cell in the agarose gel is protected from the flow of perfusion fluid yet firmly anchored by the surrounding glass capillary. The diffusion path from the perfusion fluid to the cell is only approx. 50 μm of gel. Attached cells can be injected without using agarose (Section 3.3.4).

3.3.1 Preparation of agarose medium

(a) *Medium for filling microslides*. Culture medium is boiled until the agarose powder is dissolved. Heat-susceptible media such as those containing calcium and phophate, or proteins, must not be boiled, so boil the agarose at double concentration in distilled water and add to double-strength medium after cooling to *c*.37 °C. Medium for filling microslides contains 1% Sigma VII agarose which is kept in a 5-ml tube in a block heater at 37 °C. A 1-cm overlay of liquid paraffin reduces loss of CO_2.

(b) Medium for suspending cell stocks (Section 3.3.3) is made likewise, but using Sigma IX agarose which will not gel above 25 °C.

(c) *Modifications of media to aid microinjection*. The composition of the medium is, of course, dictated by the cell type being used. We use Williams' Medium E for liver cells and bicarbonate–CO_2 buffered

Ringer's solution for heart cells. It is advisable to add BSA (fatty acid free) to the medium to help membrane sealing during microinjection. We found the BSA concentration to be particularly critical in microinjecting chromaffin cells (6); too much BSA prevented micropipette entry, while absence of BSA prevented resealing and the cell degranulated. The ideal level was 0.5 mg ml^{-1}.

3.3.2 Setting up a microslide

The optically-flat glass capillaries (Microslides, Camlab) have a path length (floor to ceiling) of 0.1 mm, and are cut up in lengths of approx. 4 mm (see *Figure 2*, a), acid washed and stored in absolute alcohol. A supply of

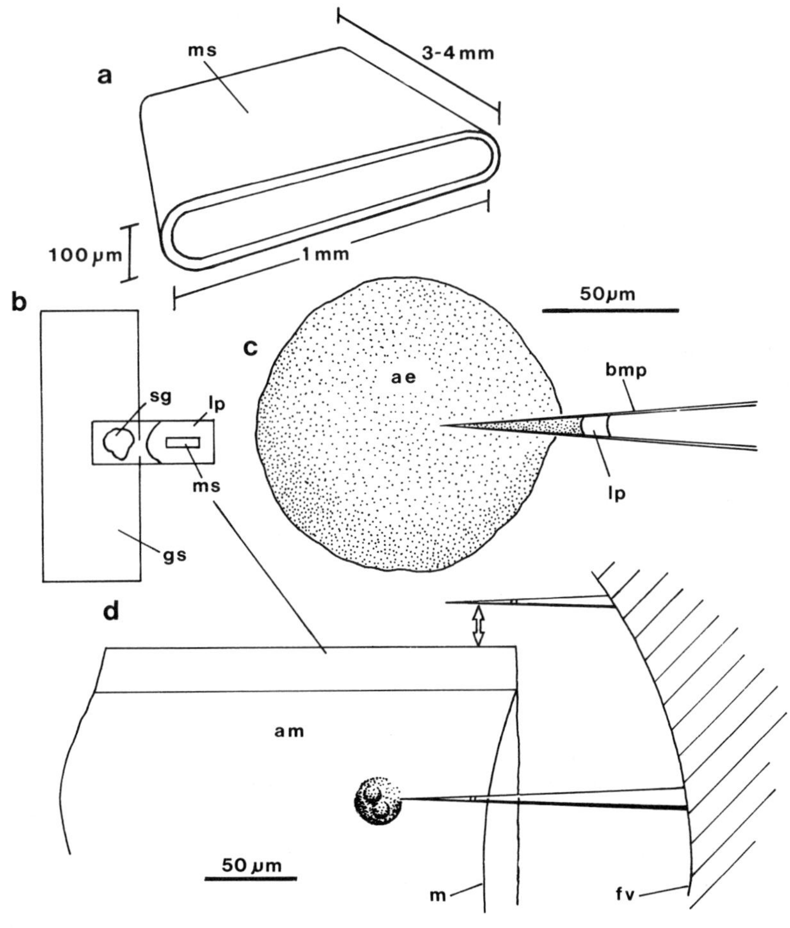

Figure 2. Cell injection. ae = aequorin, am = agarose containing medium, bmp = borosilicate micropipette, cs = coverslip, fv = field of view, gs = glass slide, lp = liquid paraffin, m = meniscus, ms = microslide, sg = silicone grease.

rectangular pieces of coverslip (22 mm × 7–9 mm) is stored in air. To set up a cell, a 3-in × 1-in glass slide is daubed with a spot of silicone grease half-way along one long side and the end of a coverslip piece (rinsed in alcohol and flamed) is pushed on to the grease so that the free end is cantilevered out and slightly tilted downwards (see *Figure 2*, b). This end is placed in the field of view of a stereomicroscope (×32 to ×80). A microslide is flamed and placed on the end of the coverslip. A blunted micropipette with a tip bore of about 50 μm is filled with enough warm agarose-containing medium (see Section 3.3.1) to fill the microslide. A small amount of this is first placed alongside the microslide so that it flows by capillarity between the microslide and coverslip: this anchors the coverslip into place. The rest is used to just fill the microslide. Two drops of liquid paraffin are dropped on to the microslide to prevent evaporation. The stage of the stereomicroscope should be held at 30–37 °C using a flow of warm air or a lamp to prevent premature gelling of the agarose, and the stock tube of agarose-medium is held in a block-heater.

3.3.3 Selecting a cell for transfer into the microslide

(a) The dish of stock cells. Most of our cells are freshly isolated from tissues and can be prevented from attaching to their dish by a thin underlay of gelled Sigma VII agarose prepared in culture medium. A concave layer of this gel holds approx. 0.5 ml of dilute cell suspension made up in medium containing 1% type IX agarose, to add viscosity to the medium. The dish is overlayered with about 3 mm of liquid paraffin and kept in an incubator at 37 °C.

(b) When a cell is needed, the dish is put on the stage of a second stereomicroscope (magnification 28× to 180×) alongside the one with the microslide and a cell of healthy appearance is sucked up into the tip of a blunt (50 μm) pipette containing medium with type IX agarose to a length of 4 or 5 mm with the meniscus almost into the full width of the bore.

(c) Suction is applied to the pipette to generate a flow of medium containing a cell into the pipette. As soon as a cell rushes into the tip the pipette is immediately and rapidly lifted up into air; the cell will come to rest 0.1 or 0.2 mm into the pipette and the suction can be released.

(d) The cell is then expelled into the microslide under the adjacent microscope (80×). A tendency for the microslide to overflow with medium can be reduced by leaving the menisci in the microslide somewhat concave when filling it. The cell should be positioned approx. 50 μm from the meniscus and ideally to one side of the mouth of the microslide (*Figure 2*, d).

(e) The microslide containing the cell is then placed in a refrigerator at about 4 °C for 5 min or so. This gels the agarose, which (Sigma VII) will not remelt at 37 °C for several hours. It is important that the agarose is gelled promptly after putting the cell in the microslide since any delay can allow the cell to start to attach to the glass. This is no problem until the microslide is placed in perfusion medium, whereupon the gel expands slightly. This tends to place a shearing force on the cell leading to large excursions in the aequorin signal a few minutes after the microslide is put in the perfusion chamber.

(f) The cell can then be placed in an incubator at 37 °C until needed for injection.

3.3.4 Setting-up attached cells in microslides

Cells that attach to glass (e.g. fibroblasts) can be put in microslides without agarose and microinjected. It is a good idea to inject while the cells are in the process of spreading since the cytoplasm is thicker, and continued spreading is a sign of lack of injection injury. In microslides without agarose problems can arise in the perfusion chamber as small bubbles get drawn through the microslide, often destroying the cells in the process.

3.4 Microinjection

Microinjection of mammalian cells about 25 μm across (e.g. hepatocyte, cardiac myocyte) is not difficult, but small cells (e.g. chromaffin cells, approx. 12 μm across) require more effort while others (e.g. HL60 leukaemic cells, approx. 12 μm) are very difficult to impale. Microinjection of a cell takes about 5 min, including filling the pipette, and the success rate with heart and liver cells is easily 50%. Since an injected cell is often used for an experiment lasting an hour or two, microinjection is not a limitation on productivity. After microinjection, cells can be left for hours without significant loss of acquorin; signals have been detected from an amoeba three days after microinjection. It is fortuitous that this foreign protein has such a long half-life in cells.

Protocol 3. Construction of microscope and micromanipulator

1. An inverted microscope (Nikon Diaphot) with Nomarski lenses is used, giving adequate image size for a hepatocyte using a 40× LWD objective and 15× eye-pieces. The objective lens must allow a cover-glass correction of 0.23 mm. A good clearance of about 2.5 cm between the cell and condenser lens is desirable.
2. A Leitz mechanical micromanipulator is mounted on a 13.0-cm-high plinth to the right of the microscope (presenting a problem for left-handed operators), with the pillar of the manipulator set 7.0 cm from the side of the microscope chassis. (*Figure 3*).
3. The microscope is set on four 5-mm-thick pieces of metal (to bypass the rubber feet) glued to a heavy baseplate (slate slab) which also supports

Protocol 3. *Continued*

the micromanipulator. This baseplate sits on two tyre inner tubes and blocks of foam rubber to provide vibration-damping.

4. A piece of alloy of about 5mm thickness is used to replace the Diaphot's stage. A surface mounting mechanical stage is fitted with the controls by the left hand. The right-hand side of the stage, facing the manipulator, is cut away so that there is no metal between the lens and the manipulator, this gives free access for the micropipette which is held horizontally on the micromanipulator (see *Figure 3*).

Protocol 4. Construction and use of the pressure injection apparatus

1. The micropipette, which is pulled to give a length of 9 cm from tip to butt, is clamped in the brass holder shown in *Figure 3* and *4*. The chuck on the end of the holder clamps the pipette rigidly while the rubber seal (from a Drummond 'Microcap') provides a gas-tight seal. The holder fits into the clamp of the Leitz micromanipulator.

2. The pipette holder is connected by flexible nylon tubing to the solenoid valve and pressure regulator shown in *Figure 3* and *4*.

3. The handle on the regulator is actuated by the left hand of the operator, and moves through an arc to raise and lower the gas pressure provided by the regulator valve R2. During injection this relieving regulator provides pressures of between 0.5 to 4 atm as indicated on gauge G1.

4. The handle on regulator R1 can be screwed in to give up to 0.5 atm pressure, seen on gauge G2.

5. The system is supplied with nitrogen or air at 4 atm.

(a) To inject cell; after fitting a pipette, regulator R1, is set to give approx. 0.3–0.5 atm pressure, which is transmitted through the solenoid valve to the pipette to prevent entry of medium into the aequorin in the tip of the pipette.

(b) After impaling the cell (see *Protocol 5*) the foot switch is depressed allowing pressure from the lever-controlled regulator to enter the pipette.

(c) The lever is moved in an arc to expel the aequorin from the pipette tip. Holding the lever in this position while repeatedly opening and closing the solenoid valve can produce pressure surges to help to free a blocked tip.

(d) This apparatus was custom-built at little cost. Commercially available equipment which has press-buttons rather than a lever is available (e.g. Digitimer) but may not be so amenable to generating rapid, variable changes in pressure.

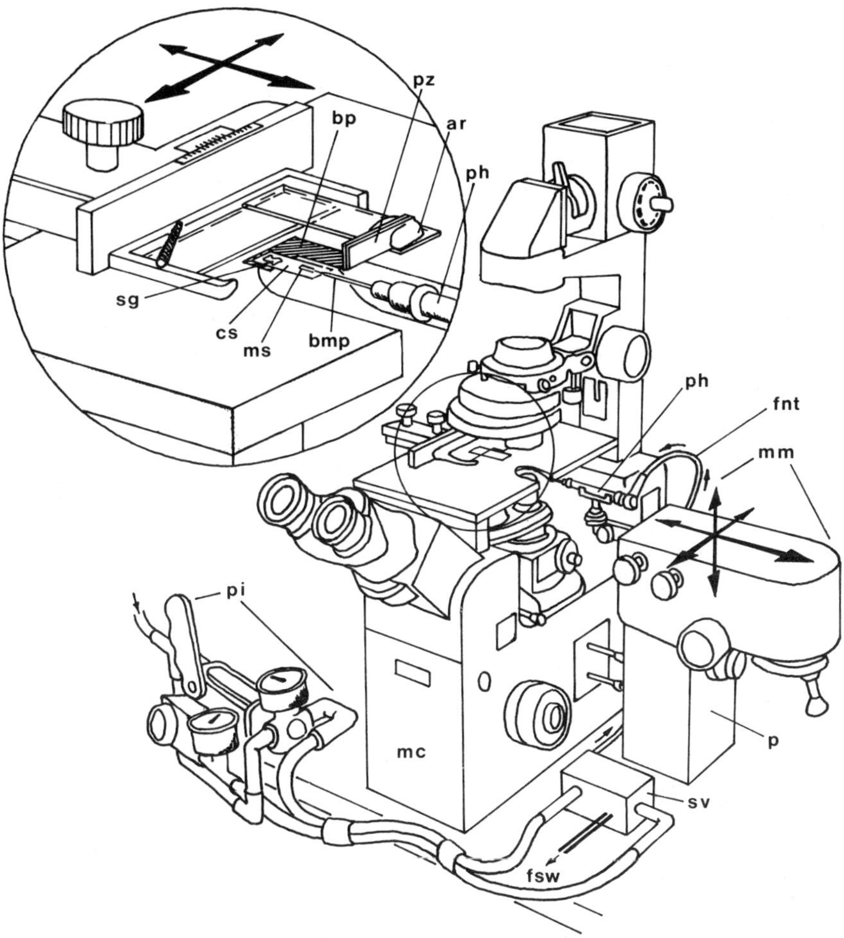

Figure 3. Cell injection apparatus. ar = 'araldite' (epoxy resin), bmp = borosilicate micropipette, bp = brass platform, cs = coverslip, fnt = flexible nylon tubing, fsw = foot switch, mc = microscope chassis, ms = microslide, mm = micromanipulator, p = micromanipulator plinth, ph = pipette holder, pi = pressure injection system, pz = piezoelectric bender element, sg = silicone grease, sv = solenoid valve.

Protocol 5. Impaling a cell

1. Take the coverslip holding the microslide and cell (Section 3.3.3) from the incubator and transfer it to the greased platform attached to the piezoelectric bender element on the microscope stage (*Protocol 3* and Section 3.4.1; *Figure 3*).
2. Align the cell with the left-hand side of the field of view using the 40×

Protocol 5. *Continued*

objective. Switch to the 10× objective; one edge of the microslide should be within the field of view without moving the mechanical stage. Focus on the outer edge of the glass wall. (*Figure 2*, d)

3. Fill a micropipette tip with aequorin (Section 3.2.3; *Figure 2*, c), back-fill it with buffer [Section 3.4.1 (b)] and mount it on the micro-manipulator (*Protocols 3* and *4*). With the joystick set fully over to the right and centred, advance the pipette over the droplet of liquid paraffin covering the microslide using the coarse *X*, *Y* controls. Lower the pipette tip until it contacts the oil. Do this under direct vision.
4. Using the 10× lens, reposition the pipette tip alongside the edge of the microslide and adjust the vertical position until the tip is parfocal with the outer edge of the microslide (*Figure 2*, d). You now know that the tip is midway between the floor and ceiling of the microslide. Using the coarse *X–Y* controls reposition the tip so that it is pointing at the cell but not quite penetrating the agarose/oil meniscus.
5. Switch to the 40× objective and focus on the equator of the cell.
6. Carry out instruction (i) of *Protocol 4*, and set *c*.0.3 to 0.5 atm. gas pressure in the pipette.
7. Using the joystick and fine vertical adjustment, advance the pipette tip into the field of view. Adjust the vertical position of the tip until it is parfocal with the cell equator. Advance to the cell and slightly (about 1 or 2 μm) dimple its surface (*Figure 2*, d).
8. Penetrate the cell surface using either the piezoelectric cell-mover or the dielectric procedure (Section 3.4.1).
9. Press the foot switch and move the lever of the pressure-controller (*Protocol 4*) to expel aequorin into the cell. A swirl of cytoplasm near the pipette tip indicates a good injection (in a hepatocyte). Press the button to apply the full 45 V reverse-voltage to the piezoelectric bender (Section 3.4.1). to remove the cell from the pipette. Withdraw the pipette completely. Watch the cell for a minute or so for signs of damage (blebbing or a plume of cytoplasm extending into the track the pipette made in the agarose), and then return the microslide on its coverslip to the incubator.
10. To improve control over the volume injected, it is possible to expel much of the aequorin column into the medium prior to penetrating the cell. By estimating the volume of the cell and using an eye-piece graticule to estimate the length (and hence volume) of the aequorin column, a rough estimate of injection volume can be made. Aim for 1% or less of the cell's volume.

11. Mammalian cells may be more tolerant of impalement if the microscope stage is held at 30–37 °C using radiant heat rather than flowing air (which will vibrate the microslide).

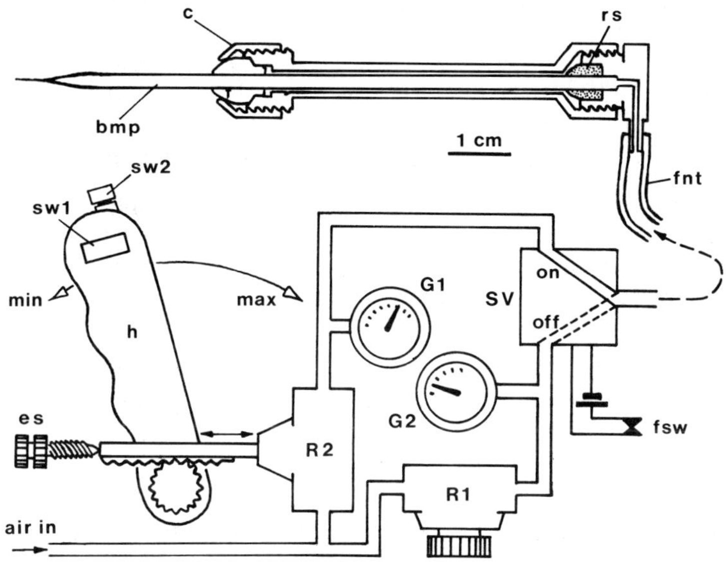

Figure 4. Pressure injection system. bmp = borosilicate micropipette, c = chuck, es = end stop, fnt = flexible nylon tubing, fsw = foot switch, G1 and G2 – gauges 1 and 2, h = handle, R1 and R2 = regulators 1 and 2, rs = rubber seal, SV = solenoid valve, sw1 and sw2 = switches 1 and 2 (actuating 'throw on' and 'throw off' of piezoelectric bender element).

3.4.1 Construction of devices to aid impalement

(a) *Piezoelectric cell-mover.* The piece of coverslip holding the microslide (Section 3.3.3) is held by a film of silicon grease on a light-weight brass platform soldered to the free end of a cantilevered piezoelectric bender element ('Bimorph' PZT-5B) (Morgan Matroc, Vernitron Division, Southampton, UK). The bender element's position in relation to the microscope objective is controlled by the mechanical stage and the cell is arranged to be at the focal point of the lens (*Figure 2d*). This arrangement allows the piezoelectric motion of the bender element to be used to rapidly throw the cell on to the pipette tip, making rapid, short movement of the micromanipulator joystick unnecessary. A voltage-reversing switch fitted into the handle of the pressure-controlling lever actuates the piezoelectric element, via a variable supply (battery box) providing up to 15 V in 1.5-V increments. The movement of the cell is optimally about 2 to 5 μm. A second non-latching switch provides a full 45 V in the reverse direction to allow the cell to be rapidly jerked completely away from the cell after microinjection.

(b) *Dielectric penetration*. Some cells (e.g. oocytes) can be difficult to penetrate despite the rapid piezoelectric movement. Because our pipettes are not filled with 3 M KCl we cannot use the electrophysiologists' trick of 'buzzing', in which a rapidly oscillating voltage is applied from an amplifier to the electrode tip to promote dielectric breakdown of the cell membrane and easy electrode entry. Instead we back-fill each aequorin pipette with old injection buffer solution, inevitably leaving an air gap behind the oil and aequorin in the tip. The pipette holder is electrically insulated from the manipulator clamp with a rubber sleeve and connected to the centre electrode in the snout of a domestic piezoelectric gas lighter. Gently squeezing the trigger of the gas lighter generates an electrostatic field at the tip of the pipette such that a cell membrane, that has first been dimpled by pushing with the micromanipulator, will be penetrated by the pipette tip. This process is more benign than mechanical prodding, is reproducible, does not result in electroporation of the cell, and the device costs next to nothing to make.

4. Apparatus for signal detection

The cell in its microslide is held in a stainless-steel perfusion chamber immediately in front of a low-noise photomultiplier. Using photon-counting, a signal will be detected from a single cell, providing all the precautions taken here are adhered to. There is no room for compromise since every aspect of the technique has been improved to the physical limit.

4.1 Construction of perfusion chamber and photomultiplier housing

Perfusion chamber (*Figure 5*). The microslide is held in a polished cavity approx. 8 × 4 × 2 mm depth in a disc of 316 stainless steel. The chamber is sealed with a circular coverslip on a film of liquid paraffin. Small-bore tubes (approx. 1 mm) at each end of the chamber allow perfusion at about 100 μl min^{-1}, controlled by suction from a variable flow peristaltic pump. The plumbing allows media to be changed by syringe suction from a flask to a 'T' piece mounted below the chamber's water jacket. The water jacket is held at 37 °C by a thermocirculator. The perfusion chamber is held together with epoxy resin (Araldite), simplifying machining and assembly.

The microslide chamber is positioned immediatley below the photocathode of the photomultiplier leaving about 3 mm air space to prevent disturbance of the electrical field of the tube by the steel cup (shims are used to adjust the separation if spuriously high count rates should occur). The photomultiplier is cooled to approx. 0 °C by a constant-temperature incubator, but there is a

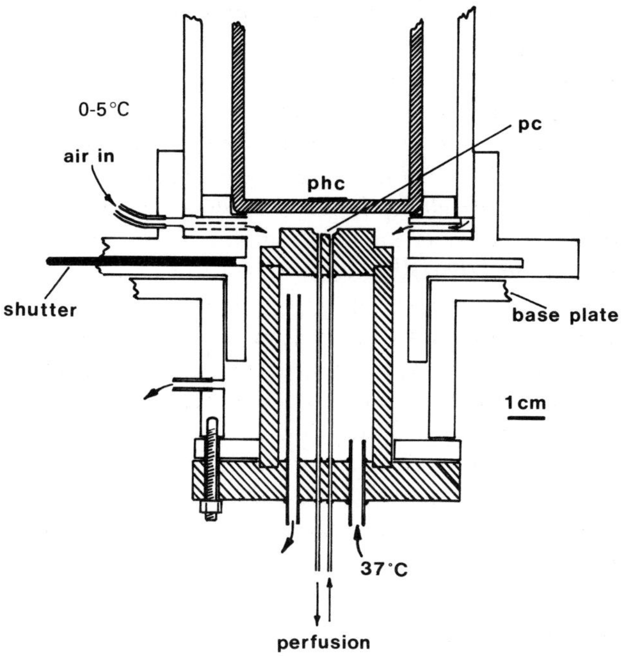

Figure 5. Perfusion chamber and photomultiplier housing. pc = polished chamber, phc = photocathode.

tendency for the photocathode to be warmed-up by the warm-water jacket of the perfusion chamber. Convection of heat is therefore prevented by a flow of cold air across the face of the photomultiplier and down the sides of the water jacket, provided by an aquarium aerator pump sitting in the cold cabinet. The need to put the warm cell chamber close to the cooled tube prevents the use of commercial, compact cooled housings which use a 1-in-thick double window to provide a frost-free window. The lack of a light guide is intentional since large volumes of transparent material raise the Cerenkov count rate from the cosmic ray background and tend to phosphoresce if exposed to blue or UV light. Even the coverslip which covers the cell is 'dark adapted', and red light is used to illuminate the cup while loading the microslide. This avoids phosphorescent signals, since a coverslip exposed to office strip lights for 10 min will generate 10 c.p.s. phosphorescent signal in this apparatus.

To minimize phosphorescence, the photomultiplier is never exposed to light, even when switched off. The use of polished stainless steel rather than transparent Perspex chambers is likewise dictated by the need to minimize background light. Epoxy adhesive and Sylgard are chemiluminescent and must also be avoided in constructing cell chambers (except for internal joints).

The counting geometry is designed to reflect as much light from the cell as possible on to the photocathode. The cell chamber is polished with jeweller's rouge. The proximity of the cell to the photomultiplier means that a 1-cm-diameter photocathode is ideal. The usual 2-in-diameter photocathode would not collect more light but background counts would be raised in proportion to the cathode area, while smaller diameter photocathodes (around 3 mm, 1 mm) would not capture all the light from the cell.

4.2 Choice of photomultiplier

We use EMI type 9789 photomultipliers, with a 10-mm-diameter bialkali photocathode and selected for low dark count. Tubes are requested from the manufacturer with a dark count rate of less than 30 c.p.s. when operated at room temperature at a voltage giving a gain of 5000 A $lumen^{-1}$. At this gain, the detection of blue-green (480 nm) photons will be determined by the photocathode's quantum efficiency (0.15 at this wavelength), and will be maximal. When cooled to 0 °C and having been kept in the dark for at least two weeks we find that most tubes perform appreciably better than quoted by the manufacturer. We look for a dark count rate (at 5000 A $lumen^{-1}$) of about 1 to 4 c.p.s. Cooling has no effect on quantum efficiency, but does reduce thermal generation of electrons. Cooling bialkali tubes below 0 °C is not useful; the photocathode takes on a high resistance and counting becomes erratic. The housing is an EMI type B2, with a dynode chain specified for use with 9789 tubes in the photon counting mode (i.e. high first-stage gain). An RF shield is factory-fitted under an insulating sleeve. Problems with corona discharge noise are rarely encountered, but are recognized from a large, varying, spiking background count; removing the tube and wiping it with an acetone-moistened paper tissue is usually the answer. A build-up of moisture in the cooled incubator can lead to corona noise; defrosting the cooled incubator weekly ensures a dry cabinet.

4.3 Amplifier–discriminator

A photomultiplier needs only two wires. One supplies high voltage, the other (the anode) carries the current generated as electrons flow up the dynode chain. One electron released from the photocathode by a photon generates a burst of, typically, 10^8 electrons lasting, typically, 4 nsec. A fast pre-amplifier, usually positioned adjacent to the photomultiplier housing, amplifies the current pulses and passes them to an amplifier–discriminator a few feet away. The discriminator rejects slow fluctuations in current and converts each brief, single photon-induced pulse, into a standard pulse suitable for triggering rate-meters, scaler, etc. The dead-time between pulses should be no more than 10 nsec, allowing a maximal count rate of 10^8 c.p.s. Since aequorin signals rarely exceed 10^4 c.p.s. photon counting is suitable for following all signals. (Some commercial systems record analogue currents once count rates start to approach the photon counting bandwidth limit.) The pulses generated by the discriminator can then be fed to rate-meters, scaler, computer, etc. In the system shown in *Figure 6* the amplifier–discriminator is EMI type. C604A.

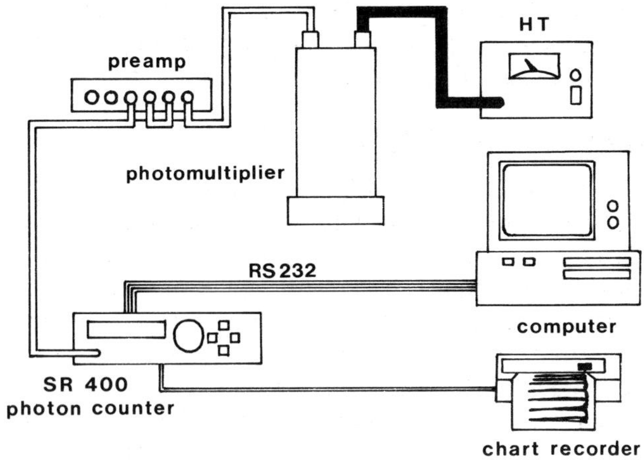

Figure 6. Schematic diagram of counting system.

4.4 High voltage

The HT supply (EMI model PM28B) provides regulated voltage for the tube, which is set to a voltage which, from the manufacturers data, gives a sensitivity of 5000 A lumen^{-1}, usually around 1200 to 1500 V for a 9789 tube. The HT must be switched off whenever the incubator door housing the photomultiplier is opened. There is no simple way to safety interlock, so position the red tall-tale lamp and HT switch close to the incubator door as a reminder. We find no problems with stability, etc., arising from switching the HT on and off several times per experiment.

4.5 Rate-meter, scaler, and chart recorder

The pulses generated by the discriminator can be fed to pulse counting ratemeters and scalers of the type more commonly used for counting nuclear particles. We have used Nuclear Enterprise NIM units; however, these units utilize different pulses from the output of the EMI discriminator and a pulse-convertor may need to be made. The output from the rate-meter is fed to a chart recorder for on-line recording of the signal. Resting signals are usually recorded with 10 c.p.s. full-scale 33 sec time-constant, 5 mm min^{-1} chart speed. But the rate-meter should embrace sectors of about 100, 300, 10^3 c.p.s., and time-constants down to approx. 1 sec^{-1} to enable capture of oscillatory Ca changes at chart speeds of up to 20 mm min^{-1} or so.

It is, however, virtually impossible to capture on-line a publishable trace of aequorin signals since the large dynamic range makes frequent scale-switches a necessity. Using a rate-meter with log output is one solution, but this makes subsequent normalization of the signal (see below) almost impossible to carry out manually.

4.6 Computerized data acquisition

The pulses generated by the discriminator which correspond to single photons of aequorin luminescence are ideal for capture digitally by a microcomputer. We use a commercial Photon Counter (SR 400; Stanford Research.) to capture the discriminators' output pulses and direct them to 50-msec-long bins in one of the two banks of on-board memory. The host computer then interrogates one memory buffer every second, carrying out a transfer via an RS232 interface. The computer has sufficient RAM to record 20 samples sec^{-1}, 65 000 counts per sample, for approx. 3 h. A slow data file is collected, at 1 sample sec^{-1}. The SR400 can also provide a chart recorder signal for on-line recordings. EMI supply a photon-counting board that fits the expansion-slots in a micro-computer, at considerable saving in cost.

4.7 A minimal computerized photon counting system

Figure 6 shows a computer-hosted system based on an Opus (IBM-AT-compatible) microcomputer and SR photon counter. This system provides a chart recorder output from a D to A in the micro which is programmed to be linear from 0–10 c.p.s. (for recording resting signals) and then logarithm from 10 to 1000 c.p.s. Time-constants are under software control from 1 to 100 sec. EMI components supply discriminated pulses to the SR photon counter. Software for interfacing the Opus to the SR photon counter and for displaying data on screen and via the D to A is available from Dr K. S. R. Cuthbertson (University of Liverpool). The use of EMI's C660 photon-counting board is under evaluation.

5. Signal normalization and calibration

5.1 Predicted photon count rates from single cells

The total number of photon counts recorded from a single hepatocyte is typically 300 000 while for a cardiac myocyte counts of *c*.2 000 000 are not unusual. At resting free Ca levels of approx. 200 nM, about 10^{-6} of this aequorin will be consumed per second, generating 'resting signals' of approx. 0.3–2 c.p.s. superimposed on a background count of about 1 to 4 c.p.s. This resting signal is detectable over approx. 10 min recording at a long time-constant (about 30 sec). If Ca_f rises to 10^{-6} M the signal from a hepatocyte will be about 10^3-fold faster (approx. 500 c.p.s.) clearly detectable above background after only a fraction of a second.

5.2 Normalization of signals

No attempt is made to determine how much aequorin is injected into a cell, except that it is of the order of 1% of cell volume. An indeterminate percentage of the amount injected will have lost its activity in storage and even more will be lost by Ca contamination entering the pipette tip during injection. Rather, we determine the total aequorin content of the cell by discharging all the aequorin at the end of the experiment while the photomultiplier is still counting. Distilled water is superfused over the cell to induce lysis and discharge of the remaining aequorin. The signals can then be normalized by calculating c.p.s. divided by total counts remaining; this is ideally carried out by microcomputer. The normalized signals can then be calibrated in free Ca by comparison with normalized signals obtained from Ca–EGTA buffers of the same Mg_f, ionic strength, pH, and temperature as believed to pertain in the cytoplasm. Attempts to inject concentrated Ca buffers into the cytoplasm of mammalian cells to carry out *in situ* calibration have not been attempted.

5.3 *In vitro* calibration of aequorin luminescence

Ca–EGTA buffers, containing about 10mM EGTA, 10 mM Hepes (pH 7.2) and isotonic KCl, are made up at 37 °C and used to determine how fast aequorin is consumed at free Ca concentrations from 10^{-7} to about 10^{-5} M. (See Chapter 7, this volume, for buffer instructions.)

5.4 *In vitro* aequorin measurements

A simple luminometer is used to record aequorin signals upon mixing small aliquots with Ca–EGTA buffers (*Figure 7*). The apparatus is placed in a dry 37 °C incubator because the photomultiplier noise is insignificant compared with the signals. Since aequorin batches can differ somewhat in their Ca-sensitivity, ideally a fresh curve should be produced for each vial of aequorin opened. In use, the plastic tube is rinsed with dilute EGTA (1 mM) and aspirated more or less dry. A small droplet, (approx. 5 μl of a dilution made from the aequorin solution left over from droplets prepared for filling pipettes) is placed on the bottom of the tube under a small drop of oil. After filling the syringe with 0.5 ml of pre-warmed buffer, the incubator door is shut and the photomultiplier and recording apparatus switched on. The sudden flood of buffer on to the small aequorin droplet ensures rapid mixing. At high Ca concentrations (e.g. 5×10^{-6} M and above) the aequorin will be more or less completely consumed during a 10 min recording. However, at lower Ca concentration it is necessary to hasten the discharge of aequorin by adding droplets of 10–100 mM $CaCl_2$ via the syringe and tubing. Avoid using dilute $CaCl_2$ to preserve more or less the same volume in the cup, or else counting geometry may alter.

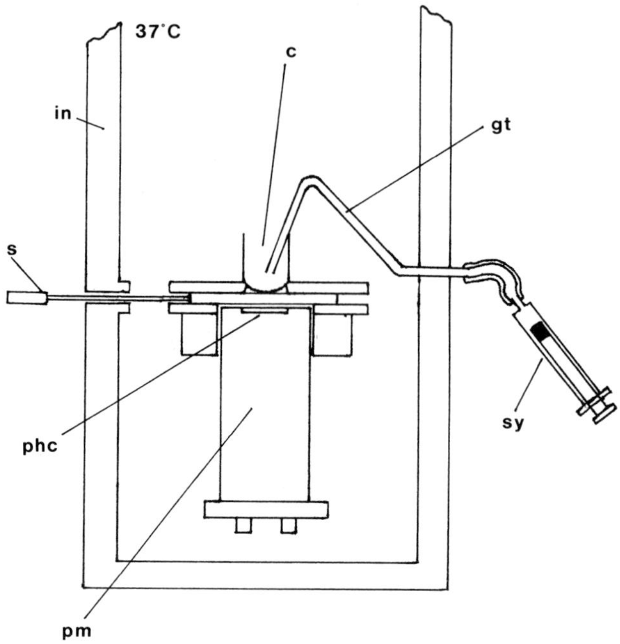

Figure 7. Simple luminometer used for *in vitro* calibration of aequorin in Ca buffers. c = cuvette (base of 1-cm diameter plastic test-tube), gt = glass tubing (approx. 1 mm diam), phc = photocathode, pm = photomultiplier, s = shutter, sy = syringe (1 ml).

5.5 Calibration curves

Figure 8 shows two curves obtained *in vitro* for mammalian cell conditions (37 °C, 150 mM KCl, Mg_f 1 mM and 5 mM) (see ref. 7). A family of curves showing the effects of pH (little effect), ionic strength. Mg, as shown in ref. 18. These can only be used to obtain a rough idea of the effects on signals; each set of conditions determines a unique curve, which will also be influenced to some degree by the batch of aequorin. A calibration curve for recombinant, homogeneous aequorin is not yet available.

5.6 Data presentation

The data obtained from *in vitro* calibration can be incorporated into the data-analysis (off-line) programme in the host microcomputer, enabling the computed fractional rate of consumption to be plotted as free Ca. Because of the large dynamic range of the original signals (approx. 0.5 c.p.s. to 1000) we use a program that varies the time-constant of the plot, between 10 or 30 sec for 'resting signals' (about 0.5 c.p.s.) but switching to 1 sec when the signal exceeds a pre-set rate (e.g. 5 or 10 c.p.s.). In this way, transient rises in free

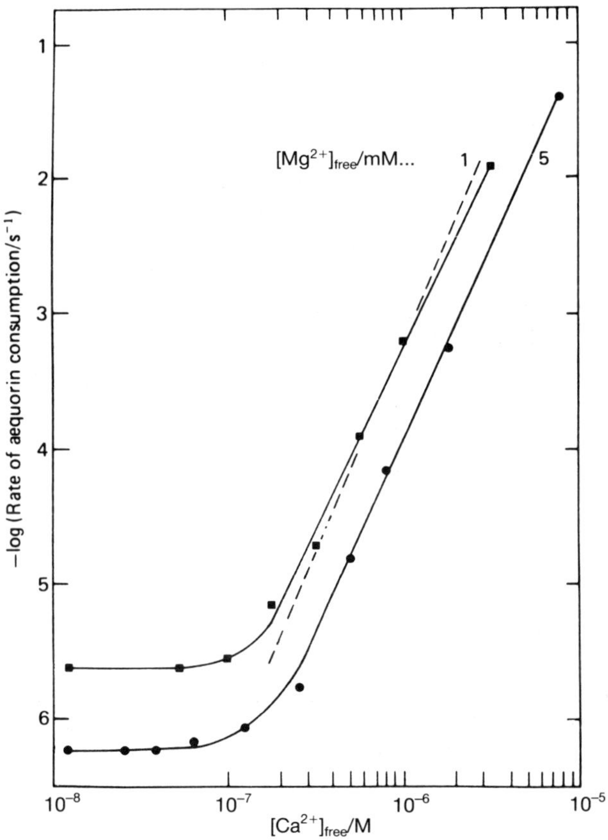

Figure 8. Relationship between the rate of aequorin consumption and $[Ca^{2+}]$ *in vitro* in 2 mM-EGTA buffers containing 150 mM KCl, 10 mM Pipes, and $[Mg^{2+}]_{free}$ as indicated, at 37 °C. (From ref. 7, with permission.)

Ca are faithfully presented without the intrusion of excessive noise in the interim periods of resting signal.

6. Imaging aequorin luminescence

6.1 Limitations

A fundamental limit on the feasibility on imaging aequorin photons is imposed by the finite amount of visual information contained in a few hundred detected photons. The question of when a number of photons appears to be a cluster or just randomly-distributed points, is not easily quantified. However, we can be fairly certain that in mammalian cells of normal size (the exception being oocytes), detectable zones of free Ca at physiological levels are unlikely to be recognized. Thus, a cell containing say 10^6 total photons, counted isotropically by a photomultiplier, when imaged by a lens, would yield only some 5×10^4 counts because of losses imposed by the numerical aperture of the lens. (About 95% of isotropically radiated light will not be collected by the lens; even a high *NA* lens will collect only 10%.) So a cell at 1 μM Ca_f would generate about 50 detected counts per second. Seen against a background of around 20 c.p.s. over the image, detection of such a cell is only just feasible. Subcellular zones, or waves of Ca_f, crossing a cell would be unlikely to be visualized. Feasibility improves in proportion to increasing cell volume and hence aequorin content, so imaging may be feasible in giant cells.

7. Simultaneous measurements of cellular parameters

7.1 Muscle tension

Because of the requirement to exclude all external light, making simultaneous recordings of other cellular parameters is not simple. Furthermore, except for really giant cells (squid giant axon, giant barnacle muscle cell) it is unlikely to be possible to record electrical potentials from an indwelling electrode in a cell because of aequorin luminescence induced at the site of impalement. However, it is possible to make tension measurements of ferret papillary muscle (19). In this system several hundred myocytes were injected with aequorin using a Pt-wire in the aequorin-filled micropipette to monitor cell impalement from the membrane resting potential. The signal will have been degraded somewhat by the large light-guide and by phosphorescence from the large, Perspex muscle bath.

7.2 Structural images

In many applications it would be useful to be able to record microscope images of illuminated cells ('Structural images') as well as recording the aequorin signals. Two ways of doing this, using a photomultiplier and a perfusable cup similar to that described above, are outlined here (see also ref. 20). In both systems the base of the perfusion chamber is sealed with a coverslip below which a 40× or 60× objective lens is focused on the cell in its microslide. The warm water jacket encloses the side of the lens and conducts heat laterally to the cell. The lens is mounted on a focusing ring (Ealing) driven by a synchronous motor and gear box. The perfusable chamber is held on a *X–Y* micrometer stage (Oriel) driven by a pair of motors and gear boxes and friction wheels. The image generated by the objective is reflected through 90 degrees and projected horizontally on to a television camera through a projector eyepiece where necessary. Illumination for the cell is provided by a plastic fibre optic (approx. 1 mm diameter) which terminates about 5 mm above the cell, between the coverslip which seals the top of the chamber and the underside of a remotely-activated shutter which protects the photomultiplier. The camera, microscope, and perfusion chamber are mounted on a base-plate which can be slid out from under the photomultiplier and shutter in order to load a microslide into the perfusion chamber (see ref. 20).

Two types of light source have been used, with two different television cameras. Initially, a CCD camera was used in conjunction with a red light-emitting diode (l.e.d.) (10 μW output at 665 nm). Clear images of myocytes, with clearly visible sarcomeres, were obtained despite the simple illuminator. By closing the shutter to protect the photomultiplier a few seconds' video recording could be made every few minutes. However, some red light did pass the edges of the shutter adding spurious signals which had to be removed from the computer recording of the aequorin signal. This system allowed monitoring of myocyte shape while perfusing anoxic medium. Later, in order to avoid interrupting the aequorin recording and in the hope of improving light collection by eliminating the shutter we have recently converted to near infra-red imaging systems. A 1300 nm laser diode (100 μW) feeds a graded fibre optic which terminates about 5 mm above the cell and is centrable over the cell. A 40× objective lens collects the image and projects it on to a 2/3-in Leddicon video tube in a proprietary chassis (Brian Reece Instruments). Reasonably clear video images are obtained. Most importantly, the laser diode produces monochromatic 1300 nm light and there is no increase in the photomultiplier signal when the laser-diode is switched on. The shutter is no longer required and we can simultaneously detect resting aequorin signals from a heart cell (approx. 1 c.p.s. above background of about 3 c.p.s., in this system) while simultaneously recording cell shape. Resolution will be somewhat degraded by the longer wavelength but sarcomeres are detectable. Some video lag may detract from seeing rapid twitches of the cell but this should be circumvented using a more intense laser diode. This system will not work with an ordinary (cheap) l.e.d.; it relies upon the extremely narrow wavelength spread of the solid state laser and the fact that a bialkali photocathode releases no electrons whatsoever when bombarded with 1300 nm photons.

8. Aequorin in cell populations

We have no enthusiasm for recording Ca_f signals from large numbers of cells. Most importantly the complicated oscillatory regulation of Ca_f that we have come to expect in agonist-stimulated cells (e.g. ref. 21.) will be obscured. Further the steep relationship between aequorin luminescence and Ca_f means that one sick cell containing several micromolar Ca_f will obscure the signal generated by several thousand cells at healthy resting levels of Ca_f. Finally, techniques for loading aequorin into cells by various methods of reversible permeabilization must be significantly disruptive to cells, especially from loss of low-molecular-weight metabolites and substrates.

Reference 7 summarizes the reversible permeabilization procedures that have been used with aequorin.

9. Recent and future developments

9.1 Modified aequorins

(a) *Isoaequorins.* Native aequorin is a mixture of several isoaequorins, which have been separated and characterized by O. Shimomura (16) and from whom samples are available. The isoaequorins differ in the Ca^{2+} concentration needed to give half-maximal light intensity. The low-sensitivity isoaequorins may find use in cells where free Ca changes are particularly great. The high-sensitivity aequorins may provide more precise records of resting free Ca levels.

(b) Coelenterazine derivatives have been synthesized, giving novel properties when reconstituted with apoprotein (ref. 22).

9.2 Aequorin transfected cells

A gene for aequorin has been cloned (12–14) and shown to produce in *Escherichia coli* apoprotein capable of being recharged by coelenterazine (14). It is very likely that by the use of a suitable mammalian expression vector that stable expression of the gene could be obtained in mammalian cell-lines. Coelenterazine has been shown to activate intracellular obelin in neutrophils loaded with obelin mRNA by reversible permeabilization (23). Thus, there are unlikely to be any constraints on using aequorin transfected cells for single-cell free Ca measurements. By circumventing the need for microinjection such an approach should provide the simplest and cheapest single-cell free Ca measurement yet devised. Once transgenic animals and plants are obtained, simple Ca_f measurements should be possible for primary cells, both *in situ* and *in vitro.*

References

1. Ridgway, E. B. and Ashley C. C. (1967). *Biochemical and Biophysical Research Communications*, **29**, 229–34.
2. Cobbold, P. H. (1980). *Nature*, **285**, 441–6.
3. Cuthbertson, K. S. R. and Cobbold, P. H. (1985). *Nature*, **316**, 541–2.
4. Cobbold, P. H. and Bourne P. K. (1984). *Nature*, **312**, 444–6.
5. Woods, N. M., Cuthbertson, K. S. R., and Cobbold, P. H. (1986). *Nature*, **319**, 600–2.
6. Cobbold, P. H., Cheek, T. R., Cuthbertson, K. S. R., and Burgoyne, R. D. (1987). *FEBS Letters*, **211**, 44–8.
7. Cobbold, P. H. and Rink, T. J. (1987). *Biochemical Journal*, **248**, 313–28.
8. Rooney, T. A., Sass, E. J., and Thomas, A. P. (1989). *Journal of Biological Chemistry*, **264**, 12859–66.

9. Kawanishi, T., Blank, L. M., Harootunian, A. T., Smith, M. T., and Tsien, R. Y. (1989). *Journal of Biological Chemistry*, **264**, 12859–66.
10. Blinks, J. R., Mattingly, P. H., Jewell, B. R., van Leeuwen, M., Harrer, G. C., and Allen, D. G. (1978). In *Methods in enzymology*, Vol. 57 (ed. M. de Luca), pp. 292–328. Academic Press, New York and London.
11. Blinks, J. R., Weir, W. G., Hess, P., and Prenolergast, F. G. (1982). *Progress in Biophysics and Molecular Biology*, **40**, 1–114.
12. Inouye, S., Hoguchi, M., Sakaki, Y., Takagi, Y., Migata, T., Iwanaga, S., and Tsuji, F. (1985). *Proceedings of the National Academy of Science, USA*, **82**, 3154–3158.
13. Prasher, D., McCann, R. O., and Cormier, M. J. (1985) *Biochemical and Biophysical Research Communications*, **126**, 1259–68.
14. Tsuji F., Inouye, S., Soto, T., and Sakaki, Y. (1986). *Proceedings of the National Academy of Sciences, USA*, **83**, 8107–11.
15. Inouye, S., Taguchi, H., Murata, M., Kakoi, H., and Goto, T. (1975). *Chemical Letters*, 141–4.
16. Shimomura, O. (1986). *Biochemical Journal*, **234**, 271–7.
17. Shimomura, O. and Shimomura, A. (1984). *Biochemical Journal*, **221**, 907–10.
18. Allen, D. G. and Blinks, J. R. (1979). In *Detection and measurement of free* Ca^{2+} *in cells* (ed. C. C. Ashley and A. K. Campbell), pp. 159–74. Elsevier/North-Holland, Amsterdam.
19. Allen, D. G. and Orchard, C. H. (1983). *Journal of Physiology* (*Lond.*) **339**, 107–22.
20. Allshire, A. P., Woods, N. M., and Cobbold, P. H. (1989). In *Methodological surveys in biochemistry and analysis*, Vol. 19 (ed. E. Reid, G. M. W. Book, and J. P. Luzio), pp. 455–61. Royal Society of Chemistry, London.
21. Berridge, M. J., Cobbold, P. H., and Cuthbertson, K. S. R. (1988). *Philosophical Transactions of the Royal Society of London*, **B320**, 325–42.
22. Shimomura, O., Musicki, B., and Kishi, Y. (1988). *Biochemical Journal*, **251**, 405–410.
23. Campbell, A. K., Patel, A. K., Razavi, Z. S., and McCapra, F. (1988). *Biochemical Journal*, **252**, 143–9.

3

The use of electrodes to study cellular Ca^{2+} metabolism

C. H. ORCHARD, M. R. BOYETT, C. H. FRY, and M. HUNTER

1. Introduction

The importance of electrodes in the study of cellular Ca^{2+} hardly needs emphasizing. Electrodes can be used to monitor Ca^{2+} flux across cell membranes and to monitor the concentration of Ca^{2+} both in the cell cytoplasm and in the extracellular space. Electrodes can also be used to manipulate intracellular Ca^{2+}. Thus, electrodes have contributed a great deal to our understanding of cellular Ca^{2+} metabolism. The length of a chapter such as this raises two main problems. First, the equipment used in the manufacture and use of electrodes varies in detail from manufacturer to manufacturer. We have, therefore, not concentrated on the details of using one particular brand of equipment but have given such information as should enable the reader to understand, and use effectively, any brand of equipment. Second, limitation of space precludes detailed discussion. We have, therefore, cited references that give further detail for the interested reader. We would particularly recommend the books by Purves (1), Thomas (2), and Thomas (3), and one by various authors (4).

2. Ca^{2+}-selective macroelectrodes

Ca^{2+}-selective macroelectrodes are electrodes that can be used to measure $[Ca^{2+}]$ in bulk solutions; for example, for the measurement of Ca^{2+} fluxes in cell suspensions.

2.1 Principles

An ion-selective electrode consists of a layer, or column, of an ion-selective ligand which separates the experimental solution from a filling solution, which contains a fixed concentration of the primary ion: in this case, Ca^{2+}. The primary ion is transported across the ion-selective ligand layer in accordance with the prevailing concentration gradient so that a potential is established and measured with respect to some stable reference level, usually a salt bridge in contact with the experimental solution. Connection to measuring devices can be achieved by placing a Ag–AgCl electrode in the filling solution of the ion-selective electrode and the salt bridge. Thus the experimental arrangement is similar to that shown for a microelectrode in *Figure 1*, E, with the Ca^{2+} selective macroelectrode taking the place of the microelectrode, and the salt bridge as the reference electrode. The arrangement may be written as the following electrochemical cell, using as an example a Ca^{2+} selective electrode filled with a solution containing 1 mM $CaCl_2$. The reference electrode is a Ag–AgCl electrode making contact with the test solution via a 3 M KCl salt bridge.

$$\underbrace{\text{Ag/AgCl/3 M KCl}}_{\text{reference electrode}}\text{/test solution/}\underbrace{\text{ion-selective membrane/1 mM CaCl}_2\text{/AgCl/Ag}}_{\text{ion-selective electrode}}$$

Each slash (/) represents a junction in the electrochemical cell where a potential can develop. The potential at the interface between the ion-selective membrane and the test solution is the one of interest and its variation as a function of concentration enables an estimate of the $[Ca^{2+}]$ in the test solution to be made. There are two objectives when using ion-selective electrodes to estimate the concentration of an ion in an unknown solution: (a) to determine the relation between concentration in the test solution and potential at this junction, and (b) to ensure that all other potentials are invariant. The first problem is solved by correct calibration procedures and will be considered in Section 2.3. The second problem is solved by paying attention to detail when constructing the rest of the electrochemical cell.

2.2 Manufacture of Ca^{2+}-sensitive macroelectrodes

The crucial step in the manufacture of ion-selective macroelectrodes is the incorporation of the ion-selective ligand in PVC. The immobilization of the ligand in a plastic ensures that stable and durable electrodes can be made readily in any form desired. For the measurement of Ca^{2+} two classes of ligand can be used. The ionic Ca^{2+} ligands can be exemplified by iso-octylphenylphosphate (IOPP) in conjunction with di-octyl phenylphosphonate (DOPP) and the neutral ligands, developed by Simon and his colleagues, by ETH 1001.

Protocol 1. To manufacture such electrodes [see Fry (5)]

1. In a fume cupboard, dissolve approximately 100 mg of high-molecular-weight PVC in 6 ml tetrahydrofuran (THF).
2. Separately, add 10 mg ETH 1001 and 2 mg K tetrakis (4-chlorophenyl

Protocol 1. *Continued*

borate) to 200 mg di-octyladipate (for a neutral ligand plastic) or 20 mg IOPP to 200 mg DOPP (for an ionic ligand plastic).

3. Mix the two solutions carefully and allow the THF to evaporate slowly over 2–3 days, leaving a clear Ca^{2+} selective plastic.
4. Prepare electrode templates by wedging a small piece of ceramic (frequentite) rod in a length of PVC tubing.
5. Redissolve the Ca^{2+} selective plastic in about 1 ml THF and dip the template into the viscous solution several times, leaving 5 min between each coating. Leave for 24 h.
6. Introduce the filling solution into the lumen of the PVC tube. For measurement of Ca^{2+} in the millimolar range use 1 mM $CaCl_2$ in 150 mM NaCl.
7. Insert a stable electrode (e.g. a Ag–AgCl electrode) into the filling solution and attach to recording apparatus.

It must be stressed that for successful use of these electrodes the other potentials in the electrochemical cell shown above are constant. The use of reversible electrodes, such as Ag–AgCl electrodes, minimizes changes in junction potentials between the metal electrode and electrolyte solutions. Procedures for preparing such electrodes are given in ref. 6. The use of a salt bridge on the reference side of the circuit is also recommended in most situations. Strong KCl solutions are often those of choice because of the small liquid junction potential between the bridge and the test solution. However, there is an inevitable leak of KCl from such bridges and in situations where this is undesirable other salt solutions should be employed. It is dangerous to place a Ag–AgCl electrode directly into the test solution as a reference connection, as such an electrode is a Cl^--sensitive electrode, so that an intervention which changes the $[Cl^-]$ would introduce an additional potential. Salt bridges may be prepared from the electrode templates as used above and the PVC tube filled with electrolyte. Alternatively, a salt bridge may be prepared by making up a 2% solution of agar in either 3 M KCl or the standard bathing solution. The solution is heated (and stirred) until it just boils. The hot solution is then injected into a plastic tube and the agar is allowed to set. The salt bridge is connected to the recording apparatus via a Ag–AgCl wire.

2.3 Calibration of Ca^{2+}-selective electrodes

Calibration of ion-selective electrodes demands interpreting the measured potential difference (p.d.) in terms of the concentration of the primary ion in the test solution. For an electrode that transports only Ca^{2+} the relation between p.d., E, and $[Ca^{2+}]$ is described by the Nernst equation:

$$E = E^0 + 2.303\ RT/zF \cdot \log_{10} \{[Ca^{2+}]_{test}\}.$$

E^0 is the potential derived from the junctions in the system so that its constancy is required for accurate determinations. The term 2.303 RT/zF is equal to approximately 30.5 mV at 37 °C—i.e. a tenfold alteration of the $[Ca^{2+}]$ will alter the measured p.d. by 30.5 mV.

Ion-selective electrodes are rarely responsive to the primary ion alone but also transport other, interferent ions, to a lesser extent. The measured p.d. is related to the primary ion (Ca^{2+}) and interferent ion (I^{n+}) concentrations by the more complex Nikolsky equation:

$$E = E^0 + 2.303\ RT/z_pF \cdot \log_{10} \{[Ca^{2+}]_{test} + \Sigma K_{pi}{}^{pot}\ [I^{n+}]^{(z_p/z_i)}\}.$$

The subscripts p and i refer to primary (Ca^{2+} in this case) and interferent ions respectively. The term $K_{pi}{}^{pot}$ is the potentiometric selectivity coefficient and reflects the relative selectivity of the electrode to the primary ion as compared to interferent ions. Note that this term is summed for all possible interferent ions; the smaller the value of $K_{pi}{}^{pot}$ the less is the contribution to the overall electrode potential from interferent ions. The measurement of $K_{pi}{}^{pot}$ is described by Ammann (7).

In practice, Ca^{2+} selective electrodes can be calibrated by placing them and a reference electrode in solutions of varying $[Ca^{2+}]$ which encompass the experimental range and containing also other electrolyes that are present in the experimental medium. The recorded p.d. can be plotted as a function of $[Ca^{2+}]$, the latter on a logarithmic scale, and the concentration of the test solution read from the calibration curve. For example, if used in solutions similar to extracellular fluid, the calibration solution should contain about 115 mM NaCl, 25 mM $NaHCO_3$, 1 mM $MgCl_2$, and 4 mM KCl and be gassed with 5% CO_2 containing gas mixture. Note also that many organic molecules also introduce a p.d., thus any change of the recorded p.d. on adding such compounds to a standard calibrating solution should be assessed if they are to be used experimentally. When recording results obtained with ion-selective electrodes the composition of the calibrating solutions should always be noted. In order to compare results obtained with various calibrating solutions a 'standard solution' should also be used against which all other readings can be compared. The next paragraph suggests such a solution, namely 1 mM $CaCl_2$ in 150 mM NaCl.

Ion-selective electrodes measure the activity of an ion in solution rather than the concentration. The two variables are related by an activity coefficient γ. Because γ depends upon many factors including the particular ion, its counter-ion ionic strength, etc., it is likely to be different in calibration and test solutions. In view of this it is not useful to express data in terms of ionic activity but in terms of concentration with respect of a standard solution.

For Ca^{2+} electrodes the unknown concentration can be referred to a 1 mM $CaCl_2$ solution in 150 mM NaCl, which may be assigned an arbitrary zero potential in the above equations.

2.4 The use of Ca^{2+}-selective macroelectrodes

Such electrodes are useful in a variety of situations from determination of Ca^{2+} in unknown solutions to the measurement of Ca^{2+} fluxes in cell suspensions. The ceramic-cast electrodes described above will respond to $[Ca^{2+}]$ as low as 10 nM in the presence of physiological concentrations of Na^+, K^+, H^+, and Mg^{2+}. Their relatively low DC resistance (1–10 MΩ) ensures that their response time is rapid (< 10 msec) and that conventional pre-amplifiers may be used to record signals. Circuitry similar to that used to record signals from electrolyte-filled micropipettes (*Figure 1*, *E* and *2* and Section 3.6) is perfectly adequate.

3. Microelectrodes

3.1 What is a microelectrode?

A microelectrode is the name commonly given to a piece of glass tubing that has been heated, and drawn to a fine tip (<1 μm; see Section 3.3.2) and then filled with an electrolyte solution (see Section 3.4). In common with previous authors, we shall refer to the unfilled electrode as a micropipette. In addition, we will be concerned with pipettes with larger tip diameters (1–2 μm) that are used for patch-clamping (Section 3.7.2), whole-cell clamping (Section 3.7.2) and cell dialysis (Section 3.7.4. *ii*). These are known as patch-pipettes (unfilled) or patch-electrodes (when filled). *Figure 1*, B, shows the names normally ascribed to the different parts of the pipette, and which we shall use in this chapter.

The electrode is normally mounted on a micromanipulator that allows fine control of the movement of the electrode (see Section 3.5), and connected to a pre-amplifier (*Figure 2* and Section 3.6). In its simplest configuration, electrodes are used to measure the voltage across a cell membrane. This is done by inserting the microelectrode into the cell, and monitoring the voltage between the microelectrode and a reference electrode (*Figure 1*, E, and Sections 2.2. and 3.6) in the extracellular fluid in the experimental bath on some sort of output device (see Section 3.6.).

Before discussing the use of microelectrodes for the study of cell Ca^{2+} metabolism, some discussion of their manufacture and use will be given. Sections 3.2 to 3.6 refer mainly to conventional microelectrodes. Details that are specific to ion selective electrodes and patch-electrodes are discussed in Sections 3.7.1 and 3.7.2 respectively.

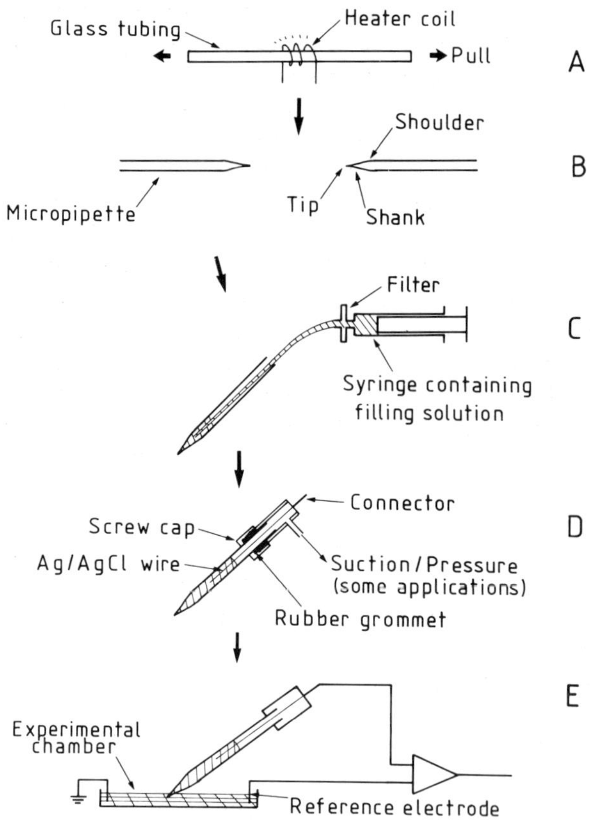

Figure 1. The stages in the manufacture and mounting of a microelectrode. The glass tubing is first heated and pulled (**A**) into two halves (**B**, which shows the names given to the different regions of the electrode). The micropipette is then filled using a syringe (**C**), before being mounted in a holder (**D**) and connected to an amplifier (**E**). See text for details.

3.2 Electrical properties of microelectrodes

In order to use microelectrodes effectively, it is important to understand their electrical properties. As shown in *Figure 2*, A, an electrode has a tip resistance (R_t), a tip potential (V_t), an electrode capacitance (C_e), and a wall resistance (R_w). Each of these properties will be considered.

3.2.1 Tip or electrode resistance (R_t)

The tip of an electrode is narrow and offers an appreciable resistance to the flow of electrical current. This resistance depends on two factors: the size of the tip and the conductivity of the electrode filling solution. Relative to its tip diameter, a patch electrode, which is filled with a solution of low ionic strength (Sections 3.4 and 3.7.2) will therefore have a higher resistance than a conventional electrode, which is filled with, for example, 3 M KCl, whilst an ion-selective microelectrode has a very high resistance as a result of the ligand in the electrode tip (Section 3.7.1). Electrode resistance is routinely measured because, for a given filling solution, it is a useful indicator of the electrode tip diameter and can be used, therefore, to choose electrodes of the appropriate tip diameter for the tissue and the type of experiment. Conventional electrodes usually have resistances between 5 and 100 MΩ. The higher resistance electrodes with sharper, finer tips are needed for impaling small cells. The lower resistance, wide tip electrodes are suitable for current injection (for voltage-clamp experiments, for example; see Section 3.7.3). An electrode of a given tip diameter can be made sharper and thus easier to impale into cells by bevelling the tip. There are various devices for bevelling electrodes (see, for example, ref. 8). The result is that the electrode tip is ground down so that it is similar in shape to the tip of a syringe needle. When an electrode is being bevelled the electrode resistance is usually continuously measured in order to monitor the bevelling process. A bevelled electrode has a lower resistance (for this reason bevelled electrodes are more suitable for current injection). During the experiment (especially when impalements are being attempted) the resistance can be checked to ensure that the electrode tip is unbroken.

The electrode resistance can be measured in a number of ways. All commercially available amplifiers have an inbuilt means to measure electrode resistance at the push of a button. The usual method is to pass a known current through the electrode tip. According to Ohm's law the passage of a current across a resistance will produce a voltage drop. From Ohm's law the resistance can be calculated from the voltage drop.

The high resistance of an electrode places a special requirement on the amplifiers used with microelectrodes (Section 3.7.1. *iii*): they must have a high input resistance, which is several orders of magnitude greater than the resistance of the electrode (see *Figure 2*).

3.2.2 Electrode resistance and the pick-up of noise

Because microelectrodes have a high resistance they act as aerials and pick up electrical interference. This results in noise at the output of the electrode, i.e. fluctuations in voltage. The commonest form of interference is mains pick-up or 'hum', i.e. noise induced by the alternating current in the mains or power cables running to and within mains powered devices such as oscilloscopes. Mains interference has the frequency of mains electricity (normally 50 or 60 Hz). The noise pick-up is greater the higher the electrode resistance. Interference induces an electrical current. When this flows across the tip resistance it results in a voltage drop, which is seen as noise. The greater the tip resistance the greater the voltage drop. Because noise can be as large as,

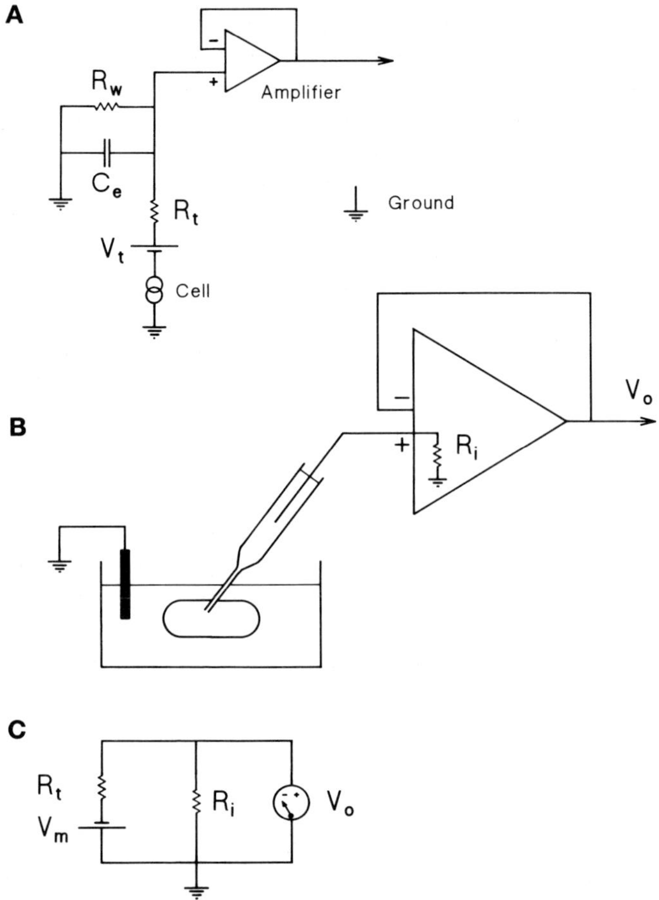

Figure 2. A: the electrical properties of a microelectrode. **B**: a typical recording arrangement showing the input resistance, R_i, of the amplifier. **C**: equivalent circuit diagram of **A**. Note that the amplifier measures the potential at its positive input with respect to ground. R_t and R_i act as a voltage divider. It follows from this that the output voltage of the amplifier $V_0 = V_m\ (R_i/R_t+R_i)$, where V_m is the membrane potential. Only if $R_i >> R_t$ will $V_0 = V_m$.

or greater than, the biological signal (and hence obscure the biological signal) steps have to be taken to eliminate it; this principally involves shielding. Usually the electrode and the tissue bath are enclosed within a Faraday cage: a metal box which is grounded; this shields the electrode from the interference. The front of the cage is often open, but if interference is a particular problem, as it is with high-resistance ion-selective microelectrodes, the cage front can be fitted with doors. Within the cage any metal objects, such as a microscope, should also be grounded. The electrode can be fitted with a small grounded shield or driven shield (this is considered further below). Because noise is principally picked up by the electrode, the connection between the electrode and the microelectrode amplifier should be as short as possible. Commercially available amplifiers have a head stage and this can be mounted within a few centimetres of the electrode (the wire connecting the electrode to the head stage should have a driven shield – see below). The bathing solution within the tissue bath is usually connected to ground via a Ag–AgCl electrode, and it is advisable to connect the Ag–AgCl reference electrode to ground via the ground connection on the microelectrode headstage.

Noise can be picked up elsewhere. For example, it can be picked up by the cable conducting the output of the microelectrode amplifier to the oscilloscope. To reduce noise pick-up by cables which are carrying small voltages, the cables should be shielded. Shielded cables consist of a central core (which carries the signal) surrounded by a shield: a copper braid connected to ground.

The connections of the various shields to ground is also important. It is best to connect each shield to a central grounding point via a low resistance connection (some workers use a heavy copper braid). Purves (1) recommends that the central grounding point should be the ground connection of the oscilloscope (normally the principal recording device). A shield should not be connected to ground by two routes; this will result in a 'ground loop'. The perils of ground loops are considered by Purves (1).

3.2.3 Tip potential (V_t)

Normally, when any type of microelectrode is immersed in the bathing solution a potential is measured from the electrode. This potential arises at the tip of the electrode and it has two components: a liquid junction potential and a true tip potential. The liquid junction potential arises because the solution in the microelectrode has a different composition to that of the bath – the cations and anions diffuse along their concentration gradients in and out of the electrode tip and they tend to do so at different speeds (because the ions have different mobilities); the two charged species become separated and a potential is set up. This potential is very small, i.e. less than a millivolt, if one uses 1 M or 3 M KCl as the electrode filling solution, because the mobilities of K^+ and Cl^- are similar. The liquid junction potential is likely to be greater if other filling solutions are used. The true tip potential arises from fixed surface charges on the glass, which act as an ion-selectivity filter at the very tip of the microelectrode where the lumen of the microelectrode is very small. This is mostly a problem with fine tipped microelectrodes (resistances greater than 30 MΩ) and is exacerbated by the use of filling solutions in which the cation/anion mobility ratio is large. Unless the overall tip potential (the sum of both components) is compensated for it will result in an incorrect measurement of membrane potential. All commercially available micro-electrode amplifiers have a control which allows the electrode tip potential to be 'backed-off' or 'offset': a voltage equal and opposite to the tip potential is added to the signal from the microelectrode. When a microelectrode is in the bathing medium and just before it is inserted into a cell, the tip potential is backed-off in this way so that the output from the microelectrode amplifier is 0 mV. What cannot be compensated for are changes in the tip potential once the electrode tip is within the cell, and Adrian (9) has shown that such changes do occur. Adrian's practical solution was only to use electrodes with tip potentials of less than 5 mV, because evidence showed that the greater the initial tip potential, the greater the change in tip potential on impalement. The tip potential can be measured in various ways but we recommend the following approach:

(a) Place a microelectrode in the bath and break off the electrode tip. Back off the recorded potential so that it reads 0 mV. There is now no tip potential (because the tip has been removed) and the sum of all other potentials has been set to zero.

(b) Replace the broken electrode with another one. The recorded potential is the tip potential of the new electrode (because the sum of all the other potentials will still be equal to zero).

On removal of an electrode from a cell it may be found that the electrode potential is no longer 0 mV because the tip potential has changed. (This is different to the situation discussed above, because the change in tip potential discussed above will reverse on withdrawing the electrode from the cell.) If the change is say greater than 5 mV the experimental recordings of membrane potential should be rejected, because one will never know whether the tip potential changed on impalement, throughout the time the electrode was in the cell, or on withdrawal of the electrode. It is possible that the smaller the initial tip potential, the smaller this change.

3.2.4 Electrode capacitance (C_e)

A capacitor consists of two electrical conductors separated by an electrical insulator. A microelectrode has a capacitance, because the glass wall of the electrode (an electrical insulator) separates two conductors: the filling and bathing solutions. An electrode is described as having a capacitance to ground, because the bathing solution surrounding the electrode is grounded. The greater the area of overlap of the filling and bathing solutions (separated by the glass wall) the greater the electrode capacitance (this is why the electrode capacitance is greater, the greater the depth of the electrode tip in a bathing solution or a tissue). Electrode capacitance is less with thick-walled electrode glass (it is inversely proportional to wall thickness). The electrode capacitance slows the response of the microelectrode, because in order to change the potential at the input of the electrode amplifier (*Figure 2*) the electrode capacitance must be first charged by a current flowing through the tip resistance (R_t). The greater the electrode capacitance or resistance the slower the response of the electrode (the time-constant of the electrode, $\tau_e = R_t \cdot C_e$). With conventional electrodes the τ_e is not usually a problem although it will, for example, slow the upstroke of an action potential. However the electrical response of ion-selective microelectrodes is very slow (τ_e is large), because of the large tip resistance of these microelectrodes (Section 3.7.1).

Because of the problems caused by the electrode capacitance, there are various means to reduce its effects. All commercially available microelectrode amplifiers have an electrode capacitance compensation control. There are various ways in which this can be adjusted to its optimal setting (the manual for the microelectrode amplifier will give details). For example, if an action potential is being recorded with a conventional microelectrode, the capacity compensation is increased to obtain the fastest possible upstroke to the action potential. Capacity compensation involves positive feedback and too much compensation results in ringing (a large voltage oscillation at the output of the amplifier). Often a small metal shield around an electrode is used to reduce the pick-up of electrical interference by the electrode or to reduce the interaction between a current passing and a voltage-sensing electrode. If the shield is grounded, however, it will increase the electrode capacitance to ground and slow the response of the electrode. To avoid this, the shield can be 'driven': it can be connected to the output of the microelectrode amplifier. Because of this, the driven shield will always be at the same potential as the electrode and therefore current flow through the tip resistance is not required to charge up the capacitance between the shield and the electrode. In the case of ion-selective electrodes, in which the electrode capacitance is a particular problem, other means have been used to reduce the effects of the electrode capacitance. The electrode can be painted to within microns of the tip with a conducting paint. If the paint is grounded by virtue of its contact with the bathing solution it eliminates the noise caused by the pick-up of electrical interference and also by fluctuations in the solution level, but it does slow the response of the electrode because the capacitance to ground is greater. The final refinement of this technique is to insulate the conducting paint from the bathing solution by painting over it a layer of nail varnish and connecting the conducting paint to the output of the electrode amplifier. This driven shield not only reduces noise pick-up, but it also greatly accelerates the time response of the electrode by eliminating the capacitance between the filling and bathing solutions.

3.2.5 Wall or shunt resistance (R_w)

The glass wall of an electrode separating the filling and bathing solutions does have some conductivity. This conductivity is very small and it has only been shown to cause problems with ion-selective microelectrodes made with thin-walled glass. Thin-walled glass has a lower wall resistance than thick-walled glass and this resistance is in parallel with the large tip resistance and the wall resistance shunts the tip resistance. The effect of this is to reduce the potential recorded by the electrode. Use thick-walled glass for ion-selective microelectrodes.

3.3 Manufacture of micropipettes

Although there are many variations, the basic method of manufacture is the same for all types of micropipettes: the central section of a piece of glass tubing is heated. When the glass becomes soft the tubing is pulled into two halves (*Figure 1*, B). However, the exact method of manufacture varies, depending on the use for which the micropipette is intended.

3.3.1 Glass

i. Types of glass tubing

Although many different types of glass tubing can, and have, been used, the most common type in use at present is borosilicate glass, which is widely available. Tubing for pulling micropipettes is available in diameters from 1.00 mm outer diameter (o.d.) to 3.00 mm o.d. The most commonly used size of glass is probably 1.5 mm o.d. This can be obtained in both thick-wall and thin-wall varieties. The advantage of the thick-wall type of glass is that when pulled, the relatively thick walls have a high wall resistance and low capacitance (Section 3.2). The disadvantage is that for a given external (pulled) diameter, the internal diameter is relatively small, making it harder to use for such applications as voltage-clamping when a large orifice is important for current passing. Conversely, thin-wall glass will have a low wall resistance and high capacitance, but can also have a relatively large orifice for a given external diameter.

Glass with several barrels (up to seven) is also available commercially for manufacture of electrodes. When pulled, one ends up with a fine tip consisting of several barrels. This can be useful for cell dialysis (Section 3.7.4. *ii*), when each barrel can be filled with a different solution or drug, or for the manufacture of ion-selective microelectrodes (Section 3.7.1), when one barrel can act as the ion-selective electrode, and the other as the reference electrode. There are two types of multi-barrel glass available. One looks like a bundle of glass tubing. The other looks like a single tube, with partitions running down the length of the tubing to divide the interior into two or four sections (theta or quad tubing). The disadvantage of this latter type is that when pulled, the central division can become thin enough to act as a low resistance pathway.

Electrode glass tubing containing glass filament is also available. This filament acts as a conduit down which filling solution can run into the tip of the pipette. This is perhaps the commonest way of filling micropipettes (see Section 3.4). However, when making ion-selective microelectrodes, non-filamented tubing is often used, because the filament may allow the back filling solution to permeate the ligand (see Section 3.7.1), thus providing a shunt pathway around the ion sensor.

ii. Preparation of electrode glass

Electrode glass is commonly available in 6-in lengths. This is prepared for pulling as follows:

(a) Cut glass into 3-in lengths.

(b) If required, fire-polish cut ends. This involves placing the cut end of the glass briefly into a hot bunsen flame to melt small glass particles on the cut surface, so that they are no longer loose to fall into the electrode and block the tip. This is normally only done when it is necessary to eject something from the electrode.

(c) Wash glass, and oven-dry, if required. (For the purposes described in this chapter this is not necessary.)

(d) Store (we use a covered beaker) for use.

3.3.2 Pulling micropipettes

Pipette pullers are, in essence, very simple. They hold the two ends of a piece of glass tubing in clamps, and heat the centre of the tubing using an electrically heated element. The two ends of the glass are pulled apart when the heated portion becomes soft, causing the glass to neck and break in the heated region, creating two (almost identical) micropipettes.

Two basic types of pipette puller are available commercially. Vertical pipette pullers, as their name suggests, hold the glass tubing vertically in two clamps. The upper clamp is fixed, the lower clamp moves to pull the glass tubing. To manufacture conventional micropipettes, this type of puller uses gravity to exert a weak pull on the glass. When the glass becomes soft it starts to neck. When the tubing has elongated by a certain amount (which can be fixed by the operator), a solenoid exerts a stronger pull which separates the glass tubing into two halves. Horizontal pullers operate on the same principle. In this case, however, the first pull is also produced by a solenoid, and so the strength of this pull can also be controlled by the operator. To manufacture wide bore pipettes ('patch-pipettes') for whole-cell clamp (see Section 3.7.2) and intracellular dialysis (see Section 3.7.4), a modified type of vertical puller is used (see Section 3.7.2 *i*).

Two types of heater coil are commonly used: either a coil of resistance wire (normally nichrome) or a strip of foil (we use platinum–iridium alloy). The advantage of nichrome wire is that it is mechanically robust, so that if it is accidentally knocked it will retain its shape. The disadvantage is that it is relatively slow to heat and to cool, and can only be run at relatively low temperatures, so that lower resistance electrodes are produced. Platinum–iridium foil, on the other hand, can be run hot, and can produce high-resistance electrodes, but is mechanically weak, so that if it is knocked it can be pushed out of shape, which can have marked effects on electrode characteristics.

ii. Factors determining electrode shape and resistance

The resistance and tip diameter that you require will vary with the application: large cells can be impaled with larger tip diameter microelectrodes than small cells! The best way of getting the electrode suited to your purposes is to experiment. Perhaps the two most important factors, however, are

- the heat of the coil—the hotter the coil the finer the tip and the longer the shank, and
- the size of the coil—for conventional microelectrodes we use a small, close-fitting coil of Pt–Ir, which produces pipettes with a relatively short shank.

However, it is best to start with the manufacturer's recommendations, and then experiment. Once the puller is set up to produce pipettes that meet your requirements, leave it alone—it can be hard to find the ideal settings again.

iii. To pull a pipette

This account refers to the pulling of a conventional micropipette in a vertical puller that has initially been set up according to the manufacturer's recommendations (*Protocol 2*).

3.4 Filling micropipettes

There are two considerations about filling a pipette. First, what to fill it with, and second, how to get the solution into the pipette.

The first of these considerations, what to put into the pipette, could be the subject of a chapter by itself. Much depends on the use to which the microelectrode will be put. The 'standard' filling solution used for conventional microelectrodes being used to monitor membrane potential is 3 M KCl. The advantage of this filling solution is that it a good electrical conductor, and that K^+ and Cl^- have similar diffusion coefficients, so that diffusion of these ions into the cell will result in only small charge separation, and so will minimize the junction potential. However, this diffusion of K^+ and Cl^- is a disadvantage because such an electrode will tend to increase intracellular $[K^+]$ and $[Cl^-]$, which may have undesirable physiological consequences (particularly in the case of Cl^-). For this reason, other electrolyte solutions, e.g. potassium acetate or potassium citrate, may be used.

For large-bore microelectrodes, e.g. those used for whole-cell voltage-clamping (Section 3.7.2), choice of filling solution becomes much more

Protocol 2. Pipette pulling

1. Place the glass tubing (we use 3-in lengths), in the lower (movable) clamp. Raise the clamp (with the tubing held in it) so that the tubing passes through the heater coil, and into the upper (immovable) clamp.
2. Adjust the height of the lower clamp so that it is as far below the heater coil as the upper clamp is above the coil.
3. Check that approximately the same length of glass is above the coil as below. If not, lower the bottom clamp, adjust the position of the glass in the clamp, and try again.
4. Fasten upper clamp.
5. Close draught excluder, which stops draughts changing the heat of the coil, thus altering electrode characteristics.
6. Turn puller on.
7. After pipette has been pulled, does the shape look all right? (If the shank is too long, it will be too flexible; too short, it will be too stiff.) If not, adjust settings and try again. If it looks all right, check its resistance (see Section 3.2.1). Is this all right? If not, try again (increase heat to increase resistance). If it is, fill it and try using it.

critical because the contents of the electrode will enter the cell cytoplasm much more easily, and may change the cytoplasmic composition markedly (a technique that can be used experimentally; Section 3.7.4). In this case (unless one wants to alter the cytoplasmic composition), it is obviously desirable to use a solution composition that mimics the intracellular environment, or at least in its most important features. Thus, the concentrations of the important ions (Na^+, K^+, Cl^-, Ca^+, and H^+) should be set to the levels thought to exist in the cell studied. In addition, an ATP regenerating system is sometimes incorporated into the solution. The final solution, therefore, will probably be a compromise between simplicity and reproducing the important features of the cell cytoplasm (which may depend on the cell type and the experiment being performed). Some examples of filling solutions used during microelectrode studies of isolated heart cells are shown in *Table 1*.

Introducing the filling solution into the tip of the micropipette can be done in several ways (see ref. 1 for a comprehensive account). Here we will concentrate on the most commonly used method: the use of electrode glass containing a glass filament to introduce an electrolyte solution into the tip. (Introducing ion sensor ligand into the tip of an ion-selective electrode poses separate problems and will be dealt with in Section 3.7.1.) Our procedure for filling such pipettes is set out in *Protocol 3*.

Table 1. Examples of electrode-filling solutions (taken from ref. 16[a] and from solutions used in our own laboratories)

Constituent	Concentration (mM)
Solution 1[a]	
K-aspartate	89.4
$MgSO_4$	7.1
Glucose	5.0
Tes (pH buffer)	30.0
KCaEGTA	0.131
K_2EGTA	0.369
K_2ATP	3.3
Na_2-creatine phosphate (CP)	3.75
$Tris_2CP$	8.25
NacAMP	0.03
Dithiothreitol	0.1
Creatine phosphokinase	30 U/ml
pH 7.1 (achieved by titration with ~11 mM KOH)	
pCa 7.0[b]	
Solution 2	
KCl	3000.0
Solution 3	
KCl	60.0
K-citrate	2100.0
Hepes (pH buffer)	5.0
pH 7.1 (achieved by titration with KOH)	
Solution 4	
As for solution 3, but with the addition of:	
EGTA	500.0

[a] Solution 1 is an example of a solution suitable for intracellular dialysis, satisfying metabolic and ionic requirements.

[b] For calculation of pCa see ref. 17. Solution 2 is perhaps the most commonly used filling solution for conventional microelectrodes. Solution 3 is one that we use for electrodes used to impale isolated cardiac myocytes, and which seems to lead to less 'blebbing' of the cell membrane than Solution 2. Solution 4 is used to buffer intracellular Ca^{2+} to low levels (see text for details) in isolated myocytes.

Protocol 3. Pipette filling

1. The electrolyte solution is drawn into a syringe.
2. A millipore filter in a holder is attached to the front of the syringe, and a fine piece of tubing attached to the front of the filter holder. The tubing is either a very narrow needle, or a piece or polyethylene tubing that has been drawn thin over a low bunsen burner flame. The diameter of the tubing should be small enough to enable it to be inserted into the micropipette. The length should be sufficient for it to reach the shoulder of the pipette. (NB: Although we routinely filter our filling solutions, it is not always necessary).
3. Pull a pipette as described in Section 3.3.2. *iii*.

Protocol 3. *Continued*

4. Introduce the tubing into the back of the pipette, until its tip lies at the shoulder of the pipette. Use the syringe to introduce the filling solution into the pipette. If necessary (e.g. if the filling solution is very expensive), a small volume (~2 μl) can be introduced. This will use the glass fibre within the pipette as a conduit to run to the tip, and electrical contact can be made to the solution (if necessary) using a long wire to run down inside the electrode to the shoulder. If the solution is cheap, the pipette is usually filled to a greater extent by keeping pressure on the syringe as the tube is withdrawn, so that it fills the micropipette as it is withdrawn.

3.5 Mounting microelectrodes

For use, the microelectrode is normally mounted on a micromanipulator, that allows fine control of the position of the microelectrode. There is a wide range of micromanipulators available, from coarse manipulators, which are suitable for use with large cells, to manipulators that have very fine control of movement, which are suitable for work on small isolated cells.

Probably the most important aspect of any manipulator is that it should be able to advance the microelectrode along the axis of the barrel of the electrode. The other aspect that should be borne in mind is that some manipulators are operated remotely either hydraulically or by stepper motors. Although expensive, these have the advantage that hand tremor is not transmitted to the electrode, in the way that it is with manipulators where the operator touches the assembly that carries the electrode.

The electrode itself is normally mounted in a Perspex holder, which has a rubber grommet, through which the electrode is pushed (*Figure 1*, D). In some holders a screw cap is then tightened down onto the grommet to hold the electrode secure (other applications, which require pressure to be applied to the back of the electrode require more secure fastening arrangements). Electrical connection to the filling solution is normally made via a Ag–AgCl wire that dips into the filling solution, and is connected to a pre-amplifer.

3.6 Recording microelectrode output

A wide range of amplifiers is now available commercially. Most amplifiers include a headstage with a suitably high input resistance, connections on the headstage for a driven shield and grounding of the tissue bath, electrode resistance measurement, a tip potential offset control, and electrode capacitance compensation (see Section 3.2). Different amplifiers are available depending on the particular use of the electrode (whether it is for iontophoresis, voltage clamp, etc., as well as for membrane potential recording). The amplifier will have extra controls for these particular functions. Details of these controls are outside the scope of this chapter.

The output of the amplifier is connected to a recording device to display and record any measurements made (be it a measurement of Ca^{2+} current or intracellular Ca^{2+}). The most commonly used output devices are pen recorders, oscilloscopes, tape recorders, and computers. A discussion of the advantages and disadvantages of each of these is beyond the scope of this chapter, but the interested reader should read refs 1 and 4.

3.7 The use of microelectrodes to study Ca^{2+}

3.7.1 Ca^{2+}-selective microelectrodes

The modification of a micropipette to endow it with ion-selective properties enables Ca^{2+} selective electrodes to be manufactured with tips of less than 1 μm in diameter. Such electrodes have found two main types of application: the measurement of [Ca^{2+}] in the cell cytoplasm, and the measurement of [Ca^{2+}] in the restricted extracellular space of multicellular preparations.

i. Manufacture

The principle of the electrode is the same as that of the macroelectrode (Section 2): a column of ion-selective ligand separates a filling solution from the test solution and the resulting p.d. is measured with respect to a suitable, stable reference. In order to achieve this, a small amount of ion-selective material is incorporated into the tip of a micropipette so that when the pipette is placed in the intracellular compartment the ligand column contacts this space. However, the ion-selective materials presently used are hydrophobic oils which make glass pipettes difficult to fill, so that the procedures of Section 3.4 must be modified:

(a) Treat the blank micropipette with a silane vapour. This appears to screen negative charges on the inside of the pipette and allow the ion-selective oil to fill the tip. Silanization is achieved by exposing the pipette to a vapour of a chloro- or amino-silane for a few minutes and then heating the electrode to about 200 °C to bake the silane to the glass wall. Many procedures to achieve this end have been described. One of the simplest is to heat gently a small quantity of silane in a bottle. Hold the electrode over the vapour for 2–3 min and then immediately bake the electrode in an oven. (The vapour of these silanes is extremely noxious so that such procedures must be performed in a well-ventilated fume cupboard.)

(b) Introduce a small quantity of ion-selective material into the electrode barrel as described in Section 3.4 and allow it to flow to the tip. If necessary use a fine whisker to tease it down to the tip. This filling technique results in a relatively long column (about 1 mm) of the ion-selective material in the electrode tip. Electrodes with shorter columns are less noisy. To obtain shorter columns, suck the ion-selective material into the electrode tip by connecting the back of the electrode to a 50-ml syringe and pulling the plunger back to the 50-ml mark (**beware!**—*All connections must be airtight*).

Carry out this process under the inspection of a dissection microscope. This technique has the advantage that the length of the column of the ion-selective material can be accurately controlled. We have used electrodes, filled in this way, with 200 μm columns.

(c) Back-fill the remainder of the electrode, as described in Section 3.4, with an electrolyte containing a $[Ca^{2+}]$ of 1 μM if used for intracellular measurements and 1 mM for extracellular measurements. Ensure that there are no air-bubbles trapped between the ion-exchange column and filling solution.

(d) Connect the electrode to the measuring system with Ag–AgCl wire.

ii. Calibration

Such electrodes have employed the neutral ligand ETH 1001 supplied in the form of a 'cocktail' by the chemical company Fluka, and are responsive to $[Ca^{2+}]$ as low as 10–100 nM in the presence of physiological quantities of interfering ions. When used to measure the very small intracellular $[Ca^{2+}]$ accurate calibration below about 10 μM requires the use of Ca^{2+} buffers, such as EGTA, to achieve adequate standard solutions. Calibration solutions should contain 5 mM EGTA as well as ions found in the intracellular space, i.e. 120 mM KCl, 10 mM NaCl, 1 mM $MgCl_2$, pH 7.0, achieved by using a buffer such as Hepes and titrating with KOH. To construct a calibration curve, the recorded p.d. is plotted as a function of the $[Ca^{2+}]$ (on a logarithmic scale).

Calibration solutions are made by adding known quantities of $CaCl_2$ to the above solution and titrating with KOH to pH 7.0. The total concentration of Ca^{2+} added (as $CaCl_2$) to achieve a desired free $[Ca^{2+}]$ is obtained from the formula:

$$\mathrm{pCa} = \mathrm{p}K + \log\,([\mathrm{CaEGTA}]/[\mathrm{EGTA}])$$

where pCa = $-\log [Ca^{2+}]$, analogous to the definition of pH. Virtually all the Ca^{2+} added binds to EGTA and so is equivalent to [CaEGTA]. The remaining unbound [EGTA] is equal to the total EGTA (5 mM in this case) minus the [CaEGTA]. The value of pK must be found by experiment for the particular solution used. For the above solution it has a value of approximately 6.44. A method describing the principle of its determination is given by Bers (10).

iii. Use

When such ion-selective microelectrodes penetrate a cell the recorded p.d. has two components. One results from the difference in $[Ca^{2+}]$ across the ion-exchange column. From this p.d., the intracellular $[Ca^{2+}]$ can be obtained, using the calibration curve (see above). The second component is the cell membrane potential, E_m. In order to arrive at the first component the value of E_m must therefore be subtracted from the total recorded p.d. and this may be achieved by separately measuring E_m with a conventional electrolyte-filled microelectrode. The outputs from the ion-selective electrode and conventional microelectrode are connected to the non-inverting and inverting inputs of a differential amplifier so that the output is a p.d. which represents the difference in $[Ca^{2+}]$ between the extracellular and intracellular space. However, the penetration of a preparation with two micro-electrodes makes the use of intracellular ion-selective electrodes difficult in this context for all but the largest cells.

The physical properties of such electrodes also impose restrictions on their use. The presence of the ion-selective column in the tip of the microelectrode ensures that the DC resistance is very high, of the order of 10^{11} Ω. This is similar to the input impedance of pre-amplifiers used to record from electrolyte-filled micro-pipettes so that if such pre-amplifiers were used a significant reduction of signal would ensue (*Figure 2*). Thus input stages with very high input impedance (10^{14}–10^{15} Ω) must be used and several are commercially available (e.g. AD 549L from Analog Devices). The high DC resistance is also associated with a significant input capacitance, derived from both the electrode and the input stage, so that the input has the configuration of a low-pass filter, limiting the frequency response of the electrode to signals less than 0.1 to 1 Hz. This causes two problems. First, if an action potential is recorded it will be greatly truncated. This causes major problems because to record the potential due to the ion under study the membrane potential (recorded by a conventional microelectrode with a fast time-response) is subtracted from the output of the ion-selective microelectrode (above). The major difference in the time-response of the two types of electrodes introduces major artefacts. Second, rapid changes of intracellular $[Ca^{2+}]$, associated for example with muscle contraction, will not be accurately recorded with such electrodes. The most successful use of these electrodes in biological systems has been the measurement of steady-state intracellular $[Ca^{2+}]$ and changes associated with slowly transporting pumps and exchangers.

3.7.2 Monitoring Ca^{2+} channels and Ca^{2+} current using patch-clamp techniques

The patch-clamp technique can be used to study Ca^{2+} flux across the cell membrane either as single-channel currents or as macroscopic whole-cell currents. With this technique an electrode with a relatively large tip (1–2 μm diameter) is placed against the cell membrane, suction is applied to the back of the electrode and a high-resistance seal (greater than a giga-Ohm) is formed between the electrode and the glass (a giga-seal).

i. Manufacture of microelectrodes

The main difference between a patch electrode and a conventional intracellular microelectrode is that the patch electrode tip must not penetrate the cell, but seal on to the cell membrane without damaging the cell. Electrodes are made from any type of glass. Electrodes are often made from haematocrit tubes made of borosilicate glass (take care to buy ones which have not been treated with the anticoagulant Li^+). Other glass types, e.g. aluminosilicate, have been used, either for their improved noise characteristics or enhanced sealing ability (11). The different stages in the manufacture of a patch-pipette are as follows:

(a) The pipette is pulled on a special type of vertical puller, in a two-stage process. An initial pull is used to give a short (1 cm or so) constricted portion. The heating coil is then allowed to cool, and the glass is recentered in the heating coil. A second (terminating) pull is then applied to separate the glass into two pipettes. The pipettes should have a short shank and wide tip to afford the smallest possible resistance. The size of the pipette tip is determined by two factors: the length of the first pull and the heat of the coil during the second pull; increasing either of these variables will give a smaller pipette.

(b) Estimate the tip size of the patch pipette from its 'bubble number': attach the pipette to a 10-ml syringe with a length of airtight, flexible tubing. Dip the tip of the pipette in a beaker of ethanol, depress the plunger of the syringe gradually and watch for bubbles appearing from the tip of the pipette. The volume in millilitres (proportional to pressure) at which bubbles just begin to appear is called the bubble number. For whole-cell recording use bubble numbers between 6 and 8, for single-channel studies use lower bubble numbers.

(c) Coat the pipette from its shoulder to within about 100 μm of the tip with a dielectric (e.g. Sylgard) to reduce pipette capacitance. Sylgard is applied to the pipette as a glutinous resin and is cured by heating (see (12) for a simple method of doing this). Although not essential for steady-state recordings, coating is neccssary to allow voltage-step experiments to be performed without saturation of the amplifier due to charging of the electrode capacitance. Since Ca^{2+} currents are both voltage- and time-dependent a pulse protocol must be used in their study; hence the pipettes must always be coated with Sylgard.

(d) Fire-polish the tip of the pipette to make it smooth. A small bead of glass is placed on a platinum filament. The pipette is brought close (within 10 μm) to the glass bead with the aid of a micromanipulator and the power supply (~1A) to the filament turned on. The pipette tip should be seen to darken slightly, indicating that fire-polishing is taking place, within about 10 sec. If this does not occur either turn up the current supply to the filament or move the pipette closer to the glass bead. If you cannot convince yourself that fire-polishing is taking place, remove the pipette and check the bubble number; it should have decreased by about half a unit.

ii. Pipette solutions

The choice of pipette solution depends on whether single-channel or whole-cell studies are to be performed. For single-channel work, ions other than Ca^{2+}, e.g. Ba^{2+}, are usually used as the current carrying species. This is because single-channel currents with Ba^{2+} are larger than those observed with Ca^{2+} and consequently more easily measured. The pipette solution should be buffered with Hepes or another organic buffer to physiological pH and the osmolality maintained at around 300 mOsM for mammalian work. For whole-cell experiments, where the cell becomes dialysed by the pipette solution, the intracellular solution is frequently made up to match the intracellular solution (see Section 3.4). One problem with whole-cell Ca^{2+} currents is that they diminish with time; this so-called 'run down' of Ca^{2+} current is also seen as the disappearance of Ca^{2+} channel activity in excised patches. This is presumably due to the loss of some intracellular molecule that sustains channel activity, and the inclusion of ATP in the pipette solution can minimize the phenomenon, suggesting that the channels must be phosphorylated to retain their activity.

All pipette solutions should be filtered (0.22 μm) before use (see Section 3.4), as should the solution bathing the cells, to enhance the probability of obtaining a giga-seal. If your glass has no internal fibre, the tip of the pipette must first be filled by dipping the tip of the pipette into pre-filtered pipette solution and applying negative pressure, with a 10-ml syringe attached to the back of the pipette, for about 5 sec. This should fill several hundred micrometres of the tip of the pipette. Next, back-fill the pipette with filtered solution (Section 3.4); this is most easily done by attaching a Millipore filter to the tip of a 10-ml syringe with a needle attached. Air-bubbles remaining in the shank of the electrode must be displaced by vigorously flicking with a finger.

iii. Obtaining a giga-seal

It is difficult to describe how to obtain a giga-seal since no one knows what forces are involved in its formation. However, most experimenters soon realize which factors aid in getting good seals. For example, vibration is incompatible with seal-formation. An air-table is a sound investment; failing this, mount the apparatus on a concrete or metal slab supported by tennis balls or an inflated cycle inner-tube. In addition, micro-manipulators should be mounted directly on the microscope to minimize bending moments between the two, thus minimizing movements between the pipette tip and the preparation.

The procedure in *Protocol 4* is then used to obtain a giga-seal.

iv. Problems with sealing

If you have problems obtaining giga-seals, try the following remedies:

(a) Make sure that all solutions are filtered.

Protocol 4. Obtaining giga-seals

1. Take care not to overfill the pipette and to keep the outside of the electrode dry and clean since fluid and/or dirt in the holder can cause tremendous electrical noise.
2. Mount the electrode in its holder. (Holders for patch-electrodes have a side arm for connection of a piece of tubing which can be connected to a pressure source for the application of pressure to the back of the electrode as described below.)
3. Apply positive pressure (~5 cm water) to the back of the electrode (a water-filled manometer is useful for this purpose) and immerse the tip of the electrode in the filtered bath solution.
4. Start to monitor the resistance of the electrode (Section 3.2). The circuit should be in the voltage-clamp mode of operation. To monitor electrode resistance a small repetitive voltage-clamp pulse will be applied. The current during the pulse will be lower, the greater the electrode resistance.
5. Bring the electrode close to the cell under visual control. Slowly advance the electrode towards the cell and watch the electrode resistance closely. As the electrode touches the cell the resistance will increase (i.e. current will decrease). Stop moving the electrode immediately this happens.
6. Release the positive pressure from the electrode and now apply gentle negative pressure (~−5 cm water) to the rear of the electrode. Once again, a manometer is useful for this purpose, although a syringe or mouth suction can also be used. With luck, the resistance will increase sharply and the current trace should now appear as an almost straight line, with small capacitative artefacts at the beginning and end of the voltage pulse. If the electrode does not seal, try increasing the suction. If a seal has not formed after about 30 sec it is unlikely that it will ever form, so try another electrode! Patch-electrodes will not seal if they are dirty and can usually be used once only.
7. Increase the size of the voltage pulse to about 10–20 mV and compensate the capacitive artefacts using the controls on the patch-clamp amplifier.

(b) Fire-polish the electrodes immediately before use and bubble the tip in ethanol to clean it. Alternatively, try freshly pulled electrodes without fire-polishing.

(c) Make sure that all vibrations are minimized. Use a manipulator with a remote control system, such as the Narishige oil-filled micromanipulator.

(d) Try aspirating fluid (and dirt) from the top of the fluid in the chamber with a Pasteur pipette or by wiping the surface film with lens paper.

(e) Try cleaning the surface of the cells with collagenase or some other digestive enzymes.

If the above fail it just may be that the cells are not suitable for patch-clamp work. Ask around: someone else may have tried the cells and found a solution. Otherwise try another preparation.

v. Single-channel recordings

After turning off the voltage pulse for electrode resistance measurement one can now begin to record single-channel currents from the region of membrane under the electrode in the 'cell-attached' mode by using a suitable voltage-pulse protocol. Most work of this type is now performed with a mini- or microcomputer, A/D converter and suitable software to both produce the pulse protocol and aquire data. A number of different combinations are available commercially and the most popular is an IBM microcomputer or one compatible with a Tecmar A/D card and p-Clamp software. The A/D card and software are available in the USA from Axon Instruments. An equivalent British system would use A/D hardware and software from Cambridge Electronic Design (CED).

The seal between the electrode and cell is extremely robust, indeed the adhesion between the electrode and membrane is greater than the cohesive properties of the membrane, so that if the electrode is withdrawn from the cell the small patch of membrane contained within the tip of the electrode detaches from the cell forming an 'inside-out' patch, so called because the membrane that faced the inside of the cell now faces the bath (outside) solution.

Another type of excised patch can be formed by first breaking the patch of membrane whilst cell-attached and then retracting the electrode from the cell. Under these circumstances a flap of membrane surrounding the electrode whilst cell-attached flips around the electrode tip and seals, forming an 'outside-out' patch, so called because the membrane that faced the outside of the cell (i.e. the extracellular face) now faces the bath solution. The patch configuration chosen allows one to test the sideness of channel agonists or antagonists.

This resealing of the detached portions of membrane outside the electrode tip and forming a vesicle can be a problem in forming inside-out patches, and manifests itself as a loss of single-channel currents or marked rounding of the currents due to the series resistance–capacitance of the outside-out patch. This outside-out portion must be disrupted. This is normally done by taking the tip of the electrode rapidly out of the bath fluid and then popping it back into the bath as quickly as possible (<1 sec if possible). This brief 'air-exposure' will usually disrupt the vesicle and allow the experiment to continue. Repeat air-exposure with increasingly longer periods in the air until the vesicle (or patch!) ruptures.

vi. Whole-cell recording

Following formation of the seal and pipette capacitance compensation, whole-cell recording can be begun by disrupting the patch of membrane across the electrode tip. This is done by applying large negative (–50 cm water) pressure pulses or by applying large (>200 mV) potentials to the patch-electrode. Watch the electrode resistance carefully at this stage. As soon as the patch breaks the resistance will fall and the capacitance will increase markedly: stop the negative pressure or voltage pulses immediately. The magnitude and time-course of the capacitive transient is determined by the series resistance of the patch-electrode and the cell capacitance (the larger the cell, the larger the capacitance). The cell will now become dialysed with the electrode solution. At this stage the electrode potential is usually set to the zero current point (corresponding to the resting potential) of the cell or some pre-determined value. Ca^{2+} current can now be studied with the use of suitable voltage-clamp protocols. Capacitance compensation can be performed either by the analogue controls of the patch clamp amplifier or by digital compensation by the computer. Whichever method is used, care should be taken that the amplifier is not allowed to saturate during the current pulses. It is advantageous to use electrodes with a low resistance, since the lower the series resistance (i.e. the larger the pipette), the smaller will be these transients.

3.7.3 Measurement of Ca^{2+} current using switch-clamp

For a detailed description of this technique see ref. 13. In the preceding section measurement of membrane current with the whole-cell voltage-clamp technique was described. A useful feature of this technique is that the cell is dialysed and the intracellular environment of the cell can be controlled. However, cell dialysis can be a disadvantage because it is also necessary to study membrane currents in the presence of a normal intracellular environment. The discontinuous or switching single-electrode voltage-clamp technique (or switch-clamp) is a useful alternative. The switch-clamp technique results in less cell dialysis because a conventional microelectrode with a smaller tip diameter is used. Because the switch-clamp involves conventional electrodes, it can also be used with cells that are unsuitable for use with patch electrodes. The switching single-electrode voltage-clamp technique was a development of the two-microelectrode voltage-clamp technique. This involved two microelectrodes: one was used to record the membrane potential and the second was used to deliver current in order to control the membrane potential. The switch clamp involves a single electrode only, which is used for both functions. It is clearly easier to impale a cell with one rather than two electrodes.

i. Principles of the switch-clamp

The principles of the switch-clamp are illustrated in *Figure 3*. The underlying principle of the technique is to switch the electrode rapidly between two

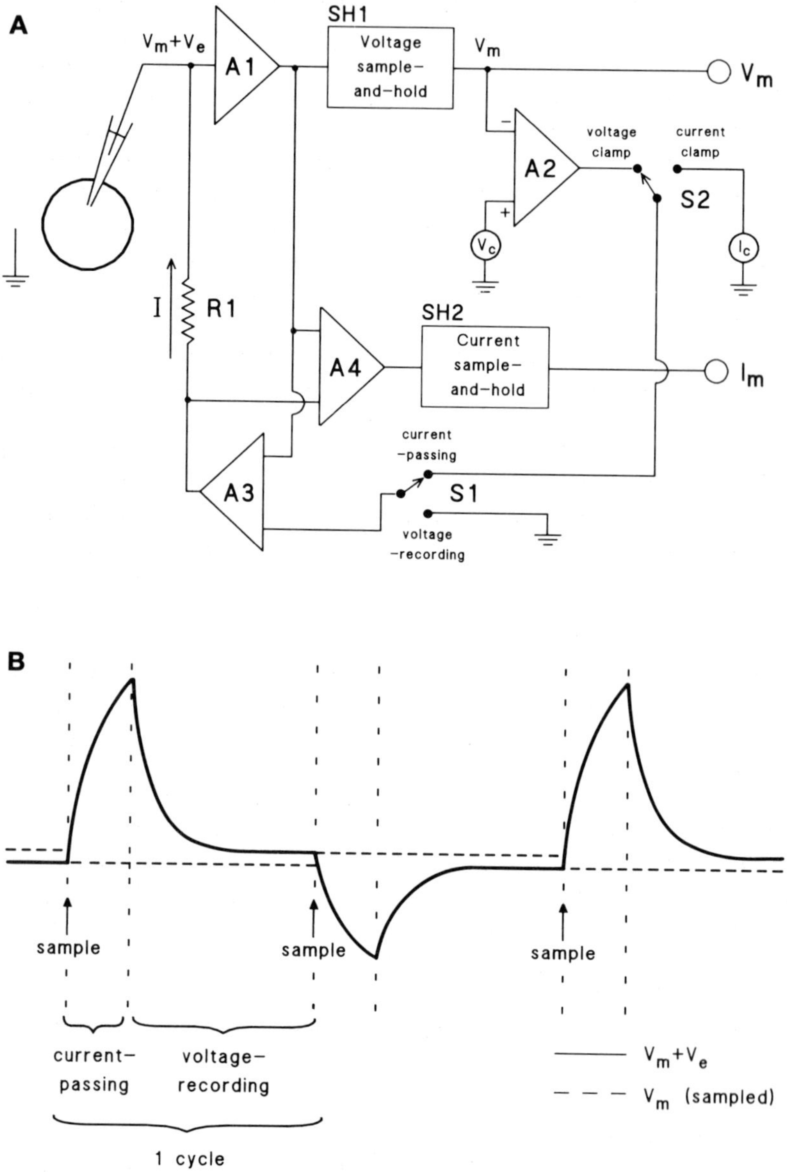

Figure 3. The switch-clamp technique (see text for details).

modes: a voltage-recording mode and a current-passing mode. The switching frequency can be controlled and in our hands is typically 5 kHz. The amplifier A1 (*Figure 3*, A) records both the membrane potential (V_m) and the potential across the tip resistance of the electrode (V_e). $V_m + V_e$ is shown in *Figure 3*, B, as the solid line. At the end of the voltage-recording part of the cycle, V_e has disappeared (see below) and the potential recorded is V_m alone. At this time, V_m is sampled and then held by a sample-and-hold device (SH1). The output of the sample-and-hold device is V_m alone and is shown in *Figure 3*, B, as the dashed line. It will be assumed that the circuit is being used for voltage-clamp (switch S2 in the left-hand position). Throughout the cycle, V_m is compared to the voltage-clamp command potential V_c by the differential amplifier A2. During the current-passing part of the cycle, the switch S1 is in the upper position and the output of A2 (the difference between V_c and V_m) is used to control the current-passing amplifier A3. A3 generates a current (I) which is applied to the cell via the resistor R1 and the microelectrode. The current is dependent only on the input from A2 (it is independent of the microelectrode resistance because of feedback, via A1, from the current-sampling resistor R1). The strength and direction of the current I is appropriate to change V_m towards V_c and is equal and opposite to the membrane current (I_m). The voltage drop across the current-sampling resistor R1 is proportional to I, and this is measured by amplifier A4. Because current is passed during part of the cycle only, a second sample-and-hold device is used (SH2); this samples and then holds I during the current-passing half of the cycle. The passage of I across the tip resistance results in a large voltage drop, V_e, and therefore the output of the electrode is no longer equal to V_m alone (*Figure 3*, B). During the voltage-recording part of the cycle no current is injected into the cell, because the input to the current-passing amplifier A3 is zero; switch S1 is moved to its lower position. V_e does not disappear instantaneously, because during the current-passing part of the cycle I charged-up the electrode capacitance. During the voltage-recording part of the cycle, V_e slowly disappears as the electrode capacitance discharges. *Only when the capacitance is fully discharged can the output of A1 be sampled and held and taken as a measure of* V_m. When the circuit is used for current clamp, switch S2 is in the right-hand position, and the current-passing amplifier is controlled by the current clamp command signal, I_c. Once again, in the current clamp mode of operation, the electrode is switched rapidly between current-passing and voltage-recording. Various switch-clamp amplifiers are available commercially; manufacturers include Axon Instruments and Dagan Corporation.

ii. Use of the switch clamp

The procedure in *Protocol 5* is based on the use of a Dagan Corporation Model 8800 with single cardiac ventricular myocytes.

Protocol 5. Use of the switch clamp

1. Pull and fill a suitable microelectrode. We use 10–20 MΩ microelectrodes filled with solution 3 (*Table 1*).
2. Fix the microelectrode into a microelectrode holder mounted on a micromanipulator. Lower the microelectrode into the bathing solution.
3. Measure the mícroelectrode resistance. Dispose of the electrode if not satisfactory.
4. Position the electrode next to the cell ready for impalement.
5. Switch the device to the current-clamp mode of operation. No current should be applied.
6. Bring the output from the electrode ('V_m') to 0 mV using the tip potential offset control.
7. During an experiment the highest possible switching frequency is used, because the switching frequency must be rapid in comparison to the signals recorded. However, during the voltage-recording part of the cycle, V_e must be allowed to decay completely as in *Figure 3*, B. If too high a switching frequency is used, $V_m + V_e$ will be sampled before V_e has decayed fully and the output of the sample-and-hold device will not be equal to V_m. *This is very important and an ignorance of this problem will result in data fraught with artefacts.* For example, with an electrode in bathing solution alone (i.e. no cell present) it is possible to apply a voltage-clamp pulse and obtain a respectable record of 'V_m' with good voltage control if the switching frequency is suitably high (try it!). If $V_m + V_e$ is inspected it will be seen that V_e is not decaying before the end of the voltage-recording part of the cycle. As a result the electrode tip is being voltage-clamped! For this reason it is important that the $V_m + V_e$ signal from amplifier A1 is monitored continuously on an oscilloscope.

   The faster the decay of V_e during the voltage-recording part of the cycle, the faster the switching frequency that can be used. The time-constant of decay is equal to $R_e \cdot C_e$; the lower the resistance of the electrode and the smaller its capacitance, the faster the decay. Therefore, chose low-resistance electrodes (providing they will still impale cells satisfactorily). To reduce C_e, set the switching frequency at its lowest value and apply a steady current (remember that the device is in the current-clamp mode of operation). Now increase the capacity compensation until the fastest decay of V_e is obtained but the $V_m + V_e$ signal does not ring (i.e. oscillate).
8. Increase the switching frequency to the highest value that still allows time for the complete decay of V_e.
9. Impale the cell. In the case of single heart cells we gently push the electrode against the cell so that the cell membrane is dimpled but not

Protocol 5. *Continued*

punctured. A further short sharp push on the electrode will hopefully cause it to impale the cell. Alternatively, the amplifier may have a buzz control; if this is pressed, the electrode tip is caused to vibrate and this may facilitate impalement.

10. The electrode properties may change after the impalement and also at other times during the experiment. Check continuously (whether in current- or voltage-clamp) that the capacity compensation and the switching frequency are at their optimal values.

11. With the voltage-clamp gain at a low value, switch the device to the voltage clamp mode of operation.

12. Increase the voltage clamp gain. To obtain good voltage control this should be as high as possible, but if it is too high the circuit will ring. Set it at its optimal value. It is possible to increase the gain further if the 'phase' control of the circuit is adjusted to its optimal value. See ref. 13 for further information about phase control.

13. To reduce noise on the V_m and I_m signals, increase the time-constant of the anti-aliasing filter (a low-pass filter between A1 and SH1) (prior to this stage, the time-constant of the anti-aliasing filter should have been set at its minimum value). If the time-constant is increased excessively, the decay of V_e will be slowed and the true value of V_m will not be recorded at the end of the voltage-recording period.

---

iii. Techniques for recording Ca^{2+} currents

The previous sections have outlined voltage-clamp techniques that can be used to monitor membrane currents from whole-cells. To measure the common L type Ca^{2+} current (measurement of the T and N types will not be considered) a 200 msec voltage clamp pulse can be applied to 0 mV from a holding potential of −40 mV. Often, the Ca^{2+} current may have to be isolated from other currents, which overlap it in time. There are many ways to do this. For example, the use of a low holding potential (−40 mV) will inactivate the Na^+ current and transient outward current; 4-aminopyridine or TEA (tetraethylammonium) can be used to block K^+ currents; replacement of K^+ by Cs^+ both extracellularly and intracellularly will also eliminate K^+ currents. Further discussion of Ca^{2+} currents is outside the scope of this chapter. Interested readers should refer to the article by Noble (14) for example.

3.7.4 Manipulation of intracellular [Ca^{2+}]

The use of single-cell preparations and microelectrodes has made it possible in recent years to use microelectrodes to change the composition of the intracellular compartment. This is normally combined with electrical measurements from the cell using the same microelectrode, making this a very powerful technique.

i. Diffusion from a microelectrode

The contents of a conventional microelectrode will leak into the cell cytoplasm after penetration of the cell membrane by the electrode. Although this can be inconvenient, it can also be used experimentally to alter the composition of the cell cytoplasm. One common use of this method is to introduce Ca^{2+} chelators into the cell cytoplasm, to lower intracellular [Ca^{2+}]. We use electrode filling solutions which include 0.5 M EGTA to stop contractions and intracellular Ca^{2+} transients in isolated cardiac cells. Presumably, therefore, chelators introduced in this way buffer bulk cytoplasmic [Ca^{2+}] (although there is some doubt as to whether Ca^{2+} close to the cell membrane is buffered). The high concentration of EGTA within the microelectrode provides a diffusion gradient for EGTA entry into the cell, but the final concentration of chelator in the cell is unknown.

A further disadvantage of this method is that it is not possible to make electrical recordings before and after buffering Ca^{2+} in the same cell (since Ca^{2+} buffering occurs as soon as the electrode penetrates the cell). To overcome this limitation, a method has recently been described using double-barrelled electrodes made from theta tubing (Section 3.3.1. *i*), in which one barrel acts as a conventional microelectrode, while the other contains a thin polythene tube through which the contents of the second barrel can be altered (15). Thus, at the start of an experiment this barrel could contain a solution similar to the cell cytoplasm. This could then be exchanged for another solution, which would diffuse into the cell cytoplasm, when required. Using this system, injection of the Ca^{2+} chelator BAPTA (125 mM) into the second barrel stopped contraction of a cardiac cell within 60 sec.

ii. Cell dialysis

The principles of cell dialysis are much the same as the method mentioned at the end of the previous section. However, instead of a conventional microelectrode, a patch-electrode (Section 3.7.2) is used. Attachment of the electrode to the cell, and obtaining access to the cell interior are the same as for whole-cell recording (Section 3.7.2, *v*). However, the wide bore of the electrode means that equilibration of the cell contents with the cytoplasm is much more complete than in the case of conventional microelectrodes (above), and so more care must be taken with the electrode filling solution (Section 3.4). However, by using a thin tube within the electrode, the contents of the electrode (and hence of the cell interior) can be exchanged much more rapidly than is possible using a conventional microelectrode.

References

1. Purves, R. D. (1981). *Microelectrode methods for intracellular recording and electrophoresis.* In *Biological Techniques* series. Academic Press, New York and London.
2. Thomas, M. V. (1982). *Techniques in calcium research.* In *Biological Techniques* series. Academic Press, New York and London.
3. Thomas, R. C. (1978). *Ion selective microelectrodes: how to make and use them.* In *Biological Techniques* series. Academic Press, New York and London.
4. Standen, N. B., Gray, P. T. A., and Whitaker, M. J. (ed.) (1987). *Microelectrode techniques: the Plymouth workshop handbook.* The Company of Biologists Ltd, Cambridge, UK.
5. Fry. C. H. (1991). *The manufacture and use of ion-selective electrodes.* Hugo-Sachs KG, Freiburg, Germany.
6. Schwan, H. P. (1964). In *Physical techniques in biological research*, Vol VI (ed. W.L. Naswtuk), pp. 75–96. Academic Press, New York and London.
7. Ammann, D. (1986). *Ion-selective micro-electrodes. Principles, design and application.* Springer-Verlag, Berlin, Heidleberg, New York.
8. Lederer, W. J., Spindler, A. J., and Eisner, D. A. (1979). *Pflügers Archiv für die gesamte Physiologie*, **381**, 287–8.
9. Adrian, R. H. (1956). *Journal of Physiology*, **133**, 631–58.
10. Bers, D. M. (1982). *American Journal of Physiology*, **242**, C404–8.
11. Rae, J. L. and Levis, R. A. (1984). *Biophysical Journal*, **45**, 144–146.
12. Corey, D. P. and Stevens, C. F. (1983). In *Single-channel recording* (ed. B. Sakmann and E. Neher), pp. 53–68. Plenum Press, New York and London.
13. Finkel, A. S. and Redman, S. J. (1985). In *Voltage and patch clamping with microelectrodes* (ed. T. G. Smith, H. Lecar, S. J. Redman and P. W. Gage), pp. 95–120. American Physiological Society, Bethesda, Maryland.
14. Noble, D. (1984). *Journal of Physiology*, **353**, 1–50.
15. Terrar, D. A. and White, E. (1989). *Journal of Physiology*, **417**, 10P.
16. London, B. and Kreuger, J. W. (1986). *Journal of General Physiology*, **88**, 475–505.
17. Fabiato, A. (1989). In *Methods in Enzymology*, Vol. 157 (ed. S. Fleischer and B. Fleischer), pp. 378–417. Academic Press, New York and London.

4

Metallochromic indicators

M. V. THOMAS

1. Introduction

Metallochromic indicators were used quite extensively for intracellular Ca measurement before the introduction of the fluorescent indicators, and they can still be useful alternatives under some circumstances. In any case the two indicator families have a great deal in common, because they measure Ca in exactly the same way, i.e. by undergoing an optically detectable change as a result of binding Ca ions.

For the metallochromic indicators the major change is an increase in absorbance at relatively long wavelengths (600–700 nm). This is a less convenient measurement than fluorescence, since the background absorbance of a sample is likely to be relatively much higher than the background fluorescence. On the other hand, the requirement for long-wavelength rather than UV light simplifies the optical system and greatly reduces the risk of damage to the sample. The major problems with the metallochromic indicators are their relatively poorer selectivity against other ions (particularly Mg and protons), and their variable stoichiometry. Both problems make calibration of the indicator responses a more difficult and uncertain task, but with carc, some reasonably reliable estimates can be made.

2. Absorbance measurements

2.1 Definition of absorbance

The absorbance A of a sample at any specified wavelength is defined as $A = \log (I_0/I)$, where I_0 and I are respectively the incident and transmitted light levels. The absorbance varies linearly with both the path length and the concentration of the absorbing species, and the molar absorbance (or extinction coefficient) is defined as the absorbance of a 1-cm path length of a 1 M solution. The Ca complexes of the metallochromic indicators have molar absorbançes of up to about 30 000 at long wavelengths.

2.2 Measurement of absorbance

The main problem with absorbance measurements is that they are relatively non-specific, in that very many factors can contribute to the total absorbance of a biological sample. However, by making absorbance measurements differentially between two appropriately spaced wavelengths, it is possible to maximize the response from the indicator, while minimizing the contributions from other sources. The wavelength pair should be as close together as possible in order to minimize these contributions, while spanning a range over which the indicator shows a significant differential absorbance change. As a further (and very desirable) refinement, one can measure the differential absorbance change between several wavelength pairs during an experiment, to verify that the differential absorbance spectrum corresponds to the response of the indicator to Ca.

3. Indicator molecules

3.1 Arsenazo-III

Arsenazo-III has been by far the most widely used indicator for biological applications. Many other metallochromic indicators have been synthesized, but if any of these are superior to arsenazo-III for Ca measurement, then nobody appears to be aware of it. All these indicators are derivatives of chromotropic acid, and differ in the two substituent groups. Other indicators include chlorphosphonazo-III and sulphonazo-III, and ref. 1 provides further chemical information on these and other indicators.

3.2 Antipyrylazo-III

The only other metallochromic indicator which has been used to any significant extent for biological Ca measurements is antipyrylazo-III (2), which has been used as an alternative to arsenazo-III when a lower affinity to Ca was desired (3).

3.3 Absorbance spectra

The absorbance of arsenazo-III in the absence of Ca is greatest at about 540 nm, and relatively small at wavelengths longer than 650 nm (see ref. 4 for spectra). Binding of Ca reduces the 540 nm absorbance and greatly increases the absorbance at long wavelength. The increase shows two peaks, centred at 600 and 650 nm, and the isosbestic wavelength (i.e. that for which there is no absorbance change) is 570 nm. Mg and protons cause broadly similar absorbance changes to Ca, but with the significant difference that there is only a single peak, centred at 600 nm (ref. 4). To maximize the discrimination against these ions, the absorbance of arsenazo-III is normally measured differentially (see below) between about 660 nm and 690 nm.

The absorbance spectrum of antipyrylazo-III is broadly similar, except that the isosbestic wavelength is 620 nm (see ref. 3 for spectrum). However, the effect of Mg is completely different, causing a decrease in absorbance at wavelengths longer than 550 nm. Mg causes no significant absorbance change at wavelengths longer than 660 nm, so this range is preferred for Ca measurement.

4. Selectivity and stoichiometry

The relatively poor selectivity and variable stoichiometry of the metallochromic indicators constitute the major problems of this method of intracellular free Ca measurement. Nevertheless, if suitable control measurements are made, it is possible to be reasonably confident about the biological data obtained using these substances. There is general agreement that the indicators respond linearly to changes in free Ca concentration in the biological range, but several other aspects have been controversial.

4.1 Stoichiometry

The various findings are best explained by the following model, although there are some disagreements and uncertainties (see ref. 5) which still do not appear to have been resolved. Arsenazo-III forms 1:1 complexes with Mg, i.e. the complex formula is MgL (L = ligand), but at least three different complexes can be formed with Ca. These are CaL, CaL_2, and Ca_2L_2. CaL_2 seems to be the dominant form under the conditions of relatively high indicator and low Ca concentrations which normally apply to biological measurements, but some workers have reported results that suggest primarily CaL stoichiometry. (The formation of Ca_2L_2 does not seem in practice to affect the linearity of the dye with respect to Ca, because it is formed only at relatively high Ca concentrations, and/or because the formation of Ca_2L_2 from CaL_2 causes little further absorbance change.)

With regard to pH, the metallochromic indicators have a relatively large number of protonation sites, and several of these appear able to take up protons around neutral pH, causing a significant pH-dependence of the absorbance spectrum. Note that Mg and protons will also both compete with Ca, thereby reducing the Ca sensitivity of the indicator as well as modifying its absorbance spectrum.

4.2 Calibration

It is unwise to attempt calibration by calculation from the many stability constants involved, and the best alternative is to calibrate the indicator under as near the **exact** conditions of use as possible, so that the results can be applied directly. This will mean using not only the appropriate ionic strength, pH and Mg concentration, but also the same **indicator** concentration, since the CaL_2 stoichiometry causes the indicator effectively to become more sensitive to free Ca as the indicator concentration is increased.

This consideration may well rule out use of standard 1-cm cuvettes, since the optical densities could be too high to allow meaningful measurements. However, it is easy to arrange for a shorter path length if narrower cuvettes are not available, e.g. by inserting transparent plastic or glass strips in the cuvette.

Another important point is that the free Ca concentrations in the calibration media must either be defined by an excess of a Ca buffer such as EGTA or (preferably) BAPTA, or measured with a Ca electrode. A problem with EGTA is that addition of Ca displaces protons from it, so a considerable excess (at least tenfold) of a pH buffer is required in order to stabilize the pH of the calibration medium if Ca salts are added to it. This problem does not occur to any significant extent with BAPTA, since its carboxylate groups are all deprotonated at physiological pH.

Clearly, it would be very desirable to calibrate the indicator response *in situ*, but this is possible for only a small number of favourable preparations, and even in these cases there can be complications, e.g. any changes in pH or Mg concentration caused by the calibration procedure will also affect the indicator response.

5. Using the indicators

5.1 Availability

Arsenazo-III and antipyrylazo-III are available from several suppliers, including Sigma. The purified forms of these substances are recommended for obvious reasons, although they are significantly more expensive. They are normally supplied as sodium salts, which are freely soluble at or near neutral pH. Stock solutions appear to be stable for many months, but refrigerator storage is recommended. The concentration is a matter of choice, but concentrations of around 10 mM are the most convenient. For injection into cells (see below), it is also advisable to include an isotonic concentration of KCl, and to check that the pH is near neutrality. A few millimolar of an appropriate pH buffer (e.g. Mops at pH 7.2) can also be included if there is any uncertainty on this point.

5.2 Indicator concentration

The metallochromic indicators are normally used at concentrations of up to a few hundred micromolar. Under these conditions, the indicators will have a significant buffering capacity (maybe tenfold or more), but in practice this does not appear to cause a significant increase in the total intracellular buffering capacity. However, it is important to check this by comparing the results obtained using a range of different indicator concentrations. The absorbance of arsenazo-III at 570 nm is independent of pH and of Mg concentrations as well as Ca, so it can be used as a reliable measure of the indicator concentration.

5.3 Introduction into cells

The number and diversity of charged groups on the metallochromic indicators precludes the possibility of introducing these substances into cells in an uncharged (membrane-permeant) form, and hence they must normally be introduced by injection through a micro-pipette. Both pressure (6–9) and ionophoretic injection (10) have been used successfully, and the injection can be monitored by the increase in absorbance at an isosbestic wavelength.

As for many other impermeant substances, a number of more exotic methods of introducing them into cells have been tried. In general, these have not proved very satisfactory, but good results can be achieved when reliable permeabilization and resealing techniques can be applied, e.g. as for mitochondria (ref. 11 and Chapter 15, this volume).

6. Instrumentation

For reliable results with the metallochromic indicators, differential absorbance measurement is practically essential. In practice, measurements at several wavelengths are very useful, to provide more information of the spectrum of the absorbance changes. Most conventional laboratory spectrophotometers are not designed for this sort of measurement, and in any event it may be necessary to build a system around the sample. The major components of such a system are described below.

6.1 Light source

Since measurements are performed at relatively long wavelengths, a standard halogen light source can be used. However, it must be driven from a well-stabilized DC power supply, as the light output is highly dependent on the drive voltage. Furthermore, the spectral content also varies, so the effects cannot be fully compensated for by differential measurement (although this will certainly help). Suitable power supplies are readily available from many manufacturers, such as Farnell in the UK. The most important practical point is to take care with the connections to the bulb. Push-fit connections can have a relatively high and variable resistance, and firm screw connections should be used instead (even if it means rebuilding the bulb-holder). If this precaution is taken, then any economical light source can be used.

There is no need to mount the source directly on to the rest of the system, as fibre optics light-guides provide a very convenient method of light transmission.

6.2 Wavelength selection

This is most conveniently done by optical interference filters, which can be relatively inexpensive. Interference filters typically have a bandwidth of around 10 nm, which is a good compromise between the conflicting requirements of a reasonable wavelength resolution and a reasonable amount of total light transmission.

There are two possible methods of implementing the wavelength selection. In the first method, the sample is illuminated with (relatively) broadband light, and the transmitted light is split between two or more photodetectors via separate optical interference filters of the required wavelengths. This method is fairly simple to set up, except that the need for a beam-splitter and multiple photodetectors complicates matters somewhat.

In the second method, the sample is illuminated sequentially with light of different wavelengths, using (for example) a series of optical interference filters in a rapidly-rotating wheel. This requires only a single photodetector, but the detection electronics is more complicated, as the signal must be separated into its individual wavelength components. I chose this method (6–9) because it allows several wavelengths to be monitored at the same time. It also has the advantage that the light intensity at the sample is lower, since the wavelength selection has been carried out prior to it, allowing relatively more intense illumination at the wavelengths of interest compared to the total light input. However, this is desirable anyway, since each wavelength is present only intermittently, and so its average level is relatively lower than for continuous illumination.

In practice, both methods can work well. Although the first method may be simpler, the second is certainly more versatile. Full constructional details of a complete system using a filter wheel are given in ref. 5, but a more sophisticated version is now commercially available from Cairn Research Ltd.

6.3 Sample illumination

In many cases the system will be assembled around a microscope, and the microscope optics will be used to illuminate and to collect light from the preparation. However, the use of miniature light-guides for either or both of these functions can be a very attractive alternative, and I have made extensive use of glass image conduit from Edmund Scientific. This is a fused bundle of optical fibres, with an overall diameter of 3.2 mm. It can be heated and then formed or drawn out to make illumination and collection probes of any desired dimensions. These can be built into the experimental chamber or positioned using a micromanipulator, as appropriate.

6.4 Light detection

Either photomultipliers or photodiodes can be used for light detection.

6.4.1 Photomultipliers

Photomultipliers are intrinsically quieter at low light levels, but they are much larger and more expensive than photodiodes, and also require a stabilized high-voltage power supply. Suppliers include Thorn-EMI Electron Tubes Ltd. This company can supply a wide variety of photomultipliers, many of which will be suitable for this application. The most important point is to use a tube with an extended red response, i.e. an S20 photocathode, as conventional photocathodes have very poor sensitivity at wavelengths longer than about 600 nm.

6.4.2 Photodiodes

Photodiodes are much cheaper and easier to use than photomultipliers, but are also noisier. However, this does not necessarily degrade the overall noise performance of the system. The ultimate limit is set by photon noise, which increases with the square root of the amount of light reaching the detector. This means that there will be a light level at which the photon noise begins to dominate the detector noise, above which there will be no advantage in using a photomultiplier. To ensure that this situation is achieved, it is best to direct the light on to a photodiode which has a relatively small active area, in order to achieve a relatively high light-intensity. The photodiode should be connected to an operational amplifier with low input current, in a virtual ground configuration with a high-value feedback resistor. The application notes of the Analog Devices AD549 operational amplifier contain a useful discussion of this subject, and are recommended for further information.

6.5 Electronics

Differential absorbance measurements can be made between two wavelengths by a logarithmic ratio amplifier. Such devices are available commercially in various forms, from individual integrated circuits to assembled instruments. However, for measurement of **small** absorbance changes, a linear differential amplifier with some means of balancing the two inputs (to produce a near-zero output) will perform just as well if not better. The reason is that logarithmic electronic circuits are inherently less accurate than linear ones, whereas a logarithmic response is approximated very well by a linear circuit so long as the signal excursions are small and an appropriate gain correction is applied.

6.6 Noise

Whatever the measurement system, the resolution will ultimately be limited by noise. Of course, some of the noise (or drift) may be generated by the sample, but it is always advisable to optimize the system noise performance. The following simple test will allow the predominant source of the noise to be identified, which is a necessary first step towards its reduction. The output from the light source should be varied over a severalfold range by using an iris diaphragm and/or a neutral density filter, and the effect on the amplitude of both the signal and the noise should be measured. If the system is behaving ideally, the noise will be photon noise, which will increase with the square root of the signal (light) level. Therefore the signal-to-noise ratio will also increase with the square root of the signal level, so there is every advantage to be gained by supplying and/or collecting more light.

The advantages of having more light will of course be even greater if the predominant noise source is the detector and its electronics, since in this case the noise will be independent of the signal level, so the improvement will be directly proportional to the amount of extra light collected. However, it also suggests that some attention should be paid to the photodetector and its electronics (see above).

There is also the possibility that at least some of the noise will be contributed by the light source, particularly on time-scales of 1 to a few seconds. This type of noise can be identified quite readily because it increases directly with the light level. It can arise because of bad electrical connections to the bulb (note previous recommendation for screw connections), intermittent short-circuits across part of the bulb filament (which can occur when the bulb nears the end of its life) or convection currents within or around the bulb. The convection currents cause a shimmering effect which is difficult or impossible to eradicate, but it can sometimes be reduced by experimenting with the bulb orientation and by ensuring that there is a smooth airflow through the light source.

References

1. Budesinsky, B. W. (1969). *In Chelates in analytical chemistry* (ed. H. V. Flaschka and A. J. Barnard, Jr.) pp. 1–91. Marcel Dekker, New York.
2. Budesinsky, B. W. (1974). *Analytica Chimica Acta*, **71**, 343–7.
3. Scarpa, A., Brinley, F. J., Jr and Dubyak, G. (1978). *Biochemistry*, **17**, 1378–86.
4. DiPolo, R., Requena, J., Brinley, F. J., Jr, Mullins, L. J., Scarpa, A., and Tiffert, T. (1976). *Journal of General Physiology*, **67**, 433–67.
5. Thomas, M. V. (1982). *Techniques in calcium research*. Academic Press, London.
6. Thomas, M. V. and Gorman, A. L. F. (1977). *Science, New York*, **196**, 531–3.
7. Gorman, A. L. F. and Thomas, M. V. (1978). *Journal of Physiology*, **275**, 357–76.
8. Gorman, A. L. F. and Thomas, M. V. (1980). *Journal of Physiology*, **308**, 259–85.
9. Gorman, A. L.F. and Thomas, M. V. (1980). *Journal of Physiology*, **308**, 287–313.
10. Miledi, R., Parker, I., and Schalow, G. (1977). *Proceedings of the Royal Society of London*, **B198**, 201–10.
11. Al-Nasser, I. and Crompton, M. (1986). *Biochemical Journal*, **239**, 31–40.

5

NMR measurement of cytoplasmic free calcium concentration by fluorine labelled indicators

JAMES C. METCALFE and GERRY A. SMITH

1. Introduction

A set of prototype indicators has been developed to measure by ^{19}F NMR the free cytosolic concentration of Ca^{2+} ($[Ca^{2+}]_i$), Mg^{2+} and Na^+ and the cytosolic pH in cells using indicators labelled with fluorine (1, 2). Although the design and use of the Ca^{2+} indicators is described here, it is worth emphasizing that the co-loading of indicators for different cations has been demonstrated which allows simultaneous measurement of the free concentration of more than one cation (1). The indicators are designed on the same principle as the fluorescent indicators for cations which they are intended to complement. The ^{19}F NMR indicators can be used in perfused tissues and organs where the use of fluorescent indicators is very restricted, whereas the fluorescent indicators can be used for single-cell measurements which are beyond the sensitivity of the ^{19}F NMR indicators. Both types of indicator can be used in cell suspensions but for most studies of this kind the fluorescent indicators are easier to use. Eventually it is expected that the ^{19}F NMR indicators will be applied to *in vivo* studies in the intact animal and be combined with spectroscopic imaging techniques which will provide $[Ca^{2+}]_i$ measurements from precisely located regions within specific tissues. It is likely that localized measurements of $[Ca^{2+}]_i$ and the free cytosolic concentration of other cations will be useful in both physiological and pathological studies.

2. Design and properties of indicators

The principle of indicator design is the same for both ^{19}F NMR and fluorescent indicators. The structure contains a component which selectively and reversibly binds Ca^{2+} when the free Ca^{2+} concentration changes and this component is attached to a spectroscopic 'reporter' group which senses the binding of Ca^{2+}. The prototype Ca^{2+} indicator (*Figure 1*) was developed from

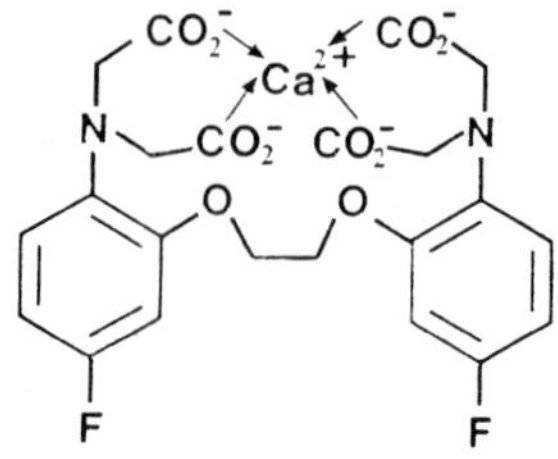

Figure 1. The prototype Ca^{2+} indicator, 5FBAPTA. The four carboxylate groups chelate a single Ca^{2+} ion in solution and the two fluorine nuclei constitute the ^{19}F reporter group. The position of the fluorine nuclei on the aromatic rings is counted anticlockwise from the oxygen substituted at the 1 position and the structure shown is therefore the 5 fluoro derivative of BAPTA.

the structure of the 1,2-bis(*o*-aminophenoxy)ethane-*N*,*N*,*N*′,*N*′-tetraacetic acid Ca^{2+} chelator (BAPTA) introduced by Tsien (3) in which the selectivity of the indicator for Ca^{2+} is determined primarily by the structural geometry of the four carboxylate groups. The two fluorine nuclei constitute the ^{19}F NMR reporter group. If 5FBAPTA is either fully uncomplexed or fully complexed with a polyvalent cation, the ^{19}F NMR spectrum consists of a single resonance because the two fluorine nuclei are in a chemically equivalent environment. The identity of a complexed cation can be determined from the position (chemical shift) of the resonance in the spectrum of the complexed indicator. For example, the resonance from the Ca^{2+}-bound indicator is 5.7 p.p.m. downfield (at 37 °C; pH 7.0) from the free indicator and this provides an unambiguous identification of the ion complexed to the indicator. In some cell systems (e.g. Ehrlich ascites tumour cells) a minor resonance due to the presence of Zn^{2+} ions in the cells has been detected (1).

The detailed chemical and ^{19}F NMR properties of 5FBAPTA which enable it to be used as an indicator of $[Ca^{2+}]_i$ have been described and reviewed elsewhere (4). Briefly, the indicator has an apparent dissociation constant (K_d) for Ca^{2+} of 0.71 μM and 0.54 μM at 37 °C and 30 °C respectively, and therefore titrates over the physiological range of $[Ca^{2+}]_i$ from 0.1 μM to 10 μM. It has a K_d for Mg^{2+} of greater than 100 mM and therefore $[Mg^{2+}]_i$ in the range from 0.1 mM to 1 mM does not affect the determination of $[Ca^{2+}]_i$. Analysis of the ^{19}F NMR spectra of solutions of 5FBAPTA partially saturated with Ca^{2+} gave a dissociation rate constant of 570 sec^{-1} and an association rate constant of 8×10^8 litre mol^{-1} sec^{-1} at 37 °C which is close to the diffusion limit. These kinetic parameters set a theoretical limit of about 10 msec to the time-resolution for changes in $[Ca^{2+}]_i$. In practice, the rate which radio-frequency (RF) pulses can be applied to the sample containing the indicator will usually be at intervals longer than 10 msec (for example, 30 msec is the shortest pulse interval we have used in perfused heart experiments to follow the Ca^{2+} transient during the cardiac cycle). The important practical consideration is that the indicator cannot be used to follow changes which occur faster than its dissociation rate for Ca^{2+} and it is therefore important to know the kinetic parameters for new indicators (see Section 6). It is already clear that for some experiments in which the range of $[Ca^{2+}]_i$ changes is large, it is useful to co-load two $[Ca^{2+}]_i$ indicators which have different K_d values. Provided the indicators have different chemical shifts for all of the resonances from the free and bound forms of the indicators, the full range of $[Ca^{2+}]_i$ changes can be followed in a single experiment. Where the titration ranges of the indicators overlap the consistency of the indicated $[Ca^{2+}]_i$ values can also be determined. This is important in the evaluation of new indicators where, for example, the apparent K_d in the cytosol might differ from the K_d in physiological media if the indicator interacts weakly with cellular components in the cytosol.

3. General features of indicator loading

The 5FBAPTA indicator is loaded into cells with the four carboxylate groups esterified with acetoxymethyl ester (AME) groups. The 5FBAPTA-AME derivative is non-polar and permeates freely into cells where endogenous esterases remove the ester groups to leave the indicator trapped in the cytosol. The rate at which the indicator accumulates in the cytosol in the completely de-esterified form is determined by several factors, including the temperature and pH of the medium, and there is no general recipe for optimal loading for different cell systems which vary in their intrinsic esterase activities (5). The rate at which the fully de-esterified indicator accumulates in the cells is also affected by the way in which the 5FBAPTA-AME is dispersed into aqueous solution. At concentrations above a few micromolar in normal physiological media the indicator forms a cloudy dispersion and it should be noted that the bubbles in gassed media can cause rapid precipitation of dispersions of 5FBAPTA-AME and the AME derivatives of other indicators.

It is important to check that all of the indicator ester is fully hydrolysed under the conditions used for loading because unhydrolysed ester may affect the biological responses to be examined. The unhydrolysed ester is difficult to detect in the NMR spectrum because it gives a very broad resonance. One way to determine the extent of hydrolysis after loading is by the use of $[^3H]$-5FBAPTA-AME (Amersham International). The indicator can be recovered by a Folch extraction and the proportions of hydrolysed and unhydrolysed indicator can be determined by scintillation counting of the aqueous and organic phases obtained from the extraction since any unhydrolysed ester is retained in the organic phase.

It is worth noting that different indicators may load at widely different rates into the same cell system and an important variable appears to be the ease with which the ester disperses in the aqueous phase. In some experiments a dispersing agent (e.g. pluronic F-127; Kodak) has been used to promote uptake of the indicator ester into the cell system. The more hydrophobic AME derivatives of indicators (e.g. fura-2-AME) are usually hydrolysed more slowly and also tend to become associated with intracellular organelles in unhydrolysed or partially hydrolysed form. For these indicators it is especially important to define loading conditions under which virtually all of the indicator is fully hydrolysed.

The demonstration of normal biological responses after loading with the indicator is, of course, an essential step in validating the use of the indicators for each application. The loading of 5FBAPTA, in common with the fluorescent indicators, will stimulate lactate production in a variety of cells as intracellular concentration of the indicator is increased (1, 6). If biological function of the cell system is impaired, it is useful to determine whether lactate production has been stimulated and whether the cellular concentration of ATP has been depleted significantly since a decrease in ATP concentration often results in the inhibition of function. For some systems a convenient way of determining whether the cellular ATP level has been affected by the loading of 5FBAPTA is by ^{31}P NMR in a switchable NMR probehead tuned for both ^{31}P NMR and ^{19}F NMR (7).

4. Thymocytes

4.1 Loading with 5FBAPTA

Typical loading conditions are summarized for murine thymocytes (8) or spleen cells which do not affect the ability of these cells to respond to appropriate mitogens with an increase in $[Ca^{2+}]_i$ and the activation of the c-*fos* and c-*myc* proto-oncogenes. Cells (30 ml) at about 2×10^7 ml^{-1} in RPMI 1640 medium buffered with 10 mM Hepes adjusted to pH 7.3 are incubated with 20 μM of 5FBAPTA-AME for 45 min at 37 °C. The cells are washed by centrifugation (500 g for 3 min) to remove unhydrolysed ester. The half-time for efflux of 5FBAPTA from thymocytes at 37 °C is approximately 4 h. The main efflux pathway is through the anion transport protein and is blocked by inhibitors of this channel (the leakage rate varies with the cell system as for the fluorescent indicators). The loading conditions described give a cytosolic concentration of the fully hydrolysed indicator of 0.5 mM to 1.0 mM estimated by the use of [3H]–5FBAPTA-AME as described above.

4.2 NMR measurements

The experiments described here refer to a Bruker wide-bore NMR spectrometer operating at 376 MHz for ^{19}F NMR. The cells ($3–5 \times 10^8$ ml^{-1}) are re-suspended in medium in the NMR sample chamber at 37 °C (*Figure 2*). The cells can be maintained in a functionally responsive state for many hours by continuous mixing using a gentle air pressure cycle and by continuous dialysis against fresh, oxygenated medium pumped through fine fibres. The rate of

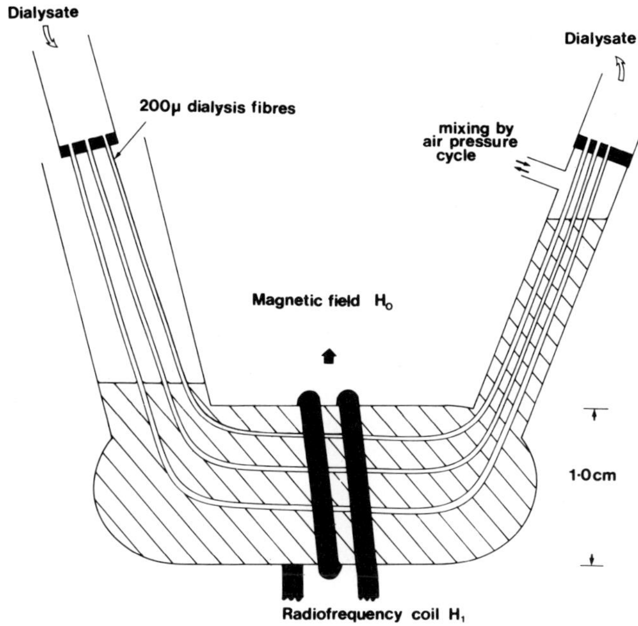

Figure 2. Sample perfusion chamber for cell suspensions. The cells are maintained in suspension by an air pressure cycle which raises and lowers the medium in the closed arm of the chamber at a rate adjusted to maintain a uniform suspension. The sample is continuously perfused with oxygenated medium through about 30 dialysis fibres (200 µm o.d.) at a rate to maintain the pH in the chamber at 7.3. The RF coil is wrapped around the chamber and the whole chamber is mounted on the probehead for insertion into the spectrometer. The temperature of the medium is monitored by a thermocouple in the effluent dialysate.

flow of the dialysis medium, determined in preliminary experiments, is adjusted to prevent acidification of the medium in the sample chamber by the lactic acid released by the cells. We have found that kidney dialysis fibre units adapted to fit into the NMR sample chamber work well for this dialysis.

Spectral accumulation for about 15 min using a RF pulse rate of 5 Hz is sufficient to generate a spectrum with a signal to noise ratio of greater than 10:1. This is adequate to yield well-defined $[Ca^{2+}]_i$ measurements in the normal titration range of the indicator. A typical spectrum of 5FBAPTA in a cell suspension consists of two resonances with the low field peak (centred at 0.0 p.p.m.) corresponding to free 5FBAPTA and the low field peak at 5.7 p.p.m. corresponding to the indicator complexed to Ca^{2+} (*Figure 3*). From the relative proportions of the two resonances, measured by integrating the areas under the peaks, the free Ca^{2+} concentration can be calculated directly from the spectrum as $[Ca^{2+}]_i = B/F \times K_d$ where B and F are the areas of the resonances from the bound and free forms of the indicator and the apparent K_d for Ca^{2+} at 37 °C is 0.71 µM.

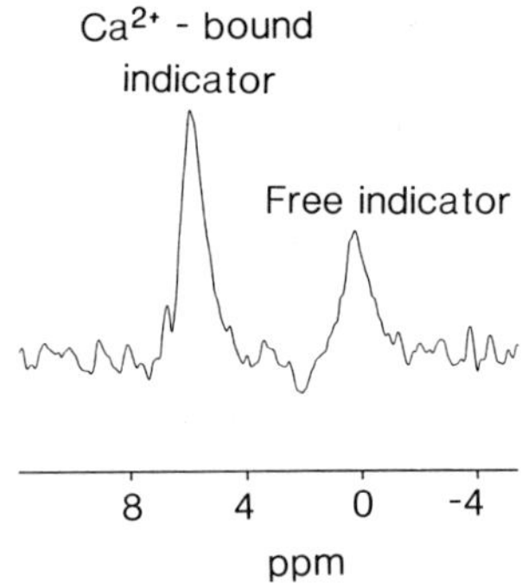

Figure 3. The spectrum of 5FBAPTA in thymocytes. The two components of the spectrum correspond to free indicator and indicator chelated with Ca^{2+} which are in slow exchange under the conditions of the experiment (37 °C; 188 MHz). In this experiment the cells have been treated with a Ca^{2+} ionophore (100 nM A23187) and the calculated $[Ca^{2+}]_i$ is 1.1 μM.

It is necessary to check that the spectral accumulation parameters do not result in significant saturation of the signal by repeating the spectral accumulation block with a slower RF pulse rate (e.g. 2 Hz). If there is a significant increase in the total integrated intensity of the spectrum from the indicator ($F + B$) at the slower pulse rate then the signal is partially saturated. The effect of signal saturation is to reduce the signal to noise of the spectrum and hence the sensitivity of the indicator. However, it has been shown that the effective T_1 relaxation times of the bound and free forms of the indicator are the same from which it follows that partial saturation of the signal will not affect the relative areas of the resonances from the two forms of the indicator and will not, therefore, affect the indicated $[Ca^{2+}]_i$ (1).

As an alternative to the use of [^{3}H]–5FBAPTA-AME, the intracellular concentration of the 5FBAPTA may be calculated using the unlabelled indicator by comparing the total integrated NMR signal ($F + B$) from the indicator in a known total intracellular volume of the cells in suspension with the corresponding signal obtained from a known concentration of 5FBAPTA in solution. For this comparison a relatively slow RF pulse rate must be used to avoid partial saturation of the signal from the aqueous solution of 5FBAPTA, since the T_1 relaxation time of 5FBAPTA in aqueous solution is typically five- to tenfold longer than the relaxation time in the cytosol.

5. Perfused hearts

5.1 Heart preparation and loading

The following protocol has been used extensively to load adult ferret hearts with 5FBAPTA. Hearts excised from deeply anaesthetized ferrets are weighed and placed in ice-cold medium (1). The aorta is cannulated within 2 min of excision and perfused retrogradely using a peristaltic pump to maintain a constant flow rate of 4 ml min^{-1} (g wet $wt)^{-1}$ with perfusion medium (Krebs Henseleit). The medium is equilibrated with 95% O_2:5% CO_2 to maintain a pH of 7.4 at 30 °C and is filtered upstream of the heart in two stages (final pore size 5 μm). It is important to test that the perfusion rate is sufficient to maintain optimal function not only of the unloaded heart but also of the perfused heart after loading with the indicator. After cannulation, the atrio-ventricular node is crushed and the heart paced at 0.2 Hz to 2.0 Hz with stimulus pulses of supramaximal voltage and 10 msec duration. Stimulation is achieved via salt bridges consisting of narrow bore plastic tubing filled with 3 M NaCl in agar in order to avoid electrical interaction with the RF coil. The aortic perfusion pressure and the isovolumic left ventricular pressure (LVP) can be monitored continuously using pressure transducers. After preparation the heart is paced at 1.0 Hz for a 60-min equilibration period. Following equilibration the aortic pressure is typically 100 to 150 mm Hg, systolic LVP is 160 to 170 mmHg and diastolic LVP is 15 to 18 mmHg.

The 5FBAPTA-AME is loaded from a 50 mM stock solution in dimethyl sulphoxide by infusion into the perfusion line at a constant rate of 200 μl h^{-1} for 30 min. The infusion is made downstream of a bubble trap, to avoid precipitating the 5FBAPTA-AME from dispersion and the 5FBAPTA-AME reaches the heart at approximately 5 μM (1).

Infusion of 5FBAPTA-AME for 30 min under the conditions described results in a progressive decrease in systolic LVP and an increase in diastolic LVP so that the LVP developed in the cycle is decreased to about 20% of the value before loading (*Figure 4*). This substantial effect on function is mainly attributable to the Ca^{2+} buffering effect of the indicator in the cytoplasm. However, the loaded heart responds to pharmacological stimuli (e.g. adrenalin) and experimental manipulation of the medium (e.g. increasing the Ca^{2+} concentration or lowering the Na^+ concentration) in qualitatively the

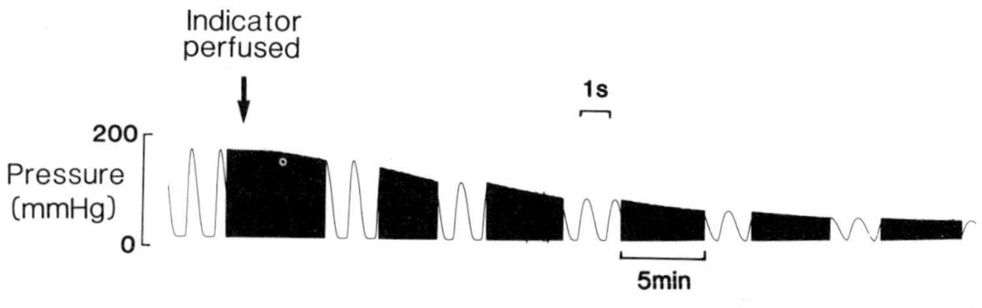

Figure 4. The effect of loading 5FBAPTA-AME on the systolic and diastolic pressure in a perfused (Langendorff) ferret heart (the details of the loading technique are in the text). The expanded time-scale sections of the trace show that as the indicator is taken up into the cells the time taken to reach maximum pressure increases and the recovery is slowed until the period of constant diastolic pressure is eliminated.

same way as unloaded hearts. Furthermore, the effect on LVP of loading the indicator is almost completely reversible by the addition of $ZnCl_2$ (50 μM) to the perfusion medium. This cation has a much higher affinity for 5FBAPTA than $Ca^{2+}$and effectively chelates to all of the indicator in the cytoplasm which eliminates its Ca^{2+} buffering capacity. The indicator has therefore provided useful new data for the perfused heart in spite of the substantial effect it has on function.

5.2 NMR measurements

Heart NMR measurements are made in a 33-mm diameter probehead in a perfusion chamber which will accommodate hearts of up to 10 g wet weight within the RF transceiver coil. The perfusion effluent is drawn off above the level of the heart to minimise the magnetic field inhomogeneity caused by susceptibility differences. The perfusion tubing is several metres long to distance the ferromagnetic perfusion equipment from the magnet. Teflon tubing is used for the line carrying the 5FBAPTA-AME to the heart since it does not absorb the ester significantly, but silicone rubber tubing is used near the probe head because the fluorine in Teflon distorts the baseline of the ^{19}F

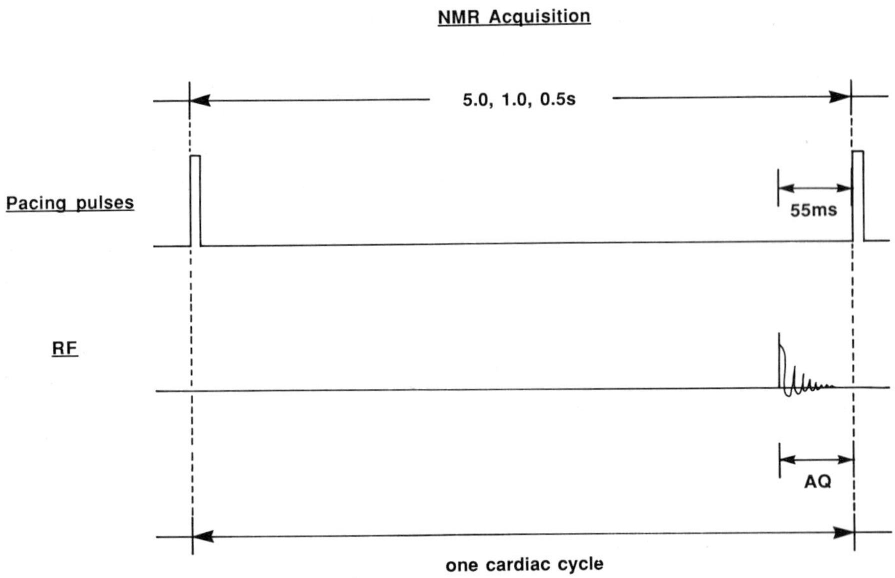

Figure 5 (a). Acquisition of the NMR spectrum from 5FBAPTA to measure end-diastolic $[Ca^{2+}]_i$ at different pacing frequencies. The diagram shows one cardiac cycle, with variable pacing at either 0.2, 1.0, or 2.0 Hz. Two pacing pulses are shown with one RF pulse at 55 msec from the end of diastole immediately before the next pacing pulse. The free induction decay is recorded during the 55 msec interval (the acquisition time, AQ) and accumulated from successive cycles.

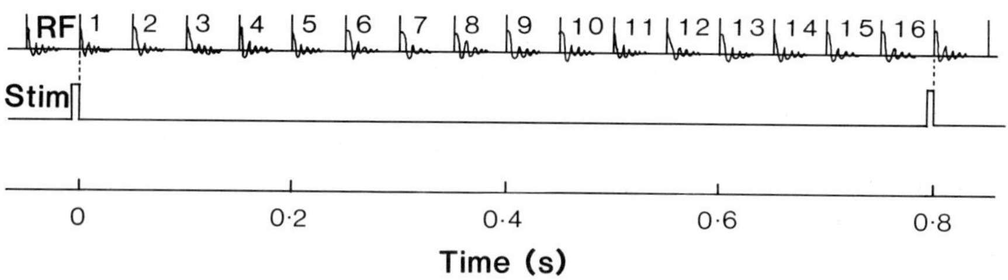

Figure 5 (b). The gating sequence used to record the $[Ca^{2+}]_i$ transient. RF pulses with a predetermined optimal flip angle are applied 16 times each cardiac cycle at intervals of 50 msec, and a pacing frequency of 1.25 Hz. The FIDs corresponding to each interval are accumulated separately, and used to calculate $[Ca^{2+}]_i$ values when adequate signal to noise has been achieved.

NMR spectrum if the Teflon tubing is too close to the RF transceiver. The lines are water-jacketed and maintained at 30 °C using a circulatory pump in the water bath in which the perfusion medium is warmed. The temperature of the medium surrounding the heart is maintained at 30 ± 0.5 °C and monitored by a thermocouple inserted in the effluent waste line. The magnet is shimmed with the heart *in situ* using the ^{1}H resonance of water.

For the loading conditions described, ^{19}F NMR spectra with a signal-to-noise ratio of greater than 10:1 can be accumulated in blocks of 4000 free induction decays (FIDs) with a RF pulse frequency of 5Hz. To measure end-diastolic $[Ca^{2+}]_i$, the stimulator is triggered from the pulse controller of the NMR spectrometer and the gating sequence shown in *Figure 5a* is used to accumulate NMR data. The FID is recorded once each cardiac cycle in the 55 msec period immediately preceding the next pacing pulse. The optimal pulse flip angle is determined for each pacing frequency.

To record cardiac Ca^{2+} transients, the gating sequence shown in *Figure 5b* is used. RF pulses at the optimal flip angle are applied 16 times each cardiac cycle (this sets the time-resolution to which changes in $[Ca^{2+}]_i$ can be determined in these experiments). The FIDs from each pulse are recorded and the data corresponding to each time-interval within the cardiac cycle are accumulated separately. Data accumulation for about 2 h is required to achieve a signal-to-noise ratio of 10:1 for the spectrum for each time interval. A typical $[Ca^{2+}]_i$ transient is shown (*Figure 6*).

6. Developments

It is clear from the effects of the 5FBAPTA indicator at the intracellular concentrations of approx. 100 μM required to obtain spectra with an adequate signal-to-noise ratio for $[Ca^{2+}]_i$ measurements that there is a substantial effect on heart function, presumably through the buffering effect of the indicator on changes in $[Ca^{2+}]_i$ through the cardiac cycle. Similar effects probably occur in

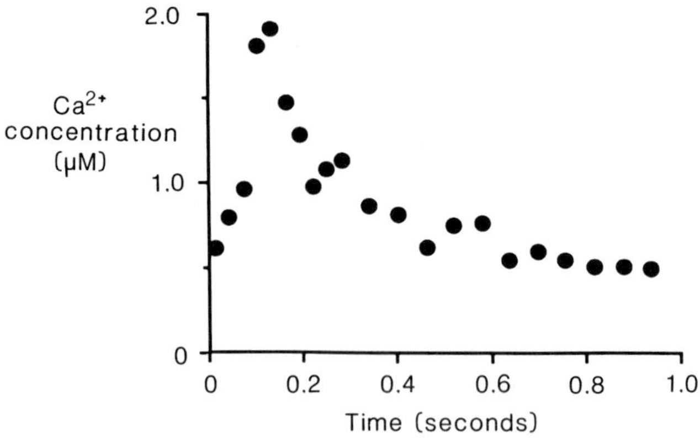

Figure 6. The $[Ca^{2+}]_i$ transient in a perfused ferret heart loaded with 5FBAPTA as illustrated in *Figure 5* and described in the text. The peak in $[Ca^{2+}]_i$ precedes the development of maximal systolic pressure and returns to the end diastolic level before the next cycle. Recent experiments with new indicators show, however, that loading with 5FBAPTA under the conditions described causes an increase in the end-diastolic $[Ca^{2+}]_i$.

the contractile cycle of isolated cardiac myocytes loaded with fluorescent indicators although an experimental analysis of functional perturbation has not been made. The main requirement for the development of the ^{19}F NMR indicators is to increase their sensitivity so that spectra can be obtained using much lower concentrations of the indicators, analogous to the improved sensitivity of fura-2 compared with quin-2 in the development of the fluorescent indicators. One way to achieve this is by the design of new indicators based on BAPTA derivatives incorporating the trifluoromethyl group and we will report soon on a new indicator of this type.

References

1. Kirschenlohr, H. L., Metcalfe, J. C., Morris, P. G., Rodrigo, G. C., and Smith, G. A. (1988). *Proceedings of the National Academy of Sciences, USA*, **85**, 9017.
2. Smith, G. A., Hesketh, T. R., and Metcalfe, J. C. (1988). *Biochemical Journal*, **250**, 227.
3. Tsien, R. Y. (1980). *Biochemistry*, **19**, 2396.
4. Metcalfe, J. C. Hesketh, T. R., and Smith, G. A. (1985). *Cell Calcium*, **6**, 183.
5. Bachelard, H. S., Bader-Goffer, R. S., Brooks, K. V., Doblin, S. D., and Morris, P. G. (1990). *Journal of Neurochemistry*, **51**, 1311.
6. Hesketh, T. R. Smith, G. A., Moore, J. P., Taylor, M. V., and Metcalfe, J. C. (1983). *Journal of Biological Chemistry*, **258**, 4876.
7. Marban, E., Kitakazi, M., Kusuoka, H., Porterfield, J. K., Yue, D. T., and Chacko, V. P. (1987). *Proceedings of the National Academy of Sciences, USA*, **84**, 6005.
8. Smith, G. A., Hesketh, T. R., Metcalfe, J. C., Feeney, J., and Morris, P. G. (1983). *Proceedings of the National Academy of Sciences, USA*, **80**, 7178.

6

X-ray microanalysis

MARIA FIORA WENDT-GALLITELLI and GERRIT ISENBERG

1. Introduction

The heart ventricular myocyte is one of those contractile cells where movements of Ca couple excitation to contraction, rhythmically every second. According to our textbooks, it is the Ca bound to troponin C (TnC) that controls the actin–myosin interaction (attachment of cross-bridges). Quantitation of this activator Ca, therefore, would help in understanding the way of contractile activation. Besides knowledge of the concentration of Ca–TnC as a function of time, one should know how and from where the activator Ca derives, and how and to where it is redistributed.

During the last two decades, a variety of different methods has addressed the above question. The outcome can be summarized in the simplified model of *Figure 1*. For mammalian ventricular myocytes, the model postulates most of the activator Ca to derive from an intracellular 'release compartment' (R) where it has been stored during diastole. Ca-release is induced by Ca-entry during excitation. Ca diffuses from the release compartment into the myofibrillar space along a gradient. Ca diffusion occurs with Ca binding to TnC. TnC competes with sites on the SR Ca-pump which have an even higher Ca affinity. Thus, after a while, Ca has sequestered out of the myofibrillar space into the 'uptake compartment' (U). Ca-influx through the sarcolemma is thought not only to trigger SR Ca-release but also to fill the SR with Ca. Vice versa, not all of the released Ca recirculates to the SR, a fraction of 20% might be extruded into the extracellular space. At steady-state beating, the sum of Ca-influx and Ca-efflux should equal.

The Ca compartments of the model are compartments in a functional sense. It is tempting to attribute them to morphological structures (*Figure 1*, B) like the sacrolemma (SL), the sarcoplasmic reticulum (SR) with its junctional and free components, and to the myofibrillar TnC. So far, the experimental evidence for such a correlation of function and structure is rather indirect. [Ca] of cell compartments has been measured analytically after their separation by fractional centrifugation (SR-microsomes). However, one does not know whether the preparative steps between mincing the tissue and Ca-measurements have destroyed the original Ca distribution. By means

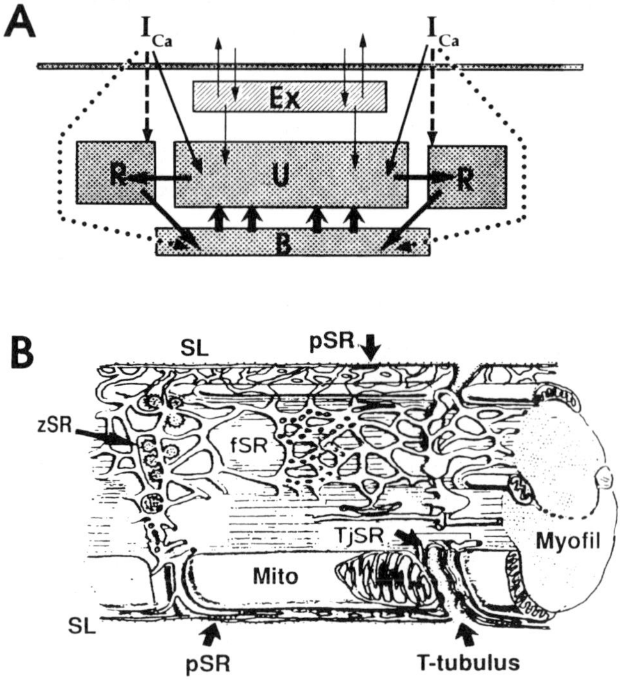

Figure 1. Schematic of functional Ca-compartments and their possible morphological counterparts. **A**: Role of functional Ca-compartments for cardiac excitation-contraction-coupling. Excitation opens sarcolemmal Ca-channels and Ca-current I_{Ca} transports extracellular Ca-ions into the cell. Most of this Ca does not bind the myofilaments (B, dotted line of I_{Ca}) but fills (solid line of I_{Ca}) the 'uptake-compartment' U. U also removes Ca from the space between the myofilaments. With some delay, Ca diffuses from U to the 'release compartment' R. At the end of diastole, R is filled with Ca. At start of systole, Ca is released from R. Activation of Ca-release is triggered by Ca-ions entering with I_{Ca} (dashed line). Released Ca binds to troponin C and activates contraction. The degree of Ca-load of R and thereby the amount of released activator Ca is modulated through Na-dependent Ca-influx and Ca-efflux ('exchanger compartment', Ex). **B**: Schematic of morphological compartments. The sarcolemma (SL) separates the extracellular (e.c.) from the intra-cellular (i.c.) space. The i.c. space is occupied by myoplasm containing bundles of myofibrils (Myofil). They are surrounded by free sarcoplasmic reticulum (fSR). At the level of the *z*-lines the SR couples to the SR of transverse tubules (TjSR). Similar couplings are seen to the external surface SL (pSR). The corbular part of zSR (arrow) does not contact the SL.

of ion-selective electrodes, luminescent proteins or Ca-sensitive dyes (1), the concentration of Ca ionized in the 'accessible water space' ($[Ca^{2+}]_i$) can be quantified in its time-course during the cardiac cycle, without disturbing the cell function too much. However, $[Ca^{2+}]_i$ is proportional to bound [Ca] only then when the probe can equilibrate with Ca during release ('downstream') and re-uptake ('upstream') sufficiently. Further, [Ca] in small intracellular compartments like the SR are usually not indicated.

A recent approach for measuring not only free but also bound Ca, and not only in the myoplasm but also in intracellular compartments, has come from the X-ray microprobe analysis (synonymous electron probe microanalysis, EPMA). When analysis is performed in a defined functional state, knowledge of Ca concentrations may connect the functional and morphological compartments together.

2. The method of X-ray microprobe analysis

2.1 The necessary equipment

EPMA analyses thin freeze-dried sections of the cell in the vacuum of the electron microscope; the concentrations are given in mmol/kg dry mass. Calcium concentrations of a given functional state are kept by rapid-freezing and cryotechniques. By means of a *cryoultramicrotom*, the frozen myocytes can be cut at temperatures between −140 and −160 °C to 100–200-nm-thick cryosections. The sections were picked up on copper grids, were *freeze-dried*, carbon-coated and finally mounted in a *specimen-holder* cooled at −100 °C. Specimen were analysed using a Philips CM12 *STEM electron microscope* equipped with a LaB_6 filament. Using the STEM-mode, the analysing beam was adjusted to 16 nm diameter at 100 kV acceleration voltage. X-ray quanta were sampled with a 30 mm^2 Link *energy dispersive Si-detector* with 145 eV resolution. A *goniometer stage* tilted the specimen by 30 degrees in order to optimize the number of counts arising from the specimen and to minimize those from the copper grid. A Link *pulse processor system* quantified the data according to the Quantem program. Finally, quantitative data was statistically evaluated with a SAS statistic package (SAS Inc., NC, USA).

2.2 Principles of energy-dispersive X-ray microprobe analysis

In the microscope, an incident high-energy beam electron can ionize an inner orbit electron of the atom. The incident and the ionized electrons leave the atom, and an electron from an outer orbit drops into the vacated inner shell. The loss of energy in this transition is emitted as *characteristic X-ray* quantum. For example, the incident beam may have 'knocked out' an electron from the K-shell of a Ca-atom; filling the ionized K-shell with a L-shell electron creates a 3.69 keV X-ray photon. Similar transitions in the shells of other elements generate X-rays of different energies. By means of a detector (see below), part of the emitted X-ray quanta can be collected. A computer sorts the number of counts according to their energy. The result is an *energy-dispersive spectrum* in which the peaks are localized at energy lines characteristic for each element (*Figure 2*).

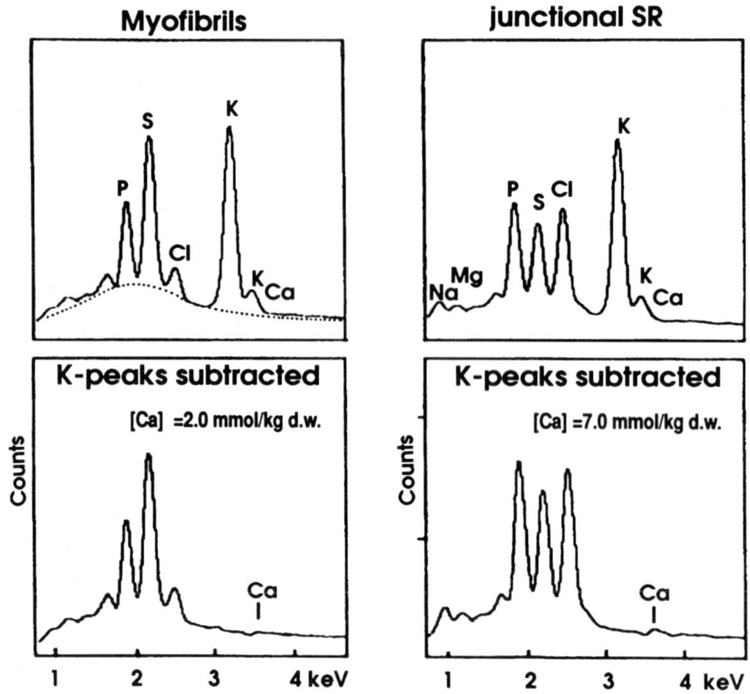

Figure 2. X-ray spectra of myoplasm and junctional sarcoplasmic reticulum (jSR). Peaks representing the elements are indicated. The *upper left* panel indicates the shape of the continuum as a curved line under the peaks. The high K-peak is $K_{k\alpha}$, the small K peak is $K_{k\beta}$. In lower panels the potassium $K_{k\alpha}$ and $K_{k\beta}$ peaks have been subtracted from the elemental spectrum in order to visualize the size of the Ca-peak. (The spectrum jSR with subtracted K-peaks, *bottom right*, has a higher gain than its counterpart in the *upper right*.) The total Ca concentration is labelled in mmol/kg d.w.

The area of the peaks is proportional to the number of counts and, therefore, to the number of atoms in the analysed volume. The incident electron also interacts with the nucleus of the atom, becomes decelerated, and the energy-loss generates the uncharacteristic X-ray bremsstrahlung. This uncharacteristic X-ray radiation constitutes the background in the energy dispersive spectrum and is called *continuum*. The number of X-rays of the continuum is directly proportional to the total mass of the analysed volume. Hence, the ratio peak/continuum is directly related to the mass fraction of an element in the analysed volume. According to the continuum method (2) data is quantified by the Hall formula

$$C_{\mathrm{a(sp)}} = \frac{(N_{\mathrm{a}}/W)_{\mathrm{sp}}}{(N_{\mathrm{a}}/W)_{\mathrm{st}}}\ C_{\mathrm{a(st)}}. \qquad \text{[eqn 1]}$$

Here C_a is the mass fraction of element a (expressed in mmol/kg total mass of the specimen under the beam); N_a is the number of characteristic counts from element a after background subtraction; W is the simultaneously observed number of counts of continuum X-rays from the analysed microvolume of the specimen (sp) or from the standard (st).

2.3 Detection, processing, and calibration of X-ray spectra

In the detector, each incident X-ray photon generates a pulse proportional to the X-ray energy. The detector is usually covered with a thin Be-window that absorbs low-energy X-rays. This absorption prevents detection of elements with atomic number less than 11 (Na) and diminishes the sensitivity for Na, Mg, and P as compared with the sensitivity for elements with higher atomic number, such as Ca. Thus, with modern Si(Li) detectors the minimal detectable concentration Na concentration is about 6 mmol/kg dry weight (d.w.).

A computer pulse-processes the characteristic and continuum X-rays and converts them into concentrations (3). As a first step, the number of events per energy level is determined and stored (multichannel analysator). The spectrum is corrected for the contributions of extraneous components from holder, grid, and film, and the X-ray continuum is subtracted. After digital filtering, the computer fits the collected spectrum for position and shape to the peaks of spectra from single elements that are stored as 'library'. Those reference files of *primary standards* stem from analysis of crystals of Na, Mg, P, S, Cl, K, and Ca salts. The continuum should be evaluated (4) in the region between 1.34 and 1.46 keV, since regions of higher energies contain large contributions determined by extraneous components such as microscope, grid, etc. In very thin sections the continuum intensities depend on the elemental composition and are proportional to the local average Z^2/A, where Z is the atomic number and A the atomic weight of the elements. Thus, equation 1 has to be corrected. The correction factor Z^2/A is 3.6 for water and approximately 3.3 for most organic compounds.

Quantitation of elemental concentrations is based on the Hall formula with the linear relation between the peak/continuum ratio and elemental concentration (5). For a biological matrix (cytosol) the relation between the peak/continuum ratio and concentration is calibrated by cryosections of *secondary standards*. Secondary (or external) standards are prepared in two ways. (a) Various weight ratios of albumin and phosvitin were mixed, the Na, Mg, P, and S content of them determined by atomic absorption. (b) In 20% dextran (200 000 mol.wt.), salt solutions of four different concentrations were prepared. 2 g dry mass of (salt plus dextran) was diluted to 10 ml with distilled and de-ionized water. The Ca-standards contained 600 mmol/kg d.w. KCl in order to have count ratios between the interfering Ca_{K_α} and the K_{K_β} peaks similar to those in mammalian cytosol (see below). Each standard was cryocut, freeze-dried, and analysed at <−100 °C (to avoid loss of S) at several points (>5). The analysis resulted in calibration curves with a 'standard factor' $f_{st} = [(P_a/W_{std}) \cdot g_{std}] \cdot C_{std}$ with P_a = peak area of the element, W = background integral, $g_{std} = n/A_w = g$ 'factor' with n = the atomic number, A_w = atomic weight, and C_{std} the concentration of the element in the standard (mmol/kg d.w.). The f_{st} values are 0.007 for Na, 0.019 for Mg, 0.023 for P, 0.028 for Cl, 0.040 for S, 0.045 for K, and 0.079 for Ca. The f_{st}-values quantify the unknown concentrations in spectra from biological sections. Since all sections were freeze-dried before analysis, the concentrations are given in mmol/kg d.w. of analysed mass.

2.4 Minimal detectable concentrations and spatial resolution

The detector has an energy resolution (full width at half maximum; FWHM) that decreases with increasing X-ray energies, it is as high as 70 eV at 0 keV FWMH but approx. 160 eV at 8 keV. FWMH also decreases when the count rates exceed 1500 Hz (number depending on detector). Analysis of the Ca peak (3.69 keV) is a problem, because the overlapping $K_{k\beta}$ (3.59 keV) has to be subtracted (see *Figure 2*, lower right). Shifts in the calibration and degradation in the resolution can introduce significant errors to the subtraction procedure. Although fitting routines including the first and second derivatives of the K peaks can correct (4) for those effects, the location of peak centroid and the energy resolution of the detector should be carefully calibrated before the start of analysis. With these precautions [Ca] can be quantified down to concentrations as low as 0.3 mmol/kg d.w. High *spatial resolution* of EPMA depends essentially from the diameter of the analysing (incident) electron beam and the thickness of the cryosection. In modern electron microscopes the minimal beam diameter can be focused to a few nanometres and is no longer limiting. However, cryosections cannot be cut thinner than 100 nm. If the free SR has a diameter of 40 nm (3), analysis in a 100-nm section contains a fringe of adjacent myoplasm (see *Figure 9* in Section 5.2). During diastole, SR has a higher [Ca] than the surrounding myoplasm, therefore SR [Ca] will be underestimated.

2.5 Statistical analysis

Statistical Poisson errors in the counts of peaks and background contribute to the statistics of the measured concentrations. Errors also arise from the subtraction of extraneous components (film, grid, and holder). If the peak-to-background ratio is close to zero, then concentrations can become negative. Negative values have to be included in statistical averaging. The statistical Poisson error of the single measurement decreases with the analysis time. The optimal analysis time is inversely proportional to the thickness of the section, thus long analysis time is necessary to bring the statistical error of the single measurement down to less than 1 mmol/kg d.w. (SD). In the case of low [Ca] the ratio peak/continuum is close to zero, and we found analysis times of 600–1000 sec to be a compromise by which Ca concentrations in the range of 0.3–0.5 mmol/kg d.w. (80–120 μM) can be detected. An improvement of the detection limit from 300 to 30 μmol/kg d.w. would require an unreasonable longer analysis time of 27 h per point.

Quantitative data was statistically evaluated with a SAS statistic package (SAS Inc., NC, USA). Nested analysis of variance for unbalanced groups took into account differences in concentrations due to cell-to-cell and animal-to-animal variance. If the overall *F* test was significant, differences between the single groups were then tested by multiple linear comparison and considered significant when $p < 0.05$.

3. Ca in ventricular myocytes: problems and requirements

3.1 Single isolated myocytes instead of multicellular preparations

Shock-freezing, cryotechniques, and EPMA have been applied to both multicellular thin trabeculae (6) and single myocytes isolated from guinea-pig ventricles (7). Both preparations have their advantages and disadvantages.

(a) In pre-loaded trabeculae force can be recorded until the very moment of the impact with the coolant. In single cells, the contractile state is monitored as unloaded shortening from slack-length (about 1.85 μm diastolic sarcomere spacing) only about 0.3 sec before the contact with the coolant.

(b) The single-cell preparation avoids the problem of cell-to-cell heterogeneity that is inherent in multicellular specimens. Due to limited cooling rate, acceptable cyropreservation is achieved only in the outermost cells of the trabecula. These cells are thin and do not bear T-tubules, i.e. may be not representative for ventricular myocardium. Further, during pharmacological inotropy by, for example, adrenalin or ouabain, the superficial cell layer (the subject of analysis) may have been intoxicated at earlier times than the inner core.

(c) In the isolated cell, analysis over the sarcolemma sharply 'jumps' from the i.c. to the e.c. elemental composition, since the adjacent bath is protein-free and has a very different continuum. Cells of multicellular tissue are surrounded by an interstitium rich in proteins where this sarcolemmal 'jump' is smeared.

(d) In single cells, the method of timed shock-freezing under voltage-clamp conditions could be developed. Combination of multicellular sucrose-gap technique with timed shock-freezing was not conceivable.

(e) The patch-electrodes, used for voltage-clamping, 'dialyse' the electrolytes into the cell, cytosolic components are diluted. Presently, we have limited the period between rupture of the membrane patch and the moment of freezing to 5 min.

3.2 Definition and preservation of the functional state

For EPMA, the original Ca-distribution in the cell is preserved by rapid shock-freezing. Within the last six years, we developed a method for shock-freezing voltage-clamped cells. The effort was inspired by the ambition to find out how Ca^{2+} ions, entering as Ca-current through the SL, fill those i.c. compartments as SR, mitochondria, and myofilaments after they have been depleted by a long rest period (7). Here, we used the voltage-clamp to optimally potentiate the contraction by means of a paired pulse protocol. Together with the contraction, the membrane current was used as a parameter defining the functional state at the moment of shock-freezing.

In general, the combination of the two methods must meet the following criteria:

(a) It should allow shock-freezing of the cell at an exact time in the functional state.

(b) The functional parameters such as membrane current and contraction should be monitored up to the very moment of freezing.

(c) The mechanical movements required for shock-freezing should not produce damage or leakage of the cell.

(d) A sufficient cooling rate should preserve the ultrastructure of the shock-frozen cell.

(e) The cell should be delivered in a system (holder) suitable for further cryocutting.

(f) The frozen cell should be found easily and reproducibly despite its small dimensions in relation to those of the container of the coolant.

4. Method for shock-freezing isolated cells under voltage-clamp

4.1 The silver holder

The cell was shock-frozen, not alone but together with the patch pipette and within a special silver holder that served as reference electrode. A 1.5-mm long taper was pressed into a 16-mm-long, 1.2-mm-wide and 50-μm thin silver foil, by means of a prickle. Slow electrolysis (1% nitric acid) opened the tip of the taper to 150–200 μm diameter and thinned the wall of the taper to 1–2 μm.

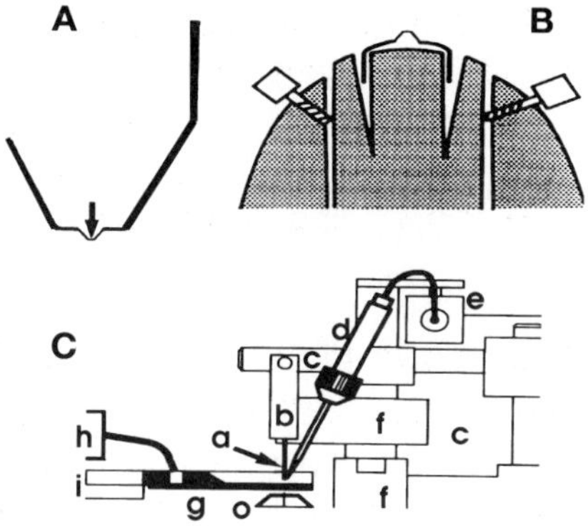

Figure 3. Schematic of the silver cell holder (**A**), its adaptation to the LKB cryoultramicrotom (**B**) and its use during electrical recordings (**C**). **A**: Arrow indicates the position of patch electrode and the myocyte on the pioloform film. **B**: For cryocutting, the silver-holder is turned by 180 degrees, thus the film with the myocyte is cut with the first sections. **C**: The silver holder (**a**) is fixed by a clip (**b**) to a Leitz-micromanipulator (**c**). The Leitz manipulator (**c**) is stable enough to bear a hydraulic Narishige-manipulator (**f**) to which the electrode-holder (**d**) and the headstage of the voltage-clamp amplifier (**e**) are mounted. The experimental chamber (**g**) is open at one end, thus, it can be removed horizontally with the holder and electrode kept still. A heater (**h**) pre-warms the bath solution to 36 °C. Chamber (**g**), microscope (objective **o**), and heater (**h**) are mounted by an independent manipulator (**i**) to a sliding table. (**A** and **B** from Wendt-Gallitelli and Isenberg, *American Journal of Physiology*, **256**, H574–83, 1989, with permission.)

The open tip of the taper was covered with a 20-nm thin transparent pioloform film which was important for several aspects of the method:

- the position of the cell on the film helped to find the cell with the cryocutting knife
- on the film, optimal contact of the coolant was provided
- through the film cell, shortening and sarcomere length could be monitored
- during the movement leading to shock-freezing, the film prevented the solution from streaming through the opening, thereby blowing off the cell.

4.2 Cell transfer into the silver holder

The experimental chamber (volume of 0.2 ml) was open at one end (compare *Figure 5*). When the cells had settled on the bottom, they were continuously superfused with a pre-warmed physiological salt solution (36 °C). A fresh silver holder was mounted on a Leitz micromanipulator. Using a hydraulic Narishige-manipulator (MO-103), the tip of the electrode was positioned 0.5 mm away to the right of the silver taper. By means of the common Leitz-manipulator, the silver holder and electrode were positioned on top of a

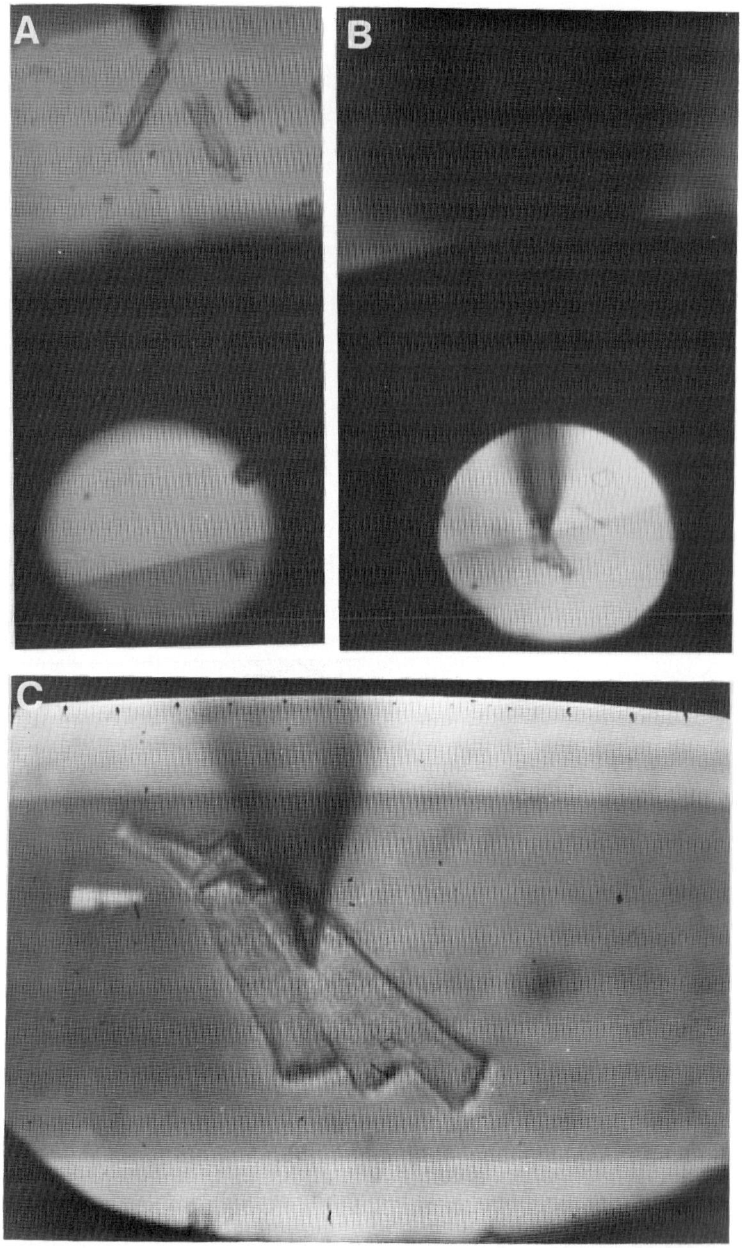

Figure 4. Transport of the ventricular myocyte by means of a patch electrode (dark shadow) into the funnel of the silver holder. Photographs were taken from the monitor of a TV microscope. **A**: position before attachment of the electrode. **B**: after transport into the silver funnel. **C**: Large magnification; the cell is positioned on the surface of the pioloform film. (From Wendt-Gallitelli and Isenberg, *American Journal of Physiology*, **256**, H574–83, 1989, with permission.)

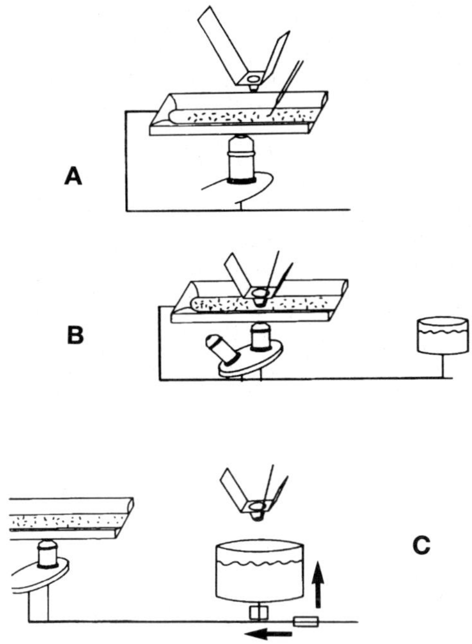

Figure 5. Schematic of transferring and shock-freezing myocytes. **A**: Initial arrangement of microscope, holder, chamber, and patch electrode. **B**: After cell transfer into the silver holder. **C**: Microscope and chamber have been moved away, and coolant is shut around the holder, the electrode, and the cell. (From Wendt-Gallitelli and Isenberg, *American Journal of Physiology*, **256**, H574–83, 1989, with permission.)

selected myocyte (*Figure 5*). The bath solution covered the holder up to the wide opening. With the Narishige-manipulator the tip of the electrode was placed on the cell and, after slight suction in the pipette, was lifted from the bottom. The GΩ-seal and cell access were performed in the usual way. When the cell had been checked to ensure normal action potentials and contractions, the cell transfer was started. With the Narishige-manipulator, the cell was moved upwards beyond the wide opening of the holder (in solution), moved across the dark shadow until seen through the holder's opening (*Figure 5*, B) and downwards until attachment to the pioloform film (32-fold objective). The cell at the tip of the pipette and on the top of the pioloform film of the silver holder is called the 'preparation'.

Success of the transfer and stability of the preparation depends crucially on avoidance of mechanical breakdowns. The flow of the superfusing bath might have to be stopped to prevent a 'blow off' of the cell from the tip. The silver holder and patch-pipette should be mounted in such a way that possible vibrations let them vibrate in unison; any 'friction' of the cell on the pioloform film damages the cell membrane and destroys the cell. Often, the thin film was barely visible, then attachment was seen by bending of the cell. Possible leakiness of the cell was monitored by membrane potential and input resistance; leaky cells had to be discarded.

4.3 Characterization of the functional state

Once the cell was laid on the film and the 'preparation' was established, stimulation and solution flow were turned on again. (The bath solution was composed of 150 mM NaCl, 5.4 mM KCl, 2 mM $CaCl_2$, 1.2 mM $MgCl_2$, 5 mM Hepes/NaOH, pH 7.4, 36 °C.) In these conditions, the myocytes had resting potentials of −85 mV and action potentials with an amplitude of 136 ± 6 mV and a duration of 210 ± 43 msec (mean ± SE from 12 cells). These values are normal, indicating that the transfer had not damaged the cell. Then, the contractions were optimally potentiated by a paired pulse protocol, a first 180 msec pulse to +5 mV being followed by a second 160 msec pulse to +50 mV and the holding potential was set to −80 mV (*Figure 6*). A pre-pulse technique was used to inactivate TTX-sensitive Na-currents (I_{Na}). The first pulse activated Ca-influx through channels (I_{Ca}), the second pulse the Ca-influx through Na:Ca-exchanger. Repolarization to −80 mV ('diastole') turned Na:Ca-exchanger into the Ca-extrusion (tail current I_{ex}).

Before shock-freezing, the pulse-pairs were applied at least ten times. Potentiation of shortening was monitored on the TV-screen and recorded on tape off line. The extent of shortening was evaluated; it peaked about 120 msec after start of the first pulse to +5 mV ('systole') to 10% (corresponding to a minimal sacromere length of 1.64 μm). Potentiation went along with an enlargement of the tail current I_{ex}, as can be expected from the enhanced cellular Ca load. In this way the functional state was characterized by electrophysiological and mechanical parameters. Cells were discarded if they responded to the paired pulsing with aftercontractions and transient inward currents (at −80 mV) suggestive of Ca-overload.

4.4 Shock-freezing: timing and possible artefacts

Shock-freezing was timed at the end of diastole (−80 mV) or at different times during systole (+5 mV). During the events that led finally to shock freezing (see below) the position of the preparation was kept absolutely still, and its surround was moved instead. 8-bar air pressure applied to pneumatic pistons slid the microscope backwards by 55 mm, thereby removing the preparation from the chamber. The movement did not disturb the electrical recordings (*Figure 6*). Recording in air was possible, since the preparation still contained the silver holder as reference electrode. The vertical movement of the coolant started 40 msec before the sliding table finally came to a stop. The preparation came in contact with the coolant 13 msec later and was shut for 20 msec in the 50-mm high layer of stirred supercooled propane

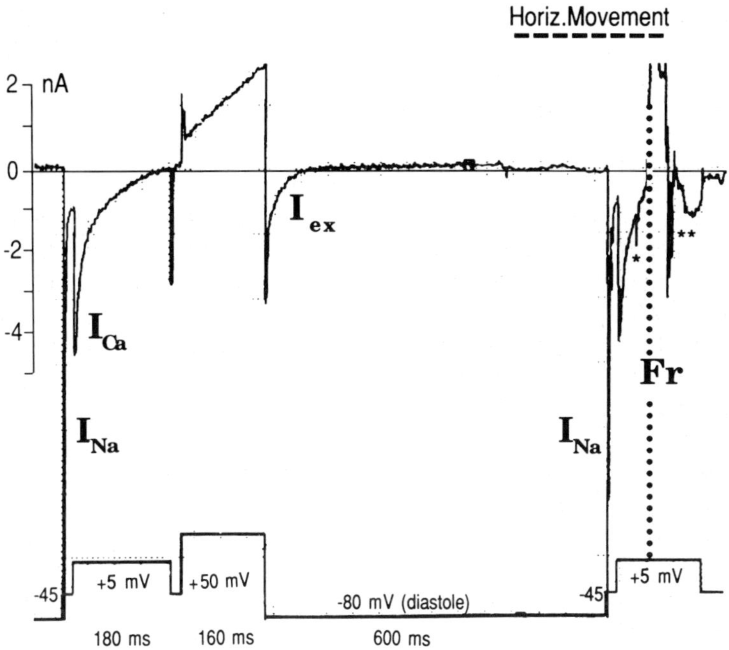

Figure 6. Net membrane current during shock-freezing. The lower panel shows the voltage-clamp protocol. During diastole, the holding potential is −80 mV. At a rate of 1 Hz, paired voltage-clamp pulses are applied. The first pulse starts with a short 20 msec depolarization to −45 mV that activates and inactivates a TTX-sensitive Na current (label I_{Na}, peak out off scale). Next, an 180 msec pulse to +5 mV activates and inactivates an inward Ca current (label I_{Ca}). After a 20 msec repolarization to −45 mV, the second pulse depolarizes the membrane for 160 msec +50 mV and induces a large outward current mostly carried by K ions. The final repolarization to −80 mV induces a large tail current due to electrogenic Na:Ca-exchange (label I_{ex}). When the preparation is moved horizontally, out of the experimental chamber and into the air, no change in membrane current is seen, and the pulse-protocol evokes similar I_{Na} and I_{Ca} as before. The operation of the magnetic valve starting the vertical movement of the piston is seen as a small capacitive current (label *). Contact with the coolant (dotted line, label Fr) is indicated by a capacitive artefact, followed by a large outward current. The abrupt stop of the movement induces vibrations which cause large current deflection (**). At the end, the current is zero. The recordings were taken from a cell superfused with 36 °C warm bath solution containing 2 mM $CaCl_2$.

(−195 °C). When the horizontal movement stopped abruptly, mechanical vibrations caused large artefacts, after which the current was zero. Then, we lifted the electrode, removed the silver holder from its clip and let it fall through the propane on to the bottom of the container. The container was removed from the piston, and the preparation stored in frozen propane until further cryocutting and analysis. At this stage of development, the success rate is about 50%. Some cells died when the giga-Ohm seal broke during the

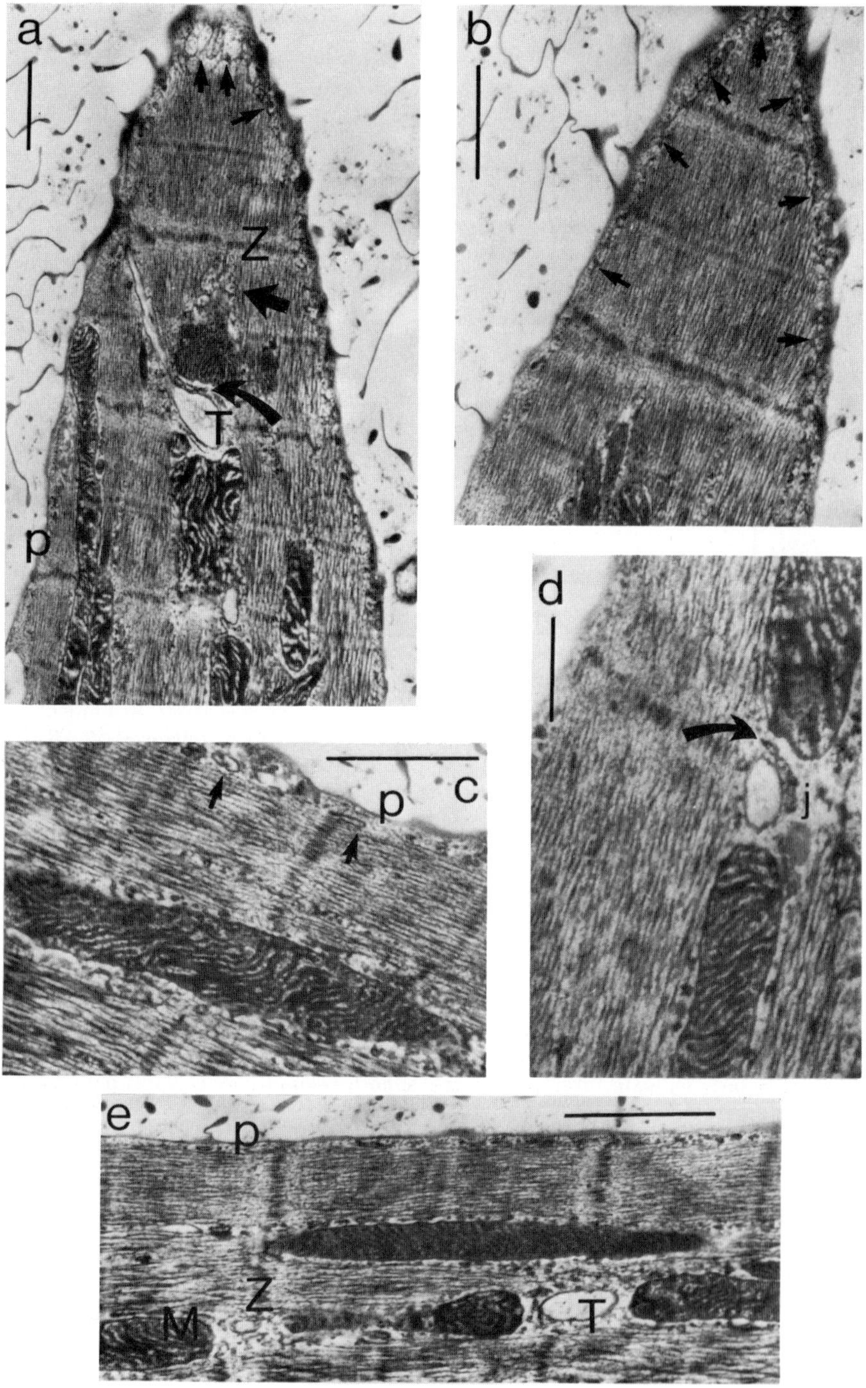
a
Z
T
p
b
d
j
p
c
e
p
Z
M
T

different phases of the cell transport. Other failures resulted when the common meniscus broke into one for the bath and one for the removed preparation.

The movement was timed by the computer through controlling the opening of magnetic valves. The pre-step from −85 to −45 mV did not trigger contraction, therefore start of depolarization to +5 was used as 'zero time' for systole. Since the pneumatic pistons had to move a mass of approx. 3 kg, the moment of contact with the coolant scattered by about 15 msec. Fortunately, the recorded membrane current monitors this contact (*Figure 6*); thus, the exact timing during systole could be evaluated a posteriori. For statistical analysis of systolic [Ca], data were summarized in groups between 15–25 msec, 30–45 msec and 60–120 msec after start of depolarization.

4.5 Freeze-substitution as control for quality of shock-freezing

In highly hydrated biological samples, the freezing rate is slowed by the insulating properties of water, which retards the diffusion of heat. Therefore, even under optimal conditions, only a 4–10-μm thin superficial layer remains free of ice-crystal damage. We used absence of cryoartefacts as an indicator of the quality of cryofixation. Freeze substitution followed by thin sectioning yielded images with sufficient contrast uniform throughout the 12-μm-thick cell; electron micrographs from a representative cell are shown in *Figure 7*. Cryoartefacts manifested themselves by disturbing the orderly geometry of the contractile material and consisted of clefts of increasing size, usually oriented along the longitudinal cell axis (lower part in *Figure 7*, c). The mitochondrial membranes became annular, and their interior became foamy in appearance. The glass electrode did not limit the quality of freezing. The absence of ice-crystals suggests that the rate of freezing was >10 000 °C/sec (8). Most of our shock-frozen cells showed a 2- to 5-μm-thick zone of minimal

Figure 7. Longitudinal sections of a ventricular myocyte shock-frozen and freeze substituted. p, sarcolemma. M, mitochondria. j, junctional SR. T, T-tubules, Z, *z*-line. (a) Minimal cryoartefacts on the left. Some distortion of myofilaments on right which are first signals of ice crystals damage (cryoartefacts). Note absence of a glycocalix on external site of plasmalemma. Bar, 1 μm; curved arrows = junctional SR; short arrows = subsarcolemmal SR; thick arrow = free network of SR. (b) Better frozen region of cell. Note on left regular pattern of thick and thin filaments, indicating absence of ice-crystal formations during freezing. Arrows indicate subsarcolemmal vesicles of SR. Bar, 1 μm. (c) Small subsarcolemmal vesicles of SR are clearly recognizable (arrows). Note well-preserved arrangement of myofilaments and structure of *z*-lines. Bar, 1 μm. (d) Junctional SR, indicated by curved arrow, is present at one side of T-tubules, Bar, 0.2 μm. (e) Minimal cryoartefacts in superficial zone indicated by perfectly parallel arrangement of filaments and preservation of superficial mitochondria. Note the sharp border at sarcolemma. Bar, 1 μm. (From Wendt-Gallitelli and Isenberg, *American Journal of Physiology*, **256**, H574–83, 1989, with permission.)

cryoartefacts. The 5-μm width of the zone is of the same order of 2–5 μm as in other preparations (9).

In regions where cryoartefacts were undetectable (*Figure 7*, b, c, e), the arrangements of the myofilaments seemed to be comparable to that known from glutaraldehyde-fixed specimens. The thick and thin filaments ran in a regular pattern, and cross-bridges with their periodicities were resolved. The architecture of the *z*-lines seemed to be undisturbed. Also, unit membranes of different organelles were well resolved. The SR was essentially similar to the corresponding images from conventionally fixed specimens. The isolated cells showed T-tubules with adjacent junctional SR (*Figure 7*, a, d). The sarcolemma was faced with vesicles of peripheral junctional SR. The interior of the SR seemed to be filled with an electron-dense material.

5. The freeze-dried cryosections

5.1 Cryosectioning

The silver holder was mounted and fixed under liquid nitrogen in an aluminum holder constructed for the cryoultramicrotome (see *Figure 3*, B). The specimens were cut at temperatures between −140 and −160 °C as thin as possible; that is, 100–150 nm. The sections included part of the silver holder as well as the bath solution surrounding the cell. The cryosections were transferred by long, thin white hairs of dalmation dogs from the knife edge to the dry copper grid (50 meshes) which had been covered with a thin pioloform film and carbon-coated (≤4 nm) previously. It was important to carbon-coat the grids *immediately before* cryocutting, otherwise the extremely small sections were blown away from the grid by electrostatic charges. The grid was covered with another grid prepared identically. The sandwich with the cryosections was transferred to a heavy brass container cooled with the liquid nitrogen cooling the cryochamber. At the end of the cryocutting day, the cooled grid-container was transferred into the apparatus, where freeze-drying was performed overnight at 10^{-6} Torr. After freeze-drying, the lid of the grid container was removed and the sandwiches were carbon-coated for final analysis.

5.2 The freeze-dried section in the electron microscope

Figure 8 shows cryosections of ventricular myocytes. The cooled and covered sections cannot be chemically stained, neither before nor after freeze-drying; the contrast of unstained cryosections is correspondingly low. Nevertheless, one can identify mitochondria and distinguish them from the myofibrils. In some sections, *z*-lines, I-, and A-bands are recognizable, and also T-tubules to which the dense membraneous structures of the junctional SR sometimes appose.

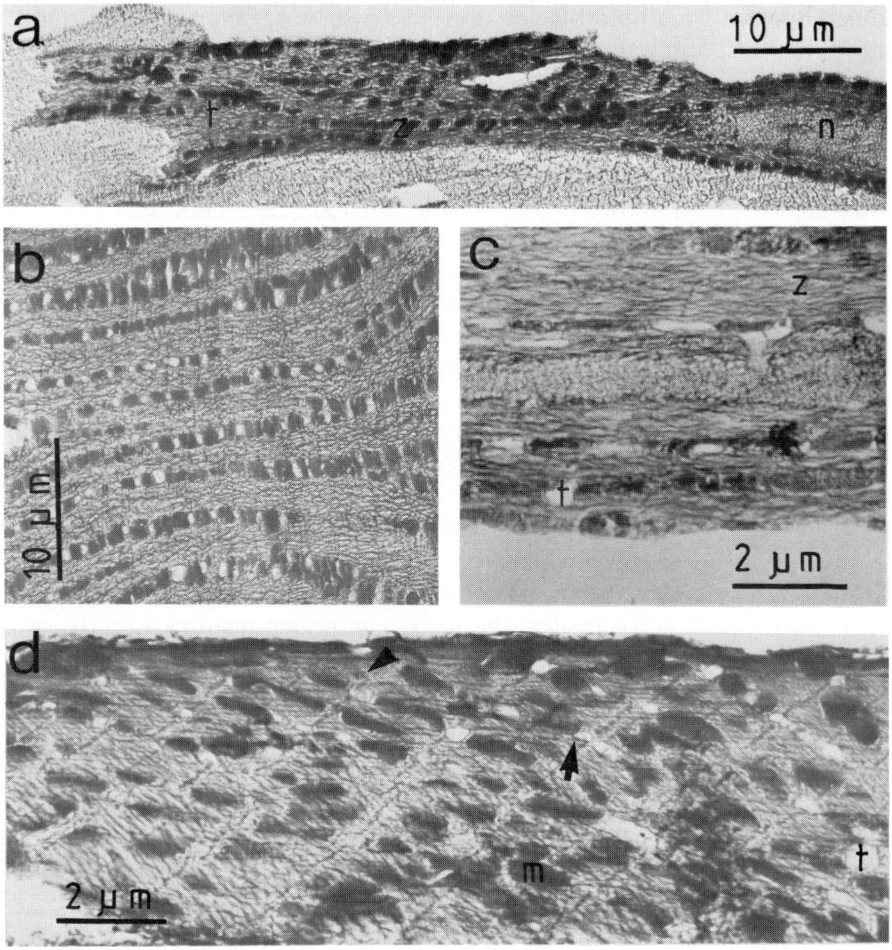

Figure 8. Cryosections in diastole and systole. (a) Low magnification, part of a longitudinal cryosection obtained from a myocyte frozen during diastole. Nucleus (n), T-tubules (t), sarcomeres with *z*-lines and mitochondria are recognizable. The cell is surrounded by the frozen bath solution. (b) Similar magnification of a myocyte frozen during systole (40 msec). The cryosection is not longitudinal, since the cell, by contracting, moves its borders from the horizontal plane of the pioloform film. Therefore, structures such as *z*-lines and junctional SR become difficult to recognize. (c) and (d) Longitudinal cryosections of myocytes frozen during diastole. At higher magnifications, junctional SR (arrow) and corbular SR (arrowhead) can be recognized in better frozen regions of the section.

The contrast of an image from an unstained section results from the fact that the water content of morphological cell compartments, such as cytoplasm, mitochondria, and other membrane-bound organelles, differs (generally, the 5-nm lipid bilayers have a lower water content). By freeze-drying, the different water content is transformed in a different thickness. To illustrate these changes, we consider a 120-nm-thick hydrated cryosection (*Figure 9*, A). Most likely, freeze-drying does not significantly modify the lateral extensions of the cell. (For example, the diastolic sarcomere length measured in the cryosection between two *z*-lines is 1.85 μm, which is undistinguishable from the 1.85 μm in the isolated living myocyte.) Thus, we postulate freeze-drying to essentially reduce the thickness of the section. Since freeze-drying removes the water quantitatively, reduction of thickness is proportional to the water content of the compartment before freeze-drying. In the *mitochondria*, the water content is about 67% (10). Thus, freeze-drying would reduce a 120-nm section to 40 nm (*Figure 9*, B1). The adjacent *myofibrillar space* (myoplasm), where the water content is 75% (11) would shrink from 120 nm to 30 nm. The water content of the *sarcoplasmic reticulum (SR)* is not known. However, the system of 40-nm-wide tubules (12) bears two unit membranes which may shrink less. Thus, the section including SR-tubules is thought to have a slightly larger thickness than the myoplasm (e.g. 33 nm).

Unknown hydration values, as in the SR, can be estimated on the basis of the known mitochondrial water; the X-ray continuum measured in SR, or in

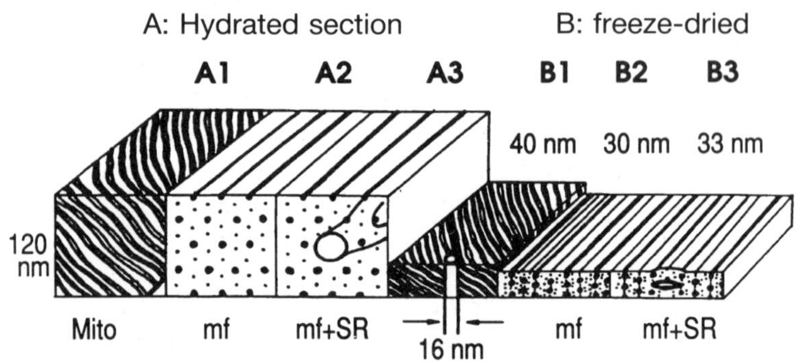

Figure 9. Schematic showing the volume-loss of a cryosection due to freeze-drying. **A**: The 120-nm-thick hydrated section goes through mitochondria (A1), myoplasm with myofibrils (A2) and myoplasm containing longitudinal SR (A3). **B**: Reduction of volume by loss of water is thought to reduce exclusively the height of the section. The mitochondrial water content is assumed to be 67%, here the section shrinks to 40 nm (B1). In the myoplasm, the water content is 75%, and the section shrinks to 30 nm (B2). The freeze-dried section has a different total mass in mitochondria and myoplasm which provides contrast in the electron microscope. When a tubulus of free SR is included in the myoplasm, lipid bilayers with low water content may shrink less than the myoplasm; however, the effect on total mass and on contrast is small (B3: 33 nm).

the A-band, was compared with the continuum in adjacent mitochondria using the same dose of electrons (i.e. the same count time and beam intensity). For example, the continuum was 10 700 counts in the mitochondria, 9000 in the SR, and 8000 in the myofibrillar space. In reference to 67% water in the mitochondrial volume, we estimate for total Ca in the mitochondria a conversion factor 0.24 ± 0.02. For the SR the conversion factor was 0.28 ± 0.07; the large S.E.M. is probably due to the fact that the lumen of the SR can be collapsed (high continuum from the lipids) or open (low continuum) (19).

In the microscope, the beam electrons are absorbed according to the mass, e.g. thicker parts of the section absorb the electron to a larger extent. The space inside the transversal tubules is 'empty'. Absorption by mitochondria is higher than by the myoplasm. In the case of the SR-tubules, possible differences in thickness are low: too low to reveal the border of these structures. Thus, the difficulty in morphological SR-identification superimposes to the problem of Ca-quantification (Section 2.4). The problems have largely hampered quantification of [Ca] in the network of free SR.

5.3 Water loss and definition of concentrations

EPMA measures the concentrations in mmol/kg dry weight (d.w.) of the analysed compartment. Most of us are used to considering concentrations given as molarity (mM, related to the volume of the compartment) or molality (related to the kilograms of water in the compartment; mmol/kg water). Interconversion of these concentrations would be possible if the water content were known.

As a matter of fact, the water content of the analysed compartment is not known. In principle, it could be quantified by EPMA: hydrated cryosections, transferred into the microscope, have to be measured in regard to the continuum. In the hydrated section, the bremsstrahlung originates largely from H and O atoms. After freeze-drying in the vacuum of the microscope, this component should have disappeared and reduced the continuum when it is measured again. Thus, the comparison of the continuum before and after freeze-drying can quantify the original water content (2). In heart ventricular cells, these measurements are missing up to now. Instead, we have to trust in analytical estimates of 75% intracellular water (11) and 67% for the mitochondrial water (10). This uncertainty superimposes when we transform the concentrations into molarity or molality.

5.4 Elemental composition of the compartments (diastole)

In the 'whole-cell configuration', the patch-electrodes have access to the intracellular space. The diffusional contact through the 2-μm-wide tips is thought to equilibrate electrolytes of the electrode solution with those in the myoplasm. The efficacy of such a 'cell dialysis' can be tested by a comparison between the concentrations. The electrode solution was composed of 140 mM KCl, 5 mM Na_2ATP, 5.5 mM $MgCl_2$, 20 μM EGTA, 10 mM Hepes, 5 mM KOH, pH 7.4; since there were no macromolecules, molarity (mM) and molality (mmol/kg water) are equal.

The *myoplasm* occupies 58% of the volume of rat ventricular cells (13), and the myofibrils 46.7 vol.%. When the myoplasm was analysed by EPMA, the electron beam (16 nm diameter) was located on the overlapping myofilaments (A-band). The elemental concentrations were homogeneous, i.e. those measured near the tip of the electrode did not significantly differ from others analysed 50 μm away at the cell edge. The myoplasmic K-concentration was 444 ± 22 mmol/kg d.w. Diffusion distributes K in the myoplasmic water, hence, the concentrations were transferred to molality. Since in the myoplasm, 3 kg water goes along with 1 kg d.w. (see *Figure 9*), the 444 mmol/kg d.w. were divided by three, which results in a K molality of 148 mmol/kg myoplasmic water, a concentration which corresponds nicely to the 145 mM K^+ in the electrode solution. Similarly, for Na, the 43 ± 5 mmol/kg d.w. corresponds to 14 mmol/kg myoplasmic water, which is close to the 12 mM Na^+ in the electrode. The similarity suggests that K and Na do equilibrate under the given conditions. For chlorine, the 152 ± 19 mmol/kg d.w. is converted into 51 mmol/kg myoplasma water. This concentration is only one-third of the Cl concentration in the electrode solution, indicating that Cl does not freely equilibrate. The measured 36 ± 2 mmol/kg d.w. Mg is transformed to 12 mmol/kg myoplasma water, a value more than twice as high as the 5.5 mM in the pipette. The 12 mmol/kg myoplasma water can be compared with 0.5 mM free $[Mg^{2+}]_i$, suggesting that 95% of total myoplasmic Mg is bound.

In *Mitochondria*, EPMA measured for K, Na, Cl, and Mg concentrations that appeared to be only 60% of those in the myoplasma. However, the water content is lower is mitochondria than in myoplasm (see above). Transformation into molality (mmol/kg mitochondrial water) brings the concentrations of K from 278 ± 16 mmol/kg d.w. to 132 mmol/kg water, of Na from 25 ± 5 mmol/kg d.w. to 14 mmol/kg water, of Cl from 103 ± 12 mmol/kg d.w. to 51 mmol/kg water, and of Mg from 21 ± 1 mmol/kg d.w. to 10 mmol/kg water. Thus, in terms of molality, mitochondria and myoplasma contain those elements such as K, Na, Mg, and Cl in nearly identical concentrations. However, the [Ca] (0.6 mmol/kg mitochondrial water) is lower and the [P] (203 mmol/kg mitochondrial water) is higher in mitochondria, about twice as high as in the myoplasm (0.9 and 113 mmol/kg myoplasmic water, respectively).

For the *junctional SR*, the water content is not known, but it can be estimated (p. 150). When we compare the elemental concentration in mmol/kg d.w. between jSR and myoplasm, the jSR had a high concentration of sodium (73 ± 8 mmol/kg d.w. or 27 mmol/kg SR water), phosphorus (388 ± 13 mmol/kg d.w. or 144 mmol/kg SR water), and chlorine (203 ± 13 mmol/kg d.w. or 75 mmol/kg SR water), whereas the concentration of sulphur (284 ± 15 mmol/kg d.w. or 101 mmol/kg SR water) was lower than in the myoplasm. The concentration of magnesium (35 mmol/kg d.w. or 13 mmol/kg SR water) and potassium (428 mmol/kg d.w. or 158 mmol/kg SR water) did not differ.

6. Time-course of Ca during the cardiac cycle

In this last section, we give an example for Ca-analysis by EPMA. The example gives the total [Ca] and its change during the contraction of a guinea-pig ventricular myocyte, potentiated by a 1 Hz paired voltage-clamp (see Section 4.3). The potentiated contraction might be compared with the *in vivo* situation (14) where the heart beats at a frequency of 3 Hz or more.

6.1 The myofibrillar space

At the *end of diastole*, the myoplasm contained a total [Ca] of 2.6 ± mmol/kg d.w. ($n = 62$, 19 cells). The number can be connected into a *molality* of 0.87 mmol/kg myoplasmic water. We are confident that this [Ca] is not contaminated by [Ca] of i.c. SR compartments. It is extremely unlikely that a cryosection longitudinal to the sarcomeres would have crossed the compartment of free SR. Further, any membrane-bearing compartment would have distorted the texture of the myofilaments seen in the cryosection. Finally, SR-[Ca] should decrease during systole, whereas a systolic increase in myofibrillar [Ca] was measured.

In identical experiments, we measured $[Ca^{2+}]_i$, the concentration of the Ca ionized in the intracellular water, by microspectroflourimetry (indo-1) to 0.18 μM (15). The comparison of total [Ca] with $[Ca^{2+}]_i$ indicates that 99.97% of total [Ca] is not free but bound. The ratio of bound to free Ca is about tenfold higher than <99% which Fabiato estimated from the published intracellular Ca-buffers (16). The discrepancy highlights how important it is to measure the total [Ca] directly. In the myofibrillar space, Ca is mostly bound to proteins that occupy a significant fraction of the analysed volume; therefore it seems to be better to relate the total [Ca] to the volume where the concentration is analysed—that is, the total [Ca] of 2.6 mmol/kg d.w. should be transformed it to a *molarity* of 0.65 mM (division by 4-litre vol. per kg d.w., see *Figure 9*). As ligands binding Ca in the myofibrillar space, TnC can be excluded since the cells were relaxed, calmodulin (25 μM) and Ca, Mg-sites of myosin (150 μM per litre compartment) remain to be considered. Both ligands together can bind 0.4 mM Ca. Therefore, an additional yet unknown Ca ligand is suggested.

During *systole*, myoplasmic [Ca] increased. 20 msec after start of the clamp step to +5 mV, total [Ca] was 5.5 ± 0.3 mmol/kg d.w. or 1.4 mM. 20 msec later it was 4.6 ± 0.5 mmol/kg d.w. (1.15 mM), and at the time of peak contraction (120 msec after start of depolarization) total [Ca] had fallen to 3.1 mmol/kg d.w. (0.8 mM). $[Ca^{2+}]_i$ had a similar time-course, i.e. reached its peak of 0.9 μM within 20 msec and rapidly decayed afterwards. Between diastole and the

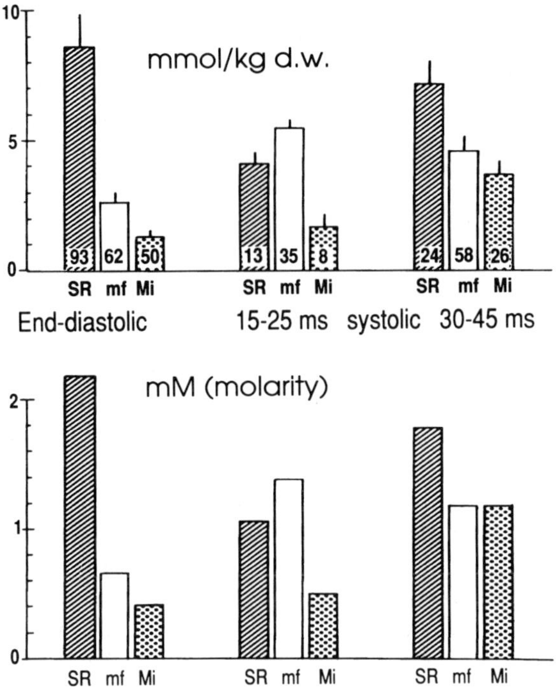

Figure 10. Time-course of total Ca concentration during diastole and different periods of systole. Analysis of Ca in the myofibrillar space, the mitochondria and the sarcoplasmic reticulum (level of the *z*-lines). Mean values from 30 myocytes (± SEM; number = no. of spectra) shock-frozen after potentiation with the paired-pulse protocol (*Figure 6*). The number of spectra is indicated. **A**: total [Ca] in mmol/kg d.w. *Bottom*: The concentrations of total [Ca] have been converted to mmol per l analysed compartment.

initial 20 msec of systole, total [Ca] increased by 2.9 mmol/kg d.w. or 0.72 mM. This increment is about 30 times higher than can be expected from Ca-influx through channels, evidencing the necessity of Ca-release from i.c. stores (SR). Due to excessive binding, the increment in total Ca is about 1000 times higher than the increment in free $[Ca^{2+}]_i$. The ratio of the increment of total to the increment in free $[Ca^{2+}]_i$ suggests a 'Ca buffering power' of 1 mM for pCa unit (15). The analysed myofibrillar space (A-band), contains about 300 μM TnC, enough to bind up to 900 μM Ca. It is interesting to note that total [Ca] falls at a time when the shortening of the myocyte still continues. Hence, time to peak of contraction seems to be controlled, not by the period during which Ca has bound to TnC, but by the kinetics of cross-bridge cycling.

The above results demonstrate that EPMA can evaluate a 'time-course' of total myofibrillar [Ca]. In electrophysiological experiments or experiments with fluorescent Ca-probes a time-course can be measured easily and from the same cell. With EPMA, the effort of a time-course is tremendous. Since shock-freezing finishes the experiment, a next cell is required for each point of time. Due to biological scattering between the different myocytes data from several cells have to be averaged for a significant conclusion. The new information on total [Ca] bound locally to the myofilaments and total [Ca] stored in compartments seems to justify the effort.

6.2 SR-Ca-content

During diastole total SR [Ca] was 8.6 ± 1.3 mmol/kg d.w. The same cells had total myoplasmic [Ca] of 2.6 ± 0.4 mmol/kg d.w. This difference is statistically significant. Thus, during diastole, there is a significant Ca-gradient between the SR and the myofibrillar space, as expected for a Ca-loaded release-compartment. 20 msec after start of depolarization, during early systole, total SR [Ca] had fallen to 4.1 ± 0.4 mmol/kg d.w., whereas total myoplasmic [Ca] had increased to 5.5 ± 0.3 mmol/kg d.w. These differences in [Ca] between SR and myoplasm are no longer significant. That is, the gradient between the SR and the myoplasm had disappeared within the short period of 20 msec. The rapid fall of stored Ca is expected for a compartment that 'releases' its Ca. 20 msec later, SR [Ca] has already increased to 7.2 ± 0.9 mM as if Ca, rapidly unbound from TnC, were sequestered into the SR again.

Qualitatively, our results support the identification of the SR (analysed on the level of the *z*-lines) with functional 'release-compartment'. They seem to suggest that the hypothesis of a separate 'uptake-compartment' (*Figure 1*) is not necessary. For quantitative interpretation, however, the diastolic SR [Ca] is too small in order to take account for the increment in total myoplasmic [Ca]. This can be demonstrated with a simple estimate: Due to a lack of knowledge we postulate that the decrease in SR [Ca], measured by us over the *z*-lines, holds true for both junctional and free SR. Then the SR volume fraction is 3% (9). The volume fraction, taken by the myofibrils where Ca is supposed to bind, is 47%. In such a way, the decrement in SR [Ca] of 4.5 mM would increment myofibrillar [Ca] by only 72 μM, whereas 720 μM was measured. In part, this discrepancy reflects the problem of analysing [Ca] in SR tubules and cisterns with dimensions smaller than the thickness of the cryosections (*Figure 9*, B3); during the diastole, when the SR-[Ca] is high but myoplasmic [Ca] is low, the SR [Ca] is probably underestimated (compare p. 138). Also, there is no morphometry on the volume fraction of guinea-pig ventricular SR. In addition, the question whether all parts of SR or only junctional SR releases activator Ca is not unequivocally answered either.

6.3 Mitochondrial Ca

In mitochondria, total [Ca] was always lower than in the myoplasm. At the end of diastole, total Ca was 1.3 ± 0.2 mmol/kg d.w. or 0.4 mM. This value is higher than the 0.5 ± 0.3 mmol/kg d.w. which we detected in previous studies on unstimulated myocytes (7). During early systole, 20 msec after start of depolarization, mitochondrial Ca increased to 1.7 ± 0.4 mmol/kg d.w. or 0.5 mM, a value not significantly different from diastole but significantly lower than the myoplasmic Ca measured at the same time. 20 msec later, mitochondrial [Ca] increased to 3.7 ± 0.5 mmol/kg d.w. or 1.2 mM. This value is about three times beyond the diastolic one (significance $p < 0.05$). It resembles the 1.2 mM total myoplasmic Ca measured simultaneously in the same cells. At the peak of contraction, mitochodrial [Ca] had fallen to near the diastolic values (1.7 ± 0.8 mmol/kg d.w. or 0.5 mM).

The result of mitochondria taking part in the beat-to-beat changes of i.c. [Ca] points again to the importance of shock-freezing being timed on a millisecond scale. That is, freezing the cells at the peak of contraction would not have discovered the large [Ca] change that occurs 30–45 msec after start of depolarization. The Ca-fluxes between myoplasm and mitochondria occur on a time-scale much faster than previously suggested (17), we suggest that in living cells the mitochondrial Ca-transport is enhanced by 'co-factors' of the surrounding myoplasm which may be lost during tissue fractionation. Since mitochondria occupy about 36% of the cell volume (13) they constitute a large Ca-buffer during the normal cardiac cycle, i.e. the mitochondria certainly contribute to the space to which Ca has 'rapid access'. The Ca-fluxes have a direction such that they cannot contribute to the activator Ca, instead they damp the myoplasmic Ca-transients. The changes in matrix-[Ca] are thought to modulate mitochondrial enzymatic activity (18).

References

1. Beuckelmann, D. J and Wier, G. (1988). *Journal of Physiology*, **405**, 233–55.
2. Gupta, B. L. (1989). In *Electron probe analysis*, Vol. 4 (ed. K. Zierold and K. H. Hagler), pp. 199–212. Springer Series in Biophysics. Springer-Verlag, Berlin and Heidelberg.
3. Somlyo, A. V., Shuman, H., and Somlyo, A. P. (1989). *Methods in enzymology*, Vol. 172 (ed. S. Fleischer and B. Fleischer), pp. 203–29. Academic Press, Orlando, Florida.
4. Kitazawa, T., Shuman, H., and Somlyo, A. P. (1983). *Ultramicroscopy*, **11**, 251–62.
5. Hall, T. A. (1979). *Journal of Microscopy*, **117**, 145–63.
6. Wendt-Gallitelli, M. F. (1986). *Basic Research in Cardiology*, **80**, Suppl. 1, 25–32.
7. Wendt-Gallitelli, M. F. and Isenberg, G. (1989). *American Journal of Physiology*, **256**, H574–83.

8. Heuser, J. E., Reese, T. S., Dennis, M. J., Jan, L. I., and Evans, L. (1979). *Journal of Cell Biology*, **81**, 275–300.
9. Nassar, R. Wallace, N. R., Taylor, I., and Sommer, J. S. (1986). *Scanning Electron Microscopy*, **1**, 309–28.
10. Werkheiser, W. C. and Bartley, W. (1957). *Biochemical Journal*, **66**, 79–91.
11. Polimeni, O. I. (1974). *American Journal of Physiology*, **227**, 676–83.
12. Page, E. and Upshow-Early, J. (1977). *Circulation Research*, **40**, 355–67.
13. Page, E. (1978). *American Journal of Physiology*, **235**, C147–58.
14. Bond, M., Jaraki, A. R., Disch, C. H., and Healy, B. P. (1989). *Circulation Research*, **64**, 1001–12.
15. Wendt-Gallitelli, M. F. and Isenberg, G. (1990). *Pflügers Archiv für die gesamte Physiologie*, **415**, R50.
16. Fabiato, A. (1983). *American Journal of Physiology*, **245**, C1–14.
17. Carafoli, E. (1985). The homoiostasis of Ca in heart cells. *Journal of Molecular and Cellular Cardiology*, **17**, 203–13.
18. McCormack, J. G., Browne, H. M., and Dawes, N. J. (1989). *Biochimica et Biophysica Acta*, **973**, 420–7.
19. Wendt-Gallitelli, M. F. and Isenberg, G. (1991). *Journal of Physiology*, **435**, 349–72.

7

Ca^{2+} buffering: computer programs and simulations

ALEXANDRE FABIATO

1. Introduction

The free calcium concentration ([free Ca^{2+}]) inside most animal cells is in the nanomolar (nM) to micromolar (μM) range. When one wants to make up a solution simulating the intracellular milieu, it is not possible to obtain such a low [free Ca^{2+}] by simply adding a small amount of calcium to the solution.* In fact, the total calcium contamination of the other salts contained in the solution would be in excess to the [free Ca^{2+}] that should be obtained. Thus, it is necessary to use Ca^{2+} buffers that are made of a ligand in equilibrium with a metal, which in this case is Ca^{2+}. Each complex between a metal and a ligand presents an equilibrium of the type:

$$M + L \rightleftharpoons ML$$

where M is the free metal, L the free ligand, and ML the complex. Such an equilibrium is characterized by a stability constant, or binding constant, or association constant, K_{ML} defined as:

$$K_{ML} = [ML]/[M] \cdot [L]$$

where [ML], [M], and [L] are the concentrations of the complex, the free metal, and the free ligand respectively.

A first degree of complexity is because these ligands which bind Ca^{2+} also bind H^+, so that instead of having a single complex between a given metal and a given ligand, several complexes are formed corresponding to various degrees of protonation. Each of these complexes has its own stability constant. Thus, apparent stability constants have to be calculated to characterize the multiple equilibria among one metal, one ligand, and the H^+ ion at a given pH value.

* All throughout this article, calcium is written in full, or abbreviated as Ca without ionic charges in the equations, whenever it is necessary to emphasize that total rather than free, i.e. ionic, Ca^{2+} is referred to.

A second degree of complexity is because the solution also contains other metals and ligands. Some of them are required for maintaining the free concentrations of other ions such as Mg^{2+} that are also important for physiological experiments. Thus, two or more of the metals contained in the solution bind to a given ligand and, conversely, more than one of the ligands bind to a given metal. Accordingly, equations for multiple equilibria should be solved to calculate the total concentrations of metals and ligands necessary to obtain specified free metal concentrations.

The solutions used for most experiments must have not only specified free concentrations of metals but also a specified ionic strength. The ionic strength ($\Gamma/2$) is defined as:

$$\Gamma/2 = \Sigma C_i z_i^2/2$$

where C_i is the concentration of the ion, and z_i is the charge of the ion. The calculation of $\Gamma/2$ has to take into account all the ionic species corresponding to all the equilibria among ligands, metals, and H^+ ions. Thus, complex calculations are needed for computing not only the total concentrations of metals and ligands in a solution but also the $\Gamma/2$ of this solution.

In other cases, instead of maintaining $\Gamma/2$ constant, the total concentration of a major monovalent cation has to be maintained constant. Alternatively, some investigators prefer to maintain the ionic equivalent (I_e) which is similar to the ionic strength but does not take into account the valence of these ions (1). Thus,

$$I_e = \Sigma C_i/2.$$

A major problem in interpreting the studies using Ca^{2+} buffers is that there is a considerable amount of disagreement in the literature over the stability constants for the complexes between the most commonly used metals and ligands. To evaluate this problem quantitatively, one needs to be able to do the operation opposite to that done in calculating the composition of a solution; i.e. to compute the free metal concentrations from specified total concentrations of metals and ligands. It is also necessary to do this computation from total concentrations to free concentrations for the calculation of the free ionic concentrations that are likely to exist within intact cells.

For the past fifteen years, we have developed calculator and computer programs that accomplish these calculations. Programmable calculators and computers permit extensive use of loops and tests of comparison, which makes it possible to use successive approximations for the calculations. This permits many operational calculations for which equations could not be solved explicitly. First, we reported these calculations on a hand-held calculator, the Texas instrument TI-59 (2). Subsequently, I have reported a Fortran version of the same program using an IBM-PC (3), as well as some ancillary programs written in BASIC and Pascal. These programs are currently used by a large number of investigators who kindly gave me extensive feedback that permitted many improvements of these programs. Dr Rashid Nassar from Duke University and Dr Michael R. Berman from Johns Hopkins University are two examples of colleagues who provided me with much help and feedback, for which I am most grateful. Accordingly, the first aim of this article is to update these computer programs. What is changed is not, of course, the basic computation but mostly the computation of the stability constants for which new measurements or corrections are always available. In addition, some colleagues have kindly developed computer programs inspired by our programs and, accordingly, I shall briefly mention the availability of these programs for the benefit of interested users.

Initially, the first Ca^{2+} buffer used was ethylenediamine tetraacetic acid (EDTA). It binds Mg^{2+} as well as Ca^{2+}. This explains the interest in ethyleneglycol-bis(*β*-aminoethyl ether)-*N*,*N*,*N'*,*N'*-tetraacetic acid (EGTA) which binds Ca^{2+} much more strongly than Mg^{2+} (4). Indeed, EGTA has been used in most of the recent experiments in which a low and controlled [free Ca^{2+}] was needed. However, EGTA presents some difficulties. First, it binds H^+ ions, which may be a major disadvantage: thus, when there is an increase of [free Ca^{2+}] resulting, for instance, from a Ca^{2+} release, Ca^{2+} displaces the H^+ ion from EGTA. This results in a lowering of the pH that may not always be buffered by the pH buffer. The lowering of pH in turn decreases the apparent affinity constant of EGTA for Ca^{2+} and, thereby, causes the [free Ca^{2+}] to be higher than expected. A second major difficulty with EGTA is its low on-rate, and extremely low off-rate, of binding to Ca^{2+} (5–7). This presents major problems in experiments which require a rapid mixing of two solutions at two [free Ca^{2+}] values in the aim of reaching an intermediary [free Ca^{2+}]. Thus, a computer simulation is reported to show the transient surge of [free Ca^{2+}] preceding the reaching of the final steady-state concentration which is observed during such a rapid mixing (8). This constitutes Section 2 of this chapter.

Recently, Dr Roger Tsien (9) has described new Ca^{2+} buffers which are highly selective for Ca^{2+} against not only Mg^{2+} but also H^+ and which have much higher on- and off-rates of Ca^{2+} binding than EGTA. The advantages of these new Ca^{2+} buffers are reviewed in Section 3 of this chapter. An example of their use is demonstrated by a method for measuring the rate of (total) calcium release from the sarcoplasmic reticulum of skinned fibres (i.e. of muscle fibres from which the sarcolemma has been removed).

Finally, Section 4 describes a new class of compounds. These are caged compounds (10) that permit a rapid increase (or decrease) of [free Ca^{2+}] because their affinity for Ca^{2+} suddenly decreases (or increases) upon application of a flash of light of a sufficient intensity. The potential uses and limitations of these caged compounds are described in this fourth section of the article.

2. Update on the stability constants and the computer program

2.1 Update on the stability constants

Recently, Drs Simon M. Harrison and Donald M. Bers (11) have described an improved method for adjusting the H^+ and Ca^{2+} association constants of EGTA for temperature and Γ/2. They used a semi-empirical Debye–Hückel formalism for the Γ/2 adjustment, and Van't Hoff isochores for the temperature adjustment of the stability constants. They verified by Ca^{2+} electrode measurements that their results closely described the influence of Γ/2 and temperature on the apparent stability constant of the CaEGTA complex. This appears to be the best correction for temperature and Γ/2 currently available. Thus, after careful consideration I thought that it was necessary to incorporate this method of correction of the stability constants into the ancillary program termed STACONS (3) which permits the calculation of the stability constants with appropriate corrections for temperature and Γ/2.

I am most grateful to Dr Donald Bers for communicating to me the empirical equations that he uses to describe the influence of Γ/2 for 0 to 0.5 M and temperature (T) for 0 to 40 °C on the activity coefficient of protons (γ_H). The association constants compiled by Martell and Smith (12) are stoichiometric and as such they use the H^+ concentration [free H^+] rather than the H^+ activity (10^{-pH}). This problem is discussed in detail on pp. 464 and 465 of ref. 13. Thus, for a given pH, it is crucial to use the appropriate γ_H which depends upon temperature and Γ/2. It must also be noted that these corrections require an accurate knowledge of the absolute stability constants and enthalpy of complexation (ΔH) as given in ref. 12.

The empirical equations that Dr Bers has derived from the data of Harrison and Bers (11) to describe the effects of Γ/2 and T on the H^+ and Ca^{2+} association constants of EGTA are the following:

$$\gamma_H = 1.45045 \cdot \exp(-B \cdot \Gamma/2) + 0.063546 \cdot \exp(-43.97704 \cdot \Gamma/2) + 0.695634$$

where

$$B = 0.522932 \cdot \exp(0.0327016 \cdot T) + 4.015942.$$

Dr Bers has found that these equations constitute a good approximation of the original data reported by Harned and Owen (14).

Thanks to the collaboration of Drs Harrison and Bers, these equations will be incorporated into the new version of what used to be my program STACONS that will be distributed. Then STACONS is no longer only mine but also to be credited to Drs Harrison and Bers. This modified program will not be accompanied by additional explanation, since the rationale of Harrison and Bers has been thoroughly explained in their article (11). However, if additional clarifications were required, these could be obtained from Dr Simon M. Harrison at the Department of Physiology, University of Leeds, Leeds LS2 9JT, UK, Dr Donald Bers at the Department of Biomedical Sciences, University of California, Riverside, Riverside, CA 92521–0121, USA, or Alexandre Fabiato at the Department of Physiology, Medical College of Virginia, Richmond, VA 23298–0551, USA.

2.2 New versions of the main computer program

As already indicated (3), Dr Rashid Nassar (Department of Physiology, Duke University Medical Center, Durham, North Carolina 27710, USA) has developed Pascal programs that run on an IBM PC and reproduce some of the programs reported in ref. 2. These programs are simple and very useful. They are recommended for investigators who would like to have a starting point to further modify the programs in Pascal. Other adaptations of these calculator programs by Dr Michael Berman and Dr Richard A. Meiss are also mentioned in ref. 3. Very worthwhile also is the text editor (AscEdit) that Mr Nassib Nassar has developed for the data input in SPECS, the main Fortran program described in ref. 3.

An even more complete work has been recently done by Dr Michel Vivaudou (Laboratoire de Biopysique Moléculaire et Cellulaire, CENG, 85X, Avenue des Martyrs, 38041 Grenoble, France. Tel: 76 88 48 67 FAX: 76 88 51 55). He has written a program that is an extensive PASCAL version of SPECS (the main computer program described in ref. 3, as previously said) usable on an IBM-PC or any IBM-PC-compatible computer. He has called the program ALEX. It contains a number of original features. ALEX is written in Turbo Pascal, version 5.5, whereas SPECS is written in Fortran. The Pascal language permits a more structural design which facilitates error detection and updates. Because of the use of a more modern language, the structure of the input file used in ALEX has been considerably relaxed so that the spacing between input data is not as critical as in SPECS. The only requirement is that names and numbers should be separated by at least one space. The names can contain either upper- or lower-case letters. The numbers can be entered in any format recognized by a common pocket calculator. Despite this relaxation, ALEX still recognizes correctly the SPECS input file format, which should allow for extensive cross-testing of the two programs. ALEX also calculates the volume of the stock solutions for making up the experimental solutions. This very valuable program is available with detailed instructions from either Dr Michel Vivaudou at the address indicated above or from Alexandre Fabiato at the Department of Physiology, Medical College of Virginia, Richmond, VA 23298–0551, USA. [Telephone: (804) 786–9563 FAX: (804) 371–7382.] We would suggest that new users request both SPECS and ALEX since they are both distributed free of any charge. This would help both them and us in the cross-testing of the two programs under a variety of conditions.

3. Computation of the instantaneous change of [free Ca^{2+}] during the rapid mixing of two solutions buffered at different free calcium concentrations

The effect of the low on- and off- rates of the binding of Ca^{2+} to EGTA on the equilibration of the [free Ca^{2+}] after rapid mixing of two solutions at different [free Ca^{2+}] values is demonstrated by the computer-generated curves shown in *Figure 1*. The algorithm for this computer program (about which Dr George D. Ford from my Department gave me some kind advice) is very general and can be useful for interpreting many experiments consisting in the rapid mixing of two solutions at different [free Ca^{2+}] values. However, the magnitude of the problem can only be demonstrated on a practical example. Thus, the computer simulation shown in *Figure 1* corresponds to a particular experimental situation. In this particular case (8) the aim was to obtain a progressive change of [free Ca^{2+}] from pCa 7.00 to pCa 6.15 (where pCa is $-\log_{10}$ [free Ca^{2+}]) by mixing two solutions at different pCa values in 10 msec. The solutions were assumed to contain 0.068 mM total EGTA and the

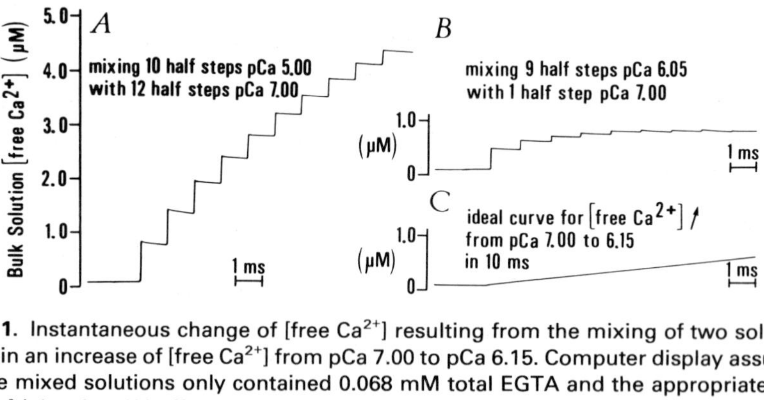

Figure 1. Instantaneous change of [free Ca^{2+}] resulting from the mixing of two solutions to obtain an increase of [free Ca^{2+}] from pCa 7.00 to pCa 6.15. Computer display assuming that the mixed solutions only contained 0.068 mM total EGTA and the appropriate [total calcium] (plus the pH buffer and KCl to adjust Γ/2 to 0.170 M). The computations assumed that the rate constants for binding of Ca^{2+} to EGTA were $K_{on} = 2 \times 10^6\ M^{-1}\ sec^{-1}$ and $K_{off} = 0.8\ sec^{-1}$. (Reproduced from part of Figure 13 in ref. 8, with permission of the *Journal of General Physiology*.)

appropriate [total calcium] to obtain the desired [free Ca^{2+}] values, plus a pH buffer and KCl to adjust $\Gamma/2$ at 0.170 M but no other divalent cation or Ca^{2+} buffer.

The solutions were delivered into the experimental chamber by microsyringes moved by stepping motors. The experiment started when a volume of solution at pCa 7.00 displaced by a given number of half-steps of one of the stepping motors had already been injected into the chamber. The calculations are not affected either by the fact that the motors were operated in the half-step mode or by the definition of the volume displaced by each half-step (52 picolitres). The experiment consisted of injecting, in the pre-existing volume of solution at pCa 7.00, a volume of solution at lower pCa. The number of half-steps of each of the stepping motors needed to inject the volumes of the two solutions in the chamber will be used to quantify these volumes; for instance, we shall refer to 10 half-steps of solution at pCa 5.00 rather than to 520 picolitres of this solution.

The ideal pattern increase of [free Ca^{2+}] would have been that shown in *Figure 1*, C, which assumed an instantaneous reaching of the steady-state equilibrium at the time of each half-step of microinjection of the high [free Ca^{2+}] solution. Under these conditions, a progressive increase of [free Ca^{2+}] from pCa 7.00 to pCa 6.15 would have been obtained. Instead, because of the low on- and off-rates of Ca^{2+} binding to EGTA, each half-step of microinjection of the high [free Ca^{2+}] solution produced an initial surge of [free Ca^{2+}] followed by a slow decline. The next half-step of microinjection occurred much before this decline was complete and there was summation of this effect for the subsequent half-steps of microinjection. Thus, at the end of the series of half-steps of microinjection of solution at higher [free Ca^{2+}], the [free Ca^{2+}] in the chamber was higher than the final steady-state value of pCa 6.15. After the end of the microinjection series, it took several tens of milliseconds for the final steady-state [free Ca^{2+}] to be reached. The amplitude of maximum surge of [free Ca^{2+}] above the steady-state value varied with the [free Ca^{2+}] level in the high [free Ca^{2+}] solution. Hence, computations indicated that a steady-state [free Ca^{2+}] of pCa 6.15 could be obtained either by mixing ten half-steps of solution at pCa 5.00 into a pre-existing volume of 12 half-steps of solution at pCa 7.00 (*Figure 1*, A) or by mixing nine half-steps of solution at pCa 6.05 with one half-step of solution at pCa 7.00 (*Figure 1*, B), but in the former case (*Figure 1*, A) the amplitude of the surge of [free Ca^{2+}] was far greater because the high [free Ca^{2+}] solution had a much higher [free Ca^{2+}] value than in the latter case (*Figure 1*, B).

The algorithm for this computer program gave the instantaneous [free Ca^{2+}] as follows:

$$[\text{free Ca}^{2+}] = \frac{(2K_{\text{on}}C + B - \lambda)\ (B + \lambda)\ \exp^{-\lambda t} - (2K_{\text{on}}C + B + \lambda)\ (B - \lambda)}{2K_{\text{on}}[2K_{\text{on}}\ C + B + \lambda - (2K_{\text{on}}C + B - \lambda)\ \exp^{-\lambda t}]}.$$

In this equation, t is the time in seconds, and K_{on} is the on-rate constant of Ca^{2+} binding to EGTA in units of litres per mol times reciprocal seconds (M^{-1} sec^{-1}). The constant λ is the pseudo-rate constant of this product of two exponentials (a function with a time-course resembling that of a tangent) and its value is given by:

$$\lambda = \{[K_{off} + K_{on} ([\text{total EGTA}] - [\text{total Ca}])]^2 + 4\ K_{off}K_{on}\ [\text{total Ca}]\}^{1/2},$$

where the total concentrations, represented by brackets, are in molar (M) units and K_{off} is the off-rate constant of Ca^{2+} binding to EGTA in reciprocal seconds (sec^{-1}). The constant B depends upon the on-rate constant (K_{on}), the off-rate constant and the total concentrations as:

$$B = K_{off} + K_{on}\ ([\text{total EGTA}] + [\text{total Ca}]).$$

C is a constant that depends on the [free Ca^{2+}] in the low and high [free Ca^{2+}] solutions and on the volume or number of half-steps of these two solutions ($n_{low} + n_{high}$) as:

$$C = ([\text{free } Ca^{2+}{}_{low}]n_{low} + [\text{free } Ca^{2+}{}_{high}]n_{high}/(n_{low} + n_{high}).$$

As previously indicated these equations are very general. However, for demonstrating how their application to the particular problem (8) taken as example resulted in the computer-generated curves shown in *Figure 1*, the following explanations are required. They explain how the values of K_{on} and K_{off} were selected for these particular conditions. Comparing the data obtained at pH 7.00 by Smith *et al.* (5) and pH 6.80 by Harafuji and Ogawa (7) showed that small changes of pH do not affect K_{on} appreciably. A Q_{10} was derived from the data of Smith *et al.* (5) at 25 ° and 8 °C. This permitted the inference of K_{on} at 22 °C and pH 7.10, which was $1.58 \times 10^6\ M^{-1}\ sec^{-1}$ from the data of Smith *et al.* (5) with cacodylate pH buffer and $1.64 \times 10^6\ M^{-1}\ sec^{-1}$ from the data of Harafuji and Ogawa (7) with Tris pH buffer, but 3.51×10^6 $M^{-1}\ sec^{-1}$ for the data of Harafuji and Ogawa (7) with imidazole pH buffer (there is as yet no explanation for this discrepancy). An intermediate value of $2.00 \times 10^6\ M^{-1}\ sec^{-1}$ was used for the computations shown in *Figure 1*. K_{off} was calculated to be 0.8 sec^{-1}, the quotient of K_{on} over the apparent affinity constant for the CaEGTA complexes at pH 7.10 and 22 °C ($3.986 \times 10^6\ M^{-1}$ sec^{-1}).

4. Use of a fast Ca^{2+} buffer to compute the rate of total calcium release from the sarcoplasmic reticulum of skinned fibres

This previously unpublished method for measurement of the rate of total calcium release from the sarcoplasmic reticulum of a skinned cardiac cell is based on the buffering of the change of myoplasmic [free Ca^{2+}] resulting from Ca^{2+} release from the sarcoplasmic reticulum by a fast Ca^{2+} buffer. This method, which is applicable to the measurement of the rate of total calcium release from the sarcoplasmic reticulum of skinned fibres of skeletal or smooth muscle and from the endoplasmic reticulum of permeabilized non-muscle cells, illustrates the usefulness of the fast Ca^{2+} buffers recently developed by Dr R. Y. Tsien (9), whom I thank for his kind advice. EGTA was not appropriate for this purpose because, as explained in the preceding section, its on-rate constant is low (2×10^6 M^{-1} sec^{-1}) and, even more importantly, because its off-rate constant is very low (0.8 sec^{-1}). At [free Ca^{2+}] around or below the dissociation constant (reciprocal of the association or stability constant defined in Section 1) of the Ca^{2+} buffer, which is $1 \times 10^{-6.60}$ M for the CaEGTA buffer at pH 7.10, the effect of the off-rate is predominant over that of the on-rate in determining the rate of Ca^{2+} buffering. One of the Ca^{2+} buffers described by Dr R. Y. Tsien (9), compound 2c of Table 1 in ref. 9, was used because its dissociation constant of $1 \times 10^{-5.8}$ M is in the appropriate range for the [free Ca^{2+}] reached during Ca^{2+} release in skinned cardiac cells and because its on-rate constant, and even more its off-rate constant of binding to Ca^{2+} are much higher than those of EGTA. This compound, hereafter termed 'Br_2 BAPTA', is the *P,P'*-dibromo derivative of 1,2-bis(*o*-aminophenoxy)ethane-*N,N,N',N'*-tetraacetic acid (BAPTA). Its formula is $C_{22}H_{22}N_2O_{10}Br_2$, and its formula weight is 634.19.

The batch of Br_2BAPTA used for these experiments was graciously prepared for me by Dr E. May and Mr P. Zink, from the Department of Pharmacology of my Institution, starting from a sample of tetraethyl ester of BAPTA kindly given to me by Dr A. T. Charteris of BDH Chemical Ltd. The degree of purity of our batch of Br_2BAPTA was 95%, which was kindly verified by Dr R. Y. Tsien. This degree of purity was taken into account in the computation of the weight of powder used for making up the solutions. Like EGTA, Br_2BAPTA has a very low affinity for Mg^{2+}, but it binds much less H^+ than EGTA does (9).

The on-rate of Ca^{2+} binding to Br_2BAPTA was estimated to be 300 times that of EGTA by Dr R. Y. Tsien (i.e. 600×10^6 M^{-1} sec^{-1}). From the quotient of this on-rate constant over the affinity constant ($1 \times 10^{5.8}$ M^{-1}) the off-rate constant can be estimated to about 1×10^3 sec^{-1}.

In a skinned cardiac cell from the adult rat ventricle, a Ca^{2+}-induced release of Ca^{2+} from the sarcoplasmic reticulum was elicited by an increase of [free Ca^{2+}] in the solution from pCa 7.40 to pCa 6.25, which corresponds to the optimum trigger in this tissue (13, 15). The pMg was 2.50; the pMgATP was 2.50; the pH was 7.10; Γ/2 was 0.170 M with K^+ and Cl^- as major ionic species; the temperature was 22 °C. The amplitude of the aequorin-detected Ca^{2+} transient (see refs 8, 13, and 15 for method of detection) resulting from the Ca^{2+} release decreased when the [total Br_2BAPTA] was increased at constant [free Ca^{2+}] in the solutions (*Figure 2*). The rate of change of myoplasmic [free Ca^{2+}] (d[free Ca^{2+}]/dt) was measured at a fixed myoplasmic [free Ca^{2+}] of $1 \times 10^{-5.6}$ M for each of the transients obtained in the presence of different [total Br_2BAPTA] values. This was done by calculating the myoplasmic [free Ca^{2+}] reached at two points of the ascending phase of the Ca^{2+} transient respectively above and below that corresponding to $1 \times 10^{-5.6}$ M free Ca^{2+}. For each point the myoplasmic [free Ca^{2+}] was inferred from the ratio of the peak of the transient of aequorin light to the maximum light and the aequorin-light calibration curve (see ref. 8 for method). The difference between the two inferred myoplasmic [free Ca^{2+}] values was divided by the time-interval to obtain the rate. The accuracy of the resulting rate of change of [free Ca^{2+}] was, therefore, not affected by the nonlinearity of the response of aequorin to [free Ca^{2+}] (16).

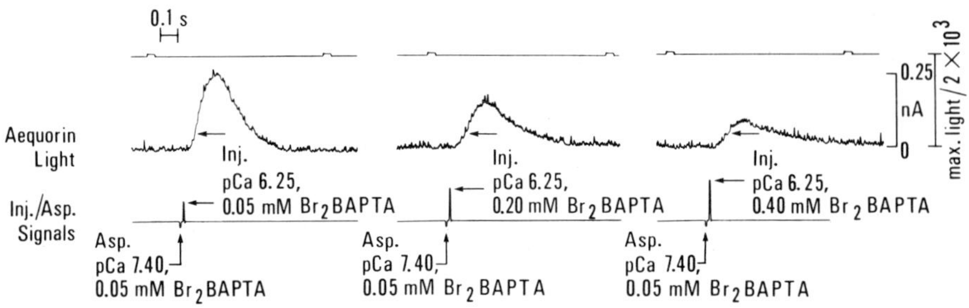

Figure 2. Method for determination of the rate of total calcium release from the sarcoplasmic reticulum. The experiment was done in a skinned cell from the adult rat ventricle. Aequorin light and the injection–aspiration signals of the microprocessors controlling the two stepping motors were recorded simultaneously. The injection signals are upward, and the aspiration signals are downward. The timings of the signals correspond to the actual timings of each injection or aspiration. The amplitude of the signals was different only to permit the distinction between the two microprocessor-controlled microinjection-aspiration syringes. The solution changes were effected by aspiration of the previous solution followed, after a brief interval, by the injection of the new solution. At the end of the experiment the maximum light was obtained by discharging all the aequorin contained in the skinned cell, and the myoplasmic [free Ca^{2+}] reached at different times during the ascending phase of the Ca^{2+} transient was derived from the ratio of the instantaneous value of the light over the maximum light and a calibration curve of aequorin bioluminescence (see ref. 8 for additional details on this methodology). Each upper horizontal arrow shows the point in the ascending phase of the calcium transient at which a [free Ca^{2+}] of $1 \times 10^{-5.6}$ M was reached. On enlargements of this area of the tracing, the times at which [free Ca^{2+}] values of $1 \times 10^{-5.7}$ M and $1 \times 10^{-5.5}$ M were reached were determined for each ascending phase of Ca^{2+} transient. Dividing the difference in concentration by the time-interval gave the rate of change of myoplasmic [free Ca^{2+}] at a [free Ca^{2+}] of $1 \times 10^{-5.6}$ M. In all solutions the pMg was 2.50, the pH was 7.10 buffered with 30 mM Bes and $\Gamma/2$ was 0.170 M with K^+ and Cl^- as major ionic species. The solutions contained 12 mM phosphocreatine and 15 U/ml creatine phosphokinase. The temperature was 22 ± 0.1 °C.

The dependence of the rate of change of myoplasmic [free Ca^{2+}] upon the [total Br_2BAPTA] permitted the inference of the rate of total calcium release from the sarcoplasmic reticulum according to the following rationale, which is derived from that briefly outlined by Drs R. Y. Tsien and T. J. Rink (17) for measuring the rate of total calcium uptake with quin-2. The derivation is somewhat different, however, because with quin-2 the Ca^{2+} buffer was also used for measuring the change of cytoplasmic [free Ca^{2+}]. I am most grateful to Dr George D. Ford, from my Department, for his help with this computation.

The conservation equation is:

$$[\text{total Ca}] = [\text{CaSR}] + [\text{free Ca}^{2+}] + [\text{CaCB}] + [\text{CaBr}_2\text{BAPTA}] \quad \text{[eqn 1]}$$

where CaSR is the calcium content of the sarcoplasmic reticulum (abbreviated as SR) whose rate of change is to be measured, CaCB is the calcium bound to cellular buffers (abbreviated as CB), and $CaBr_2$BAPTA is the calcium bound to Br_2BAPTA.

There should be conservation. Thus,

$$\frac{d[\text{total Ca}]}{dt} = 0, \text{ hence:}$$

$$\frac{d[\text{CaSR}]}{dt} + \frac{d[\text{free Ca}^{2+}]}{dt} + \frac{d[\text{CaCB}]}{dt} + \frac{d[\text{CaBr}_2\text{BAPTA}]}{dt} = 0. \quad \text{[eqn 2]}$$

From an adaptation of the rationale explained by Tsien and Rink in ref. 17, one can write:

$$\frac{d[\text{CaBr}_2\text{BAPTA}]}{dt} = K_d \times \frac{[\text{total Br}_2\text{BAPTA}]}{(K_d + [\text{free Ca}^{2+}])^2} \times \frac{[\text{free Ca}^{2+}]}{dt}, \quad \text{[eqn 3]}$$

where K_d is the dissociation constant of the Ca^{2+} binding to Br_2BAPTA.

Similarly,

$$\frac{d[\text{CaCB}]}{dt} = \frac{K_c[\text{total CB}]}{(K_c + [\text{free Ca}^{2+}])^2} \times \frac{d[\text{free Ca}^{2+}]}{dt} \quad \text{[eqn 4]}$$

where K_c is the dissociation constant of the cellular Ca^{2+} buffers.

Let us define B, the buffer power of Br_2BAPTA as the derivative of bound calcium with respect to [free Ca^{2+}].

The dissociation equation gives:

$$\text{CaBr}_2\text{BAPTA} = \frac{[\text{free Ca}^{2+}] \times [\text{total Br}_2\text{BAPTA}]}{[\text{free Ca}^{2+}] + K_d}. \quad \text{[eqn 5]}$$

The derivative of this composite function is

$$B = \frac{d[CaBr_2BAPTA]}{d[free\ Ca^{2+}]} = \frac{K_d[total\ Br_2BAPTA]}{K_d\ ([free\ Ca^{2+}])^2} . \qquad [eqn\ 6]$$

Similarly, the buffering power, C, for the cellular Ca^{2+} buffers is:

$$C = \frac{d[CaCB]}{d[free\ Ca^{2+}]} = \frac{K_b[total\ CB]}{(K_c + ([free\ Ca^{2+}])^2} . \qquad [eqn\ 7]$$

Then eqn 3 and eqn 4 become:

$$\frac{d[CaBr_2BAPTA]}{dt} = B \times \frac{d[free\ Ca^{2+}]}{dt} \qquad [eqn\ 8]$$

and

$$\frac{d[CaCB]}{dt} = C \times \frac{d[free\ Ca^{2+}]}{dt} . \qquad [eqn\ 9]$$

Let

$$R = \frac{d[CaSR]}{dt} \qquad [eqn\ 10]$$

and

$$A = \frac{d[free\ Ca^{2+}]}{dt} . \qquad [eqn\ 11]$$

Substituting eqn 8 and eqn 9 in the conservation equation, i.e. eqn 1, gives:

$$R = -A\ (1 + B + C). \qquad [eqn\ 12]$$

Dividing first by A and then by R,

$$\frac{1}{A} = \frac{-(1 + B)}{R} - \frac{C}{R} . \qquad [eqn\ 13]$$

The plot of $1/A$ as a function of $(1 + B)$ gives a straight line with a slope $= -1/R$ and intercept $= -C$. Thus, in *Figure 3* the reciprocal of the slope would give the rate of total calcium release in mmol sec^{-1} per litre of cell water accessible to aequorin and to Ca^{2+}, and the intercept would give the cellular buffering (buffering power has no unit).

From the rate of total calcium release measured in skinned cardiac cells, in mmol sec^{-1} per litre of cell water accessible to aequorin and Ca^{2+}, it is possible to infer the rate of total calcium release that would occur in the intact cardiac muscle under the same conditions. This latter rate can be expressed in mmol sec^{-1} per kilogram wet weight and computed from the result obtained in skinned cardiac cells and from the following data for the rat ventricular cell: tissue water = 77% of wet weight (18), intracellular water = 75% of total water (19), volume occupied by the mitochondria = 36% of cell volume (20) and water content of the mitochondria = 65% (21, 22).

To permit comparison with the data from fragmented cardiac sarcoplasmic reticulum, the results can also be expressed in mmol · sec^{-1} per milligram of sarcoplasmic reticular protein. This computation can be done assuming a concentration of reticular protein of 2.1 mg per kilogram tissue wet weight for the mammalian cardiac tissue (23). One must also know whether Ca^{2+} is released from all portions of the cardiac sarcoplasmic reticulum of skinned cardiac cells or only from the terminal cisternae including the extended junctional sarcoplasmic reticulum (24, 25).

The goal of this section of the article was to explain how a fast Ca^{2+} buffer

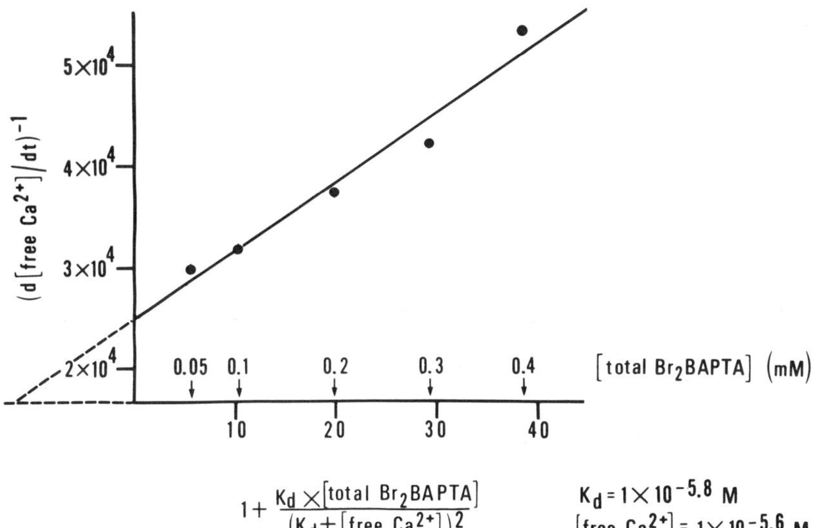

Figure 3. Computation of the rate of total calcium release from the data of experiments of the type shown in *Figure 2*. K_d is the dissociation constant of the binding of Ca^{2+} to Br_2BAPTA, which is $1 \times 10^{-5.8}$ M. The [free Ca^{2+}] was $1 \times 10^{-5.6}$ M. The reciprocal of the slope of the curve would give the rate of total calcium release in mmol · sec^{-1} per litre of cell water accessible to aequorin and to Ca^{2+}. The horizontal axis intercept of the curve would give the cellular buffering power (buffering power has no unit).

can be used for a method of computation of the rate of total calcium release from the sarcoplasmic reticulum of skinned fibres. The value of this rate is still uncertain, not only because of the limited number of experiments but also because of a number of methodological uncertainties in particular concerning the stability constants of the complexes between Ca^{2+} and Br_2BAPTA (26). An additional potential problem was that, at least in the case of *Figure 2*, increasing the [total Br_2BAPTA] not only depressed the amplitude of the Ca^{2+} transient but also increased the time taken for the decay of the Ca^{2+} transient and slightly increased its time to peak. These are the reasons for which I eventually decided not to give the value of the rate of total calcium release, which can easily be measured from the slope of the straight line of the plot shown in *Figure 3*. It corresponds to the rate I have reported many times orally.

5. Caged calcium compounds

In many experiments in cellular physiology it is desirable to rapidly change the [free Ca^{2+}] at a critical site within a cellular preparation. This cannot be achieved by rapidly changing the externally applied solutions when the preparation is of large dimensions. The diffusion delay, which is proportional to the square of the distance (27), results in an excessively slow change of [free Ca^{2+}] at the active site. With other experimental systems, the slowness of the change of [free Ca^{2+}] is related to the technique itself. Thus, for instance, the technique of sarcoplasmic reticulum incorporated in a lipid bilayer requires that the change of solutions in both chambers, including the *cis*-chamber (the chamber simulating the cytoplasmic side of the sarcoplasmic reticular membrane), be slow to avoid breaking the bilayer (28). This prevents the study of the effect of rapidly changing the myoplasmic [free Ca^{2+}] on the Ca^{2+} channel of the sarcoplasmic reticulum, whereas such a rapid change has been suggested to be essential for the study of the process of Ca^{2+}-induced release of Ca^{2+} from the sarcoplasmic reticulum (8).

These experiments requiring rapid changes of [free Ca^{2+}] in large cellular preparations or in a lipid bilayer set-up will be greatly facilitated by the recent development of caged, or photo-cleavable, compounds (10). Until now, two types of caged Ca^{2+} compounds, i.e. of compounds that permit a rapid increase of [free Ca^{2+}] when a flash of light is applied, and two caged Ca^{2+} chelators, which produce the opposite effect upon application of the flash of light, have been described. Thus, one can add any of these compounds to the experimental solutions and allow enough delay to permit them to diffuse to the desired active site on the cellular preparation or on the lipid bilayer. Then, one may apply the flash of light at the precise time when the rapid change of [free Ca^{2+}] is desired.

The two Ca^{2+} chelators have recently been described by Adams *et al.* (29).

One is a 2-nitrobenzyl monoester of BAPTA. Upon flash photolysis it releases BAPTA which chelates Ca^{2+} and thereby decreases the [free Ca^{2+}]. The other caged chelator is diazo-2, a compound in which the photolysis alters the electron withdrawing substituents in such a way that the affinity of the compound for Ca^{2+} is increased. At the time of this writing, these compounds were not yet commercially available and therefore the problems in their physiological use could not be assessed.

The two caged Ca^{2+} compounds are (i) nitr-5, which is the most advantageous of a series of compounds described by the group of Dr Roger Tsien (30–32) and can be obtained by Molecular Probes Inc., (ii) 1-(2-nitro-4,5-dimethoxyphenyl-*N*,*N*,*N*′,*N*′-tetrakis[(oxycarbonyl)methyl]-1,2,-ethanediamine (DM-nitrophen) which has been described by Dr Kaplan's group (33) and can be obtained from Calbiochem.

I am most grateful to Dr Robert M. Barsotti (Bockus Research Institute, The Graduate Hospital, 415 South 19th Street, Philadelphia, PA 19146, USA) for initiating me to the methodology and problems of caged Ca^{2+} compounds when I started to use them. The ideal qualities of a caged Ca^{2+} compound are the following: (i) The affinity for Ca^{2+} should change as much as possible, from the 10^{-8} M (or at least 10^{-7} M) range to the 10^{-3} M (or at least 10^{-4} M) range; (ii) The compound should be specific for Ca^{2+} with high specificity in particular against Mg^{2+}; (iii) The quantum yield of photolysis should be as high as possible to permit a rapid change of [free Ca^{2+}]; (iv) The photolysis products should not be toxic or have any artefactual physiological effect.

Neither nitr-5 (30–32) nor DM-nitrophen (33) presents any significant toxicity or artefactual physiological effect of their products of photolysis.

The major advantage of DM-nitrophen (33) is the large extent of the change of its Ca^{2+} affinity that is produced by the photolysis. Thus, a decrease of affinity to Ca^{2+} by a factor of about 1×10^6 is observed during flash photolysis: from the nanomolar (5×10^{-9} M) to the millimolar (5×10^{-3} M) range. This, associated with a relatively good quantum yield of 0.18, ensures that the increase of [free Ca^{2+}] will be sufficiently ample and rapid for most experiments.

The major disadvantage of DM-nitrophen (33) is that it binds Mg^{2+} as well as Ca^{2+}, like EDTA on which the DM-nitrophen molecule is based. Thus, the experiment must start under conditions of unphysiologically low [free Mg^{2+}] whereas the photolysis produces both an increase of [free Ca^{2+}] and an increase of [free Mg^{2+}]. Consequently, to what extent the combination of these increases of divalent cation concentrations will be responsible for any physiological effect that might be observed will remain unknown. For instance, I made the computation of the [free Mg^{2+}] before photolysis in the muscle experiment illustrated by Figure 4 of the article by Kaplan *et al.* (33), and found it to be 30 μM, which is much less than the 1 mM believed to be the physiological [free Mg^{2+}] in the intact skeletal muscle fibre.

The major advantage of nitr-5 is a very good selectivity for Ca^{2+} over Mg^{2+} (30–32).

The disadvantages of nitr-5 (30–32) are (i) a very low quantum yield (0.012 to 0.035); (ii) a very small decrease in Ca^{2+} affinity after irradiation: the decrease is by a factor of 40 only, which is clearly insufficient for many experiments. After photolysis, the [free Ca^{2+}] does not immediately reach a final level; there is initially a large photolysis and then some of the Ca^{2+} rebinds to non-photolysed nitr-5 or weakly rebinds to photolysed nitr-5. The first event (rebinding to non-photolysed nitr-5) is a major problem, because there is much free nitr-5 and the change in affinity to Ca^{2+} is small. In contrast, because of the higher Ca^{2+} affinity of DM-nitrophen before photolysis and its much larger change of affinity for Ca^{2+} during photolysis, this rebinding problem does not exist with DM-nitrophen. Unfortunately, the level of [free Ca^{2+}] reached during the initial higher transient cannot be computed precisely because one does not know well enough the on- and off-rate constants of Ca^{2+} binding to nitr-5 and all the binding constants of Ca^{2+} to the constituents and organelles of the cells (15), or even the exact concentration of photolysed nitr-5. This renders the quantification of the photolysis with nitr-5 very difficult.

For both nitr-5 and DM-nitrophen, the flash photolysis must be done with a light at about 350 nm wavelength, which can be produced either by a laser or by halogen flash-lamp (34). Because of the short wavelength, all optics must be in quartz. In addition, the light path throughout the preparations, or the experimental chamber containing the solution, must be short to avoid the decrease of light by absorption as explained in the following calculation. The optical density or absorbance (A) is given by the Beer–Lambert law:

$$A = C \times \varepsilon \times l$$

where C is the concentration of the substance to be photolysed and l is the path length. With 2 mM DM-nitrophen and a path length of 2 mm, one has:

$$A = 2 \times 10^{-3}\ \text{M} \times 4.33 \times 10^{3}\ \text{M}^{-1} \cdot \text{cm}^{-1} \times 0.2\ \text{cm}.$$

Therefore:

$$A = 1.7.$$

The absorbance is related to the incident (I_0) and the transmitted (I_t) light by the following formula:

$$A = \log_{10} \frac{I_0}{I_t}.$$

Thus, in our case:

$$1.7 = \log_{10}\frac{I_0}{I_t}.$$

Hence,

$$\frac{I_0}{I_t} = 50.$$

Therefore, the transmitted light will be 2% of the incident light, which would be no problem if a laser is the light source, but may be a major problem with the limited output of a halogen flash-lamp.

Acknowledgement

This work was supported by Grant No. 901207 from the American Heart Association.

References

1. Johansson, L. (1975). *Acta Chemica Scandinavica*, **A29**, 365.
2. Fabiato, A. and Fabiato, F. (1979). *Journal de Physiologie* (*Paris*), **75**, 463.
3. Fabiato, A. (1988). In *Methods in enzymology*, Vol. 157, Biomembranes, Part Q, ATP-Driven Pumps and Related Transport: Calcium, Proton, and Potassium Pumps (ed. S. Fleischer and B. Fleischer), p. 378. Academic Press, San Diego.
4. Weber, A. and Winicur, S. (1961). *Journal of Biological Chemistry*, **236**, 3198.
5. Smith, P. D., Berger, R. L., Podolsky, R. J., and Czerlinski, G. (1977). *Biophysical Journal*, **17**, 159a.
6. Smith, P. D., Liesegang, G. W., Berger, R. L., Czerlinski, G., and Podolsky, R. J. (1984). *Analytical Biochemistry*, **143**, 188.
7. Harafuji, H. and Ogawa, Y. (1980). *Journal of Biochemistry*, **87**, 1305.
8. Fabiato, A. (1985). *Journal of General Physiology*, **85**, 247.
9. Tsien, R. Y. (1980). *Biochemistry*, **19**, 2396.
10. Kaplan, J. H. and Somlyo, A. P. (1989). *Trends in Neurosciences*, **12**, 54.
11. Harrison, S. M. and Bers, D. M. (1989). *American Journal of Physiology*, **256**, C1250.
12. Martell, A. E. and Smith, R. M. (1974). *Critical stability constants*, Vol. 1, *Amino acids*, 469 pp. Plenum Press, New York.
13. Fabiato, A. (1981). *Journal of General Physiology*, **78**, 457.
14. Harned, H. S. and Owen, B. B. (1958). *The physical chemistry of electrolytic solutions*, 803 pp. Reinhold, New York.
15. Fabiato, A. (1983). *American Journal of Physiology*, **245**, C1.
16. Blinks, J. R., Wier, W. G., Hess, P., and Prendergast, F. G. (1982). *Progress in Biophysics and Molecular Biology*, **40**, 1.

17. Tsien, R. Y. and Rink, T. J. (1983). In *Current methods in cellular neurobiology*, Vol. III, *Electrophysiological and optical recording techniques* (ed. J. L. Barker and J. F. McKlevy), p. 249. Wiley, New York.
18. Polimeni, P. I. and Page, E. (1973) *Circulation Research*, **33**, 367.
19. Polimeni, P. I. (1974). *American Journal of Physiology*, **227**, 676.
20. Page, E. (1978). *American Journal of Physiology*, **235**, C147.
21. Werkheiser, W. C. and Bartley, W. (1957). *Biochemical Journal*, **66**, 79.
22. Somlyo, A. P., Somlyo, A. V., Shuman, H., Scarpa, A., Endo, M., and Inesi, G. (1981). In *Calcium and phosphate transport across biomembranes* (ed. F. Bronner and M. Peterlik), p. 87. Academic Press, New York.
23. Levitsky, D. O., Benevolensky, D. S., Levchenko, T. S., Smirnov, V. N., and Chazov, E. I. (1981). *Journal of Molecular and Cellular Cardiology*, **13**, 785.
24. Sommer, J. R., Bossen, E., and Fabiato, A. (1982). In *40th Annual Proceedings of the Electron Microscopy Society of America* (ed. G. W. Bailey), p. 136. Claitor's, Baton Rouge.
25. Fabiato, A. (1985). *Journal of General Physiology*, **85**, 291.
26. Harrison, S. M. and Bers, D. M. (1987). *Biochimica et Biophysica Acta*, **925**, 133.
27. Crank, J. (1975). *The mathematics of diffusion* (2nd edn). Clarendon Press, Oxford.
28. Smith, J. S., Coronado, R., and Meissner, G. (1988). In *Methods in enzymology*. Vol. 157, Biomembranes, Part Q, ATP-Driven Pumps and Related Transport: Calcium, Proton, and Potassium Pumps (ed. S. Fleischer and B. Fleischer), p. 480. Academic Press, San Diego.
29. Adams, S. R., Kao, J. P. Y., and Tsien, R. Y. (1989). *Journal of the American Chemical Society*, **111**, 7957.
30. Gurney, A. M., Tsien, R. Y., and Lester, H. A. (1987). *Proceedings of the National Academy of Sciences, USA*, **84**, 3496.
31. Adams, S. R., Kao, J. P. Y., Grynkiewicz, G., Minta, A., and Tsien, R. Y. (1988). *Journal of the American Chemical Society*, **110**, 3212.
32. Kao, J. P. Y., Harootanian, A. T., and Tsien, R. Y. (1989). *Journal of Biological Chemistry*, **264**, 8179.
33. Kaplan, J. H. and Ellis-Davies, G. C. R. (1988). *Proceedings of the National Academy of Sciences, USA*, **85**, 6571.
34. Rapp, G. and Güth, K. (1988). *Pflügers Archiv für die gesamte Physiologie*, **411**, 200.

8

Use of fluorescent TnC derivatives and 'caged' compounds to study cellular Ca^{2+} phenomena

C. C. ASHLEY, P. J. GRIFFITHS, T. J. LEA, I. P. MULLIGAN, R. E. PALMER, and SARAH J. SIMNETT

1. $STnC_{DANZ}$ as an indicator

1.2 Introduction

Although it has been established using the photoprotein aequorin (*Figure 1*) that the free calcium concentration within muscle cells changes rapidly upon activation (1, 2) little is known of the time-course of calcium bound to the contractile system as a result of this transient change in the free calcium.

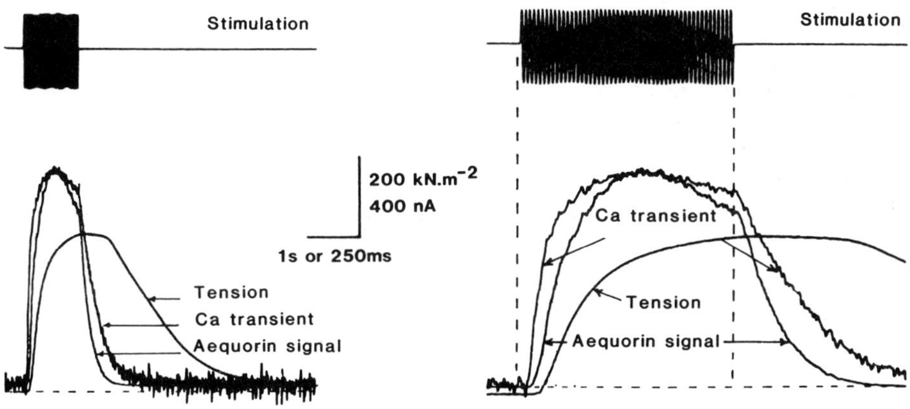

Figure 1. Aequorin-light responses, free Ca^{2+}, and tension during electrical stimulation of an intact barnacle muscle fibre. The calcium transient may be distinguished from the aequorin-light response by noting that it leads the aequorin signal during the rising phase of tension, and lags during relaxation as clearly demonstrated in the panel on the right, which shows the same record on a faster time-base. Free calcium transient was obtained from the aequorin signal by addition of the signal itself and its first derivative (see ref. 2). The 2.5th root of the waveform thus obtained was then calculated to yield the calcium transient. Temperature 12 °C. Sarcomere length 9.5 μm. Resting Ca^{2+}_i = 2 nA. (From ref. 2, with permission.)

Recently, a dansylazaridine-labelled subunit of the thin filament calcium sensor protein troponin (TnC_{DANZ}) has been described which, when excited at 325 nm shows a 2.1-fold increase in fluorescence at 510 nm, when calcium ions occupy the calcium specific regulatory sites (*Figure 2*). The affinity of rabbit skeletal TnC_{DANZ} ($STnC_{DANZ}$) (K_{Ca} 3.3 × 10^5 M^{-1}) increases at the calcium specific sites by more than 10 times when incorporated into intact troponin, but decreases by a factor of 5–10 times when whole troponin is incorporated on to the actin filament (see *Table 1*). Hence the affinity for calcium and the time-course of the consequent fluorescence change of *free* $STnC_{DANZ}$ should be similar to that of whole troponin bound to the contractile filaments provided the on-rate remains diffusion limited. However, changes in the affinity and number of these regulatory sites brought about by cross-bridge (S1) attachment will not be observed with free $STnC_{DANZ}$ and in this way the behaviour of free $STnC_{DANZ}$ may differ from that of $STnC_{DANZ}$ incorporated into the thin filaments.

Here we report upon the use of this labelled troponin C in cells and document the time-course of the fluorescence changes of free $STnC_{DANZ}$ compared to the time-course of free Ca^{2+} and isometric force development from a single muscle fibre under voltage-clamp control.

Table 1. Preliminary midpoints of titrations

	N_1 [a]	K_1 [b]	N_2 [a]	K_2 [b]	K_{ATPase} [h]
1. Tn (Tm · Tn)[c], $-Mg^{2+}$ [d]	2	8.5	2	6.3	
2. Actin · Tm · Tn[f], $-Mg^{2+}$	2	8.4	2	5.5	
3. Tn (Tm · Tn)[c], $+Mg^{2+}$ [e]	2	6.4	2	6.4	
4. Actin · Tm · Tn[f], $+Mg^{2+}$	2	6.7[g]	2	5.7[g]	6.0[i]
5. TnC_{DANZ} [h]	–	–	–	5.3	
6. $Myofibrils_{DANZ}$ [h]	–	–	–	5.6	5.4
7. TnC_{IAANS} [h]	–	6.2	–	–	
8. $Myofibrils_{IAANS}$ [h]	–	6.9	–	–	5.6
9. TnC_{AD} [h]	–	6.3	–	–	
10. $Myofibrils_{AD}$ [h]	–	6.4	–	–	5.8

[a] Number of mol Ca^{2+} bound per mol protein.
[b] Apparent affinity constant for a class of sites expressed as the – log of the midpoint of the transition.
[c] The data for Ca^{2+} binding to Tn and the Tm · Tn complex are grouped Tm · Tn are mixed in a 1:1 molar ratio.
[d] $-Mg^{2+}$ refers to the following solvent conditions: 10 mM Imidazole, pH 7.0, 150 mM KCl, 0.1 mM EDTA.
[e] $+Mg^{2+}$ refers to the following solvent conditions: 10 mM Imidazole, pH 7.0, 150 mM KCl, 0.5 mM EGTA, 2.0 mM $MgCl_2$.
[f] Actin · Tm · Tn complex mixed in a 7:1:1 molar ratio.
[g] Determined from Scatchard analysis.
[h] Solvent conditions were 120 mM Mops, pH 7.0, 90 mM KCl, 2 mM EGTA, 5 mM $MgCl_2$, and when ATPase was measured, 2 mM ATP.
[i] Heavy meromyosin (HMM) added at 1 mol HMM: 7 mol actin.
IAANS = 2-(4′-iodoacetamidoanilino)naphthalene-6-sulphonic acid; AD = acrylodan.
[Reproduced from ref. 11, with permission.]

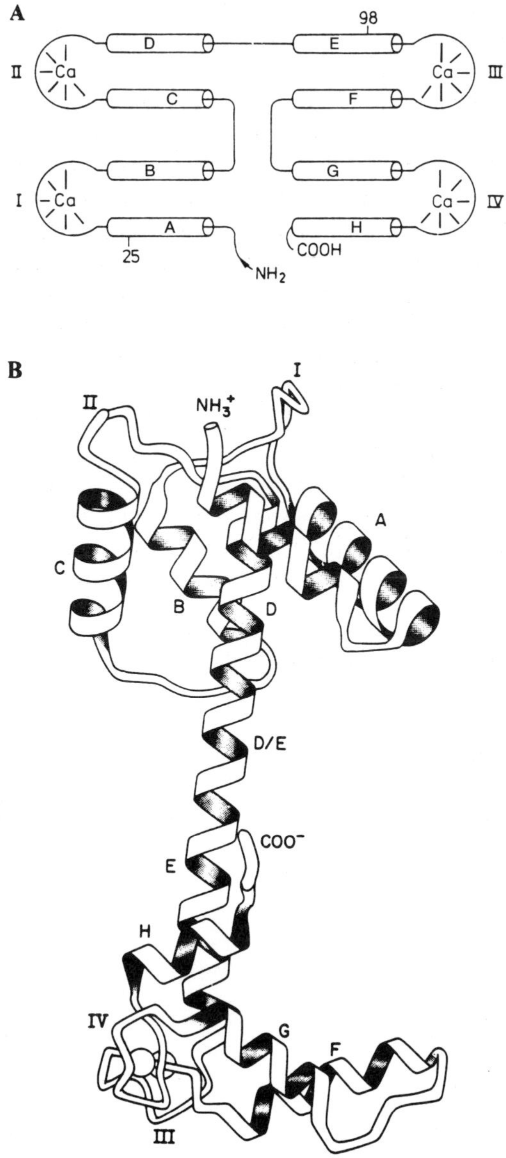

Figure 2. **A**: Schematic diagram of predicted troponin-C structure with two low-affinity specific calcium-binding domains (I and II) and two high-affinity Ca^{2+}– and Mg^{2+}–binding sites (III and IV). A–H are helical regions of the peptide chain (after Potter and Johnson, ref. 3). **B**: Structure of troponin C (ribbon representation) as derived from X-ray crystallographic studies at 2.8 Å resolution. The upper helical domain is the calcium regulatory part containing the low-affinity binding sites. The lower domain contains the unspecific Ca^{2+}–Mg^{2+} sites. (From ref. 24, with permission.)

1.2 Troponin and troponin C (TnC)

1.2.1 Ca^{2+} binding properties

The Ca^{2+}-binding parameters determined for the skeletal (S) and cardiac (C) proteins are listed in *Table 2* (from ref. 3). STnC and CTnC contain two (*Table 2*) high-affinity Ca^{2+}-binding sites ($K_{Ca} = 10^7\ M^{-1}$) that also bind Mg^{2+} competitively ($K_{Mg} = 10^3\ M^{-1}$) and have been called the Ca^{2+}–Mg^{2+} sites (Sites III and IV) (4, 5). STnC also contains two sites (*Table 1*) of lower Ca^{2+} affinity ($K_{Ca} = 10^5\ M^{-1}$) which are essentially specific for Ca^{2+} at physiological (millimolar) Mg^{2+} concentrations and have been called the Ca^{2+}-specific sites (4). CTnC, in contrast to STnC, contains only one Ca^{2+}-specific site (Site II) (*Table 2*) which is similar in affinity to the two Ca^{2+}-specific sites in STnC (5, 6). The existence of only three Ca^{2+}-binding sites on CTnC was predicted by Van Eerd and Takahashi (7) when they compared the amino acid sequences of CTnC with the sequence and structure of parvalbumin (8) and of STnC (9). Two key co-ordinating aspartic acid residues in region I were substituted for by a leucine and an alanine, so neither of these amino acids could contribute to Ca^{2+} co-ordination. Thus CTnC has essentially the same Ca^{2+}-binding properties as STnC except that it lacks the Ca^{2+}-specific site present in region I of STnC (5, 10).

Ca^{2+} binding to intact STn is quite different from binding to STnC and, in

Table 2. Ca^{2+} and Mg^{2+} binding properties of skeletal troponin (STn) and cardiac troponin (CTn) and their calcium-binding subunits STnC and CTnC

		n	Metal	K_a (M^{-1})	k_{off} (sec^{-1})
STnC	Ca^{2+}-specific	2	Ca^{2+}	3×10^5	300
			Mg^{2+}	200	~100
	Ca^{2+}–Mg^{2+}	2	Ca^{2+}	2×10^7	1–4
			Mg^{2+}	4×10^3	8
STn	Ca^{2+}-specific	2	Ca^{2+}	5×10^6	20–25
			Mg^{2+}	200	~100
	Ca^{2+}–Mg^{2+}	2	Ca^{2+}	5×10^8	0.3–0.6
			Mg^{2+}	4×10^4	5
CTnC	Ca^{2+}-specific	1	Ca^{2+}	$2–4 \times 10^5$	300
			Mg^{2+}	200	~100
	Ca^{2+}–Mg^{2+}	2	Ca^{2+}	1.5×10^7	7
			Mg^{2+}	$0.7–3 \times 10^3$	30–120
CTn	Ca^{2+}-specific	1	Ca^{2+}	$2–3 \times 10^6$	15–30
			Mg^{2+}	200	~100
	Ca^{2+}–Mg^{2+}	2	Ca^{2+}	4×10^8	0.3
			Mg^{2+}	$0.3–2 \times 10^4$	5–33

where $K_a = (k_{on}/k_{off})$

[Modified from refs 3 and 46]

general, the Ca^{2+} affinity of all four sites is increased by an order of magnitude in the absence of Mg^{2+} (*Table 2*).

Although the Ca^{2+}-specific sites were originally thought to be specific for Ca^{2+}, recent evidence suggests that Mg^{2+} binding to sites distinct from the Ca^{2+}-specific sites affects their affinity. The results of these studies (12), show that the apparent K_{Mg} for producing shifts in the Ca^{2+} affinity of these sites is approximately 200 M^{-1} in both $STnC_{DANZ}$ and STn_{DANZ} and it is likely that the effect is noncompetitive; i.e. Mg^{2+} does not bind directly to the Ca^{2+}-specific sites (3); a value for K_{Mg} of 1.2×10^3 M^{-1} was recently found for the low-affinity Ca^{2+} site on the thin filament (13). Ogawa (14), however, suggests that Mg^{2+} may bind competitively to the low-affinity sites (K_{Mg} 5.2×10^2 M^{-1}) for STnC and 4.4×10^2 M^{-1} for STn. Site-directed mutagenesis (15), has been used to produce changes in the N terminal (I, II binding sites) by a disulfide bridge and this shifts the Ca^{2+} sensitivity (DANZ fluorescence) to higher free Ca^{2+} values (16), consistent with a shift to a lower value of the affinity of these sites.

1.2.2 Kinetics of Ca^{2+} binding to STnC

The kinetic rates of calcium binding to TnC and displacement from TnC are important. Knowledge of the equilibrium binding constants alone is insufficient. It is now possible to measure some of these rate constants by using the fluorescent probe dansylaziridine (DANZ). DANZ-labelled troponin C at the methionine 25 at site I (*Figure 2*, A) exhibits a strong fluorescence upon excitation at 350 nm, when calcium occupies the calcium-specific regulatory sites (17); whilst calcium occupancy of the high-affinity sites causes a slight decrease in fluorescence. Johnson *et al.* (18) measured the change in fluorescence when DANZ-labelled STnC or STn were rapidly mixed with Ca^{2+}. Significantly, both STn and isolated STnC both bound Ca^{2+} rapidly within the mixing time of the instrument (2–3 msec) suggesting a diffusion limited on-rate. When the converse experiment was performed (that is, either calcium-saturated STn or STnC were mixed rapidly with the chelator EGTA), the DANZ fluorescence intensity decreased in a quasi-exponential manner within about 40 msec for STn and even more rapidly for STnC. The half-time of the fluorescence decay suggested a rate-constant ($k_{Ca,off}$) of approximately 23 sec^{-1} for STn, >300 sec^{-1} for STnC at 20 °C (19, 20). These measurements imply that calcium binding to these low-affinity (Ca^{2+} specific) sites is fast, approximately $K_{Ca,on} > 10^8$ M^{-1} sec^{-1} and also studies indicate that these low-affinity sites are independent in their Ca binding (20). The rate of calcium dissociation from the higher affinity Ca^{2+}–Mg^{2+} sites is slower, $K_{Ca,off}$ 5 sec^{-1}. Although the calcium on-rate to these high-affinity sites is also fast in the absence of magnesium; in the presence of a physiological free Mg^{2+} concentration (1–3 mM) (21) these sites are largely occupied by magnesium, and therefore the apparent rate of calcium binding is limited by the need to displace bound magnesium—a relatively *slow* process ($k_{Mg,off}$ 5 sec^{-1}) (22).

Consideration of the calcium binding kinetics of troponin-C (18) suggested that the low-affinity calcium-binding sites are the major regulatory sites in skeletal muscle, as Ca^{2+} binding to the high affinity (Ca^{2+}–Mg^{2+} sites) (at physiological magnesium concentrations) would be too slow to account for the observed rate of force development during a twitch, although a more prolonged elevated free Ca^{2+}, characteristic of a tetanic response, may well displace bound Mg^{2+} from these sites as well as from the Ca^{2+}–Mg^{2+} sites of the soluble calcium binding proteins (parvalbumins) present in many muscle cells at millimolar concentrations (22).

It is, therefore, likely that only the two low-affinity regulatory sites play a major role in the rapid initiation of muscular contraction. Gillis (23) has suggested on the basis of modelling that changes in the calcium occupancy of the low-affinity sites on troponin C may precede the rising and falling phases of twitch tension in vertebrate striated muscle.

1.3 Use of $STnC_{DANZ}$ in intact single fibres

Single striated muscle fibres from the arthropod (*Balanus nubilus*) were isolated and cannulated (1) and axially microinjected with a stock solution of $STnC_{DANZ}$ (13 mM nominally) in 100 mM KCl (pH 7.0). The volume injected was 0.4 μl, so that after dilution, the final intracellular concentration was in the range 50–150 μM. Fibres were stimulated by voltage-clamp with an axially inserted electrode assembly where the current electrode was 100 μm diam. platinized Pt wire and the voltage-sensing part was a 100 μm capillary filled with 0.5 M KCl and containing a 25 μm diam. bright Pt wire. The voltage-clamp circuit was standard, and force was recorded with an Akers force transducer (sensitivity 12 $mN \cdot V^{-1}$; resonance frequency 4 kHz). Fibre fluorescence was excited at 325 nm with a Liconix He–Cd laser beamed via a front-surfaced mirror on to the preparation. A section of frosted UV-compatible plastic cuvette was used as light diffuser to reduce the intensity of the laser spot and also to act as a beam spreader. The light was collected by a 5 mm × 1 m standard light-guide, passed through a filter assembly with interference filters (12.5 mm diam.), at 440, 520, and 600 nm halfband width (10 nm), and via a 5-mm rod light guide to the 10-mm sensitive face plate of an EMI 9502B photomultiplier tube arranged in an anode–ground configuration with the HV at −800 to −1000 V. The output from the photomultiplier was passed through a current-to-voltage converter with a 10 msec time-constant and the data stored on a Nicolet 1170 system with magnetic tape store.

In *Figure 3A*, the fibre autofluorescence in the absence of $STnC_{DANZ}$ is indicated upon excitation at 325 nm in a conventional spectrofluorimeter assay. The fibre fluorescence shows a maximum at more than 420 nm, followed by a gradual incline to reach a value at 600 nm, which is more than 50% of that observed at 420 nm (*Figure 3*). In the presence of injected $STnC_{DANZ}$ the fibre shows a marked additional fluorescence maximum at 520

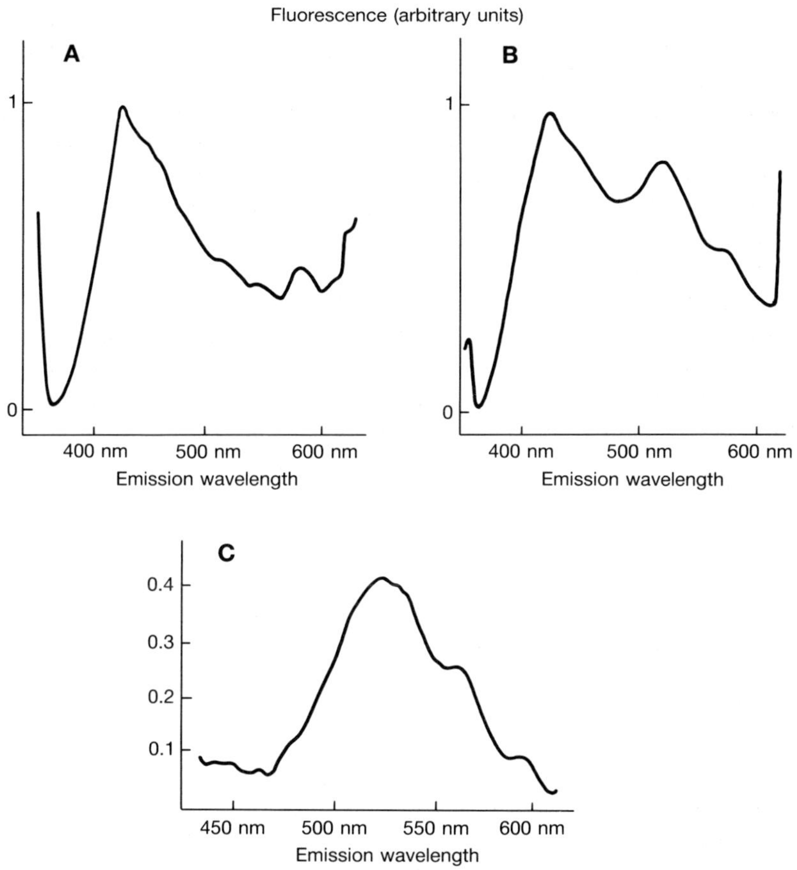

Figure 3. Fluorescence spectrum obtained from a single barnacle muscle fibre on excitation at 325 nm. **A**: Spectrum obtained from native fibre (autofluorescence). **B**: Spectrum obtained after injection of $STnC_{DANZ}$ into the same fibre. 1 h was allowed for diffusion of $STnC_{DANZ}$ from the region of injection. **C**: Subtraction of **A** from **B**. Temperature 17 °C $[STnC_{DANZ}]_i$ = 15 μM (From ref. 26, with permission.)

nm (*Figure 3B*) which is close to that observed for Ca^{2+}–$STnC_{DANZ}$ complex in free solution at the physiological Mg^{2+} ion concentration expected within intact muscle cells (25).

This result indicated that the emission peak of the Ca-dependent $STnC_{DANZ}$ signal could be assessed satisfactorily at a wavelength approx. 500 nm in intact cells, while a control wavelength for non-specific fluorescence changes could be either 600 or 450 nm, where the $STnC_{DANZ}$ signal was small. Usually either 440 or 400 nm wavelength was chosen, as these values are close to the maximum of the fibre autofluorescence and should therefore be a sensitive index of the non-specific effects if present.

Typical fluorescence signals obtained from a fibre on stimulation are shown in *Figure 4*. Records A and C were obtained before injection of the cell with $STnC_{DANZ}$. This scaled subtraction of fluorescence at 410 and 490 nm yields no net signal. The fibre was then injected with labelled troponin C and after allowing time (1–2 h) for radial diffusion of the label from the site of injection, scaled subtraction of fluorescence at 410 and 490 nm now yields a net signal. This difference signal and the associated force response upon application of a depolarizing pulse both increase with increasing stimulus intensity (*Figure 4B and D*). The time-course of the fluorescence difference signal compared to force development is clearly illustrated in *Figure 5*, which indicates that the fluorescence record rises faster than force. The $t_{1/2}$ of fluorescence rise precedes force by 55 msec ($n = 15$) and the fluorescence increase is virtually complete when force is still only 50% of its peak value, while the peak of force and fluorescence reach a maximum value at similar times of 22 °C (26).

Since the fibre difference fluorescence (490–410 nm) differs in a $STnC_{DANZ}$-injected fibre compared to that from an uninjected fibre, this suggests that the

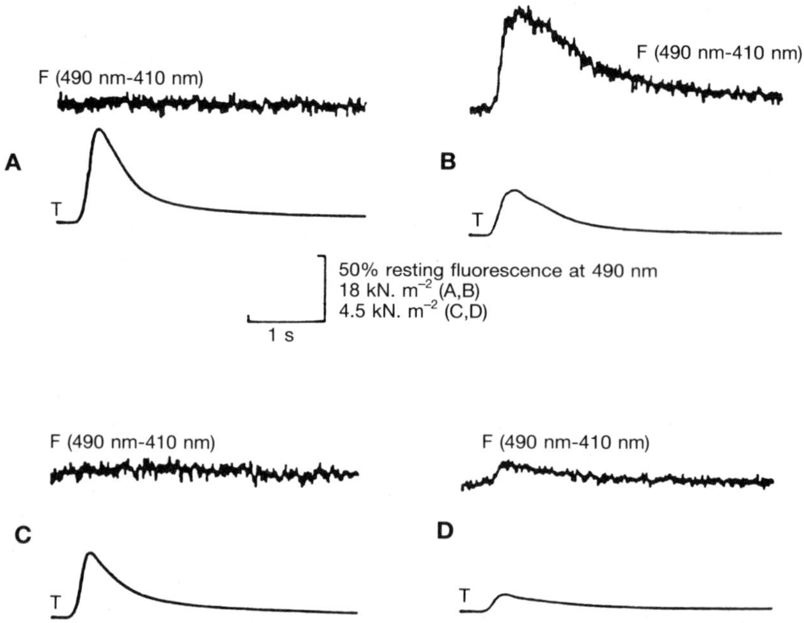

Figure 4. Force (T) and fluorescence (F) change recorded from a barnacle muscle fibre under voltage-clamp control. **A** and **C**, prior to $STnC_{DANZ}$ injection; **B** and **D**, after injection. Fluorescence records were obtained by scaled subtraction of records obtained at 410 and 490 nm on excitation at 325 nm. Depolarizing pulses, 200 msec duration. Subtractions performed on responses with similar peak force values. Temperature 22 °C. $[STnC_{DANZ}]_i$ = 122 μM. (From ref. 26, with permission.)

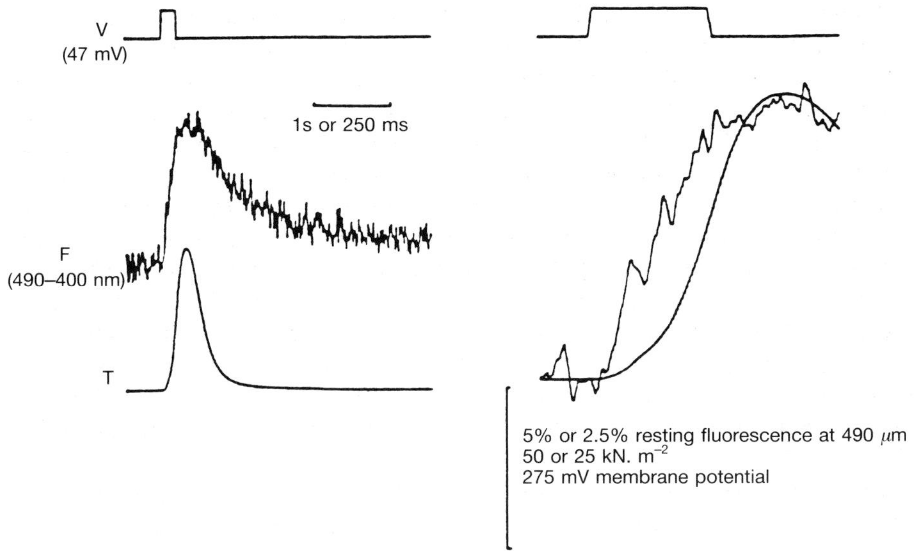

Figure 5. Time-course of force and fluorescence responses to stimulation. Left panel: membrane potential (V), fluorescence signal after scaled subtraction (F) and tension (T). Right panel: same response, but displayed on a 4 × faster time-trace. $[STnC_{DANZ}]_i$ = 50 μM. Holding potential −50 mV. Temperature 22 °C (see *Figure 4*). (From ref. 26, with permission.)

signal from the injected cell arises, most likely, from a change in $STnC_{DANZ}$ fluorescence as a result of calcium ions being bound at the 'calcium-specific' sites of this protein subunit. A possible objection to this interpretation could be that the fluorescence signal obtained at these two wavelengths may not arise from the same regions of the cell, thus rendering the scaled subtraction procedure inappropriate. For example, the incident excitation beam might not penetrate beyond the surface of the cell or alternatively the fluorescence signal from one or both measured wavelengths may be strongly absorbed before it reaches the fibre surface. Both of these possibilities would lead to a signal derived principally from the surface of the preparation. In addition, either wavelength could contain differing degrees of contribution from peripherally located fluorescent groups again leading to a signal originating primarily from the fibre surface. If the fibre is circular in cross-section, then a maximum value of 2 should be obtained from the slope of the relation of log fibre radius vs. log fibre autofluorescence if the fluorescence arises from the volume of the preparation, and a value of 1 if only from the surface (26). The results obtained indicate a similar dependence on fibre radius of fluorescence at both 520 and 440 nm, suggesting that they arise from similar regions of the cell, and that the slope for the relationship is close to the value of 2 predicted for a cell volume dependency at both wavelengths.

Finally, in an attempt to examine the likely kinetics of the $STnC_{DANZ}$

signal in response to the free calcium change, the occupancy of the low-affinity calcium binding sites following a step change in free calcium concentration was computed, on the assumption of simple competition between calcium and magnesium ions at this binding site. In this preparation, the physiological free Mg^{2+} is more than 5 mM, and the forward rate constant for magnesium binding was assumed to be (k_{Mg}) 10^4 M^{-1} sec^{-1}, while the off rate was (k_{Mg}) 50 sec^{-1}. The calcium specific sites were, surprisingly 50% occupied with magnesium (free Mg^{2+} 5 mM, pCa >7.0) (26, 27) and although the magnesium displacement reaction is likely to be rapid, it may be slow enough to produce some reduction in the apparent Ca^{2+} 'on' rate to the magnesium occupied sites compared with the rate of calcium binding to those sites which are free of magnesium for a step-change in free Ca^{2+}, especially at high levels of Ca^{2+} activation (*Figure 6*). These simulations suggest that the Ca^{2+} on rates might be reduced, at extreme values, to the same range as rate-constant values predicted from the rise time of the $STnC_{DANZ}$ fluorescence signal in the intact muscle fibre (*Figure 5*). However, there is strong evidence

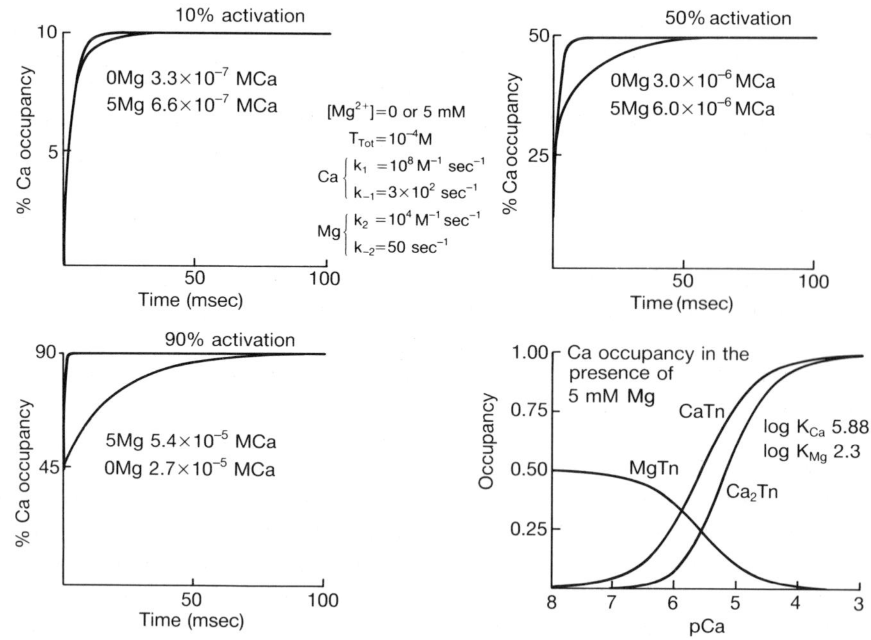

Figure 6. Computer predictions of site occupancy as a function of calcium and magnesium ion concentration. Free Ca^{2+} change as step function of suitable amplitude giving % of maximal activation in the presence or absence of Mg^{2+}. $k_{Ca,\ on} = 10^8\ M^{-1}\ sec^{-1}$, $k_{Ca,\ off} = 1.3 \times 10^2\ sec^{-1}$, $k_{Mg,\ on} = 10^4\ M^{-1}\ sec^{-1}$, $k_{Mg,\ off}$ 50 sec^{-1}. *Lower right*: equilibrium occupancy as a function of free calcium concentration [Mg^{2+}] 5 mM, pK_{Ca} 5.88, pK_{Mg} 2.3. Gear method used for numerical integration. (From ref. 26, with permission.) T_{Tot} = TnC site concentration.

to suggest that the interaction of calcium and magnesium at the low-affinity sites is more complex than simple competition, in which case the value of the 'off' rate constant for Ca^{2+} would be increased.

The data presented here indicate that electrical stimulation of $STnC_{DANZ}$-injected barnacle muscle fibres causes a net change in fluorescence at 500 nm after subtraction of non-specific effects monitored at 440 nm. This is consistent with the view that a fluorescence change would be expected at this wavelength as a result of calcium binding to the injected $STnC_{DANZ}$. Furthermore, experiments performed on uninjected fibres which were subsequently injected with $STnC_{DANZ}$ showed that this signal at 500 nm was absent before $STnC_{DANZ}$ injection. The reported dependence of fluorescence intensity upon the square of the fibre radius further supports this view, since it would appear that injected $STnC_{DANZ}$ would be excited to fluorescence throughout the fibre cross-section, and that the non-specific fluorescence at 440 nm should arise from the same region of the cell as the $STnC_{DANZ}$ signal. Localized membrane damage of injected fibres exposing the indicator to a raised (10 mM) Ca^{2+} concentration caused a roughly twofold increase in fluorescence at 520 nm after correction for non-specific changes at 440 nm, as would be expected from the *in vitro* behaviour of $STnC_{DANZ}$. These observations add further support to the notion that, in intact fibres, the fluorescence at 500 nm, after correction for non-specific changes at 440 nm, is as would be expected from the *in vitro* behaviour of $STnC_{DANZ}$ upon the binding of Ca^{2+}.

The rising phase of the fluorescence-difference signal usually precedes force development and is complete at a time when force is only 50% developed. This could suggest that the calcium bound to isolated $STnC_{DANZ}$ and putatively to the troponin on the intact thin filament, occurs substantially before the onset of force. This finding would not be in discord with observations which indicate that, in frog skeletal muscle, cross-bridge movement, measured in terms of the change in the equatorial X-ray diffraction intensity ratios ($I_{1,0}$, $I_{1,1}$), substantially precedes force development and is not in discord with the finding that, in this preparation, the free Ca^{2+} from aequorin and the first actin (A2) layer line intensity, the latter giving a measure of attached cross-bridges, also precedes force development (see later). Likewise, fibre stiffness, a measure of cross-bridge attachment, also precedes force development in single frog fibres at 5 °C. Nevertheless, the time-course of the development of the fluorescence change is slower than would have been predicted simply from the published 'on' rate constant for the calcium specific site (k_{on} 10^8 M^{-1} sec^{-1}). One possible reason for this discrepancy is that the 'on' rate for calcium to the calcium-specific sites could be limited by a relatively slow magnesium displacement reaction (*Figure 6*).

Perhaps the most interesting feature of the present results is the observation that the net change in fluorescence at 500 nm decays more slowly than force in all records so far examined. This fact is well-demonstrated in

Figure 5 where, at 22 °C, the half-time for the decay of fluorescence is on average 2.0 × that of force.

This result implies that the Ca^{2+} attached to the Ca^{2+}-specific sites of free $STnC_{DANZ}$, and also presumably to intact STnC on the thin filament, dissociates more slowly than the decline of force. This is not in discord with mechanical findings on single frog muscle fibres, where it is recognized that the exponential fall of force observed after the 'knee' of force, represents sarcomeric disorganization and does not necessarily represent the true decline of activation at the level of individual sarcomeres. This is clearly shown in experiments using the sarcomere clamp where the isometric fall of force, preceding the knee of tension, can be considerably prolonged by maintaining sarcomeric homogeneity. These experiments, if directly applicable to this muscle fibre preparation, would support the notion that this slow decay process may well involve a component or reaction which is dependent upon the relatively slow release of calcium from the contractile system.

Support for this notion also comes from the description of a prolonged 'tail' to the aequorin-light response and hence free calcium, seen both in cannulated as well as intact single barnacle (*Figure 1*) and frog muscle fibres.

Certainly in experiments where both aequorin and $STnC_{DANZ}$ have been injected into the same fibre, stimulation produces an aequorin response which reaches a peak some 100–200 msec in front of peak force and yet the $STnC_{DANZ}$ signal still shows the slower final rate of rise to peak at a time when force reaches its maximum and subsequently decays more slowly than the aequorin or force signal. In double-pulse experiments, the aequorin signal is smaller yet force is potentiated provided the 'tail' of the aequorin response is present, for time intervals less than 2 sec, yet under these conditions the $STnC_{DANZ}$ signal summates (*Figure 7*). This phenomenon cannot, in these fibres, be attributed to a low release of Ca^{2+} from sarcoplasmic binding proteins (parvalbumins) as these are absent in these muscle cells. These findings again imply that calcium is attached to the calcium sensor molecule at a time when externally recorded force is close to zero. Despite the fact that the $STnC_{DANZ}$ here is not filament bound, the binding of S1 to actin is thought to enhance Ca^{2+} binding to TnC and should therefore come off more slowly from the filament bound state (for the same on rate) than in these experiments with the non-filament bound (free) protein. Thus the rate of decay observed here should be faster than if the rate from the filament bound protein were being directly monitored.

1.4 Use of $STnC_{DANZ}$ in 'skinned' muscle fibres

Recently, it has been shown that TnC can be extracted from myofibrils by treatment with EDTA, resulting in a reduction in Ca^{2+}-dependent ATPase activity. Incubation with exogenous STnC fully restores the activity to these TnC-depleted myofibrils (28–30, 59). In the experiments described here, the

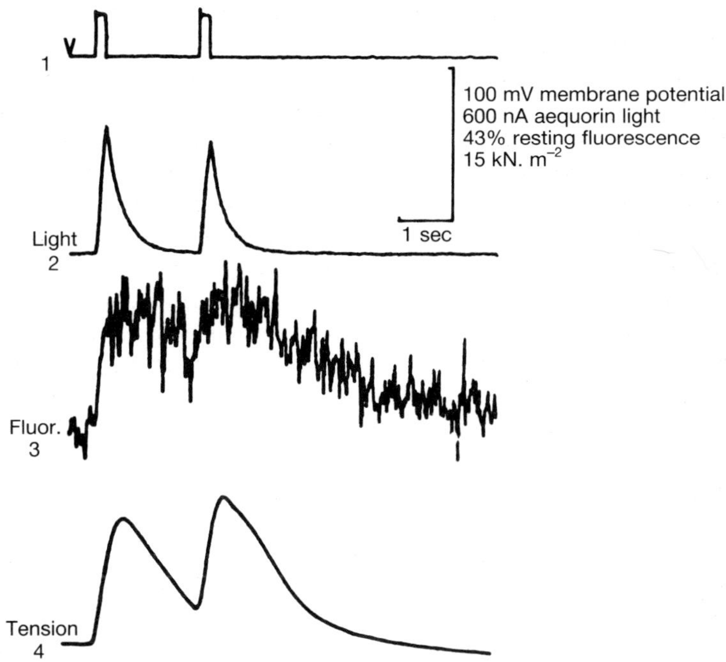

Figure 7. Force records (**trace 4**) from two identical voltage-clamp pulses (**trace 1**); aequorin-light (**trace 2**) and $STnC_{DANZ}$ fluorescence (**trace 3**) from a single barnacle muscle fibre. Deconvoluted fluorescence and luminescence records were consecutive. HP = –43 mV, V + 34 mV; 2 × 200 msec, temperature 15 °C. $STnC_{DANZ}$ approx. 10 μM. Resting luminescence: 130 nA. (From ref. 27, with permission.)

endogenous STnC was replaced by $STnC_{DANZ}$ in skinned fibres and myofibrils.

A wire hook was fixed to the end of the force transducer (AME, Horten, Norway, type AE801). The fibre bundle was wrapped around the hook and both ends glued to a needle fixed in placed about 2 mm from the end of the hook. A cylindrical cuvette with an inner diameter of about 1 mm was drawn around the fibre bundle. The fibres were bathed by passing solution through the cuvette. In this way there was no need to move the fibres in order to change the solution. Normally, the solution in the cuvette was renewed automatically every 2–5 sec.

The fibres from rabbit, fast skeletal muscle (glycerinated psoas), were skinned (Brij-treated), extracted, and reconstituted consecutively on the force transducer in the cuvette. All measurements and incubations were carried out at ambient temperature. Initially, the fibres were skinned chemically with 1% Brij 35 in relaxing solution for 20 min. Maximal stimulation of force was tested by changing the solution to pCa 4.0. Endogenous TnC was extracted from the fibres by replacing the normal relaxing solution with 2 mM EDTA–Tris, pH 7.8 and incubating 20 min while renewing the extraction solution every 15 sec. The endogenous TnC was replaced with exogenous and fluorescently labelled TnC by incubating the fibres approximately 20 min with 20 μg/ml TnC in relaxing solution. In order to stimulate the fibres to contract and relax, solutions were drawn through the cuvette from separate reservoirs. These solutions were all made to contain 7 mM EGTA, 2 mM MgATP, 1 mM Mg^{2+} (free), 105 mM K^+, 15 mM creatine phosphate, and 15 units/ml creatine phosphokinase adjusted to pH 7.00. Propionate was the major anion used and approximately 85 mM imidazole constituted the major proton buffer. The free Ca^{2+} was calculated from the total concentrations of ligands and metals using the stability constants of Smith and Martel (31). Relaxing solution contained 5×10^{-9} M Ca^{2+} (free) (pCa 8.3) and fully activating solution contained 1.0×10^{-4} M Ca^{2+} (free) (pCa 4.0).

Fluorescence intensity of the fibres was measured with a Zeiss model MPMU1 microscope. The exciting light from a xenon light-source was filtered at 340 nm (Schott SFK2) and focused on the fibres (Zeiss ULTRAFLUAR 10 objective). The epifluorescent light was collected through the objective and a 523-nm interference filter (Schott) before being converted to a voltage signal by the photomultiplier. For cardiac TnC_{IAANS}-reconstituted fibres the exciting light was filtered at 320 nm and the emission was filtered at 420 nm. Since both passed through the same objective the incident and fluorescent light had the same focus on the same part of the fibre (32).

Non-specific binding of $STnC_{DANZ}$ to EDTA-extracted fibres was also investigated, since it is possible that EDTA treatment may create additional binding sites for $STnC_{DANZ}$ other than TnC-depleted Tn complexes. Typically, fibres that were extracted with EDTA required 10–15 min incubation with $STnC_{DANZ}$ for full restoration of maximal Ca^{2+}-induced force. In *Figure 8* the bundle was incubated with $STnC_{DANZ}$ for 3.4 min (*Figure 8*, C) and again for 10 min (not illustrated, occurred between D and E of *Figure 8*.) A better signal-to-noise ratio was obtained when more $STnC_{DANZ}$ was incorporated; this increase in signal required more zero suppression and resulted in a larger maximal fluorescence change (*Figure 8*, E and F). The greater maximal Ca^{2+}-induced fluorescence change correlated with a greater maximal Ca^{2+}-induced change in tension after the second incubation with $STnC_{DANZ}$. These results can be most easily explained by a time-dependent incorporation of $STnC_{DANZ}$ into complexes regulating tension with no detectable nonspecific binding. Submaximal stimulation at pCa 5.8 elicited about the same fractional fluorescence change after the short and long incubations (*Figure 8*, D and F) (see ref. 32 for details). Thus, the extraction and reconstitution conditions do not appear to change significantly the tension responses measured isometrically in response to Ca^{2+} stimulation.

Changing the solution in the cuvette without moving the fibre made it possible to make a submaximal measurement directly by bracketing the submaximal stimulation with the relaxed and the fully stimulated conditions.

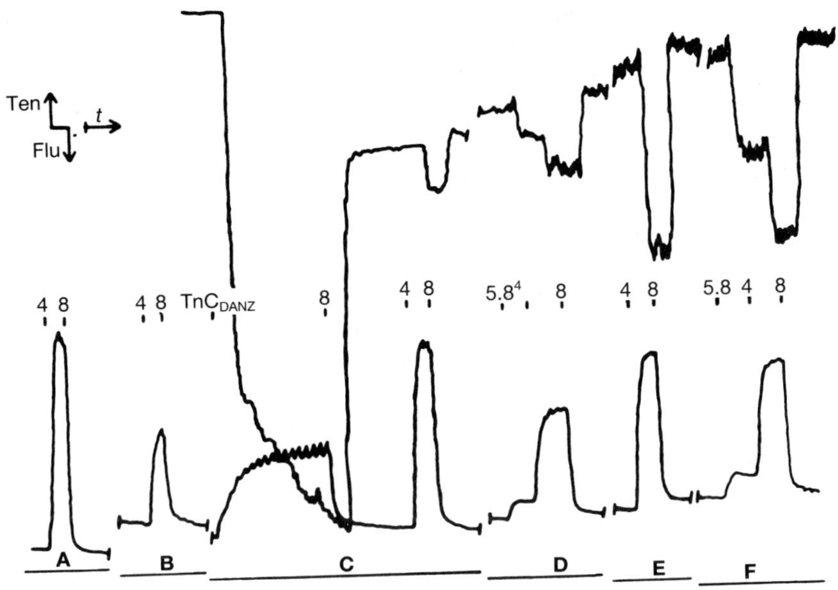

Figure 8. Tension and fluorescence record of STnC$_{DANZ}$-reconstituted psoas fibres. Shown are selected portions of a tension and fluorescence record for the same bundle of 2–3 fibres. The sequence is: **A**, a contraction of fibres after a 20-min incubation in 1% Brij 35; **B**, a contraction after a 20-min incubation with 2 mM EDTA; **C**, a 3.4-min incubation with STnC$_{DANZ}$ followed by the first contraction; **D**, submaximal and maximal contractions of the fibres 46-min and 12 contractions after STnC$_{DANZ}$ was washed out of the cuvette; **E**, the tension and fluorescence responses made immediately after reincubation of the fibres with STnC$_{DANZ}$ for 10 min; and **F**, tension and fluorescence response made 10 min later during which four contractions were made. 'Ten' (tension) and 'Flu' (fluorescence) indicate the direction of pen movement for an increase in signal, and the distance between the arrows represents the offsets by the pens. The vertical distance represents 10 mg of tension and 0.1 V of fluorescence signal except in **C**, where the tension scale remains the same, but the fluorescence scale represents 0.25 V. The chart speed and direction are indicated by *t*, which represents 1 min. The numbers 4, 8, and 5.8 represent the −log of the free Ca^{2+} concentration in the bathing solution at that point. The minimal and maximal changes in tension and fluorescence were noted at pCa 8 and 4, respectively. (From ref. 32, with permission.)

The striking result from the records of skinned fibres reconstituted with STnC$_{DANZ}$ was that the fractional fluorescence was consistently greater than the fractional tension by about 0.6 pCa units at the midpoints (*Figure 9*).

Since there are two Ca^{2+}-specific sites, different conformational states can be produced by the binding of Ca^{2+} to each site. A conformational state could occur when Ca^{2+} is bound to either site, one specific site, or both sites. Both the fractional fluorescence and tension changes were fit by making the Ca^{2+} saturation curve in between the two sets of data (*Figure 9*, one model). The fluorescence and tension data were found to be described well by the

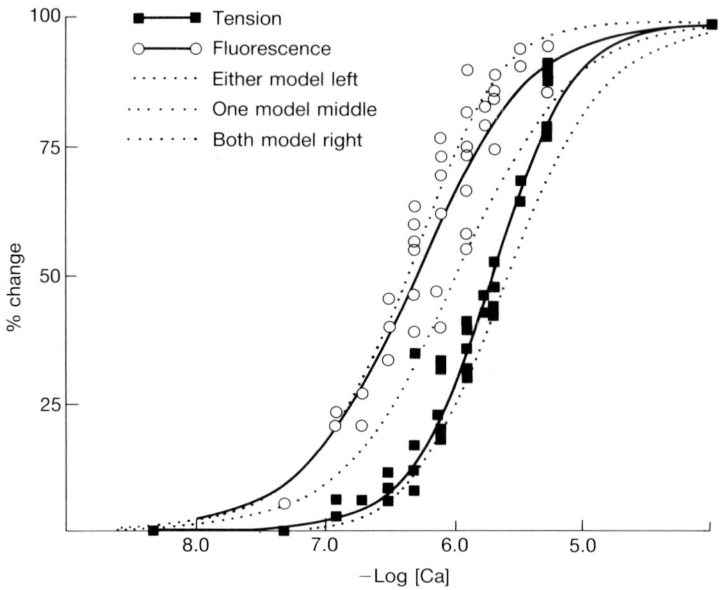

Figure 9. Steady-state fluorescence and tension changes of fibres reconstituted with $STnC_{DANZ}$. Fractional tension (■) and fluorescence (○) data which were collected in pairs were compiled from four separate experiments and four fibre bundle preparations. The procedure used for making the measurement is described in Section 1.4. The per cent maximal change was determined for all points by increasing the Ca^{2+} submaximally and then maximally as shown in *Figure 8*. The solid curves are generated with the Hill equation and the parameters appear in *Table 1*. The theoretical curves for the cumulative distributions of TnC molecules with Ca^{2+} bound at either, one particular, or both Ca^{2+}-specific sites were generated with the function

$$Y = \frac{ß[Ca^{2+}] + (K[Ca^{2+}])^2}{(1 + K[Ca^{2+}])^2}$$

where Y = fraction of TnC molecules, and K is the intrinsic association constant = $1.0 \times 10^6\ M^{-1}$ and $ß$ is equal to K for one particular site occupied with Ca^{2+}, $2K$ for either site, and zero for both sites occupied. (From ref. 32, with permission.)

cumulative distributions of TnC molecules that have Ca^{2+} bound at either and both sites, respectively. Thus, the results with fibres reconstituted with $STnC_{DANZ}$ are consistent with tension and fluorescence changes responding to two different conformational states induced by Ca^{2+} binding to the Ca^{2+}-specific sites and the affinity of the Ca^{2+}-specific sites of TnC measured in the steady-state is described by the theoretical distribution of Ca^{2+} binding to one particular site (*Figure 9*), i.e. $1 \times 10^6\ M^{-1}$ (32).

1.5 $STnC_{IAANS}$ as an indicator in intact single fibres

In experiments employing the fluorescently labelled derivative of TnC, $STnC_{IAANS}$, labelled (cys-98) at the Ca^{2+}-Mg^{2+} site (site III) (*Figure 2*), there

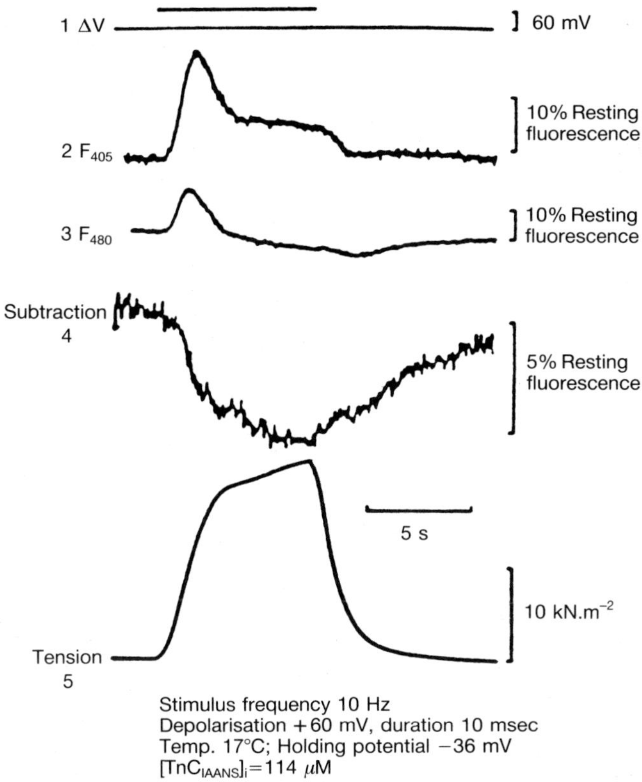

Figure 10. $STnC_{IAANS}$ fluorescence responses from an injected barnacle muscle fibre as a result of a long-duration voltage-clamp pulse. **Trace 1**, voltage; **trace 2**, 405 nm Ca^{2+}-independent; and **trace 3**, 480 nm Ca^{2+}-dependent fluorescence signal; **trace 4**, subtraction of trace 3 from trace 2; trace 5, isometric force record. (From ref. 27, with permission.)

is a fluorescence decrease, at physiological Mg^{2+} concentrations, associated with Ca^{2+} binding. The $STnC_{IAANS}$ was axially injected into single arthropod muscle fibres (27) and the fluorescence change observed at 480 nm, compared to Ca^{2+}-independent fluorescence change occurring at 405 nm. In short-duration pulse (<250 msec) experiments, produced by sinusoidal stimulation, little change in the $STnC_{IAANS}$ signal was observed despite an appreciable force transient, while a transient $STnC_{DANZ}$ signal would normally have been observed. At longer-stimulus durations (>5 sec) a marked fall in $STnC_{IAANS}$ fluorescence was observed and was followed, upon cessation of stimulation, by a return of the fluorescence to the initial values. These results are consistent with a slow Ca^{2+} loading and subsequent unloading of the Ca^{2+}-Mg^{2+} sites. The time-course of these changes is demonstrated in *Figure 10*. Since the Ca^{2+}-Mg^{2+} sites of STnC are virtually equivalent in affinity to those of the paravalbumins, the time-course of loading and unloading of free $STnC_{IAANS}$ should also give an indication of the likely *in vitro* loading of these soluble Ca^{2+} binding proteins during a force transient (27).

2. Photolabile caged Ca^{2+} compounds

The application of recently developed photorelease techniques allows the concentration of substances of interest to be changed rapidly in cellular preparations without the diffusional delays encountered using previous techniques. The procedure involves the photolysis of a biologically inactive precursor molecule within the preparation to release rapidly the desired product. One class of photolabile precursors have been created by the introduction of photolabile groups into either the chelators BAPTA (nitr-2, nitr-5, nitr-7, diazo-2, and diazo-4) or EDTA (DM-nitrophen) (33, 58), which bind divalent ions (e.g. Ca^{2+}, Mg^{2+}) with a high affinity. Upon illumination with UV light (300–400 nm) nitr-5, a caged calcium compound, photolyses releasing calcium into free solution with a time-constant of 0.3 msec. The dissociation constant (K_d) for Ca^{2+} is 0.145 μM before photolysis and 6 μM after illumination, at 0.1 M ionic strength. DM-nitrophen has an even larger change in dissociation constants (from 5 nM to 2 mM with a time-constant of <0.4 msec) as the EDTA moiety is cleaved to yield two iminodiacetic acid molecules. However, it is much less selective for calcium than nitr-5, and under physiological conditions (Mg^{2+} 1–3 mM) photolysis can cause a large release of magnesium. Diazo-2 changes its K_d from 2 μM to 0.7 μM on photolysis (time-constant 0.4 msec), allowing the free calcium concentration to be lowered rapidly to a new value (2–3 msec). These photolabile chelators provide powerful tools for fast localized control of free calcium within cellular preparations.

2.1 Caged calcium (nitr-series)

The experiments with caged ATP have demonstrated the elegant way in which the technique of flash photolysis of caged compounds can be applied to study the reaction rates of processes within skinned muscle fibres. The application of similar techniques to the study of the activation of muscle by Ca^{2+} had to await the development of a suitable compound. This required the synthesis of a calcium chelator whose affinity for calcium would fall upon photolysis. Several such photosensitive chelators have now been synthetized (33–35), the nitr-series is based upon the parent chelator BAPTA (36).

Tsien and co-workers considered the properties that would be required by an ideal caged calcium molecule for physiological investigations. In addition to the above requirements a caged calcium should:

(a) Bind Ca^{2+} prior to photolysis with a dissociation constant of 0.1–0.3 μM or lower under physiological conditions (pH 7.0, ionic strength >0.1 M), as this is the level of Ca^{2+} in resting cells (37), otherwise the chelator would not have a significant amount of calcium bound to it before photolysis.

(b) The dissociation constant for Mg^{2+} should be greater than 1 mM, so that there will not be significant amounts of magnesium bound to the compound prior to photolysis and therefore no great change in the free magnesium concentration following photolysis.

(c) After photolysis the chelator should have a dissociation constant for calcium greater than 1 μM.

(d) The release of Ca^{2+} following photolysis should be fast (>1 msec), to be useful in examining the kinetics of Ca^{2+} activated processes in cells (e.g. $k_{Ca,on}$ STnC *ca.* 10^8 M^{-1} sec^{-1}).

The first compound synthesized, nitr-1 demonstrated the feasibility of this approach (35). Unfortunately, this particular compound had a low extinction coefficient at wavelengths above 300 nm and this limited its usefulness in biological experiments. The addition of alkosyl substituents to the photosensitive nitrobenzyl group to give a nitropiperonyl group improved the absorbance at longer ultraviolet wavelengths and allowed this compound (nitr-2) to be used in neurophysiological experiments (38). Nitr-2 however, releases Ca^{2+} only relatively slowly ($k_{Ca,off}$ = 5 sec^{-1} at 20 °C) but modification of the compound by substitution of a hydrogen for the methyl leaving group gave nitr-5 and this modification considerably improved the rate of Ca^{2+} release ($k_{Ca,off}$ = 3000 sec^{-1} at 20 °C).

This compound, nitr-5, now meets most of the criteria outlined above and has been successfully used in both neurophysiological studies (39) as well as in

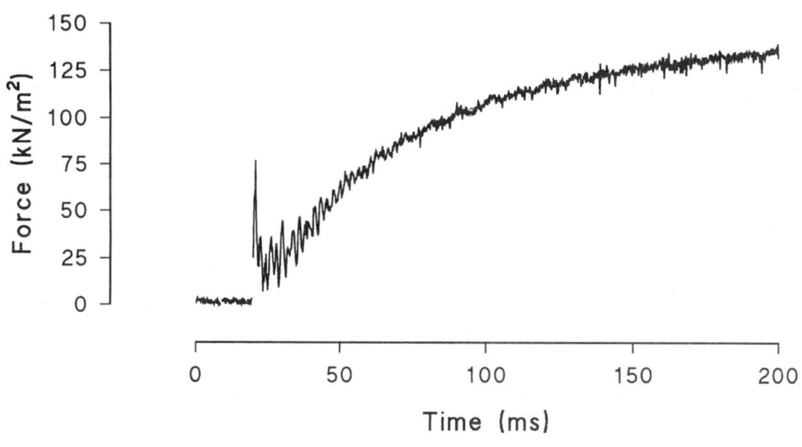

Figure 11. Laser flash photolysis of nitr-5 in a chemically skinned single fibre from frog semitendinosus muscle by a 50 nsec pulse of light at 347 nm from a frequency doubled ruby laser. Sarcomere length 2.1 μm; temperature 12 °C.

studies of muscle kinetics (40–45). Its K_d prior to photolysis is 145 nM (12 °C, I = 0.2 M) which changes to 6 μM after photolysis. The quantum efficiency has been measured (35) as 0.012 for calcium free nitr-5 and 0.035 for calcium-saturated nitr-5 (epsilon = 5500 M^{-1} cm^{-1} at 355–365 nm prior to photolysis). The only photolysis by-product is water, rather than the methanol produced on photolysis of nitr-2.

By rapidly (<1 msec) increasing the free calcium concentration within the myofibrillar space of skinned muscle using laser-flash photolysis of the photolabile calcium chelator, nitr-5, the delay imposed by diffusion in earlier pCa-clamp experiments can be overcome. Single fibres from the semitendinosus muscle of the frog, *Rana temporaria*, permeabilized with Triton X-100 (1% w/v) could be fully activated using this technique. The half-time for force development was 40 ± 2 sec (n = 14) (*Figure 11*; refs 40–43, 46) at 12 °C which is similar to that found in electrically-stimulated intact frog sartorius muscles (47), (55 msec for tetanized frog muscle at 3 °C) and in single tibialis anterior fibres (36.9 msec at 12 °C) (2), and is considerably faster than the half-time of force development seen in experiments using the pCa-clamp method, which at pCa 5.88 was 1.6 sec^{-1} at 4 °C (48). (See Chapter 7, this volume, for more detail.)

2.2 Diazo-2

2.2.1 Caged chelators

Recently diazo-2 (49), a photolabile derivative of BAPTA, has been described which upon photolysis is rapidly (>2000 sec^{-1}) converted from a chelator with a low affinity for Ca^{2+} (K_d = 2.2 μM) to one with a high affinity for Ca^{2+} (K_d = 0.073 μM).

Figure 12 shows the spectrum obtained when 20 μM diazo-2 (either calcium-free A or calcium-loaded B) was photolysed. After one flash of the frequency-doubled ruby laser, exposure of these two samples to a further series of ten flashes reduced the absorbance of calcium-free diazo-2 at 370 nm to 4% of its initial value and the absorbance of calcium-loaded diazo-2 at 330 nm to 15% of its initial value.

2.3 Use of nitr-5 and diazo-2 in frog and scallop skinned muscle fibres

Frog muscle fibres were dissected from the semitendinosus muscle of the frog *Rana temporaria.* A bundle of fibres was removed from the distal head of the semitendinosus muscle and placed in a bath containing mineral oil on a cooled stage at 4 °C. Single fibres were dissected and T-shaped aluminium clips (51) were attached to each end to give a fibre length of about 3 mm. The fibres were skinned for 10–30 min by immersion in a relaxing solution (10 mM EGTA) to which 1% Triton X-100 had been added.

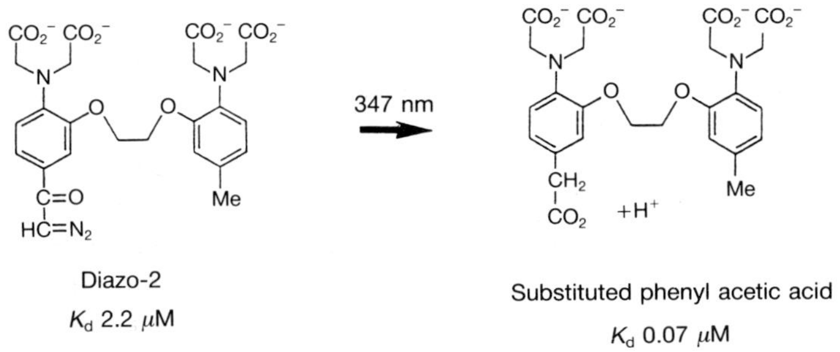

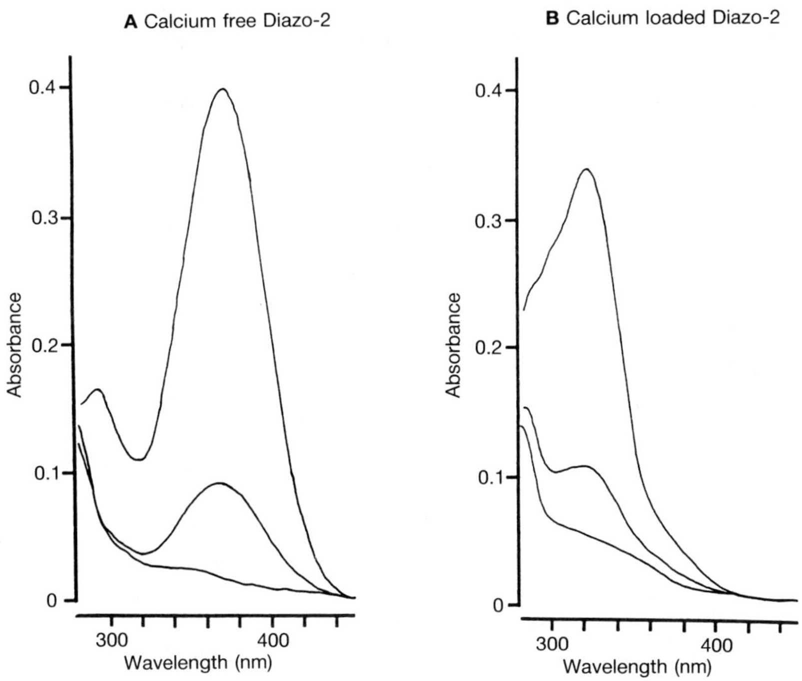

Figure 12. Spectra obtained upon photolysis of 20 μM diazo-2 in an acryl cuvette. **A**: Calcium-free diazo-2 in the presence of 1 mM EGTA. **B**: Calcium-loaded diazo-2 in the presence of 1 mM EGTA and 10 mM $CaCl_2$. In each panel the upper trace is the spectrum obtained prior to photolysis, the middle trace is the spectrum obtained after one laser flash and the lower trace is the spectrum obtained after ten further laser flashes. (From ref. 50, with permission.)

2.3.1 Tension recording

The fibres were attached to the silicon beam of a strain gauge (Akers, Norway). Measurements of the fibre width and thickness were made using a microscope mounted above the fibre, in order to calculate the cross-sectional area. Tension was continuously recorded on a chart recorder and simultaneously on a digital storage oscilloscope (Nicolet) and also on an IBM PC AT microcomputer via a CED 1401 analogue-to-digital convertor.

2.3.2 Fibre troughs

The fibres were suspended in a trough (52, 53) for the experiment. A stainless steel block was made containing five troughs each approximately 5 mm wide and 8 mm long. The troughs held, by capillarity, a volume of 12–30 μl. The block was hollow, and a cooling solution was circulated through it, so that the temperature was controlled at 12 ± 0.1 °C. The front trough had a glass floor which allowed the fibre to be illuminated by a light from below, which made the mounting and laser diffraction measurements of the sarcomere length of the fibre easier. The front of this trough was made of quartz so as to enable experiments to be performed in which the fibre could be activated by flash photolysis while still bathed in the solution, rather than an air flash.

2.3.3 Ruby laser

The light source used for the photolysis procedure was a Q-switched frequency-doubled ruby laser (Lumonics: 2 joules at 694 mnm:QSR2/6). The ruby laser consisted of a ruby rod which was optically pumped by a pair of linear flash-lamps inside a reflective cavity. Operating in the Q-switched mode the laser produces a pulse of light at 694 nm (primary) with a duration of 50 nanoseconds and a peak intensity of 100–200 megawatts.

For biological experiments with caged compounds light in the ultraviolet range is needed. This is obtained by doubling the frequency of the light emitted from the ruby laser. Frequency doubling can be accomplished by using a crystal of rubidium dihydrogen arsenate (RDA) maintained at +50 °C. Approximately 10% of the primary illuminating radiation is released from the crystal as secondary harmonic radiation at 347 nm.

The primary 694-nm and secondary 347-nm emissions from the RDA crystals are polarized at right angles, and hence the primary radiation can be attenuated by a gull-wing quartz prism beam separator, the energy of the 347-nm radiation is 200–300 mJ in the frequency-doubled mode.

2.3.4 Beam focusing

The beam of secondary radiation produced by the fibres was roughly circular in cross-section. This beam was focused to a thin horizontal beam by a plano-convex quartz lens (Ealing Electro-optics) of focal length 120 mm, placed in front of the trough. Two movable shutters in the optical path before the lens allowed the horizontal length of the fibre to be adjusted to suit the length of each individual fibre. The alignment and dimensions of the beam were checked by exposing a piece of exposed and developed photographic paper to the laser flash. The light beam left a clear image (burn pattern) on the paper and approximate adjustments could be made. The energy of the light beam could be controlled by altering the charging potential of the capacitors which drove the xenon flash-lamps of the laser. The energy of the laser flash at 347 nm used in these experiments was normally in the range of 70–100 mJ and was measured for each flash using an on-line energy meter.

2.3.5 Experimental protocol

After skinning was complete the fibres were placed in a relaxing solution, the sarcomere length was adjusted to 2.3 μm and the dimensions of the fibre were measured. The fibres were then transferred to the nitr-5 or diazo-2 solution. The level of free Ca^{2+} in the diazo-2 or nitr-5 had previously been adjusted to produce either a maximal contraction or just threshold respectively prior to illumination. After the Ca^{2+} had reached equilibrium, the trough containing the caged-compound was lowered pneumatically from the fibre, suspending the fibre in air. After 600 msec the laser flashed, and the fibre was allowed to reach a steady state before being placed in a trough of relaxing solution. By keeping the fibre at a temperature of 12 °C, which is the temperature of the dew point, the temperature of the fibre remained constant (54).

Finally, experiments with the caged chelator, diazo-2 provided rates of relaxation in single skinned frog fibres which are at least as fast as those seen *in vivo* whilst the fall in free calcium is likely to be much faster ($t_{½}$ <4 msec) (50, 55–57) than even the fast phase of the relaxation *in vivo* at 12 °C ($t_{½}$ = 70–80 msec), while in myosin-regulated scallop striated muscle the rates achieved are far faster than intact muscle (ref. 57; *Figures 13* and *14*). It is now possible to use this method to study how rapidly the TnC-bound Ca^{2+} falls, using fluorescently labelled derivatives such as $STnC_{DANZ}$ (Section 1.1) and to determine which are the steps which limit relaxation *in vivo*.

Acknowledgement

The SERC, BHF, and MRC are thanked for financial assistance. Our thanks to Dr J. D. Potter for fluorescent TnC derivatives and to Drs Steven Adams and Roger Tsien for caged calcium chelators.

References

1. Ashley, C. C. and Ridgway, E. B. (1970). *Journal of Physiology*, **209**, 105–30.
2. Griffiths, P. J., Duchateau, J. J., Maeda, Y., Potter, J. D., and Ashley, C. C. (1990). *Pflügers Archiv für die gesamte Physiologie*, **415**, 554–65.
3. Potter, J. D. and Johnson, J. D. (1982). *Calcium and cell function*, Vol. II. (ed. W. Cheung), pp. 145–73. Academic Press, New York.
4. Potter, J. D. and Gergely, J. (1975). *Journal of Biological Chemistry*, **250**, 4628–33.
5. Holroyde, M. J., Robertson, S. P., Johnson, J. D., Solaro, R. J., and Potter, J. D. (1980). *Journal of Biological Chemistry*, **255**, 1168–93.

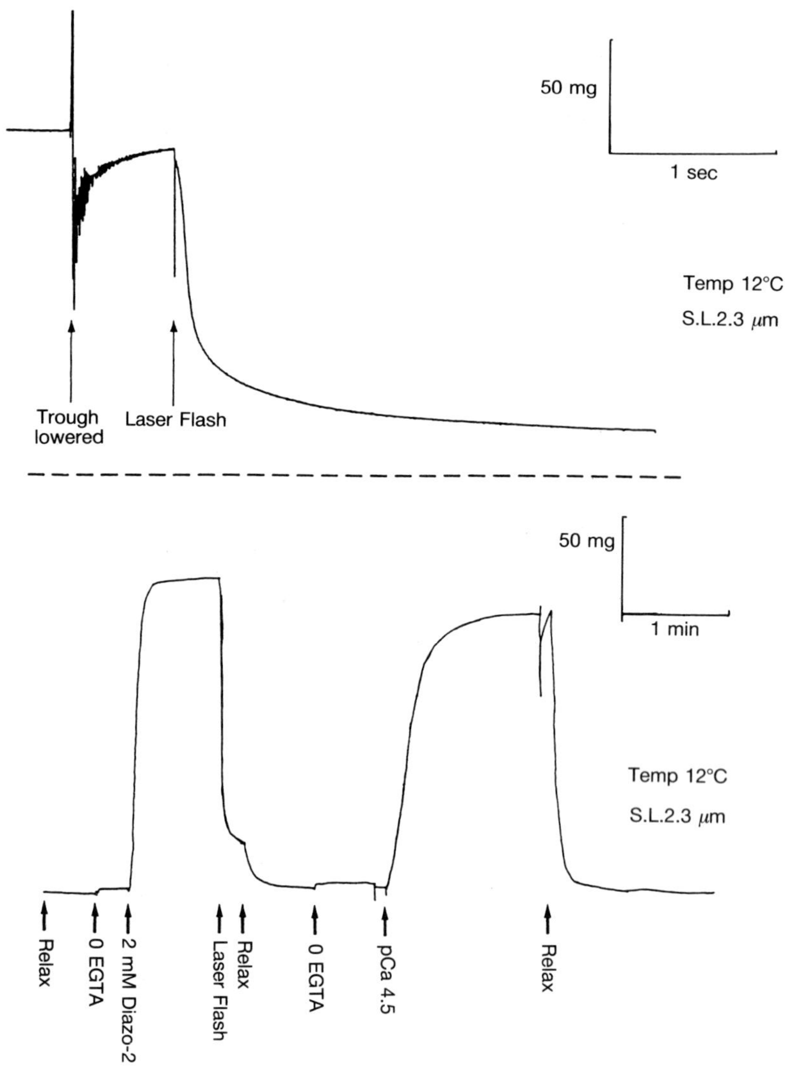

Figure 13. The upper panel shows the tension record of a single skinned frog muscle fibre which has been equilibrated in 2 mM diazo-2–0.95 mM $CaCl_2$(pCa < 5.7) and then exposed to a brief flash of light from a frequency-doubled ruby laser (347 nm). The lower panel shows the same tension recording on a slower time-scale allowing the development of tension in diazo-2 to be seen and compared with the tension developed in an activating solution, pCa 4.5. (From ref. 50, with permission.)

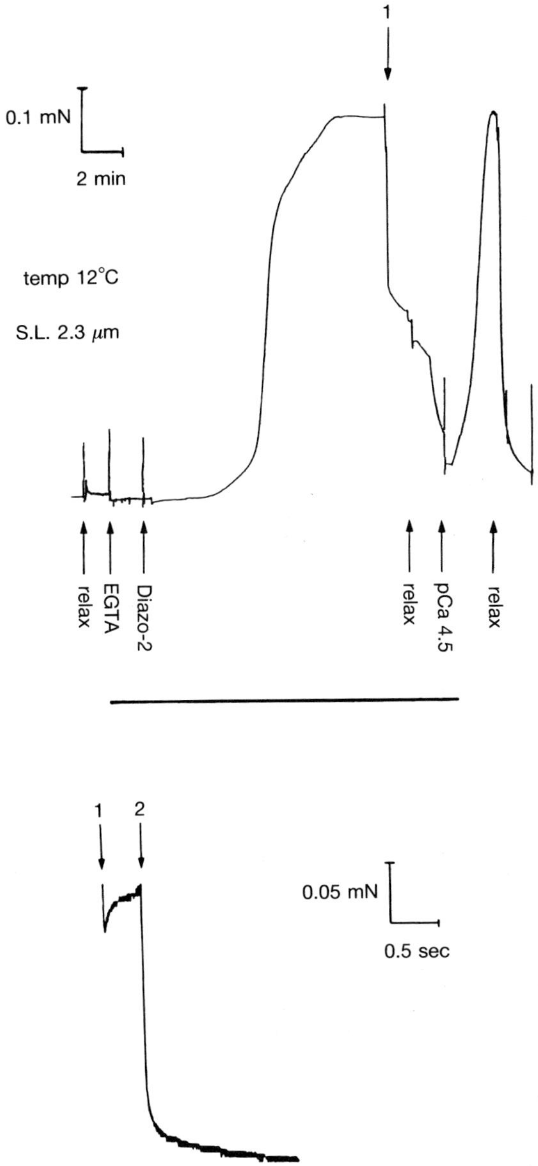

Figure 14. *Above*: The tension record from scallop skinned muscle fibre bundle which was allowed to come to equilibrium in diazo-2–$CaCl_2$(pCa ~5.0) and then flashed in air with a pulse of light (347 nm) from a frequency-doubled ruby laser. 1 = trough lowered and laser flash. This is followed by a contraction in activating solution (pCa 4.5) showing that the contraction in diazo-2 was maximal. Just prior to flashing the speed of the chart recorder was increased 10 times. Diameter 209 μm, length 3.0 mm, $t_{1/2}$ = 14 msec. *Below*: The same record shown on a faster time-scale. 1 trough lowered; 2 = laser flash. Diazo-2 = 2 mM. (From ref. 57, with permission.)

6. Johnson, J. D., Collins, J. H., Robertson, S. P., and Potter, J. D. (1980). *Journal of Biological Chemistry*, **255**, 9635–40.
7. Van Eerd, J. P. and Takahashi, K. (1976). *Biochemistry*, **15**, 1171–80.
8. Kretsinger, R. H. and Nockolds, C. E. (1973). *Journal of Biological Chemistry*, **248**, 3313–26.
9. Collins, J. H., Greaser, M. L., Potter, J. D., and Horn, M. J. (1977). *Journal of Biological Chemistry*, **252**, 6356–62.
10. Leavis, P. C. and Kraft, E. L. (1978). *Advances in Biochemistry and Biophysics*, **186**, 411–15.
11. Zot, H. G., Iida, S., and Potter, J. D. (1983). *Chemica Scripta*, **21**, 133–6.
12. Potter, J. D., Robertson, S. P., and Johnson, J. D. (1981). *Federation Proceedings*, **40**, 2653–6.
13. Zot, H. G. and Potter, J. D. (1987). *Journal of Muscle Research and Cell Motility*, **8**, 428–36.
14. Ogawa, Y. (1985). *Journal of Biochemistry* (Tokyo), **97**, 1011–23.
15. Putkey, J. A., Sweeney, H. L., and Campbell, S. T. (1989). *Journal of Biological Chemistry*, **264**, 12370–8.
16. Grabarek, Z., Tan, R. Y., Wang, J., Tao, T., and Gergely, J. (1990). *Nature*, **345**, 132–5.
17. Johnson, J. D., Collins, J. H., and Potter, J. D. (1978). *Journal of Biological Chemistry*, **253**, 6451–548.
18. Johnson, J. D., Charlton, S. C., and Potter, J. D. (1979). *Journal of Biological Chemistry*, **254**, 3497–502.
19. Rosenfeld, S. S. and Taylor, E. W. (1985). *Journal of Biological Chemistry*, **260**, 242–51.
20. Rosenfeld, S. S. and Taylor, E. W. (1985). *Journal of Biological Chemistry*, **260**, 252–61.
21. Hess, P., Metzger, P., and Weingart, R. (1982). *Journal of Physiology*, **329**, 173–88.
22. Gillis, J. M., Thomason, D., Leferve, J., and Kretsinger, R. H. (1982). *Journal of Muscle Research and Cell Motility*, **3**, 377–98.
23. Gillis, J. M. (1985). *Biochimica et Biophysica Acta*, **811**, 97–145.
24. Herzberg, O. and James, M. N. G. (1985). *Nature*, **313**, 653–9.
25. Ashley, C. C. and Ellory, J. C. (1972). *Journal of Physiology*, **226**, 653–74.
26. Griffiths, P. J., Potter, J. D., Coles, B., Strang, P., and Ashley, C. C. (1984). *FEBS Letters*, **176**, 144–50.
27. Griffiths, P. J., Potter, J. D., Maeda, Y., and Ashley, C. C. (1988). *Advances in Experimental Medical Biology*, **226**, 113–29.
28. Cox, J. A., Comte, M., and Stein, E. A. (1981). *Biochemical Journal*, **195**, 205–11.
29. Zot, H. G. and Potter, J. D. (1982). *Journal of Biological Chemistry*, **257**, 7678–83.
30. Güth, K. and Potter, J. D. (1987). *Journal of Biological Chemistry*, **262**, 13627–35.
31. Smith, R. M. and Martell, A. E. (ed.) (1974). *Critical stability constants*, I, pp. 1–65. Plenum Press, New York.
32. Zot, H. G., Guth, K., and Potter, J. D. (1986). *Journal of Biological Chemistry*, **261**, 15883–90.

33. Kaplan, J. and Ellis-Davies, G. (1988). *Proceedings of the National Academy of Sciences, USA*, **85**, 6571–5.
34. Adams, S. R., Kao, J. P. Y., and Tsien, R. Y. (1986). *Journal of General Physiology*, **88**, 9a–10a.
35. Adams, S. R., Kao, J. P. Y., Grynkiewicz, G., Minta, A., and Tsien, R. Y. (1988). *Journal of the American Chemical Society*, **110**, 3212–20.
36. Tsien, R. Y. (1980). *Biochemistry*, **19**, 2396–404.
37. Ashley, C. C. and Campbell, A. K. (1979). *Detection and measurement of free Ca^{2+} in cells*. North-Holland, Amsterdam.
38. Tsien, R. Y. and Zucker, R. S. (1986). *Biophysical Journal*, **50**, 843–53.
39. Gurney, A. M., Tsien, R. Y., and Lester, H. A. (1987). *Proceedings of the National Academy of Sciences, USA*, **84**, 3496–500.
40. Ashley, C. C., Barsotti, R. J., Ferenczi, M. A., Lea, T. J., and Mulligan, I. P. (1987). *Journal of Physiology*, **394**, 24P.
41. Ashley, C. C., Barsotti, R. J., Ferenczi, M. A., Lea, T. J., Mulligan, I. P., and Tsien, R. Y. (1987). *Journal of Physiology*, **390**, 144P.
42. Ashley, C. C., Barsotti, R. J., Ferenczi, M. A., Lea, T. J., and Mulligan, I. P. (1988). *Journal of Physiology*, **398**, 71P.
43. Ashley, C. C., Barsotti, R. J., Ferenczi, M. A., Lea, T. J., and Mulligan, I. P. (1989). *Methodological surveys in biochemical analysis*, Vol. 19 (ed. E. Reid, G. M. W. Book, and J. P. Luzio), p. 131–2. Royal Society of Chemistry, London.
44. Lea, T. J., Fenton, M. J., Potter, J. D., and Ashley, C. C. (1990). *Biochimica et Biophysica Acta*, **1034**, 186–94.
45. Lea, T. J. and Ashley, C. C. (1990). *Journal of Physiology*, **427**, 435–53.
46. Ashley, C. C., Lea, T. J., and Mulligan, I. P. (1991). *Quarterly Reviews of Biophysics*, **24**, 1–73.
47. Kress, M., Huxley, H. E., Faruqi, A. R., and Hendrix, J. (1986). *Journal of Molecular Biology*, **188**, 325–42.
48. Moisescu, D. G. (1976). *Nature*, **262**, 610–13.
49. Adams, S. R., Kao, J. P. Y., and Tsien, R. Y. (1989). *Journal of the American Chemical Society*, **111**, 7957–68.
50. Mulligan, I. P. and Ashley, C. C. (1989). *FEBS Letters*, **255**, 196–200.
51. Goldman, Y. and Simmons, R. M. (1984). *Journal of Physiology*, **350**, 497–518.
52. Goldman, Y., Hibberd, M. G., McCray, J. A., and Trentham, D. R. (1982). *Nature*, **300**, 701–5.
53. Ferenczi, M. A., Goldman, Y., and Simmons, R. M. (1984). *Journal of Physiology*, **350**, 519–43.
54. Ferenczi, M. A. (1986). *Biophysical Journal*, **50**, 471–7.
55. Ashley, C. C., Mulligan, I. P., and Palmer, R. (1990). *Journal of Physiology*, **426**, 31P.
56. McCray, J. and Trentham, D. R. (1989). *Annual Review of Biophysical Chemistry*, **18**, 239–70.
57. Palmer, R. E., Mulligan, I. P., Nunn, C., and Ashley, C. C. (1990). *Biochemical and Biophysical Research Communications*, **168**, 255–30.
58. Kaplan, J. (1990). *Annual Reviews in Physiology*, **52**, 897–914.
59. Collins, J. H., Thiebert, J. L., Francois, J.-M., Ashley, C. C., and Potter, J. D. (1991). *Biochemistry*, **30**, 702–7.

9

Calmodulin

ANTHONY R. MEANS, INDRANI C. BAGCHI, MARK F. A. VANBERKUM, and COLIN D. RASMUSSEN

1. Introduction

A variety of mechanisms exist to maintain intracellular Ca^{2+} homeostasis (1). The markable regulation of this ion has evolved in a manner so that Ca^{2+} serves as the only intracellular second messenger of its class. Along with cyclic AMP, Ca^{2+} controls a wide variety of intracellular processes. Analogous to cyclic AMP, Ca^{2+} most frequently executes its regulatory roles after binding with high affinity to intracellular Ca^{2+} binding proteins. Calmodulin is the primary example of such a receptor protein owing to its ubiquitous nature and wide ranging effects.

Calmodulin utilizes a linear sequence of 12 amino acids to bind Ca^{2+} (2). The three-dimensional structure formed to co-ordinate Ca^{2+} has been called an 'E-F hand' or 'calmodulin-fold' (3). A superfamily of proteins utilizes this configuration to bind Ca^{2+} but calmodulin is the only member that subserves such a plethora of functions. About 30 enzymes and structural proteins have been identified that bind calmodulin with high affinity (4). The Ca^{2+}–calmodulin complex can also regulate the intracellular concentration of other second messengers such as cyclic AMP and inositoltrisphosphate (IP_3). Isoforms of adenylyl cyclase and phosphodiesterase are calmodulin-dependent and the enzyme that phosphorylates IP_3 also utilizes calmodulin as a regulatory subunit. Thus it is not surprising that calmodulin is an essential protein and disruption of the single gene in yeast and fungi is lethal (5).

The purpose of this chapter is to detail methods that can be used to study calmodulin and its functions. We described purification and assays of this protein, detection of its target enzymes and strategies to produce and mutate calmodulin in both bacterial and animal cells. Each procedure is currently utilized in our laboratories and is discussed in the context of implementation to a variety of biological problems.

Because of widespread interest in calmodulin research many useful reagents are now available commercially. Purified calmodulin, anti-calmodulin, calmodulin–sepharose or -agarose, ^{3}H-propionylated calmodulin, ^{125}I-calmodulin, complete radioimmunoassay kits, and several calmodulin-deficient target enzymes are available from major biochemical suppliers such as Amersham, Boehringer-Mannheim Biochemicals, Calbiochem-Behring, and Sigma Chemical Company. For laboratories that plan extensive calmodulin research it may be more economically feasible to produce at least some of their reagents by the protocols described in this chapter.

2. Purification of calmodulin

Because of its highly conserved primary sequence, any tissue which affords good yields of calmodulin may be used as a source for the pure protein. Although calmodulin can be purified to homogeneity from rat testis by the following technique with a yield of 200 mg/kg (wet weight), use of other tissues may require modifications to this scheme.

Protocol 1. Calmodulin purification

1. *Homogenization.* Frozen testes (1 kg) are obtained from Pel-Freeze Biological Inc. They are thawed in 2 vol. of deionized water and homogenized for three 1-min bursts with a Brinkman Polytron (PT-40; setting 6) with 5-min cooling periods in ice.
2. *Heat-treatment.* The homogenate is heated to boiling in 200-ml beakers in a microwave oven (approximately 90 sec), and rapidly cooled in a methanol dry-ice bath. The agglutinated suspension is centrifuged at 20 000 *g* for 30 min and the supernatant solutions pooled. The resultant supernatant solution is brought to 10 mM Imidazole, 1 mM EGTA pH 6.1 (buffer A) and dialysed against the same buffer with several changes for 18 h at 4 °C.
3. *Ion-exchange chromatography.* The dialysed solution is applied to a phosphocellulose column (2.5 × 10 cm) equilibrated with buffer A. Activator (calmodulin) protein and the protein kinase inhibitor pass unretarded and can be directly applied to a DEAE-cellulose column (1.5 × 40 cm) equilibrated with buffer A. Unbound protein is then washed from the column (approx. 500 ml buffer A to an effluent $A_{280} = 0.05$) at which time a 1000-ml linear gradient (0–0.4 M NaCl) is initiated and 10 ml fractions collected. Fractions containing calmodulin activity are pooled and concentrated by lyophilization.
4. *Ultrogel chromatography.* The concentrated DEAE-fractions are applied to a 2.5 × 90 cm column packed with Ultrogel 44. Up to 2% of the bed volume can be applied (20 ml) yet the flow rate should be maintained at 10–15 ml/h and 5-ml fractions collected.

Table 1. Purification of calmodulin from 1 kg rat testis

	Total units	Total protein (mg)	Specific activity (U/mg)	Purification (fold)	Yield per cent
Crude homogenate	3.0×10^6	–	2392	33	100
Heat-treatment	2.8×10^6	1254	2230	68	93
Phosphocellulose	3.0×10^6	1152	2604	79	100
DEAE-cellulose	2.2×10^6	97	22 680	687	73
Gel filtration	2.1×10^6	96	21 875	663	70

Table 2. Summary of rat testis calmodulin properties

Property	Technique	Value
MW	SDS-gel	18 000 ± 1000
Sedimentation coefficient	Sucrose gradient	1.9S
Ca^{2+} binding	Equilibrium dialysis	4 mol/mol protein
α-helical content ($+Ca^{2+}$)	CD	67%
$1\%_{280}$	Spectrophotometry	2.1
pI	Isoelectric focusing	3.85
AA composition	AA analysis	lacks cysteine and tryptophan

Table 1 shows the results expected from a typical purification sequence while *Table 2* represents the physical properties of rat testis calmodulin.

3. Use of phosphodiesterase activation as a bioassay

A convenient assay to use to detect calmodulin during its purification relies on its ability to activate specific forms of cyclic nucleotide phosphodiesterases.

These enzymes catalyse the hydrolysis of the 3′–5′ diester bond of 3′–5′ cyclic nucleotides to the 5′-monophosphate nucleotide.

$$\text{cAMP} \xrightarrow{\text{Phosphodiesterase}} 5'\text{-AMP}$$

The most sensitive and therefore most commonly employed assay procedures involve the hydrolysis of tritiated cyclic nucleotides with subsequent separation of substrate and reaction product via various forms of chromatography followed by scintillation counting. When considering a PDE assay one should select a rapid, reliable, inexpensive assay, which requires minimal time. One of the most widely used assays involves hydrolysis of ^3H cyclic nucleotide by the following reactions.

3.1 Radiometric assay

$$
\begin{array}{ccccc}
 & & & & P_i \\
 & & & & \uparrow \\
{}^3\text{H-cAMP} & \rightarrow & {}^3\text{H-5}'\text{AMP} & \rightarrow & {}^3\text{H-adenosine} \\
 & \text{PDE} & & \text{VENOM} & \\
{}^3\text{H-cGMP} & \rightarrow & {}^3\text{H-5}'\text{GMP} & \rightarrow & {}^3\text{H-guanosine} \\
 & & & & \downarrow \\
 & & & & P_i
\end{array}
$$

3.1.1 Substrate preparation

Although radiolabelled cyclic nucleotides are specified by the vendor as being greater than 99% pure as determined by paper chromatography, the preparations contain as much as 10% ^3H_2O. To reduce background levels, this contamination requires the following purification of the nucleotides before use in the subsequent assays:

(a) Pack Bio-Rad AG 2 × 8 (200–400 mesh) column (0.8 × 5 cm).
(b) Wash with 5 ml 40 mM Tris (pH 7.5).
(c) Apply 0.200–1.0 ml ^3H cyclic nucleotide.
(d) Wash with 10 ml 40 mM Tris.
(e) Wash with 5 ml 0.3 N HCl collecting 0.5-ml fractions.
(f) Add 100 litres 0.5 M Tris-base to peak fractions.
(g) Immediately adjust pH to 7.0–8.0 with 1.0 M NaOH.
(h) Store at −20 °C in 40% ethanol (repurify every 4 weeks or when backgrounds increase).

3.1.2 Extinction coefficients (*E*) for cAMP and cGMP (10^{-3}M)

$$[1]\ \text{cAMP}\ E_{265\text{nm}} = 14.0 \quad [2]\ \text{cGMP}\ E_{252\text{nm}} = 13.7$$

Make up substrate solution combining the desired amounts of unlabelled and radioactive cyclic nucleotides, read on spectrophotometer, and calculate concentration using the following equation:

$$\text{OD} = \text{conc. (mmol/litre)} \times E.$$

Remove aliquots and count on the scintillation counter. Determine specific activity as follows:

$$\text{(c.p.m./litre)/(mol/litre)} = \text{c.p.m./mol} = \text{specific activity (SA)}.$$

3.1.3 Separation of reaction products by the batch assay

Figure 1 illustrates the necessity of characterizing the resin system used in the batch assay. Parts A and B demonstrate that although the 50–100 mesh resin provides greatest nucleotide (^{14}C-guanosine) recovery, the blank or background values are unacceptably high (13 and 8% respectively). Note also the necessity of purifying the substrate in reducing high blanks. Parts C, D, and E demonstrate that although there is a reduction in product recovery (68 to 52%) the use of 200–400 mesh resin significantly reduces the blanks (to approximately 2%). As described earlier, we have elected to employ a combination of conditions.

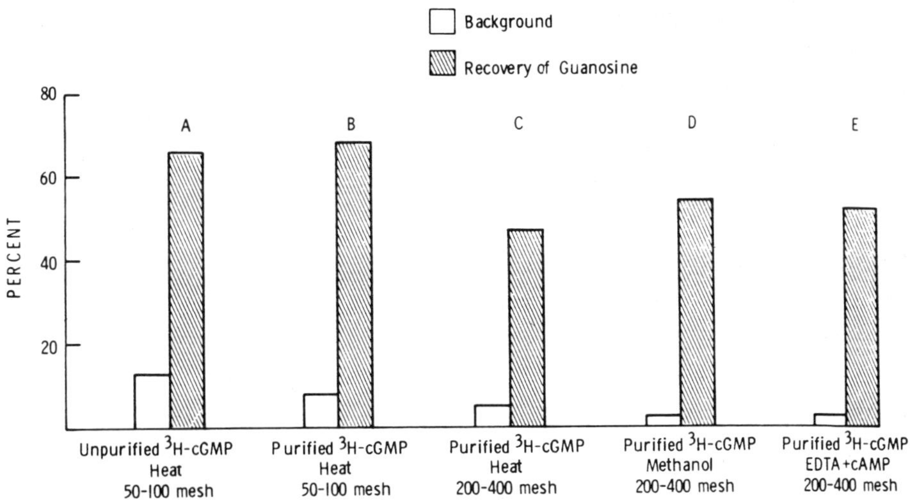

Figure 1. Characterization and refinement of batch PDE assay. Assays performed as described in text by altering the mode of termination and resin mesh. Results are average of triplicates which did not vary by more than 8%.

Protocol 2. PDE assay

Assay mixture

- 50 μl 50 mM MOPS (pH 7.0) containing 12 mM Mg acetate, 20 mM 2-mercaptoethanol, and 0.4 mM $CaCl_2$ (in excess of EGTA) 10 μl ^{3}H cyclic nucleotide (2.0×10^5 c.p.m., 2×10^{-4} M)
- 100 μl enzyme solution
- 40 μl distilled water

Protocol 2. *Continued*

Procedure

1. Initiate reaction by adding either sample or substrate followed by vortexing.
2. Incubate 5–30 min at 30 °C (optimal time and enzyme concentration selected for hydrolysis of 10–25% of the total substrate, see below).
3. Terminate reaction by adding a 200 μl solution containing 20 mM EDTA and 10 mM cAMP.
4. Add 25 μl (0.50 mg/ml) Snake Venom (*C. atrox*); incubate 15 min.
5. Add 100 μl 10^{-2} M unlabelled nucleoside (adenosine or guanosine).
6. Add 1.0 ml settled Bio-Rad AG2X8 (200–400 mesh) Resin (equilibrated in 40 mM Tris–HCl, pH 7.5).
7. Vortex twice (5 min interval).
8. Centrifuge (1000 *g*) for 5 min.
9. Count 100 μl in 2.0 ml of 10% BBS-3 Spectrofluor

Calculations

- 200 Rxn mix
- 200 stopping solution
- 25 venom
- 100 nucleoside
- 600 aqueous resin volume

- 1100 Total volume

$$\frac{\text{(Total c.p.m.-background c.p.m.)} \times 2 \text{ (50\% recovery)} \times 11.0,}{\text{Spec. act. (c.p.m./mol)} \times \text{(min of incubation)} \times \text{sample vol. (ml).}} = \frac{\text{mol hydrol.}}{\text{ml min}}$$

The ^{3}H-nucleotides can be separated on individual Bio-Rad AG-2 × 8 columns but this is a time consuming procedure (6). Alternatively, the reaction products may be separated by TLC (PEI cellulose) using a 1 M ammonium acetate: 95% ethanol (3:7) solvent system. The reactions are spotted and chromatographed with carrier nucleotides and final products identified with a short-wave (254 nm) mineral light. The cellulose spots are scraped from the plates and counted with scintillation fluid.

3.1.4 Assay reliability

No matter which method is employed for the separation of reaction products, the assay reliability should be confirmed periodically by performing enzyme-dependent and time-dependent curves in order that legitimate reaction rates can be assumed in subsequent studies. When assaying PDE, the enzyme reaction should never consume greater than 30% of the total substrate present.

3.1.5 General comments

The most widely employed PDE assay is the 'Batch' assay due to the ease of separation of the charged (cAMP or cGMP) substrate from the neutral (adenosine or guanosine) reaction product via the anion exchange resin. Unfortunately, recovery of product (adenosine or guanosine) is not complete. Thus, this system should be characterized using carbon-14 nucleoside in order to determine actual recovery. As shown in *Figure 1*, ^{14}C guanosine recovery is approximately 50%; therefore, a correction factor of 2 is included in the final calculations. The other radiometric assays (TLC and column) provide accurate determinations of PDE activity. However, because each involves rather time consuming steps, they are not generally the assay methods of choice unless relatively few samples are being analysed.

3.2 Preparation of calmodulin-deficient phosphodiesterase

Protocol 3. Calmodulin-deficient PDE preparation

1. Sacrifice 10 adult rats and remove brains.
2. Homogenize in 3 vol. 10 mM Tris (pH 7.5) containing 10 mM 2-mercaptoethanol, 5 mM EGTA, 1 mM $MgCl_2$, 0.05% NaN_3
3. Sonicate in 20 ml batches for 20 sec.
4. Centrifuge at 100 K *g* for 1 h.
5. Apply sample to a 2.5 × 30 cm column packed with DE-52 cellulose (Whatman) and equilibrate with 10 mM Tris (pH 7.5) containing 10 mM 2-mercaptoethanol, 1 mM EGTA, 1 mM $MgCl_2$, 0.05% NaN_3
6. Wash with column buffer (200 ml) then begin a linear salt gradient 0–0.4 M $(NH_4)_2SO_4$ in column buffer (200 ml to 200 ml).
7. Flow rate should be approximately 0.5 ml/min; collect 3-ml fractions.
8. Assay even-numbered fractions using batch assay (100 μl sample; 10^{-5} M cAMP) with a 10 min pre-incubation period.
9. Repeat assay except include 10 μg calmodulin preparation (see below) with a 10 min pre-incubation period (initiate reaction by adding substrate).
10. Plot column profile and determine the location of calmodulin-deficient PDE and pool fractions stimulated 5- to 20-fold by calmodulin.
11. Add glycerol to a final concentration of 50%. Prepare 0.5–1.0 ml aliquots. Freeze on dry ice.
12. Store frozen at −70 °C or below.

Protocol 3. *Continued*

13. Thaw rapidly immediately before use. Do not refreeze. Discard the unused portion of the aliquot.

This procedure will yield a fairly crude phosphodiesterase preparation that is suitable for assaying calmodulin. Several affinity chromatography techniques have been developed that yield nearly homogeneous preparations of enzyme. Calmodulin-dependent phosphodiesterase can be extensively purified by calmodulin-sepharose chromatography (7, 8). Further purification on Cibacron blue agarose has been used to remove minor contaminants present in some preps (9). Chromatography on a column composed of a conformation-specific monoclonal antibody against Ca^{2+}-calmodulin/phosphodiesterase has been used to rapidly isolate enzyme from brain and heart (10). Calmodulin independent phosphodiesterase can be purified in two steps using chromatography on a cyclic-nucleotide-sepharose column (11).

3.3 Calmodulin bioassay

While calmodulin regulates several enzyme systems, its regulation of a specific form of cyclic nucleotide phosphodiesterase affords a very good assay of this protein. Calmodulin-deficient phosphodiesterase is prepared as described in the preceeding section and used as follows.

3.3.1 Assay mixture

Previously defined reaction mixture (minus cAMP) for phosphodiesterase. Include 20 μg of calmodulin-deficient PDE except in tube 1, and perform assays in triplicate.

Tube	Calmodulin solution
1	0μl
2	0
3	5
4	10
5	20
6	30
7	50
8	100

Pre-incubate at 30 °C for 10 min, then initiate reaction with the addition of 10 μl of cAMP (2×10^{-4} M; 2×10^{5} c.p.m.).

Incubate for 30 min at 30 °C and follow one of the assay procedures (preferably the 'Batch') previously defined.

3.3.2 Tabulation of data

Tube	c.p.m.	Net c.p.m. (basal activity)	% Maximal stimulation
1	138	–	–
2	860	Basal activity	0
3	1176	316	4
4	1490	630	8
5	3467	2607	33
6	6153	5293	67
7	8750	7890	100
8	8760	7900	100

When these data are plotted as PDE activity as a function of microlitres of solution added, the calmodulin concentration can then be quantitated. One unit of calmodulin is defined as the amount (nanograms) of heat-treated extract required to stimulate activator-deficient PDE to a 50% maximal stimulation.

3.3.3 Quantitation of calmodulin activity

Calmodulin activity is commonly expressed as one unit equalling the amount of protein which stimulates CaM-deficient PDE 50% of its maximal stimulation. Because enzyme preparations vary considerably from preparation to preparation (activation may range from 2- to 25-fold stimulation), comparisons of inter-experiment not to mention interlaboratory values are quite difficult. It is therefore convenient to titrate each preparation with pure, homogeneous calmodulin, thereby converting units to nanograms of calmodulin. Henceforth, tissue concentrations of calmodulin can be expressed in universal terms, nanograms per unit weight of tissue.

4. Calmodulin radioimmunoassay

Since quantitation of calmodulin levels by the phosphodiesterase assay requires calcium-dependent binding of biologically active calmodulin, such an assay would be susceptible to inhibition by other calmodulin binding proteins, and only able to evaluate the 'biologically active' form of the protein. A radioimmunoassay, on the other hand, is predicated on the highly specific antigen–antibody reaction and, since it would not require calcium-dependent binding, would obviate these problems. The following is a discussion of the development of a radioimmunoassay to calmodulin. We will first discuss the procedure of radiolabelling the protein, then the production of antibody, and finally the assay itself.

4.1 Radioiodination of calmodulin

While several different methods are available for labelling a protein with ^{125}I, the conjugation method first described by Bolton and Hunter (12) is most amenable for labelling calmodulin to high specific radioactivity with essentially no loss in immunological or biological activity (13).

One mCi of ^{125}I-Bolton–Hunter reagent, (^{125}-I-labelled *N*-succinimidyl-3 (4-hydroxyphenol) propionate (1600 Ci/mmol supplied in benzene from Amersham) is evaporate to dryness under a stream of nitrogen in a conical tube. The walls of the tube are washed with 200 µl of benzene (desiccated with calcium hydride) and the benzene removed as before. Ten microlitres of 0.5 mg/ml calmodulin in 125 mM borate buffer pH 8.4 is applied directly to the residue at the bottom of the conical tube and incubated for 30 min at 4 °C. The reaction is then terminated by the addition of 150 µl of 200 mM glycine in the same buffer and a subsequent 15 min incubation at 4 °C. The sample is quantitatively removed and applied to a Sephadex G-25 (Pharmacia) column (0.7 × 20 cm) previously equilibrated with 0.05 M phosphate, 0.05% NaN_3, 0.25% (w/v) gelatin, 0.1 M NaCl (pH 7.5). The reaction tube is washed with 50 µl of 200 mM glycine in 125 mM borate buffer pH 8.4 which is added to the column. The sample is chromatographed, and thirty 1-ml fractions collected. Five-microlitre aliquots of each fraction and the reaction vessel are counted for ^{125}I in a Searle gamma-counter. ^{125}I-labelled calmodulin eluted from the column in the V_O is readily separated from the protein-free ^{125}I/labelled Bolton–Hunter reagent eluting in the V_I. In order to determine the total ^{125}I conjugated to the 5 µg of calmodulin, the total c.p.m. present in the V_O is corrected for recovery of ^{125}I-calmodulin on the G-25 column (routinely 90%). Since only protein bound ^{125}I-labelled Bolton–Hunter reagent will remain attached to the reaction vessel (in the case of calmodulin <1%) the c.p.m. present in the vessel are added to the corrected V_O value. This total radioactivity is then divided by the total mass of 5 µg to generate the total c.p.m./µg calmodulin. This procedure routinely conjugates 1.5 mol of ^{125}I/mol of protein resulting in a specific activity of approximately 2400 Ci/mmole or 1.56×10^8 c.p.m./µg.

4.2 Production of antiserum and purification of antibody

The specificity of any radioimmunoassay is dependent on the quality of the antiserum available. While there are various methods available to make proteins more antigenic, these procedures generally result in a loss in sensitivity to the native form of the protein. For this reason we have endeavoured to generate antibodies against the native protein. While we initially reported the production of monospecific antibodies in goat to native calmodulin (14), the goat antiserum was inappropriate for use in a radioimmunoassay. When sheep were used instead of goats, antisera were produced which were satisfactory for use in the development of a radioimmunoassay for calmodulin (13). The following is the protocol used in sheep.

4.2.1 Immunization schedule

Electrophoretically pure calmodulin (10 mg in 1 ml of 0.9% saline) is emulsified with an equal volume of complete Freund's adjuvant. Ten milligrams of protein is injected subcutaneously at multiple sites in the lower back of an adult sheep. Subsequent booster injections of 2 mg calmodulin each are administered at 42, 56, and 70 days, and every 60 days thereafter. Test bleeds are obtained at 2-week intervals following the administration of each booster injection. Upon removal, the blood is allowed to clot, and the serum obtained by centrifugation at 6000 *g* for 20 min at 4 °C. The serum from each bleed (approximately 500–700 ml) is divided into 100 ml aliquots and can be stored at −20 °C for at least 18 months. These immunization procedures and serum collection are performed for us by Bethyl Laboratories Inc.

4.2.2 Purification of antibody

100 ml of serum is thawed at 4 °C and brought to 50% saturation with solid ammonium sulfate. The suspension is slowly stirred for 2 hours at 4 °C followed by centrifugation at 10 000 *g* for 30 min. The precipitate is resuspended in approximately 50 ml of 125 mM borate buffer, 75 mM NaCl, pH 8.4 and dialysed extensively against several changes of the same buffer for 18 h at 4 °C. Ten millitres of this solution are then applied to a 10-mg calmodulin-sepharose affinity column prepared by the method of Dedman *et al.* (17). The column is washed with the borate buffer until the $A_{280} < 0.005$, at which time 200 mM glycine, pH 2.7, is applied to the column.

Those fractions which originally passed through the calmodulin-affinity column (affinity flow-through) should be pooled, reapplied to the affinity column, and processed as before. This procedure should be repeated until no protein is eluted from the column by the glycine buffer (usually three cycles). The final affinity flow through is devoid of antibodies to calmodulin and can be used as control serum.

The fractions which are eluted with the glycine buffer and exceed an A_{280} of 0.100 are pooled and dialysed against 125 mM borate, 75 mM NaCl, pH 8.4. This pooled sample consists of antibodies to calmodulin to be used in the radioimmunoassay. Specificity studies of the affinity-purified antibodies have demonstrated that only calmodulin is immunoprecipitated from a total *in vitro* translation product (13).

4.3 Development of the radioimmunoassay

4.3.1 Anti-calmodulin dilution

Since the effective antibody titre fluctuates from one preparation to the next, it is important to generate an antibody dilution curve for each preparation of calmodulin-affinity-purified antibody. In so doing, the correct concentration of affinity-purified antibody can be determined for use in the radioimmunoassay.

A series of dilutions are made from the stock of calmodulin-affinity-purified antibody (starting concentration is usually between 200 to 300 μg IgG/ml). The dilutions are prepared with an appropriate amount of affinity run through as carrier to ensure that a constant IgG concentration of 800 μg/ml is maintained throughout the dilution series. The assay is performed in 12 × 75 mm culture tubes by first adding 300 μl of radioimmunoassay buffer (125 mM borate, 1 mM EGTA, 75 mM NaCl, 10 μg/ml bovine serum albumin, pH 8.4), followed by the addition of 100 μl of the same buffer containing 10 000 to 15 000 c.p.m. ^{125}I-calmodulin. One hundred microlitres of the appropriate antibody dilution (in duplicate) is then added and the samples vortexed. Non-specific background is determined by substituting 100 μl of the affinity flow through (800 μg/ml) for anti-calmodulin. The samples are incubated for 18 h at 25 °C. The length of incubation should be tested since we have found that 90% of the total binding observed at 18 h has occurred after only 5 h. Although the sheep anti-calmodulin preparations are precipitating, separation of bound from free ^{125}I-calmodulin is facilitated by the addition of protein A (13). We use the commercial fixed cell preparation of *Staphylococcus aureus*, Pansorbin, available from Calbiochem. Ten microlitres of a 10% solution of Pansorbin is added, the assay tubes are vortexed and incubated for 30 min at 25 °C. The tubes are centrifuged at 1000 *g* for 10 min in a Beckman TJ6 centrifuge. The supernatant solution is decanted and while in the inverted position, the mouth of the tube is blotted on absorbent paper. The pellet is washed twice with 1 ml of radioimmunoassay buffer and processed as before. The relative amounts of ^{125}I present in the final pellets are determined on a Searle gamma-counter.

The data are obtained by subtracting the non-specific background (routinely less than 100 c.p.m.) and dividing each duplicate set by the total c.p.m. of ^{125}I-CaM initially added to each assay tube. The antibody dilution which precipitates between 30 to 40% of the total ^{125}I-CaM is used for all subsequent assays.

4.3.2 Preparation of standards

Initially, standards should be prepared from homogeneous calmodulin. Since calmodulin can be readily purified to homogeneity, the preparation is relatively simple. A starting stock solution of calmodulin is made in radioimmunoassay buffer devoid of bovine serum albumin at a concentration of 500 μg/ml. This final concentration can be routinely verified spectrophotometrically if rat testis calmodulin is used since $E = 0.21$ (1 mg/ml). The most accurate method for assuring reproducible concentration is to calculate the latter following amino acid analysis of a sample of the pure protein. From this stock a series of dilutions are made using the radioimmunoassay buffer. The dilution series should initially range from 10 μg/ml to 1 ng/ml. However, once the limits of the assay are determined a more limited series of concentrations is required.

4.3.3 Sample preparation

Tissues to be assayed are collected on ice and homogenized in 125 mM borate buffer, 5 mM EGTA, 75 mM NaCl pH 8.4 (5:1, v/w) with a Brinkman polytron homogenizer (5 × 10 sec bursts using a PT-10 generator at a setting of 4.5). Tissue culture cells as well as other isolated cell preparations can be homogenized by sonication in the same buffer, using a Brownwill Biosonik IV Sonicator. Following homogenization, appropriate aliquots are removed for total protein and DNA determinations. Homogenates are brought to a rapid boil in a microwave oven (less than 1 min) or heat-treated for 3 min at 90 °C and rapidly cooled to 4 °C by immersion in ice. The heat-treated samples are centrifuged at 10 000 *g* for 30 min and the supernatant solutions are used for the assay. Preparation of samples by this method routinely yields quantitative recovery of calmodulin. When samples cannot be heat-treated because of other considerations, 1% Triton X-100 should be included in the homogenization buffer to assure total release of membrane-bound calmodulin. Prior to assay, samples should be diluted to a calmodulin concentration of 50 to 100 ng/ml. Since calmodulin levels are routinely 0.1 to 1% of the total protein, this can be accomplished to a first approximation by considering the total protein content of each sample.

4.3.4 Radioimmunoassay

Initially, 200 to 300 μl of radioimmunoassay buffer is added to the entire series of 12 × 75 mm culture tubes used in the assay. Since the final reaction volume is 500 μl, the actual volume of buffer per tube will be dependent on the total volume of the other components added. Both standards and samples are added to their appropriate tubes in 100 μl volumes, followed by the addition of 100 μl of a fixed number of c.p.m. ^{125}I-calmodulin (10 000–15 000) in radioimmunoassay buffer. Finally, the reaction is initiated by the addition of 100 μl of the appropriately diluted anti-calmodulin (the amount required to achieve 30–40% binding of the tracer). Non-specific background is determined by substituting 100 μl of the affinity run through (800 μg/ml) for the anti-calmodulin. Samples are processed as described above for the antibody dilution curve, and total radioactivity in each tube obtained. *Figure 2* depicts a typical standard curve generated by the above procedure.

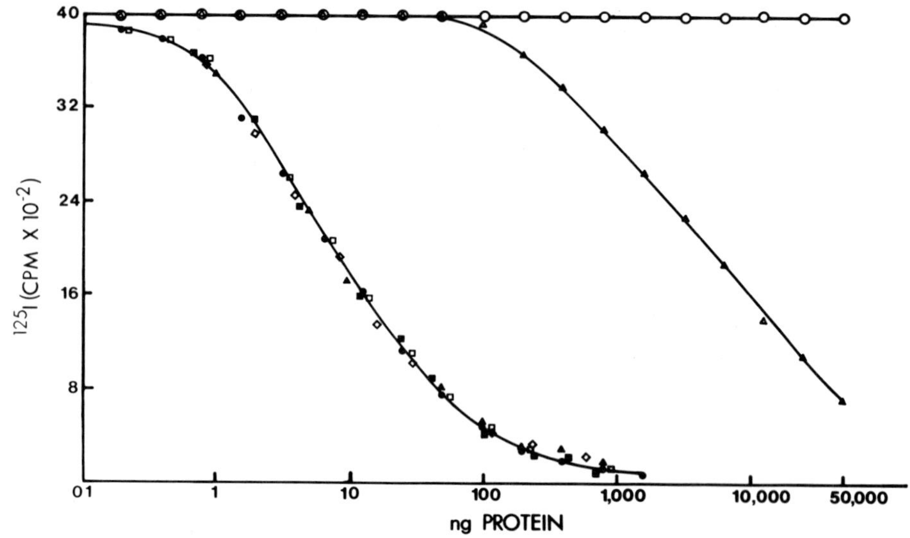

Figure 2. Relative immunological cross-reactivity of homologous calcium-binding proteins. Homogeneous calmodulins from rat testis (●), bovine brain (□), *A. hypogaea* (■), troponin C (△), parvalbumin (○); heat-treated supernatant solutions from the electroplax of *E. electricus* (◇) and *C. reinhardii* (▲).

The highly conserved nature of this protein is clearly demonstrated in this figure, since the same competition curve is generated by calmodulins purified from such diverse phyla as higher plants (the peanut) to mammals (cow brain). In addition, the same competition curves are generated whether calmodulin is in its purified state or in a heat-treated tissue extract. Because of this fact, standards made from heat-treated extracts may be substituted for the pure protein standards once their calmodulin levels have been determined with the pure protein standards. Since the determination of calmodulin levels by the radioimmunoassay is predicated on comparing the degree of competitive binding by a heat-treated sample to a standard curve, it is important to ascertain the degree to which other proteins may cross-react with the antibody. As shown in *Figure 2*, only troponin C, a protein with 50% sequence homology and biological cross-reactivity to calmodulin, exhibits immunological cross-reactivity (at a 665-fold greater concentration than calmodulin). The data presented in this figure were calculated by the log-logit computer program based on the assay statistics derived by Midgley *et al.* (15) and Duddleston *et al.* (16). A typical assay has an intra-assay variability of 3% and an inter-assay variability of 5%. The limit of detection is approximately 15 pg and the assay sensitivity is 115 pg. Non-specific background is generally less than 100 c.p.m., but varies depending on the gamma-counter used.

5. Immunoprecipitation of radiolabelled calmodulin

Once antibody is made against calmodulin, it is possible to use it to immunoprecipitate radiolabelled calmodulin from a complex mixture of protein. This procedure is particularly useful in the analysis of calmodulin in radiolabelled tissue culture cells or *in vitro* translation products.

5.1 Buffers

- Pi/NaCl: 0.9 mM $CaCl_2$, 2.7 mM KCl, 1.5 mM KH_2PO_4, 0.5 mM $MgCl_2 \cdot 6H_2O$, 137.0 mM NaCl, 8.1 mM $Na_2HPO_4 \cdot 7H_2O$
- PSDT: 10 mM sodium phosphate, 150 mM NaCl, 1 mM EGTA, 100 μM appropriate unlabelled amino acid, 1% Triton X-100, 1% deoxycholate, pH 7.5

5.2 Sample preparation

5.2.1 Radiolabelled tissue-culture cells

After labelling with a radioactive amino acid, cell monolayers should be washed four times with 10 ml each of Pi/NaCl containing 100 μmol of the unlabelled amino acid used for labelling. Cells are then removed from the dish by scraping with a rubber policeman. The cells are centrifuged in a Beckman Microfuge for 2 min, and the pellets are washed three times with Pi/NaCl containing 100 μM unlabelled amino acid used for labelling. Final cell pellets are resuspended in 1 ml of PSDT buffer at 4 °C and homogenized by sonication. Aliquots can be removed for total protein and DNA determination. The remaining sample is heat-treated at 90 °C for 5 min followed by rapid cooling in an ice bath. All samples are then centrifuged at 10 000 *g* for 30 min at 4 °C, and the supernatant solutions are removed for analysis.

5.2.2 *In vitro* translation

Samples to be assayed from *in vitro* translation should be mixed with an equal volume of 2× PSDT buffer. The samples are then heat-treated at 90 °C for 5 min and rapidly cooled in an ice bath. Cell samples are centrifuged at 10 000 *g* for 30 min at 4 °C and the supernatant solution removed for analysis.

5.3 Immunoprecipitation procedure

Calmodulin is immunoprecipitated by sheep antibody to calmodulin. Prior to incubation with the anti-CaM, 100 μl of the radiolabelled sample should be incubated with 25 μl of a 10% solution of Pansorbin (Calbiochem) for 30 min at 25 °C. The solution is then centrifuged in a Beckman Microfuge for 2 min, and the supernatant solution is removed for analysis. In order to determine subsequent recovery of calmodulin, 1000 c.p.m. of [^{125}I] calmodulin is added to each sample to be assayed. One hundred microlitres of affinity-purified anti-calmodulin (1 mg/ml) in PSDT buffer is incubated with up to 100 μl of labelled heat-treated sample. Regardless of the sample volume, each reaction should contain 1000 c.p.m. of [^{125}I]-calmodulin. Samples are incubated at 25 °C for 18 h, after which time 25 μl of a 10% solution of Pansorbin is added and the sample is incubated for an additional 30 min. Each suspension is then carefully layered over 500 μl of 1 M sucrose in PSDT and centrifuged in a Beckman Microfuge for 4 min. The supernatant solution is carefully aspirated, and the pellets are washed twice with 1 ml of PSDT. The pellets are resuspended in 100 μl of SDS sample buffer and heated for 5 min at 95 °C. The samples are now ready for analysis. The amount of [^{125}I]-calmodulin present in each sample is determined using a gamma-counter. This value is then divided by the 1000 c.p.m. originally present and multiplied by 100 in order to determine the percentage recovery of calmodulin in each sample. This value can then be used to adjust the total immunoprecipitable counts to 100% recovery.

6. ^{125}I-calmodulin-overlay procedure

A useful technique has been developed which allows the identification of calmodulin binding proteins in crude cell and tissue homogenates. This technique is referred to as the ^{125}I-calmodulin-overlay procedure. Since it was first reported by Glenny and Weber (17) this technique has been used by many investigators to identify calmodulin-binding proteins from a variety of samples. In this procedure, proteins separated on SDS-polyacrylamide gels or non-denaturing gels are incubated with ^{125}I-CaM. The incubation is performed in the presence and absence of Ca^{2+} in order to ascertain whether the interactions are Ca^{2+}-dependent. Following the incubation, the calmodulin-binding proteins are identified by autoradiography as those protein bands to which ^{125}I-CaM bound. Alternatively, the proteins can be transferred to nitrocellulose filters and the filters incubated with ^{125}I-calmodulin (18). This protocol can be performed in less time than incubation of the gel, but may result in decreased binding to specific proteins. Using a two-step electrophoretic transfer can prevent significant loss of lower M_r binding proteins (19). ^{125}I–calmodulin overlays have also been used in molecular biological applications to screen λgt11 cDNA libraries for clones producing calmodulin-binding proteins (20) and to verify that an isolated clone contains a calmodulin-binding region (21, 22). When this technique is coupled with the 'erasable Western blot' protocol (23) one can probe electrophoretically separated proteins with a specific antibody, detect the reactive product, then remove the primary and secondary antibodies and probe the same blot with ^{125}I-protein.

6.1 Assessing binding

The ^{125}I-calmodulin-overlay procedure is useful for the preliminary screening for calmodulin-binding proteins (24). In many cases, this binding is proportional to the amount of protein present in the band. However, a great deal of caution must be used in the interpretation of these data since the properties predicted from observing binding to an electrophoresed protein may be quite different from those found when binding is assessed by other independent techniques. This, in fact, appears to be the case with respect to ^{125}I-calmodulin-binding to phosphorylase (25). Equal amounts of purified myosin light chain kinase (MLCK) and glycogen phosphorylase bind equivalent amounts of ^{125}I-calmodulin in a Ca^{2+}-dependent manner. Since the binding to these two proteins was equivalent at 0.2–2.5 mM ^{125}I-calmodulin, this would suggest that phosphorylase and MLCK bind calmodulin with a similar high affinity. Calmodulin does activate phosphorylase, but only twofold, compared to a total calmodulin-dependence of MLCK. The K_a for calmodulin activation of phosphorylase is 1000-fold higher than the K_a of MLCK. The calmodulin activation of MLCK is completely Ca^{2+}-dependent, but there is no affect of Ca^{2+} on the phosphorylase activation. Furthermore, troponin C will activate phosphorylase but not MLCK.

Despite these precautions, use of the ^{125}I-protein overlay can indicate which calmodulin-binding proteins should be further studied with functional assays or specific antibodies for immunoprecipitation or the Western blot analysis (26).

6.2 Mapping calmodulin-binding domains

Another useful application of this technique is mapping the calmodulin-interacting domain of a calmodulin-binding protein. This approach has been used successfully to construct a model for MLCK (27). Proteolytic fragments of MLCK were analysed for their ability to bind ^{125}I-calmodulin. Digestion with *S. aureus* V8 protease produced a M_r = 70 000 ^{125}I-calmodulin-binding fragment that contained both phosphorylation sites recognized by cAMP-dependent protein kinase. Treatment with chymotrypsin produced a M_r = 95 000 calmodulin-binding fragment that contained only the calmodulin modulated phosphorylation site. Inclusion of calmodulin–sepharose in the MLCK protease digestion reactions permitted isolation of the calmodulin-binding fragments and determination that the 70 kd polypeptide retains MLCK activity. This analysis of ^{125}I-calmodulin-binding to fragments, generated from a series of proteolytic digestions of MLCK, permitted assignments of the calmodulin-binding domain relative to the portions of the molecule that contain the catalytic site and the two residues that are phosphorylated by cAMP-dependent protein kinase. These latter two functional domains are necessarily determined by different protocols.

Protocol 4. Protocol for overlay

The following is a detailed outline of the procedure as used in our laboratory. Samples can be electrophoresed by many techniques.

1. Following electrophoresis, gels are removed and fixed in 40% MeOH: 10% acetic acid for 30 min at 25 °C.
2. Rinse 5 times with distilled water.
3. Wash 18 h in 10% EtOH at 25 °C to remove SDS.
4. Wash 30 min more in 10% EtOH.
5. Rinse five times with distilled water.
6. Incubate 10–15 min in 0.1 M imidazole.
   - Imidazole stock solution (1.0 M): 34.05 g recrystallized Imidazole. Bring to 500 ml with distilled water. Adjust pH to 7.0.
   - 0.1 M imidazole: 50 ml 1.0 M imidazole, 450 ml distilled water.
7. Incubate 10–15 min in either Soln. G. with Ca^{2+} or Soln. G with EGTA. Soln. G:
   - 80 ml 1.0 M imidazole
   - 59.64 g KCl
   - 4.00 g BSA (200 g non-fat dry milk can be substituted for BSA)
   - 5 ml 20% NaN_3
   - bring to 4 litres with distilled water. Adjust pH to 7.0
   - add 40 ml of 0.1 M $CaCl_2$ for Soln. G with Ca^{2+} *or*
   - add 4 ml of 1.0 M EGTA for Soln. G with EGTA
   - store in cold room
8. Place gels in Sears 10-in food sealing bags and seal bag on Sears Counter-Craft Seal-and-Save. Slit through one side of bag in a corner with a razor blade. Add 30 ml of either Soln. G with Ca^{2+} or Soln. G with EGTA. Add 20 000 000 c.p.m. of ^{125}I-CaM. Exclude all air possible by pressing on bag. Seal corner with a Beckman Nalgene Heat Sealer.
9. Rock sealed bags for 18 h at 4 °C on labquake.
10. Remove gels from sealed bags into Sears 30 × 20 cm plastic superseal boxes and wash in Soln. G with either Ca^{2+} or EGTA. Change wash every 4 h, from each wash take a 1-ml aliquot to count. Count for 2 min. Record counts.
11. Continue to change wash until counts are at or near background (until 1-ml count is within 10 counts of empty carrier tube).
12. Rinse gels five times for 1 min with distilled water to remove BSA. If the gel is to be scanned for an accurate determination of protein in the

Protocol 4. *Continued*

bands, it should be washed for 12 h in 40% methanol, 7% acetic acid. This will remove BSA that will otherwise contribute to a dark-blue background. This treatment does not affect the ^{125}I-calmodulin bound.

13. Stain gels in glass basin for 1 h with 0.05% Coomassie Brilliant Blue in 40% methanol and 7% acetic acid.

14 Destain gels by repetitive washes with 7% acetic acid, 3% glycerol, 30% methanol. Gels can be scanned on a suitable densitometer for protein determination.

15. Place destained gels on Whatman chromatography paper and wrap in Bio-Rad permeable wrap (Cat. No. 165–0964). Dry gels on Bio-Rad gel dryer model No. 1125B.

16. Subject dry gels to autoradiography and identify those bands to which ^{125}I-CaM bound.

17. The autoradiograms can be scanned to quantitate the relative amount of ^{125}I–Calmodulin bound.

7. Purification of CaM binding proteins by affinity chromatography

Since CaM serves as an intracellular Ca^{2+} receptor, it is the interaction with target proteins that transduces the Ca^{2+} signal. Many of the interactions occur with high affinity and only when Ca^{2+} is bound to CaM. This property has led to a generalized scheme of purifying CaM binding proteins from a variety of cells and tissues by CaM-Sepharose affinity chromatography. Following some preliminary chromatography steps, the solution containing the CaM binding protein is passed through an affinity resin containing covalently attached CaM in the presence of Ca^{2+}. After extensive washing at high ionic strength to remove adventiously bound proteins, the specifically bound molecules are eluted with EGTA. Because of this affinity chromatography step, a significant level of purification of CaM binding proteins is achieved in a single step, allowing immense time-saving over less selective conventional techniques. In this section, a detailed purification scheme of a Ca^{2+}/CaM-dependent enzyme, myosin light chain kinase (MLCK) is presented. The enzyme has been purified to near homogeneity from a number of tissue sources including smooth muscle, skeletal muscle, and cardiac muscle (28–35). Recently, a partial cDNA clone of the enzyme has been expressed in bacteria and the expressed protein was isolated from bacterial extracts (34). Here we will outline the scheme of purification of this Ca^{2+}/CaM-dependent protein from both animal tissue and bacteria. The same procedure with only minimal modifications can be used for the purification of other CaM binding proteins as long as the interaction is Ca^{2+}-dependent.

7.1 Purification of MLCK from chicken gizzard

The purification procedure described in *Protocol 5* is a modification of the method of Adelstein and Klee (29).

Protocol 5. Purification of MLCK

1. *Homogenization.* Fresh chicken gizzards are used for the purification of myosin light chain kinase. Immediately after the chicks are sacrificed, the gizzards are removed and packed in ice. After removing all fat and connective tissue, the smooth muscle is placed in a beaker immersed in ice. The muscle is first ground in a meat-grinder at 4 °C. From 500 g of ground gizzard muscle, usually 30–50 mg of the enzyme is obtained. The ground muscle is suspended in buffer A (1:4, w/v) and homogenized in a Waring blender. The homogenate is then centrifuged at 10 000 *g* for 15 min. After resuspending the pellet in the same volume of buffer B, the suspension is centrifuged for 15 min at 10 000 *g*. The resulting pellet is washed again as described previously.

   Buffer A: (Homogenization):

   - 20 mM Hepes pH 7.0
   - 40 mM NaCl
   - 5 mM EGTA
   - 0.5% Triton X-100
   - 1 mM $MgSO_4$
   - 0.02% NaN_3
   - 1% (w/v) soybean trypsin inhibitor
   - 1% (w/v) pepstatin
   - 0.2 mM PMSF

   Buffer B: (Wash): Same as buffer A, except that the Triton X-100 is deleted.

2. *Extraction of MLCK.* The washed pellet is resuspended in 4 vol. of buffer C. The suspension is stirred gently for 20 min, centrifuged at 10 000 *g* for 30 min and then filtered through glass wool. The supernatant is made 40% saturated with $(NH_4)_2SO_4$, stirred on ice for 20 min and then centrifuged for 30 min at 10 000 *g*. To the supernatant more $(NH_4)_2SO_4$ is added to make it 60% saturated. This results in the precipitation of MLCK and the protein is sedimented at 10 000 *g* for 30 min.

   Buffer C: (Extraction):

   - 50 mM Hepes pH 7.5
   - 60 mM NaCl
   - 5 mM EGTA
   - 2 mM EDTA
   - 1 mM DTT

Protocol 5. *Continued*

- 0.02% NaN_3
- 25 mM $MgSO_4$
- 5% (w/v) leupeptin
- 1% (w/v) antipain
- 1% (w/v) pepstatin
- 0.2 mM PMSF
- 1% (w/v) soybean trypsin inhibitor

3. *DEAE–Sephacel chromatography.* The pellet obtained after ammonium sulfate fractionation is first resuspended in buffer D and then dialysed against the same buffer overnight. The dialysed sample is then applied to a DEAE-Sephacel column equilibrated with buffer D. This is followed by washing of the column with buffer D containing 20 mM NaCl until the A_{280} is less than 0.05. The enzyme is then eluted with a linear gradient of 20 mM to 400 mM NaCl in buffer D. The main peak of MLCK activity is found at about 200 mM salt.

   Buffer D: (DEAE):
   - 50 mM Hepes pH 7.5
   - 1 mM DTT
   - 1 mM EGTA
   - 0.1 mM PMSF
   - 1% (w/v) leupeptin

4. *Affinity chromatography.* The fractions containing MLCK activity are pooled and dialysed overnight against buffer E containing 1 mM EGTA. Just before loading on to the column, Ca^{2+} is added to the sample to give a final concentration of 0.5 mM. The sample is then applied to a CaM-sepharose 4B column. The column is washed with buffer E containing 0.5 mM $CaCl_2$ and 0.2 M NaCl until the (A_{280}) is less than 0.05. Finally, the enzyme is eluted with buffer E containing 0.2 M NaCl and 1 mM EGTA. The fractions containing MLCK activity are pooled, divided into aliquots, and stored frozen at −70 °C.

   Buffer E: (CaM-sepharose):
   - 50 mM Hepes pH 7.5
   - 1 mM DTT
   - 1 mM PMSF

Protocol 6. Preparation of CaM-sepharose

Calmodulin-sepharose 4B affinity resin can be prepared using the method described by March *et al.* (35).

1. Swell 15 g of CNBr-activated sepharose 4B on a sintered glass funnel and wash extensively with 1 mM HCl (200 ml/g).

Protocol 6. *Continued*

2. Dissolve lyophilized CaM in 0.1 M $NaHCO_3$, 1 mM $CaCl_2$, pH 8.3 (coupling buffer).
3. Wash the swollen CNBr-sepharose with:
   (a) 1 litre of 0.1 M $NaHCO_3$, pH 9.5
   (b) 1 litre of Millipore water
   (c) (c)1 litre of 0.1 M $NaHCO_3$ + 1 mM $CaCl_2$, pH 8.3 (coupling buffer)
4. Transfer the moist cake to a plastic bottle containing 1 vol. of coupling buffer.
5. Add CaM (2 mg of CaM/ml CNBr-sepharose 4B) to the slurry.
6. Incubate overnight on a shaker at 4 °C.
7. Pour the slurry into a sintered glass funnel and drain to a moist cake.
8. Resuspend in 1 vol. of 1 M ethanolamine, drain the slurry to a moist cake and resuspend the resin again in 1 vol. of 1 M ethanolamine.
9. Incubate overnight on a shaker at 4 °C.
10. Wash the gel with three cycles of:
    (a) 1 litre of 0.1 M $NaHCO_3$, pH 9.5
    (b) 1 litre of 0.1 M sodium acetate, pH 4.0
11. Wash the resin thoroughly with Millipore water containing 0.02% sodium azide.
12. Store as 1:1 slurry at 4 °C.

7.2 Purification of MLCK expressed in bacteria

Over-expression of the cloned cDNA in bacteria provides a convenient source for the purification of this enzyme. The protocol described in Section 7.2.1 has been developed in our laboratory (34).

7.2.1 Bacterial cell lysis and isolation of MLCK

Bacteria are first transformed with an expression vector containing the MLCK cDNA. Following induction of gene expression by the method of choice, the cells are pelleted by centrifugation. The pellet is resuspended in 1.5 vol. of 25% sucrose, 0.5 M Tris (pH 8.0). Lysozyme (3 mg/ml) to 15% of total volume and 0.25 M EDTA (pH 7.0) to 2% of total volume are then added. The suspension is shaken at 37 °C for 20 min. This is followed by addition of an equal volume of 2% Brij (a non-ionic detergent) in 0.05 M Tris–HCl (pH 8.0). The mixture is again shaken at 37 °C for 20 min. followed by sonication for 15 sec. It is then centrifuged at 10 000 *g* for 15 min. The pellet is discarded and the supernate is assayed for MLCK activity. Solid $(NH_4)_2SO_4$ is slowly added to the supernatant to 40% saturation. The resulting solution is stirred for 20 min and then centrifuged. The protein pellet is discarded and more $(NH_4)_2SO_4$ is added to the supernate to achieve 60% saturation. As before, the solution is stirred for 20 min and centrifuged. The supernate is discarded and the pellet containing MLCK activity is dissolved in a minimal volume of buffer A containing 20 mM NaCl. The protein solution is dialysed for 16 h against the same buffer and loaded on to a DEAE-cellulose followed by a CaM-sepharose column as has been described previously in step 3 and step 4 of *Protocol 5*.

Buffer A:

- 50 mM Hepes pH 7.5
- 1 mM DTT
- 1 mM EGTA

8. Bacterial expression of calmodulin

Bacterial expression can provide a versatile, economic method to produce large quantities of easily purified protein. However, this method may not always be useful since some of the common post-translational modifications that are common in eukaryotic cells cannot be reproduced in bacteria. In other cases, the protein produced may remain in the particulate fraction or in inclusion bodies (36) which usually prevents the isolation of functional protein. For CaM and some CaM-dependent enzymes, bacterial expression has proven to be a very effective tool for the study of structure–function relationships (37, 38). The vectors that we have utilized as well as the expression and isolation procedures are outlined below.

Bacterial expression vectors are created using standard recombinant DNA techniques to combine known bacterial promotors with a cDNA copy of CaM. An expression vector is a prokaryotic plasmid that contains:

(a) an *E. coli* origin of replication;

(b) a selectable marker gene (usually an ampicillin resistence gene); and

(c) a prokaryotic promoter that has proper promoter elements to bind an RNA polymerase, an operator that binds regulatory proteins, a Shine–Dalgarno box that serves to position the ribosome, and an initiation codon.

Detailed explanations for these requirements and their position effects can be found in (39). Furthermore, a number of plasmid expression vectors using different prokaryotic promoters are now commercially available from such companies as Pharmacia, Promega Corporation, or Boehringer-Mannheim Biochemicals.

For the expression of CaM, our laboratory routinely utilizes two different vectors: pCaMN1 which contains a tac promoter (a hybrid promoter utilizing elements of the trypotophan and lactose operons), and pCaMP$_L$ which utilizes the leftward promoter of lambda phage. The construction of pCaMN1 is detailed in Putkey *et al.* (39). pCaMP$_L$ was constructed by inserting a *Bgl*II/*Nco*I DNA fragment encoding the P$_L$ promoter of lambda into the *Bam*HI/*Nco*I sites of pCaM23 (39). This creates a vector that has a functional P$_L$ promoter immediately upstream of the CaM cDNA and a non-functional *tac* promoter 5′ of the P$_L$ promoter. Both of these vectors therefore, have an *Nco* I site encompassing the initiation codon which also results in the presence of a concensus initiation sequence (CCATGG) (40). Each vector has advantages and disadvantages. The P$_L$ promoter is an extremely efficient promoter that under the right conditions generates 20–25 mg of CaM/litre of bacteria by a simple heat-induction procedure (38). It therefore provides a simple economical method to produce large quantities of pure protein that is often required in physical chemistry experiments. However, the high rate of expression of the P$_L$ promoter tends to destabilize plasmids carrying this promoter. To overcome this deficiency the plasmid must be maintained in one bacterial strain, MM294cI$^+$, which carries a constitutively active repressor but transforms poorly, making molecular biology experiments more difficult. For protein expression, the plasmid must then be transferred to another bacterial strain (N5151 or AR68) which carries a lambda lysogen that encodes a temperature-sensitive repressor (cI$_{857}$) (41). Overall, these are minor inconveniences if only a few constructs are required but if numerous CaM mutants are being produced the *tac* promoter vector, pCaMN1 may be more desirable. This vector system allows one to utilize more conventional bacterial strains such as JM103, JM109, or DH5α which have markedly better transformation efficiencies. This system requires isopropylthiogalactoside (IPTG) to be used for protein induction which is fairly expensive and the yield of CaM is usually only 1–3 mg/litre bacteria. This is, however, more than sufficient protein for functional analysis of the mutants in enzyme activation assays.

8.1 Growth of bacteria and induction

8.1.1 *tac* promoter

Protocol 7. Expression of calmodulin using the *tac* promoter

1. Maintain a stock of pCaMN1 in JM109 on LB plates containing 50 μg/ml ampicillin. LB = Luria broth (45).
2. Use 5–8 colonies to inoculate 50 ml LB + amp (50 μg/ml) and incubate overnight at 37 °C. This should form a dense culture.
3. Inoculate 1 litre LB + amp (50 μg/ml) with 20 ml of the overnight culture and grow at 37 °C for 2–3 h with vigorous shaking. When the OD at 600 nm is approximately 0.4, induce CaM expression by adding 1.0 M IPTG to a final concentration of 2 mM.

Protocol 7. *Continued*

4. Incubate the bacteria for 6–8 h at 37 °C.
5. Harvest the bacteria by centrifuging at 3000 *g* for 30 min in 1 litre bottles. Eventually all of the bacteria are collected into one 200-ml bottle by another spin at 3000 *g* for 30 min. This bacteria pellet can be frozen at −70 °C for 1–2 months before isolation of the protein.

8.2 P_L promoter

As discussed above, all recombinant DNA work must be done in the MM294cI$^+$ strain, which carries a constitutively active repressor. MM294cI$^+$ bacteria can be grown at 37 °C with no problem, and standard transformation protocols can be used (42). The pCaMP$_L$ plasmid must then be transferred to N5151 bacteria for protein expression (41). Usually, one of the short-cut transformation procedures of Hanahan is utilized (Table 4 of ref. 42). Furthermore, since lambda repressor regulates its own expression, N5151 bacteria must always be grown at 30 °C even when they do not contain a plasmid with a P_L promoter. Do not use N5151 bacteria shifted to 42 °C (or ones grown at 37 °C accidentally) to make glycerol stocks or to initiate overnight cultures. It has been observed that if the basal level of CaM expression is too high in the N5151 (for unknown reasons) the bacteria grow very slowly and usually do not grow to saturating density in an overnight culture. If this happens, throw the culture away and start again using fresh N5151 bacteria picked from a plate. The procedure is as follows (41).

Protocol 8. Expression of calmodulin using the P_L promoter

1. Maintain a stock of pCaMP$_L$ in N5151 on LB plates containing 50 μg/ml ampicillin.
2. Use 5–8 colonies to inoculate 50 ml LB + amp (50 μg/ml) and incubate overnight at 30 °C. This should form a dense culture.
3. Inoculate 1 litre of LB + amp (50 μg/ml) with 40 ml of the overnight culture and grow at 30 °C for approx. 5 h with vigorous shaking to keep the culture well aerated. Monitor the A_{600}, and when it reaches 0.9–1.1 in the culture, protein expression can be induced. For some proteins it is better to induce expression while the bacteria are in mid-log phase A_{600} = 0.4–0.6.
4. Induce protein expression by adding an equal volume of LB (1 litre) that was at 68 °C to the culture. Mix well as the hot LB is added, and the final temperature should be approx. 42 °C. Re-divide the induced culture into two equal aliquots of 1 litre and incubate for 90 min at 42 °C with vigorous shaking.

Protocol 8. *Continued*

5. The bacteria are then harvested as described above for the *tac* promoter in *Protocol 7*, step 5.

8.3 Purification of expressed CaM

The procedure set out in *Protocol 9* works well for isolating CaM from bacteria, utilizing either expression system (37, 38). The ease of this procedure should be compared to that for isolation of CaM from tissue as described previously.

Protocol 9. Purification of expressed CaM

1. Resuspend the pellet of bacteria from 1 litre of original culture (before induction) into 20 ml of sucrose solution: 2.4 M sucrose, 40 mM Tris–HCl, pH 8.0, 0.10 mM EDTA. Swirl the sucrose solution gently at 4 °C for 15–30 min. to allow the pellet to resuspend properly. A Teflon rod can also be used to help resuspend the pellet. If using a frozen pellet of bacteria, allow the pellet to thaw in the presence of sucrose solution and then swirl at 4 °C.
2. The bacteria are incubated for a minimum of 2 h at 4 °C. For CaM, a very stable protein, an overnight incubation is acceptable.
3. Add 4 vol. (80 ml) of MOPS solution (50 mM Mops, pH 7.5, 100 mM KCL, 1 mM EDTA, 1 mM DTT and 200 μg/ml lysozyme). The DTT and lysozyme should be added fresh. Mix well with vigorous shaking and incubate at 4 °C for a minimum of 8 h. This is most often done overnight.
4. Centrifuge for 2 h at 4 °C in a Ti45 rotor at 40 000 r.p.m. to pellet the chromosomal DNA and particulate matter.
5. Save the clarified supernatant; this supernatant can also be frozen at −70 °C for 1–2 months if necessary.
6. Add an equal volume of buffer A (10 mM Tris–HCl, pH 7.5, 1 mM $CaCl_2$) to the clarified supernatant and then add 1.0 M $CaCl_2$ to a final concentration of 6 mM. Add the calcium very slowly with stirring or a precipitate may form.
7. Load the sample on to a phenyl-sepharose (Pharmacia) column equilibrated in buffer A at room temp. Since the large supernatant sample must be kept at 4 °C to minimize protease activity and the sample must be at room temp. when it enters the column, the sample is warmed to room temp. just before it enters the column by using an extra-long tubing that is immersed in a beaker of water. Wash the sample through the column with buffer A.

Protocol 9. *Continued*

8. Wash the column with buffer A with 0.5 M NaCl and monitor at A_{280}. A small peak should come off the column. The salt is removed by washing the column in buffer A until A_{280} goes to zero.
9. Elute CaM from the phenyl-sepharose with 10 mM Tris–HCl, pH 7.5, 1 mM EDTA. Monitor the fractions by taking absorbance readings at 277 nm. Collect the peak and add $CaCl_2$ to 1.5 mM for long-term storage. The CaM obtained at this point is usually 80–85% pure.
10. The purification is completed by a single HPLC run, using a DEAE 5PW (Waters, Protein Pak DEAE 5PW, Cat. No. 088044) column at a flow-rate of 0.5 ml/min and the following buffers and linear gradient conditions. CaM elutes from the column as the major peak at about 24 min.

    Buffer A:
    - 10 mM Tris–HCl, pH 7.5
    - 0.1 mM DTT
    - 1 mM EDTA

    Buffer B: buffer A with 1.0 M NaCl

Gradient:	%A	%B
initial	100	0
10 min	70	30
25	65	35
30	50	50
33	0	100
35	0	100
45	100	0

9. Site-directed mutagenesis

The basic strategy for site-directed mutagenesis is to introduce defined alterations in the nucleotide sequence of a known piece of DNA using a single-stranded oligonucleotide primer to synthesize a mutant complementary DNA strand using a ssDNA template. The resultant dsDNA is transformed into *E. coli*, and the resulting progeny (parental and mutant) are analysed. Theoretically, half the progeny should be mutant and half wild-type. In practice, the yield of mutant progeny is significantly lower. The procedures below, based on the method described by Kunkel (43), provide for selection against the non-mutagenized (template) strand of the ds DNS thereby enhancing the yield of mutant progeny (up to 90% mutants). With these yields it is possible to identify desired mutants simply by nucleotide sequencing of a managable number of isolates.

9.1 Design of mutagenic oligonucleotides

In our experience, certain factors are important when designing mutagenic oligonucleotides. When synthesizing oligonucleotide where a single position is mutated, the mutant position should be flanked by nine or more nucleotides to ensure efficient hybridization. In addition, if possible the 3′-most nucleotide should be either a G or C to ensure tight binding of the 3′-end of the oligonucleotide, essential for synthesis of the complementary strand by T4 DNA polymerase. Recently it has been shown that entire regions of a DNA sequence can be mutated randomly by deliberate misincorporation of nucleotides by 'doping' the phosphoaramidites used during chemical synthesis of the oligonucleotide (44). This method is useful when it is not known which mutations might be informative, but where a selection for a mutant phenotype exists.

9.2 Preparation of ssDNA templates

Buffers

LB medium (per litre): 10 g NaCl, 10 g bacto-tryptone, 5 g bacto-yeast extract.

TE buffer: 10 mM Tris–HCl (pH 8.0), 1 mM EDTA.

PEG buffer: No. 1. 25% polyethylene glycol-8000, 3M NaCl., *or* No. 2. 20% polyethylene glycol-8000, 3.75 M ammonium actetate.

Kinase buffer (5×): 0.5 M Tris–HCl (pH 8.0), 50 mM $MgCl_2$, 25 mM DTT, 2 mM ATP.

Annealing buffer (10×): 200 mM Tris–HCl (pH 7.4), 20 mM $MgCl_2$, 500 mM NaCl.

T4 dilution buffer: 100 mM potassium phosphate (pH 7.0), 5 mM DTT, 50% glycerol. T4 DNA polymerase is not stable when diluted so only dilute as much as is required for immediate use.

Synthesis buffer (10×): 175 mM Tris–HCl (pH 7.4), 37.5 mM $MgCl_2$, 5 mM dithiothreitol (DTT), 4 mM of each deoxynucleoside-triphosphate, 7.5 mM ATP.

Stop buffer: 10 mM Tris–HCl (pH 8.0), 10 mM EDTA.

Protocol 10. Preparation of ssDNA from M13

1. Standard recombinant DNA techniques are used to clone the DNA fragment to be mutated into an M13 phage vector (45). M13mp18 and mp19 are commonly used to allow production of ssDNA template from

Protocol 10. *Continued*

opposite DNA strands (46). *E. coli* strains JM109 or DH5αF′ are routinely used to prepare M13 templates.[a] Bacteria are transformed with the dsRF form of the recombinant M13 clone, selected by formation of plaques on LB agar plates, and characterized by standard procedures (45). A phage stock is prepared from the supernatant of the M13 infected isolate by removal of the infected bacteria by centrifugation and can be maintained for several months at 4 °C. This phage preparation is used to prepare U-rich DNA by the following procedures.

2. Inoculate 2 ml of LB medium with two or three colonies of BW313 (dut^- ung^-) *E. coli*, and grow for 2–3 h until an $A_{600} = 0.5$ is reached.
3. Take 50 μl of the BW313 culture and add 1–2 μl of supernatant from an M13 phage culture. Allow to stand for 30 min at 37 °C. Then add 5 ml of LB medium and grow overnight at 37 °C.
4. For the subsequent optimal recovery of mutants, repeat the infection of BW313 *E. coli* using 1–2 μl of the supernatant produced in step 3. This results in the enrichment of U-rich DNA in the ssDNA template.
5. Centrifuge the overnight culture at 3000 *g* for 20 min, remove 1.5 ml of supernatant and place in a microcentrifuge tube.
6. Centrifuge at 12 000 *g* for 5 min and remove the *top* 1.2 ml of supernatant.
7. Centrifuge again for 5 min at 12 000 *g* and remove the *top* 1 ml of supernatant. This procedure ensures the removal of all dsDNA-containing bacteria which could otherwise contaminant the mutagenesis experiment with unmutated transformation competent DNA.
8. To the 1 ml of supernatant add 0.2 ml of PEG buffer No. 1 and let stand at room temp. for 20 min.
9. Pellet phage particles by centrifugation at 12 000 *g* for 10 min.
10. Resuspend phage pellet in 50 μl of TE buffer.
11. Extract with phenol and phenol–chloroform (45). To the extracted supernatant add 5 μl 3 M sodium acetate (pH 7) and 150 μl 95% ethanol to precipitate the ssDNA. Collect the precipitate by centrifugation and resuspend in 15 μl TE buffer. Typically, 4 μl of this preparation are used per mutagensis reaction.

[a] When using JM109, cultures must be started from bacteria grown on M–9 minimal medium plates (45). Growth of JM109 on minimal medium selects for a marker required for proline biosynethsis carried on an episome. This episome also carries the F-factor which is an absolute requirement for M13 infection. In the DH5αF′ strain, the F-factor is integrated into the genome so that selection is not required and bacteria may be grown in LB medium.

Protocol 11. Preparation of ssDNA from phagemid clones

1. The DNA to be mutated is cloned by standard techniques (45) into a suitable phagemid vector (e.g. pTZ18, pTZ19, pGEM3Zf(–); phagemids are available from companies such as Stratagene, Promega Biotech). Phagemids are pUC-based plasmids into which the origin for synthesis single-stranded DNA from the phage f1 has been incorporated. The phagemid replicates as a typical ds circular DNA plasmid using the ColE1 origin until the host bacterium is infected with a helper phage.
2. Transform the recombinant phagemid into the *E. coli* strain BW313 (dut^- ung^-). Transformed colonies are selected by growth on LB-agar plates supplemented with 100 μg/ml ampicillin.
3. Prepare an overnight culture of phagemid-containing bacteria by picking individual ampicillin-resistant colonies from a fresh plate. Inoculate 1–2 ml of LB medium containing 100 μg/ml ampicillin and incubate with shaking at 37 °C overnight.
4. Inoculate 10 ml of LB medium with 50 μl of the overnight culture. Shake vigorously at 37 °C for 30 min in a sterile 50-ml tube.
5. Add helper phage (R408) (47) at a multiplicity of infection of 10 (about 2×10^9 phage). Continue shaking for 13 h at 37 °C.
6. Centrifuge the culture at 5000 *g* for 10 min to remove bacteria. Carefully remove supernatant (which contains the phage particles) and repeat centrifugation until a bacterial pellet is no longer produced (usually twice).
7. Precipitate the phage by adding 0.25 vol. of PEG buffer No. 2 to the supernatant. Let stand on ice 30 min, then centrifuge 15 min at 12 000 *g*.
8. Phenol–chloroform extract phage and precipitate ssDNA with ethanol (45). Dissolve the DNA in 50 μl of TE buffer. The ssDNA is now ready for use as a template for the synthesis of a complementary strand primed by an oligonucleotide containing the desired mutation(s).

9.3 Mutagenesis procedure

Protocol 12. Phosphorylating oligonucleotides

Chemically synthesized oligonucleotides lack the 5′-PO_4 that will eventually be required to covalently close the dsDNA after oligonucleotide-directed synthesis of the complementary strand is complete, and therefore must be phosphorylated with T4 polynucleotide kinase prior to annealing.

Protocol 12. *Continued*

Set up the following reaction:

- 200 pmol oligonucleotide
- 4 μl kinase buffer (5×)
- 5 units T4 polynucleotide kinase
- water to 20 μl

1. Mix well and incubate at 37 °C for 45 min.
2. Stop the reaction by heating at 65 °C for 10 min.
3. Dilute the kinased oligonucleotide to 2 pmol/μl.

Protocol 13. Annealing the oligonucleotide and synthesis of the complementary strand

Set up the following reaction:

- 0.5 pmol uracil-rich ssDNA template[a]
- 10 pmol mutagenic oligonucleotide[a]
- 1 μl 10 × annealing buffer
- Add water to bring final vol. to 10 μl

1. Place the reaction in a water bath at 65–70 °C. Allow the annealing reaction to cool to 30 °C over a 45-min period. After this, place the reactions on ice.
2. Add 1 μl 10× synthesis buffer, 1 μl T4 DNA ligase (2–5 units), 1 μl T4 DNA polymerase (diluted in T4 dilution buffer to 1 unit/μl).
3. Incubate on ice for 5 min to stabilize the primer, then at 25 °C for 5 min, and finally to 37 °C for 90 min.
4. To stop the reaction add 90 μl of stop buffer. The reaction is stable at −20 °C for at least one month for subsequent use in transformation.
5. Transform competent *E. coli* (e.g. JM109, DH5αF′) with 10 μl of the reaction mixture and select for ampicillin resistant colonies (for phagemids) or phage plaques (for M13).
6. Isolates are then picked and analysed for the presence of the desired mutation by nucleotide sequencing.

[a] These amounts are given as starting guide-lines. The optimal ratios of oligo:template will vary widely depending on the oligonucletide and template DNA and will need to be determined empirically to ensure production of the desired product. Keep in mind that high oligo concentrations may interfere with subsequent ligation of the dsDNA product or may result in non-specific hybridization to the template DNA.

9.4 General comments

Once the desired mutation in a CaM or other cDNA has been created, the cDNA can then be ligated into one of the bacterial or eukaryotic expression vector systems described in this chapter for functional analysis. In combination with these expression systems, the ability to create site-specific mutations provides a powerful approach to dissecting the structure–function relationships of CaM and other proteins for which cDNAs have been isolated.

10. Eukaryotic CaM expression vectors

10.1 Expression vector for use in mouse C127 cells

10.1.1 Vector design

We have recently designed a plasmid (pMT-X; *Figure 3*) that makes possible the regulatable expression of virtually any cDNA or cDNA fragment in mouse C127 cells in either a sense (to produce a protein product) or antisense (to inhibit production of a gene product) orientation. This plasmid contains the hMT-IIa promoter (48), inducible by Zn, serum, glucocorticoids or progesterone, unique *Nco*I, *Bam*HI, *Sma*I and *Eco*RI sites for directional cloning of cDNAs, and the 3′-end of the parvalbumin gene for termination of transcription and polyadenylation. If the cDNA contains or can be mutated by site-directed mutagenesis to contain an *Nco*I site (CCATGG) at the initiator ATG, then the initiation codon is in the same context as would be the hMT mRNA normally transcribed from this promoter. In addition, the *Hind*II site at the 5′-end of the hMT-IIa promoter has been mutated to a *Not*I restriction site, to allow the ligation of the bovine papilloma virus (BPV)

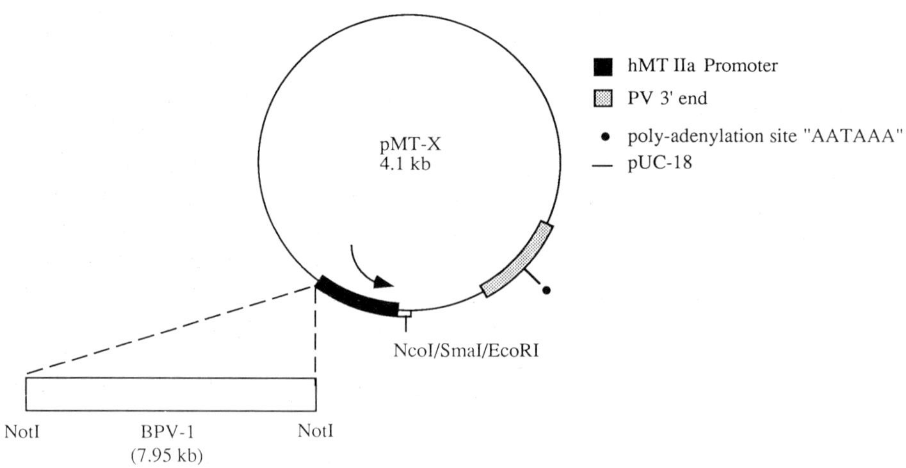

Figure 3.

genome which was modified to contain *Not*I ends. The inclusion of BPV in the vector allows the expression plasmid to be maintained as a nuclear episome, and selection for stable transformants by a focus formation assay (49–52).

10.1.2 Cell culture

The culture of mouse C127 cells has been previously described (52).

10.1.3 Transfection

i. Buffers

(a) 2 × HBS: 50 mM Hepes, (pH 7.1), 1.5 mM sodium phosphate, 280 mM NaCl.

(b) 250 mM $CaCl_2$.

Both solutions should be made fresh and sterile-filtered prior to use.

ii. Procedure

Amounts given are for transfection of one, 100-mm diameter culture plate at 70% confluence. Usually four plates are transfected with each expression plasmid to be studied.

(a) To 0.5 ml of 2 × HBS add:

(b) 15 μg salmon sperm DNA

(c) 5 μg expression plasmid DNA*

Mix well, then add 0.5 ml of 250 mM $CaCl_2$. The solution should become cloudy upon addition of the $CaCl_2$. Let stand 20–40 min at room temp. until a fine precipitate forms. Remove the culture medium from the cells and rinse once with 5 ml Hanks' balanced salts solution (or other isotonic salt solution compatible with your cultured cell type). Add the DNA/$CaPO_4$ precipitate and swirl gently to ensure even coverage of the plate. Let sit 30 min at room temp. after which 10 ml of culture medium is added. The cells are then incubated for 5 hrs. at 37 °C.

After 5 h remove the medium and add 1 ml of 1 × HBS/25% glycerol. Let sit for 60 sec, then rinse *very gently* three times with 10 ml of fresh culture medium. Finally, add 10 ml fresh medium and incubate overnight at 37 °C. The next day, remove the cells from the dish by trypsinization (or whatever method is best for your cell type) and replate into four new dishes (a 1:4 split). The cells are then fed twice weekly with fresh medium. After 10–14 days transformed foci should appear and can be selected using cloning rings as previously described (52). If a selectable marker is employed, then medium in which only cells expressing the marker gene will survive is used to culture cells. Once colonies of surviving cells appear they can be selected using cloning rings and further characterized.

* If co-transfecting a plasmid containing a selectable marker, then this should be 0.5 μg of the plasmid with the marker gene, and 4.5 μg of expression plasmid DNA.

10.1.4 Characterization of cell lines

After the initial selection of stable transformants, cell lines are characterized for the presence and organization of the transfected DNA by Southern blot hybridization, and correct expression of the gene included in the expression plasmid by Northern blot hybridization (mRNA) and either Western blot or radioimmunoassay (protein).

10.2 Nucleic acid isolation and characterization

10.2.1 Buffers

(a) TMN: 10 mM Tris–HCl (pH 8.0), 1.5 mM $MgCl_2$, 140 mM NaCl.

(b) DNA lysis buffer: 10 mM Tris–HCl (pH 8.0), 10 mM EDTA, 10 mM NaCl, 0.5% sodium dodecyl sulfate (SDS).

(c) RNA lysis buffer: TMN, 0.5% NP-40.

(d) TE/RNase: 10 mM Tris–HCl (pH 8.0), 1 mM EDTA, 20 μg/ml RNase A (prepared according to ref. 45).

(e) Denaturing buffer: 0.5 M NaOH, 1.5 M NaCl.

(f) Neutralizing buffer: 0.5 M Tris–HCl (pH 8.0), 1.5 M NaCl.

(g) Hybridization solution: 5 × SSPE, 5 × Denhardt's solution (52), 50% deionized formamide, 1% SDS, 100 μg/ml yeast tRNA.

10.2.2 DNA isolation and analysis

Isolated cells (usually 5×10^6) are resuspended in 50 μl TMN buffer, and 450 μl of DNA lysis buffer is added. The solution is pipetted vigorously 20 times using a P200 (Gilson) to shear high-molecular-weight DNA. The sample is then incubated at room temp. for 10 min. Next, 100 μg of proteinase K is added and the sample incubated at 37 °C for 4 h. Protein is removed by extraction with phenol–chloroform (45), and the DNA precipitated by the addition of 50 μl 3 M sodium acetate (pH 5.0), and 1 ml 95% ethanol, followed by centrifugation at 12 000 *g* for 10 min in a microcentrifuge. The pellet is resuspended in 100 μl of TE/RNase and incubated for 30 min at 37 °C to digest RNA.

Ten micrograms of DNA is digested with the appropriate restriction enzyme, and the sample resolved by agarose gel electrophoresis (45). After electrophoresis, the DNA is first denatured for 30 min with gentle agitation in 250 ml denaturing buffer, followed by a 30-min incubation in neutralizing buffer. The DNA is transferred to filters as described previously (52, 45). After transfer, the filter is air-dried for 30 min then baked at 68 °C for 1–2 h.

After baking, the filter is pre-hybridized in a Seal-a-Meal bag for 1 h at 37 °C in 10–20 ml of hybridization solution. Hybridization is for 16–20 h using 5–10 ml of hybridization solution containing $2\text{–}4 \times 10^6$ c.p.m./ml of radiolabelled probe prepared by the method of Feinberg and Vogelstein (53).

Filters are washed five times in 200 ml 2 × SSC (45), 0.1% SDS for 5–10 min each wash, followed by two washes of 15 min each in 0.1 × SSC, 0.1% SDS at 50 °C. Autoradiography is typically for 2 h, using Kodak X-AR film at −70 °C with intensifying screens.

10.3.2 RNA isolation and Northern blot analysis

To determine if the transfected expression plasmid is producing the expected mRNA, the following procedures can be used.

Cells are lysed in 200 μl of RNA lysis buffer for 10 min in ice with periodic vortexing. Nuclei are pelleted by centrifugation for 2 min in a microcentrifuge, and the supernatant fluid retained. The sample is extracted twice with phenol–chloroform (45), and once with chloroform–isoamyl alcohol. RNA is precipitated by the addition of 25 μl of 3 M sodium acetate (pH 5.0) and 750 μl of 95% ethanol, and the precipitate collected by centrifugation as for DNA. The pellet is washed with ethanol, air-dried, and resuspended in 100 μl of sterile water. Yields are calculated by determining the A_{260} for a 10 μl portion of the sample diluted into 1 ml of water.

RNA is resolved on formaldehyde-containing agarose gels and transferred to Nytran, or an equivalent filter (52). Pre-hybridization and hybridization (including washes) using a radiolabelled probe are the same as is used for Southern blots (Section 10.2.2).

10.2.4 Analysis of protein expression

The proper expression of proteins in mouse C127 cells can be determined by the use of either Western blot analysis (54) or radioimmunoassay, as described earlier in this chapter. Both have been used successfully in this system (49, 51, 55).

10.3 Expression vectors for use in *Aspergillus nidulans*

We have recently cloned and characterized the CaM gene of the filamentous fungus *A. nidulans*. Calmodulin in this organism is structurally and functionally more similar to vertebrate CaM than is CaM from other lower eukaryotes such as budding or fission yeast (56). *Aspergillus* has been genetically well-characterized, and recent studies have established a transformation system that combines genetic analysis and molecular biology. As such this organism promises great utility for assessing CaM-dependent function *in vivo*.

10.3.1 Vector design

The availability of both constitutive and inducible promoters in *A. nidulans* permits the design and construction of expression plasmids for use in this organism (57). The plasmid pCR-X contains the *A. nidulans* CaM gene promoter, unique *Nco*I and *Sma*I restriction sites, the 3′-end of the *A. nidulans*

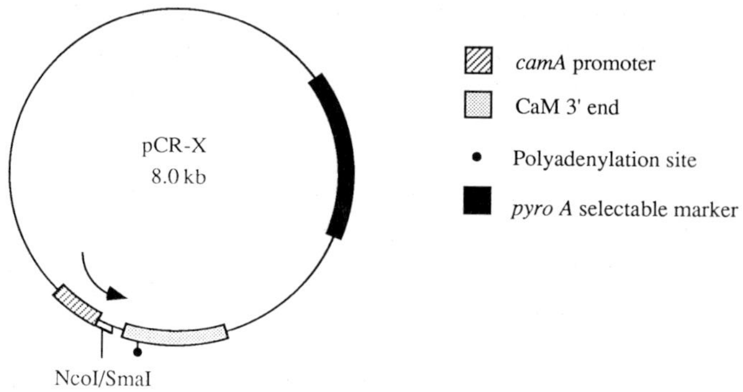

Figure 4(a).

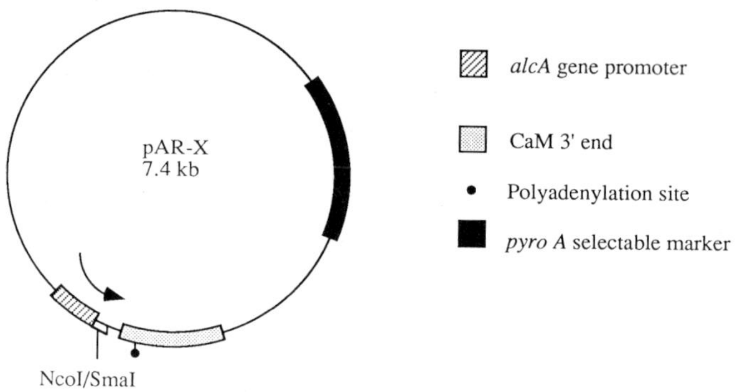

Figure 4(b).

CaM gene to provide transcription termination signals, the *pyro*A gene of *A. nidulans* as a selectable marker, all placed in the vector pGEM3Zf(–) (*Figure 4a*). In the plasmid pAR-X, the CaM promoter has been replaced with the regulatable promoter from the *alcA* gene of *A. nidulans* (*Figure 4b*). This promoter is repressed by glucose or acetate in the culture medium, de-repressed by glycerol and induced by threonine or ethanol. Vectors based on this promoter are capable of expressing protein at up to 5% of total protein.

10.3.2 Transformation of *Aspergillus nidulans*

Transformation of *A. nidulans* with plasmid DNA is performed following the procedure of Osmani *et al.* (58). Transformation occurs predominantly by homologous recombination of DNA sequences in the expression plasmid with the corresponding region in the *A. nidulans* genome (58, 59). *Aspergillus* grows as a multinucleate syncitium. Therefore, stable transformants can be selected as heterokaryons containing both normal and transformed nuclei. This is important when using a pAR-X based expression plasmid since it allows the expression plasmid to be successfully integrated and maintained in the genome without being expressed, an important feature in instances where over-expression of a gene product might have lethal effects.

10.3.3 Buffers

(a) YGUU medium: 0.5% yeast extract, 2% glucose, 5 mM uridine, 10 mM uracil, supplemented with trace elements (60) and vitamins as required by the particular strain to be transformed. For the pCR-X and pAR-X vectors, pyrodoxine HCl is omitted from the medium. If the pAL series of vector (57) is used, uridine and uracil are omitted.

(b) Lytic mix: 0.4 M $NH_4(SO_4)_2$, 50 mM potassium citrate (pH 6.0), 0.5% yeast extract, 1% sucrose, 20 mM glucose, 50 mg bovine serum albumin, 50 mg Novozyme 234 (Novo Industri, Danbury, CT; prepared according to Osmani *et al.* (58)), 25 mg Driselase and 25 μl Glucuronidase (both from Sigma). All components are combined and the mixture is centrifuged for 5 min at 2000 *g* to sediment insoluble material.

(c) Protoplast wash buffer: 50 mM potassium citrate (pH 6.0), 0.4 M $NH_4(SO_4)_2$, 0.5% sucrose.

(d) Protoplast storage buffer: 0.6 M KCl, 50 mM $CaCl_2$, 10 mM Tris–HCl (pH 7.5).

(e) PEG buffer: 25% PEG-8000, 100 mM $CaCl_2$, 0.6 M KCl, 10 mM Pipes (pH 6.8).

(f) YAS top agar: YGUU medium, 1 M sucrose, 0.8% agar.

(g) YAS plates: YGUU medium, 0.1 M sucrose, 1.5% agar.

10.3.3 Procedure

To prepare transformation competent protoplasts, 1×10^8 freshly prepared conidia (61) are placed in 10 ml YGUU medium and germinated at 32 °C for 5.5–6 h (until a germ tube is visible in greater than half the conidia). Conidia are then harvested by centrifugation (3 min, 500 *g*). The pelleted conidia are resuspended in 10 ml of Lytic mix and digestion of the cell wall allowed to proceed for 2 h at 32 °C with gentle agitation (150 r.p.m. in shaking water bath). The resultant protoplasts are washed twice in Protoplast wash buffer, centrifuged as above and resuspended in 0.4 ml Protoplast storage buffer. Transformation efficiency increases if protoplasts are allowed to sit overnight in ice prior to transformation.

For transformation, 100 μl of protoplasts are added to 2 μg of expression plasmid DNA, followed by 50 μl of PEG buffer. After 20 min in ice, an additional 1 ml of PEG buffer is added, the mixture mixed gently by inversion, and allowed to stand an additional 30 min at room temp. Aliquots (100–200 μl) are added to 3 ml YAS top agar (at 45 °C), mixed well and plated on to YAS plates pre-warmed to 37 °C. After the top agar has hardened (30 min at room temp.) the plates are incubated at 37 °C (32 °C if a temperature-sensitive strain is used) for two days to allow growth of transformed heterokaryons.

10.3.4 Selection and characterization of strains

As in cultured cells, the correct integration and expression of the expression plasmid are characterized by Southern and Northern blot analysis and Western blot or radioimmunoassay.

10.4 Isolation and analysis of DNA and RNA from *A. nidulans*

10.4.1 Buffers

(a) DNA isolation buffer: 50 mM Tris–HCl (pH 8.0), 100 mM EDTA, and 2% SDS.

(b) RNA isolation buffer: 50 mM Tris–HCl (pH 8.0), 0.3 M NaCl, 5 mM EGTA, 2% SDS.

(c) 20 × RNA buffer: 0.4 M boric acid, 4 mM EDTA, pH adjusted to 8.3.

(d) RNA sample buffer: 900 μl deionized formamide, 300 μl formaldehyde, 180 μl 20 × RNA buffer.

10.4.2 Procedure

Mycelia grown overnight in liquid culture are harvested on Miracloth (Calbiochem), washed with water, the excess fluid removed, wet weight determined and then placed in a microcentrifuge tube with holes punched in the cap. The amounts of buffers indicated below are for samples of 100–200 mg wet weight. Freeze the sample on dry ice, then lyophilize overnight. After completely lyophilized, grind the mycelia in the microfuge tube to a fine powder using a pestle that fits 1.5-ml Eppendorf tubes.

10.4.3 DNA isolation

Transfer the powdered mycelia to a tube containing 0.5 ml of DNA isolation buffer. Incubate at 65 °C for 30 min followed by treatment with 100 μg proteinase K for 16 h at 37 °C. Debris is removed by centrifugation in a microfuge for 1 min, followed by extraction (phenol–chloroform three times). DNA is recovered by ethanol precipitation, and RNA is removed by treatment with RNase A as described above followed by phenol-extraction and ethanol precipitation. Usually 1–3 μg of genomic DNA is digested with the appropriate restriction enzyme and analysed by Southern blot hybridization as described above.

10.4.4 RNA isolation

Powdered mycelia are placed in a microfuge tube containing 500 μl of RNA isolation buffer and 500 μl phenol–chloroform. Immediately vortex at full speed for 1 min. Separate the phases by centrifugation for 5 min at 12 000 *g*. Repeat the phenol–chloroform extraction two times, and extract once more with chloroform–isoamyl alcohol. Collect the RNA by ethanol precipitation and resuspend in 100 µl of sterile water. The yield is determined spectrophotometrically and 5–10 µg are resolved by gel electrophoresis. Northern transfer and hybridization to a radiolabelled probe are as described above.

10.4.5 General comments

This section has outlined methods for the use of expression vectors in two different systems, mouse C127 cells, and the genetically manipulable fungus *A. nidulans*. The vectors described should be readily adaptable to the expression of genes other than CaM or CaM-binding proteins. Studies in which the described vectors have been utilized have indicated that they provide a powerful approach to understading CaM functions *in vivo*. Where reagents are available to characterize the expression of other genes, these systems should also prove useful in assessing the effect of changes in levels of specific proteins on cellular functions in eukaryotes.

References

1. Carafoli, E. (1987). *Annual Reviews of Biochemistry*, **56**, 395–433.
2. Babu, Y. S., Sack, J. S., Greenhough, T. J., Bugg, C. E. Means, A. R., and Cook, W. J. (1985). *Nature*, **315**, 37–40.
3. Kretsinger, R. H. (1980). *Advances in Cyclic Nucleotide Research*, **11**, 1–26.
4. Means, A. R. (1988). *Recent Progress in Hormone Research*, **44**, 223–62.
5. Rasmussen, C. D. and Means, A. R. (1989). *Trends in Neurosciences*, **12**, 433–8.
6. Beavo, J., Hardman, J., and Sutherland, E. (1970). *Journal of Biological Chemistry*, **245**, 5649–55.
7. Klee, C. B., Crouch, T. H., and Krinks, M. H. (1979). *Biochemistry*, **18**, 722–5.
8. Kincaid, R. L., Manganiello, V. C., Odya, C. E., Osborne, J. C., Stith-Coleman, I. E., Danello, M. A., and Vaughan, M. (1984). *Journal of Biological Chemistry*, **259**, 5158–66.
9. Shenolikar, R. S., Thompson, W. J., and Strada, S. J. (1985). *Biochemistry*, **24**, 672–8.
10. Hansen, R. S. and Beavo, J. A. (1982). *Proceedings of the National Academy of Sciences, USA*, **79**, 2788–92.
11. Martins, T. J., Mumby, M. C., and Beavo, J. (1982). *Journal of Biological Chemistry*, **257**, 1973–9.
12. Bolton, A. E. and Hunter, W. M. (1973). *Biochemical Journal*, **133**, 529–39.
13. Chafouleas, J. G., Dedman, J. R., Munjaal, R. P., and Means, A. R. (1979). *Journal of Biological Chemistry*, **254**, 10262–7.
14. Dedman, J. R. Welsh, M. J., and Means, A. R. (1978). *Journal of Biological Chemistry*, **253**, 7515–21.
15. Midgley, A. R., Jr., Niswender, G. D., and Rebar, R. W. (1969). *Acta Endocrinologica*, Suppl. **142**, 247.

16. Duddleson, W. G., Midgely, A. R., Jr., and Niswender, G. D. (1972). *Computers and Biomedical Research*, **5**, 205.
17. Glenny, J. R. Jr. and Weber, K. (1980). *Journal of Biological Chemistry*, **255**, 10551–4.
18. Flanagan, S. D. and Yost, B. (1984). *Analytical Biochemistry*, **140**, 510–19.
19. Otter, T., King, S. M., and Witman, G. B. (1987). *Analytical Biochemistry*, **162**, 370–7.
20. Sikela, J. M. and Hahn, W. E. (1987). *Proceedings of the National Academy of Sciences, USA*, **84**, 3038–42.
21. Guerriero, V., Russo, M. A., Olson, N. J., Putkey, J. A., and Means, A. R. (1986) *Biochemistry*, **25**, 8372–81.
22. Hanley, R. M., Means, A. R., Ono, T., Kemp., B. E., Burgin, K. E., Waxham, N., and Kelly, P. T. (1987). *Science*, **237**, 293–7.
23. Kaufmann, S. H., Ewing, C. M., and Shaper, J. H. (1987). *Analytical Biochemistry*, **16**, 89–95.
24. Slaughter, G. R. and Means, A. R. (1987). *Methods in enzymology*, Vol. 139 (ed. A. R. Means and P. M. Conn), pp. 433–44. Academic Press, New York and London.
25. Slaughter, G. R. and Means, A. R. (1985). *Biochemical and Biophysical Research Communications*, **126**, 295–303.
26. Simmen, R. C. M., Dunbar, B. S., Guerriero, V., Chafouleas, J. G., Clark, J. H., and Means, A. R. (1984). *Journal of Cell Biology*, **99**, 588–93.
27. Foyt, H. L., Guerriero, V., and Means, A. R. (1985). *Journal of Biological Chemistry*, **260**, 7765–74.
28. Dabrowska, R., Sherry, J. M. F., Aromatorio, D. K., and Hartshorne, D. J. (1978). *Biochemistry*, **17**, 253–8.
29. Adelstein, R. S. and Klee, C. B. (1981). *Journal of Biological Chemistry*, **256**, 7501–9.
30. Pires, E. M. V. and Perry, S. V. (1977). *Biochemical Journal*, **167**, 137–46.
31. Yagi, K., Yazawa, M., Kakiuchi, S., Ohshima, M., and Venishi, K. (1978). *Journal of Biological Chemistry*, **253**, 1338–40.
32. Walsh, M. P., Vallet, B., Autric, F., and Demaille, J. (1979). *Journal of Biological Chemistry*, **236**, 12136–44.
33. Wolf, H. and Hofmann, F. (1980). *Proceedings of the National Academy of Sciences, USA*, **77**, 5852–5.
34. Bagchi, I. C., Kemp, B. E., and Means, A. R. (1989). *Journal of Biological Chemistry*, **264**, 15843–9.
35. March, S. C., Parikh, I., and Cuatrecasas, P. (1974). *Analytical Biochemistry*, **60**, 149–52.
36. Reinach, F. C., Nagai, K., and Kendrick-Jones, J. (1986). *Nature*, **322**, 80–3.
37. Putkey, J.A., Ono, T., VanBerkum, M. F. A., and Means, A. R. (1988). *Journal of Biological Chemistry*, **263**, 11242–9.
38. VanBerkum, M. F. A., George, S. E., and Means, A. R. (1990). *Journal of Biological Chemistry*, **265**, 3750–6.
39. Putkey, J. A., Donnelly, P. V., and Means, A. R. (1987). In *Methods in Enzymology*, Vol. 139 (ed. A. R. Means and P. M. Conn), pp. 303–17. Academic Press, New York and London.
40. Kozak, M. (1984). *Nucleic Acids Research*, **12**, 857.

41. Watt, R. A. Shatzman, A. R., and Rosenberg, M. (1985). *Molecular and cellular biology*, **5**, 448–56.
42. Hanahan, D. (1986). In *DNA cloning*, Vol. I: *A practical approach*. (ed. D. M. Glover), pp. 109–35. Academic Press, New York and London.
43. Kunkel, T. A. (1985). *Proceedings of the National Academy of Sciences, USA*, **83**, 488–92.
44. Hutchison, C. A. III, Nordeen, S. K., Vogt, K., and Edgell, M. H. (1986). *Proceedings of the National Academy of Sciences, USA*, **84**, 710–14.
45. Maniatis, T., Fritsch, E. F., and Sambrook, J. (1982). *Molecular cloning: A laboratory manual.* Cold Spring Harbor Laboratory Press, Cold Spring Harbor, New York.
46. Messing, J. (1983). In *Methods in enzymology*, Vol. 101 (ed. R. Wu, L. Grossman,and K. Moldave), pp. 20–78. Academic Press, New York and London.
47. Russel, M., Kidd, S., and Kelley, M. R. (1986). *Gene*, **45**, 333–8.
48. Slater, E. P., Cato, A. C. B., Karin, M., Baxter, J. D., and Beato, M. (1988). *Molecular Endocrinology*, **2**, 485–91.
49. Rasmussen, C. D. and Means, A. R. (1987). *EMBO Journal*, **6**, 3961–8.
50. Sarver, N., Gruss, P., Law, M. F., Khoury, G., and Howley, P. M. (1981). *Molecular and Cellular Biology*, **1**, 486–96.
51. Rasmussen, C. D. and Means, A. R. (1989). *EMBO Journal*, **8**, 73–82.
52. Rasmussen, C. D., Simmen, R. C. M., MacDougall, E. A., and Means, A. R. (1987). *Methods in Enzymology*, Vol. 139 (ed. A. R. Means and P. M. Conn), pp. 642–54. Academic Press, New York and London.
53. Feinberg, A. P. and Vogelstein, B. (1983). *Analytical Biochemistry*, **137**, 266–7.
54. Towbin, H., Staehelin, T., and Gordon, J. (1979). *Proceedings of the National Academy of Sciences, USA*, **76**, 4350–4.
55. Rasmussen, C. D. and Means, A. R. (1989). *Molecular Endocrinology*, **3**, 588–96.
56. Rasmussen, C. D., Means, R. L., Lu, K. P., May, G. S., and Means, A. R. (1990). *Journal of Biological Chemistry*, **265**, 13767–75.
57. Waring, R. B., May, G. S., and Morris, N. R. (1989). *Gene (Amsterdam)*, **79**, 119–30.
58. Osmani, S. A., May, G. S., and Morris, N. R. (1987). *Journal of Cell Biology*, **104**, 1495–1504.
59. May, G. S. (1989). *Journal of Cell Biology*, **109**, 2267–74.
60. Cove, D. J. (1966). *Biochimica et Biophysica Acta.*, **113**, 51–6.
61. Clutterbuck, A. J. (1974). In *Handbook of genetics*, Vol. 1 (ed. R. C. King), pp. 447–510. Plenum Press, New York.

10

Annexins: calcium-dependent phospholipid-binding proteins

CATHERINE M. BOUSTEAD and JOHN H. WALKER

1. Introduction

The effects of calcium as an intracellular second messenger are mediated by its interaction with calcium-binding proteins. Recently, a new family of calcium-binding proteins, the annexins, has been identified in a wide variety of species and tissues (for reviews see refs 1–4). Annexins are abundant proteins (up to 0.5% of cell protein) which bind reversibly to phospholipids at micromolar calcium concentrations, and inhibit the action of phospholipase A_2 *in vitro*. Certain annexins are also substrates for serine and tyrosine kinases, and some exhibit calcium-dependent binding to F-actin. Structurally, the annexins are characterized by a variable N-terminal tail, followed by four (in the case of annexins I–V and VII–VIII) or eight (annexin VI) repeats of a highly conserved 70 amino acid sequence. All annexins contain at least one calcium-binding site, but unlike the calmodulin family of proteins, they do not contain the amino acid sequence corresponding to the 'E-F hand' type of calcium-binding site.

The annexins were identified independently by groups studying various different aspects of cellular structure and function, often those regulated by calcium, such as exocytosis, the cytoskeleton, inflammation, and blood coagulation. As a result, the majority of annexins have each received several different names. A common nomenclature which numbers the annexins I to VIII has now been proposed (5) and will be used throughout this chapter (see *Table 1* and *Figure 1*). Methods for the isolation and purification of annexins will be described, together with assay procedures for their main properties.

2. Isolation of annexins

Most methods for the isolation of annexins utilize their reversible calcium-dependent binding either to membrane phospholipids or to tissue cytoskeleton fractions. The three approaches described in *Protocols 1–3* are the basis of most of the variations used in different laboratories. These methods may be

Table 1. Nomenclature of annexins[a]

Annexin	kd	pI	Previous nomenclature	Reference[b]
I	38	6.5	Lipocortin I	6
			p35	7
			Calpactin II	8
			Chromobindin 9	9
			GIF (Glycosylation-inhibiting factor)	10
II	Heterotetramer		Lipocortin II	11
	2 × 36	7.2	p36	12
	2 × 10		Calpactin I	8, 13, 14
			Chromobindin 8	9
			Protein I	15
			PAP IV (Placental anti-coagulant protein)	16
III	35	5.9	Lipocortin III	17
			PAP III	16
			35-alpha Calcimedin	18
IV	32	5.6	Lipocortin IV	17
			Chromobindin 4	9, 19
			Protein II	20
			PAP II	16
			35-beta Calcimedin	18
			Endonexin I	21
			32.5K Calelectrin	22
			PP4-X	23
V	34	5.0	Lipocortin V	17
			PAP I	24
			35-gamma Calcimedin	18
			Endonexin II	25
			PP4	23
			Anchorin CII	26, 27
			Calphobindin I	28
			IBC (Inhibitor of blood coagulation)	29
			VAC-alpha (Vascular anti-coagulant)	30, 31
			Chromobindins 5 and 7[c]	9
			p32/34[c]	32
VI	70	5.8	Lipocortin VI	17
			p68	33, 34
			p70	35
			73K	36
			Chromobindin 20	9
			67K Calcimedin	37
			67K Calelectrin	38
			Protein III	39
			Calphobindin II	28
			Synhibin	40
VII	47	7.0	Synexin	41
VIII	37	5.4	VAC-beta	42

[a] Annexins have also been described in ref. 43 (*Torpedo*), ref. 44 (*Drosophila*), and refs 45 and 46 (plants), but have not yet been incorporated into the nomenclature given here.
[b] References are to the amino acid or cDNA sequence where possible, or to a useful entry point into the literature.
[c] Bovine tissues appear to contain two closely related polypeptides.

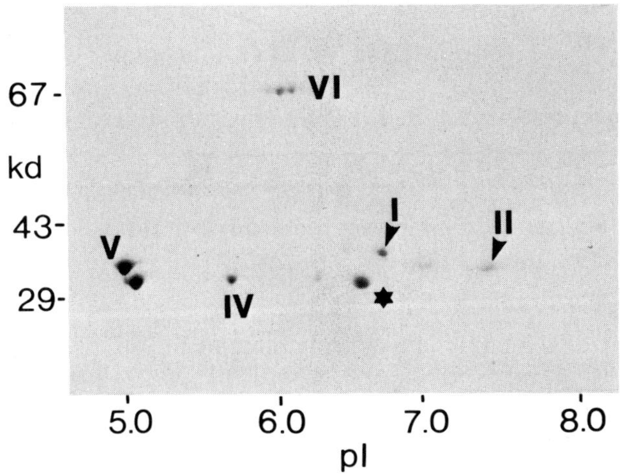

Figure 1. Two-dimensional gel electrophoresis of bovine annexins. The positions of annexins I, II, IV, V (note that two closely related proteins are found in bovine tissues), and VI are marked. The identity of the polypeptide marked ★ is not yet clear (9, 17, 18).

Protocol 1. Preparation of annexins using calcium-precipitation (32)

A. *Procedure*

All steps should be performed at 4 °C.

1. Homogenize 50 g tissue (fresh or frozen) in a Waring blender (4 × 30 sec, with 30 sec cooling between homogenization steps) in 250 ml of 0.15 M NaCl, 5 mM EGTA, 0.25 mM PMSF, 10 mM Hepes, pH 7.4.
2. Centrifuge for 30 min at 40 000 *g*.
3. Collect the supernatant and add $CaCl_2$ from a 1 M stock solution to give a final concentration of 6 mM (1 mM excess over the EGTA). Stir and leave for 15 min on ice.
4. Centrifuge for 30 min at 40 000 *g*.
5. Wash the pellet with 100 ml of 0.15 M NaCl, 1 mM $CaCl_2$, 10 mM Hepes, pH 7.4, by resuspension and centrifugation at 40 000 *g* for 30 min.
6. Repeat step 5 with 100 ml of 1 mM $CaCl_2$, 10 mM Hepes, pH 7.4.
7. Resuspend the pellet in 15 ml of 10 mM EGTA, 10 mM Hepes, pH 7.4, and centrifuge for 1 h at 100 000 *g*. The yield can be increased by a further wash of the final pellet.

B. *Variation*

This method has been modified for use with plant cells, the main difference

Protocol 1. *Continued*

being an increased concentration of EGTA at step 1 (10 mM) and the addition of phospholipid (1 mg/ml) and a higher calcium concentration (15 mM final concentration, i.e. 5 mM excess over the EGTA) at step 3 (45, 46).

Protocol 2. Preparation of annexins from a cytoskeleton fraction (15, 39)

All steps are performed at 4 °C.

1. Homogenize 50 g tissue in a Waring blender in 250 ml of 100 mM KCl, 1 mM $CaCl_2$, 1 mM NaN_3, 0.5 mM DTT, 0.25 mM PMSF, 10 mM imidazole-HCl, pH 7.4 (buffer A). Add Triton X-100 to a final concentration of 1.0% and stir.
2. Centrifuge for 30 min at 25 000 *g*.
3. Wash the pellet once in buffer A containing 1.0% Triton X-100, and then twice in the same buffer without Triton X-100, in each case by resuspension of the pellet and centrifugation for 30 min at 25 000 *g*.
4. Extract the final pellet with 20 vol. of 100 mM KCl, 5 mM EGTA, 1 mM NaN_3, 0.5 mM DTT, 10 mM imidazole-HCl, pH 7.4, and centrifuge at 100 000 *g* for 1 h.
5. Precipitate protein from the supernatant by slowly adding ammonium sulfate to give a final saturation of 75%, ensuring that the pH remains at 7.4. After 30 min at 4 °C, collect the precipitate by centrifugation at 40 000 *g* for 30 min. Resuspend the pellet in a suitable volume (e.g. 5 ml) of 10 mM imidazole-HCl, pH 7.4, containing 1 mM NaN_3, and 0.5 mM DTT, and dialyse against several changes of the same buffer.

Protocol 3. Preparation of annexins using phenyl-Sepharose (79)

A. *Procedure*

1. Homogenize tissue at 4 °C in 5 vol. of 0.15 M NaCl, 2 mM EGTA, 2 mM PMSF, 10 mM Hepes, pH 7.4. Alternatively, acetone powders can be used by extracting 1 g of powder with 15 ml of buffer for 1 h at 4 °C.
2. Centrifuge for 1 h at 100 000 *g*.
3. To the supernatant, add $CaCl_2$ to a concentration of 3 mM (1 mM excess over EGTA).
4. Load on to a phenyl-sepharose column (10 g tissue/100 ml of resin) pre-equilibrated with 0.15 M NaCl, 0.5 mM Ca^{2+}, 10 mM Hepes, pH 7.4.
5. Continue to elute with the same buffer until the eluate is protein-free.

Protocol 3. *Continued*

6. Elute with 0.1 mM Ca^{2+}, 10 mM Hepes, pH 7.4, until the eluate is again protein-free.

7. Elute the specifically Ca^{2+}-bound proteins with 10 mM EGTA, 10 mM Hepes, pH 7.4.

B. *Variation*

An alternative is to use high salt (1 M NaCl) buffers in the later stages of the procedure to limit the leaching of hydrophobically bound proteins from the column (18).

---

successfully applied to most tissues. We have found *Protocol 1* to be particularly useful for the isolation of annexins IV, V, and VI, whereas *Protocol 2* gives good yields of annexins I and II, as well as annexins IV and VI. *Protocol 3* describes the purification of annexins from extracts of tissues and acetone powders by using the hydrophobic affinity chromatography resin phenyl–Sepharose in the presence of calcium. Unlike calmodulin, which exposes hydrophobic sites as a function of calcium concentration, annexins probably bind indirectly to the phenyl-Sepharose via lipids present in the initial tissue extract (18, 43, 44).

Annexins have a wide distribution and most tissues examined to date contain at least one member of the family. Useful tables are available of the distributions in cell-lines and tissues of annexins I (11, 17, 49, 50), II (11, 17, 39, 49, 50), III (17), IV (39, 50), V (17, 50), and VI (36, 39, 50, 51, 52). *Table 2* provides a list of the main tissue sources for annexins and references to the isolation methods used.

2.1 Protease inhibitors

For the preparation of intact annexins IV, V, and VI it is usually sufficient to add the serine protease inhibitor phenylmethylsulphonylfluoride (PMSF) to the homogenization buffer at a final concentration of 0.25 mM from a stock solution of 100 mM in ethanol. This should be done immediately before use, as PMSF has limited stability in aqueous buffers. Other annexins, notably annexins I and II, are susceptible to proteolysis at the amino terminal end (57, 76, 77). In the case of annexin II, proteolysis at the amino terminal end of the 36 kd subunit can remove the binding site for the 10 kd subunit. The use of further protease inhibitors is therefore advisable. Since little is known of the endogenous proteases which act on annexins, it may be necessary to add a range of inhibitors. Some examples of mixtures which have been used are (a) aprotinin (1 μg/ml), leupeptin (1 μg/ml), iodoacetic acid (4 mM), benzamidine (1 mM) and PMSF (1 mM) for annexin I from human placenta (57), and (b) PMSF (0.1 mM), with leupeptin, chymostatin, and soybean trypsin inhibitor (10 μg/ml each) for annexin II from bovine aorta (58). A useful list of protease inhibitors is provided in ref. 78.

Table 2. Tissue sources of annexins

Annexin	**Tissue/cell**	**Reference**
I	Lung (bovine)	49, 53
	(porcine)	54, 55
	(rat)	54
	Spleen (human)	56
	Placenta (human)	11, 49, 51, 55, 57
	A431 cells	8
II	Aorta (bovine)	58
	Intestinal mucosa (porcine)	15
	Lung (bovine)	49, 53, 59
	Mammary gland (mouse)	60
	Placenta (human)	11, 49, 51
	Spleen (human)	56
	A431 cells	8
	Fibroblasts (chicken, transformed with Rous sarcoma virus)	61
III	Intestinal mucosa (bovine)	17
	Liver (rat)	18
	Lung (bovine)	53
	Mononuclear cells (human blood)	62
	Peritoneal lavage (rat)	17
	Placenta (human)	16, 63
	Spleen (human)	56, 63
IV	Adrenal medulla (bovine)	9, 22
	Brain (porcine)	65
	Intestinal mucosa (porcine)	15
	Lens (bovine)	66
	Liver (bovine)	9, 22
	(human)	56
	(porcine)	39
	(rat)	18
	Lung (bovine)	53
	Mammary gland (mouse)	60
	Placenta (human)	16, 63
V	Aorta (bovine)	31
	Brain (porcine)	65
	Chondrocytes (chicken)	67
	Intestinal mucosa (bovine)	17
	Kidney (rat)	68
	Lens (bovine)	66
	Liver (human)	56
	(rat)	18
	Lung (bovine)	32, 53
	Mononuclear cells (human blood)	62
	Peritoneal lavage (rat)	17
	Placenta (human)	16, 29, 56, 57, 63, 69
	Spleen (human)	56
VI	Adrenal medulla (bovine)	9, 22
	Aorta (bovine)	70
	Brain (porcine)	65

	Gizzard (chicken)	71
	Intestinal mucosa (bovine)	17, 36
	Liver (bovine)	9, 22
	(human)	56
	(porcine)	39
	(rat)	71
	Lung (bovine)	53, 62
	(porcine)	72
	Lymphocytes (porcine)	73
	Mammary gland (mouse)	60
	Mononuclear cells (human blood)	62
	Placenta (human)	51, 56, 63
	Spleen (human)	56
	(mouse)	34
	Thymus (mouse)	34
	Tonsils (human)	33
VII	Adrenal medulla (bovine)	74
	Liver (bovine)	75
VIII (recombinant)	Human placental cDNA in *E. coli*/pRH 292	42
Calelectrin	Electric organ (*Torpedo*)	43
Plant annexins	Suspension culture cells (Tomato)	45, 46

3. Further purification of annexins

Once a preparation of annexins has been obtained, a preliminary identification of the proteins is possible by using two-dimensional gel electrophoresis. The positions of the major annexins have been well characterized (ref. 9; *Figure 1*), and the technique is particularly valuable since as many as six different annexins in the 32–38 kd range may be present in any one tissue.

The individual annexins can be purified using standard protein purification techniques. A useful initial separation is achieved by anion exchange chromatography on DEAE-cellulose (Whatman) or Mono Q (Pharmacia).

Protocol 4. Ion-exchange chromatography using DEAE-cellulose (32)

1. Dialyse the annexin preparation (up to approximately 7.5 mg) for 16 h against 5 litres of 20 mM Hepes, pH 7.4.
2. Prepare a 1.0 × 5.0 cm column of DEAE-cellulose (DE-52, Whatman, Maidstone, UK) according to the manufacturer's instructions, and equilibrate in the same buffer.
3. Apply the protein to the column at a flow rate of 12 ml/h and collect the flow-through fractions.
4. Once all the unbound protein has been eluted, apply a linear gradient of 0–0.3 M NaCl in 20 mM Hepes, pH 7.4. Continue to use a flow rate of 12 ml/h and collect 1 ml fractions.

A typical separation of annexins using this method is shown in *Figures 2* and *3*. Annexin I and annexin II (both monomer and 90 kd complex) pass through the column without being retarded. Annexin IV (and annexin III if present) elutes at approximately 0.08 M NaCl, and annexin V (which exists in two forms in bovine tissue as shown in *Figures 1* and *3*) and annexin VI elute at approximately 0.23 M NaCl.

The further purification of the higher-molecular-weight annexins (annexin I 90 kd complex and annexin VI) from the 32–38 kd proteins is readily achieved using gel filtration on a suitable resin such as Ultrogel AcA 44 (Pharmacia). A standard procedure is to apply protein to a 1.6 × 70 cm column of Ultrogel AcA 44 in 0.1 M NaCl, 1 mM EGTA, 1 mM NaN_3, 50 mM Tris–HCl pH 7.4 at a speed of 6 ml/h and collect fractions of 1 ml. An alternative method for the separation of annexin VI from annexin V is to use Affigel Blue–agarose, which binds annexin VI but not annexin V (80). Annexin I and annexin II monomer may be separated by using FPLC ion-exchange chromatography on Mono S (Pharmacia) (11, 53). Both proteins are retarded at pH 6.0 in 50 mM Mes buffer and can be eluted with a gradient of NaCl. An alternative method is to use a hydroxylapatite column (Bio-Rad), applying the proteins in 10 mM imidazole pH 7.5 and eluting with a 0–200 mM linear gradient of $NaPO_4$, pH 7.5 (49). Annexins III and IV can be separated using FPLC with a Mono Q column. Protein is dialysed against 40 mM Tris pH 8.5 containing 0.5 mM DTT and applied at a flow rate of 1 ml/min to a column of Mono Q previously equilibrated in the same buffer. The column is developed with a 200 ml linear gradient of 0–0.15 M NaCl in the same buffer. These conditions can also be used to separate the two closely related forms of annexin V found in bovine tissues (53).

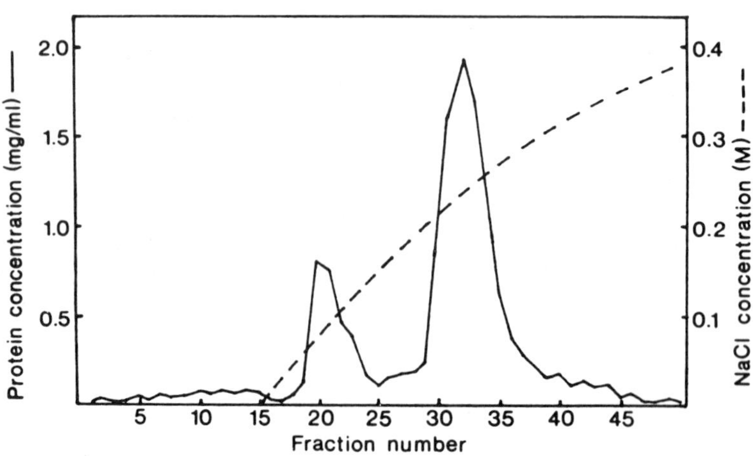

Figure 2. DEAE–cellulose ion-exchange chromatography of bovine lung annexins.

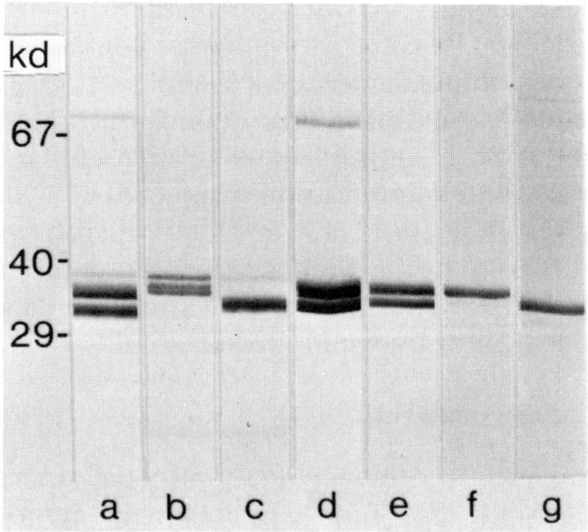

Figure 3. Isolation and purification of bovine lung annexins. Coomassie Blue stained 10% SDS-PAGE of: (a) final supernatant of the lung preparation; (b) unbound fraction from the DEAE-cellulose column containing annexins I and II; (c) protein eluting at 0.08 M NaCl from DEAE–cellulose (containing annexin IV); (d) protein eluting at 0.23 M NaCl from DEAE–cellulose (annexin VI, 70 kd, and annexin V, 34 kd, plus the 32 kd bovine protein closely related to annexin V); (e) annexin V and the related 32 kd protein after gel filtration to remove annexin VI; (f) and (g) annexin V and the related 32 kd protein separated by further anion-exchange chromatography.

4. Preparation of antibodies

4.1 Polyclonal antisera

The annexins are all highly antigenic, enabling polyclonal antisera to be readily prepared using conventional procedures. A standard schedule is to inject New Zealand White rabbits on four occasions at fortnightly intervals with 100 μg of purified annexin emulsified with an equal volume of Freund's complete adjuvant for the first injections and with an equal volume of Freund's incomplete adjuvant on subsequent occasions. Injection is made subcutaneously and intra-muscularly at several sites on the back of the rabbit. The antiserum should give a satisfactory titre one week after the final injection. In some cases it may be necessary to adsorb a polyclonal antiserum raised to one annexin with other annexins to ensure complete specificity. Nevertheless, we have always found the major specificity of our antisera to be against the annexin used for immunization providing highly purified antigen was used.

4.2 Anti-peptide antibodies

Amino acid sequence data is available for most annexins. It is therefore possible to produce synthetic peptides for immunization. The high degree of amino acid sequence identity between different members of the annexin family (40–70%) makes it possible to synthesize consensus peptides corresponding to the regions of greatest sequence similarity. Using this approach, anti-consensus peptide antisera have been raised which recognize more than one annexin (63, 81). Such antisera have been useful in identifying new annexins in mammals (63), amphibia and insects (81).

Antibodies to synthetic peptides may also be prepared to ensure complete specificity to a particular annexin. For example, this approach has been used successfully for annexin I (82) using antibodies raised to a synthetic peptide corresponding to a region of the amino terminal tail.

4.3 Monoclonal antibodies

Several groups have prepared monoclonal antibodies specific to annexin I (83, 84, 85) and annexin II (77, 86, 87). Interestingly, monoclonal antibodies to annexin I have been identified which recognize the Ca^{2+}-bound form, but not the Ca^{2+}-free form (88). Clearly, these may be of use in defining the calcium-binding site of annexin I.

5. Phospholipid binding assays

One of the key features of annexins is their ability to bind to acidic phospholipids in the presence of calcium. Lipid-binding can be studied using a number of different techniques. The following centrifugation assay is typical of a method which has been widely used to investigate both the phospholipid specificity and calcium-dependence of lipid-binding (32, 56, 89). Sucrose-loaded liposomes of defined phospholipid composition are mixed with annexin in the presence of varying free calcium concentrations. Following incubation, the lipid-bound annexin is separated from the unbound protein by spinning down the liposome pellet. Certain phospholipids, notably phosphatidic acid and phosphatidylethanolamine, do not readily form stable liposomes when used alone. However, liposomes can be prepared from these lipids, and from phosphatidylinositol or phosphatidylserine, as 1:1 mixtures with phosphatidylcholine. Liposomes made from phosphatidylcholine alone do not bind any annexin so far described at calcium concentrations below 1 mM.

Protocol 5. Phospholipid binding assay

A. *Preparation of liposomes* (90)

1. Add 1–3 mg of the required phospholipid dissolved in 0.5–1.0 ml of 1:2 chloroform-methanol to a 2-litre flat-bottomed Erlenmeyer flask, and allow it to spread evenly over the bottom.

Protocol 5. *Continued*

2. Pass a stream of nitrogen through the flask to allow the solvent to evaporate slowly, without disturbing the solution.
3. Pass water-saturated nitrogen through the flask for approximately 15 min until the cloudy appearance of the phospholipid layer disappears as water is absorbed.
4. Carefully pour degassed 240 mM sucrose (10 ml per 1.5 mg of phospholipid) down the side of the flask and allow the phospholipids to swell under nitrogen for approximately 2 h without disturbing the flask. The sucrose concentration used here is chosen to be approximately iso-osmolar with the buffer to be used in the binding assays.
5. Gently swirl the flask to yield a suspension of thin-walled phospholipid vesicles. The vesicles remain stable to pipetting, centrifugation and resuspension of the pellets for approximately 12 h at room temp. after which time they become more fragile.

B. *Liposome binding assay* (32)

1. Collect the liposomes by adding 2 volumes of buffer A (100 mM KCl, 2 mM $MgCl_2$, 1 mM EGTA, 20 mM Hepes, pH 7.4) or buffer B (buffer A plus 2 mM $CaCl_2$ to give 1 mM free Ca^{2+}) and centrifuging for 10 min at 12 000 *g* in a microcentrifuge. Other concentrations of calcium can be used together with EGTA to generate a range of free calcium concentrations (see Chapter 7, this volume).
2. Wash the liposomes twice more in the required buffer by resuspension of the pellet and centrifugation at 12 000 *g* for 10 min, to equilibrate the liposomes in the buffer and remove any liposomes which do not sediment under the conditions of the assay.
3. Incubate liposomes corresponding to 150 μg of phospholipid with 10 μg of annexin for 15 min at 20 °C in a total volume of 500 μl of either buffer A or buffer B.
4. Centrifuge for 10 min at 12 000 *g*.
5. Collect the supernatants, and precipitate the protein by adding an equal volume of 20% trichloroacetic acid, leaving the samples on ice for 15 min and centrifuging at 12 000 *g* for 15 min at 4 °C. Remove the supernatant and mix the precipitate with SDS sample buffer.
6. Wash the phospholipid-containing pellets once in 500 μl buffer A or B, then extract with 300 μl acetone at −20 °C for 30 min to remove excess lipid. Centrifuge at 12 000 *g* for 15 min at 4 °C, carefully remove the supernatant and allow any remaining traces of acetone to evaporate before mixing the pellet with SDS sample buffer.
7. Analyse equal proportions of the supernatants and pellets from the assay

Protocol 5. *Continued*

by 10% SDS-PAGE. The percentage of protein in the pellet can be determined by densitometric scanning of the polyacrylamide gel. Alternatively, protein assays can be used to determine the percentage of annexin in the supernatant and pellet fractions.

C. *Variations on the assay*

When only limited quantities of annexins are available, the assay can also be performed using ^{125}I-labelled annexin (89). Calcium-dependent binding of annexins to liposomes formed by sonication (56, 91), to liposomes of mixed lipid extracted from a particular tissue (43), or to membrane preparations such as chromaffin granule membranes (22), can be demonstrated using similar techniques.

5.1 Other methods of measuring phospholipid-binding

Calcium-dependent binding of annexins to lipids can also be measured using columns of lipid immobilized on phenyl-Sepharose (48) or glass beads (55). Sucrose flotation gradients (91, 92, 93) or gel filtration on Sepharose 4B (69) have been used to separate free from liposome-bound annexin in the presence and absence of calcium. Further methods for studying the interaction of annexins with phospholipids include the Langmuir trough (94), the automated ellipsometer (42) and a fluorescence quenching assay (64).

5.2 Measurement of calcium-dependent aggregation of liposomes and membranes

In addition to binding to phospholipids, many annexins are also able to promote the aggregation of liposomes or membrane vesicles in the presence of calcium. Aggregation is usually measured by monitoring the increase in absorbance at 540 nm of a suspension of liposomes or membranes in the presence of annexin and calcium over a range of concentrations (74). It is also possible to observe the aggregation of fluorescently-loaded liposomes by fluorescence microscopy (49).

6. Inhibition of phospholipase A_2

All annexins examined to date show the ability to inhibit phospholipase A_2 in the *in vitro* assay given below. The mechanism of inhibition and its physiological significance are not yet clear, but it may be due to the ability of annexins to bind to the substrate phospholipid rather than to a direct interaction of annexins with phospholipase A_2 (95, 96, 97).

Protocol 6. Assay for inhibition of phospholipase A_2 (98)

A. *Reagents*

1. The substrate for the reaction is ^{3}H-oleic acid labelled *Escherichia coli* membrane phospholipids. These are prepared by incorporating ^{3}H-oleic acid (10 Ci/mmol, Amersham, UK) into their medium during their exponential growth phase. The labelled *E. coli* are then killed by autoclaving, recovered by centrifugation, and washed repeatedly in 200 mM Tris–HCl pH 8.0 containing 10 mg/ml fatty-acid-free bovine serum albumin (BSA). When the washings contain negligible radioactivity, the substrate may be stored at 4 °C in 200 mM Tris–HCl pH 8.0 containing 0.1% NaN_3 until required. Before use in the assay, the *E. coli* should be washed twice in 200 mM Tris–HCl containing 10 mg/ml fatty acid free bovine serum albumin. The suspension of *E. coli* for use should contain approximately 3×10^5 c.p.m./ml.
2. Buffer A: 70 mM $CaCl_2$, 700 mM Tris–HCl, pH 8.0.
3. Buffer B: 2 mM $CaCl_2$, 20 mM Tris–HCl, pH 8.0.
4. Phospholipase A_2, e.g. porcine pancreatic phospholipase A_2 (Sigma), 500 ng/ml in buffer B. 50 ng is used for each assay.
5. Annexin at a suitable range of concentrations. 1–5 μg per assay may provide a good starting point.
6. 2 M HCl.
7. 100 mg/ml fatty-acid-free BSA (Sigma) in water.
8. Scintillation fluid, e.g. ACS, Amersham.

B. *Procedure*

1. For each experimental point, performed in duplicate, 100 μl of sample is mixed with 50 μl buffer A and 100 μl of diluted phospholipase A_2 and made up to a total volume of 350 μl with buffer B. Controls are also prepared in triplicate by omitting the PLA_2 (blank value) or the sample (100% value).
2. Pre-incubate for 10 min at 20 °C.
3. Start the reaction by adding 50 μl of *E. coli* substrate and allow to proceed for 10 min at 20 °C.
4. Stop the reaction by adding 50 μl of 2 M HCl and 100 μl of 100 mg/ml fatty-acid-free BSA to each tube. The BSA will bind the liberated fatty acids.
5. Centrifuge the tubes for 10 min at 10 000 *g* in a bench microcentrifuge at 20 °C.

Protocol 6. *Continued*

6. Equal aliquots of the supernatants containing the liberated ^{3}H-oleic acid are mixed with scintillant and subjected to liquid scintillation counting. After subtracting the blank value to adjust for non-specific release, PLA_2 activity is expressed as a percentage inhibition compared to the 100% value.

6.1 Alternative methods of measuring phospholipase A_2 inhibition

Inhibition of phospholipase A_2 may also be measured using a defined phospholipid substrate (such as l-stearoyl-2-[1′-^{14}C] arachidonyl-sn-glycero-3-phosphocholine), rather than bacterial membrane phospholipids (99). References to further methods of measuring the inhibition of phospholipase A_2 are given in ref. 96.

7. Actin-binding assay

Several of the annexins, notably annexins, I, II, and V, have been shown to bind to F-actin in a calcium-dependent manner *in vitro*. At high calcium concentrations, annexins I and II also bundle actin filaments (15, 49).

Protocol 7. F-actin-binding assay

A. *Reagents*

1. F-actin is prepared from a suitable source, e.g. rabbit skeletal muscle acetone powder, as described (100).
2. Binding buffer: 100 mM KCl, 2 mM $MgCl_2$, 1 mM NaN_3, 0.5 mM dithiothreitol, 0.1 mM ATP, 10 mM imidazole-HCl, pH 7.4, containing either 1 mM EGTA or 1 mM $CaCl_2$, or Ca^{2+}/EGTA buffers covering a range of calcium concentrations.

B. *Procedure*

1. Incubate F-actin (e.g. 0.5 mg/ml), annexin (e.g. 0.1 mg/ml) or actin-annexin mixtures in binding buffer (100 μl) for 1 h at 25 °C.
2. Centrifuge for 20 min at 150 000 *g* (Beckman airfuge or Beckman TL-100 centrifuge).
3. Collect the supernatants and pellets, mix with SDS sample buffer, and analyse equivalent amounts of each by SDS-PAGE. If necessary the supernatants can be concentrated by precipitation with an equal volume of 10% TCA (see *Protocol* 5, B, step 5).

Protocol 7. *Continued*

C. *Variations*

1. The assay can also be performed using ^{125}I-labelled annexin, and the radioactivity in the pellet determined in a gamma-counter (89). This may be useful if a low ratio of annexin to actin is used.
2. To monitor F-actin bundling, the same assay can be performed but using lower-speed centrifugation in a bench microcentrifuge (12 000 *g* for 15 min) to collect F-actin bundles (101). The bundles of F-actin may be examined by negative staining the resuspended sample and viewing by electron microscopy.
3. F-actin binding of annexins may also be measured using chromatography of F-actin linked to Sepharose-4B (102).

8. Conclusions

Annexins are clearly important cellular proteins which have attracted considerable interest. At present, the precise functions of annexins are obscure. Nevertheless, it seems unlikely that this can remain true much longer. Solution of the crystal structure of annexin V is imminent and genetic approaches should soon enable annexin expression to be eliminated in *Hydra* and in cultures of higher eukaryotic cells. These developments should further our knowledge of annexins as major calcium-regulated proteins of the cell.

References

1. Geisow, M. J., Walker, J. H., Boustead, C. M., and Taylor, W. (1987). *Bioscience Reports*, **7**, 289–98.
2. Crompton, M. R., Moss, S. E., and Crumpton, M. J. (1988). *Cell*, **55**, 1–3.
3. Klee, C. B. (1988). *Biochemistry*, **27**, 6645–53.
4. Burgoyne, R. D. and Geisow, M. J. (1989). *Cell Calcium*, **10**, 1–10.
5. Crumpton, M. J. and Dedman, J. R. (1990). *Nature*, **345**, 212.
6. Wallner, B. P., Mattaliano, R. J., Hession, C., Cate, R. L., Tizard, R., Sinclair, L. K., Foeller, C., Chow, E. P., Browning, J. L., Ramachandran, K. L., and Pepinsky, R. B. (1986). *Nature*, **320**, 77–81.
7. Fava, R. A. and Cohen, S. (1984). *Journal of Biological Chemistry*, **259**, 2636–45.
8. Glenney, J. R. (1986). *Proceedings of the National Academy of Sciences, USA*, **83**, 4258–62.
9. Creutz, C. E., Zaks, W. J., Hamman, H. C., Crane, S., Martin, W. H., Gould, K. L., Oddie, K. M., and Parsons, S. J. (1987). *Journal of Biological Chemistry*, **262**, 1860–8.
10. Iwata, M. and Ishizaka, K. (1987). *Proceedings of the National Academy of Sciences, USA*, **84**, 2444–8.
11. Huang, K-S., Wallner, B. P., Mattaliano, R. J., Tizard, R., Burne, C., Frey, A., Hession, C., McGray, P., Sinclair, L. K., Chow, E. P., Browning, J. L.,

Ramachandran, K. L., Tang, J., Smart, J. E., and Pepinsky, R. B. (1986). *Cell*, **46**, 191–9.
12. Radke, K. and Martin, G. S. (1979). *Proceedings of the National Academy of Sciences, USA*, **76**, 5212–16.
13. Kristensen, T., Saris, C. J. M., Hunter, T., Hicks, L. J., Noonan, D. J., Glenney, J. R., and Tack, B. F., (1986). *Biochemistry*, **25**, 4497–503.
14. Saris, C. J. M., Tack, B. F., Kristensen, T., Glenney, J. R., and Hunter, T. (1986). *Cell*, **46**, 201–12.
15. Gerke, V. and Weber, K. (1984). *EMBO Journal*. **3**, 227–33.
16. Tait, J. F., Sakat, M., McMullen, B. A., Miao, C. H., Funakoshi, T, Hendrickson, L. E., and Fujikawa, K. (1988). *Biochemistry*, **27**, 6268–76.
17. Pepinsky, R. B., Tizard, R., Mattaliano, R. J., Sinclair, L. K., Miller, G. T., Browning, J. L., Chow, E. P., Burne, C., Huang, K-S., Pratt, D., Wachter, L., Hession, C., Frey, A. Z., and Wallner, B. P. (1988). *Journal of Biological Chemistry*, **263**, 10799–811.
18. Kaetzel, M. A., Hazarika, P., and Dedman, J. R. (1989). *Journal of Biological Chemistry*, **264**, 14463–70.
19. Hamman, H. C., Gaffey, L. C., Lynch, K. R., and Creutz, C. E. (1988). *Biochemical and Biophysical Research Communications*, **156**, 660–7.
20. Weber, K., Johnsson, N., Plessman, U., Van, P. N., Soeling, H-D., Ampe, C., and Vandekerckhove, J. (1987). *EMBO Journal*, **6**, 1599–1604.
21. Geisow, M. J., Fritsche, U., Hexham, J. M., Dash, B., and Johnson, T. (1986). *Nature*, **320**, 636–8.
22. Sudhof, T. C., Ebbecke, M., Walker, J. H., Fritsche, U., and Boustead, C. (1984). *Biochemistry*, **23**, 1103–9.
23. Grundmann, U., Abel, K-J., Bohn, H., Loebermann, H., Lottspeich, F., and Kuepper, H. (1988). *Proceedings of the National Academy of Sciences, USA*, **85**, 3708–12.
24. Funakoshi, T., Hendrickson, L. E., McMullem, B. A., and Fujikawa, K. (1987). *Biochemistry*, **26**, 8087–92.
25. Kaplan, R., Jaye, M., Burgess, W. H., Schlaepfer, D. D., and Haigler, H. T. (1988). *Journal of Biological Chemistry*, **263**, 8037–43.
26. Pfaffle, M., Ruggiero, F., Hofmann, H., Fernandez, M. P., Selmin, O., Yamada, Y., Garrone, R., and Von der Mark, K. (1988). *EMBO Journal*, **7**, 2335–42.
27. Von der Mark, K., Hofmann, C., Pfaeffle, M., and Gropp, R. (1990). *Trends in Biochemical Sciences*, **15**, 12.
28. Yoshizaki, H., Mizoguchi, T., Arai, K., Shiratsuchi, M., Shidara, Y., and Maki, M. (1990). *Journal of Biochemistry*, **107**, 43–50.
29. Iwasaki, A., Suda, M., Nakao, H., Nagoya, T., Saino, Y., Arai, K., Mizoguchi, T., Sato, F., Yoshizaki, H., Hirata, M., Miyata, T., Shidara, Y., Murata, M., and Maki, M. (1987). *Journal of Biochemistry* (*Tokyo*), **102**, 1261–73.
30. Maurer-Fogy, I., Reutelingsperger, C. P. M., Pieters, J., Bodo, G., Stratowa, C., and Hauptmann, R. (1988). *European Journal of Biochemistry*, **174**, 585–92.
31. Reutelingsperger, C. P. M., Kop, J. M. M., Hornstra, G., and Hemker, H. C. (1988). *European Journal of Biochemistry*, **173**, 171–178.
32. Boustead, C. M., Walker, J. H., and Geisow, M. J. (1988). *FEBS Letters*, **233**, 233–8.

33. Crompton, M. R., Owens, R. J., Totty, N. F., Moss, S. E., Waterfield, M. D., and Crumpton, M. J. (1988). *EMBO Journal*, **7**, 21–7.
34. Moss, S. E., Crompton, M. R., and Crumpton, M. J. (1988). *European Journal of Biochemistry*, **177**, 21–7.
35. Geisow, M. J. and Walker, J. H. (1986). *Trends in Biochemical Sciences*, **11**, 420–3.
36. Glenney, J. R. (1986). *Experimental Cell Research*, **162**, 183–90.
37. Smith, V. L. and Dedman, J. R. (1986). *Journal of Biological Chemistry*, **261**, 15815–18.
38. Sudhof, T. C., Slaughter, C. A., Leznicki, I., Barjon, P., and Reynolds, G. A. (1988). *Proceedings of the National Academy of Sciences, USA*, **85**, 664–8.
39. Shadle, P. J., Gerke, V., and Weber, K. (1985). *Journal of Biological Chemistry*, **260**, 16354–60.
40. Pollard, H. B. and Scott, J. H. (1982). *FEBS Letters*, **150**, 201–6.
41. Burns, A. L., Magendzo, K., Shirvan, A., Srivastava, M., Rojas, E., and Alijani, M. R. (1989). *Proceedings of the National Academy of Sciences, USA*, **86**, 3798–802.
42. Hauptmann, R., Maurer-Fogy, I., Kryster, E., Bodo, G., Andree, H., and Reutelingsperger, C. P. M. (1989). *European Journal of Biochemistry*, **185**, 63–71.
43. Walker, J. H. (1982). *Journal of Neurochemistry*, **39**, 815–23.
44. Johnston, P. A., Perin, M. S., Reynolds, G. A., Wasserman, S. A., and Sudhof, T. C. (1990). *Journal of Biological Chemistry*, **265**, 11382–7.
45. Boustead, C. M., Smallwood, M., Small, H., Bowles, D. J., and Walker, J. H. (1989). *FEBS Letters*, **244**, 456–60.
46. Smallwood, M., Keen, J. N., and Bowles, D. J. (1990). *Biochemical Journal*, **270**, 157–61.
47. Geisow, M. J., Childs, J., Dash, B., Harris, A., Panayotou, G., Sudhof, T., and Walker, J. H. (1984). *EMBO Journal*, **3**, 2969–74.
48. Edwards, H. C. and Booth, A. G. (1987). *Journal of Cell Biology*, **105** 303–11.
49. Glenney, J. R., Tack, B., and Powell, M. A. (1987). *Journal of Cell Biology*, **104**, 503–11.
50. Hayashi, H., Owada, M. K., Sonobe, S., Kakunaga, T., Kawakatsu, H. and Yano, J. (1987). *FEBS Letters*, **223**, 267–72.
51. Hayashi, H., Owada, M. K., Sonobe, S., and Kakunaga, T. (1989). *Journal of Biological Chemistry*, **264**, 17222–30.
52. Owens, R. J., Gallagher, C. J., and Crumpton, M. J. (1984). *EMBO Journal*, **3**, 945–52.
53. Khanna, N. C., Helwig, E. D., Ikebuchi, N. W., Fitzpatrick, S., Bajwa, R., and Waisman, D. M. (1990). *Biochemistry*, **29**, 4852–62.
54. De, B. K., Misono, K. S., Lukas, T. J., Mroczkowski, B., and Cohen, S. (1986). *Journal of Biological Chemistry*, **261**, 13784–92.
55. Ando, Y., Imamura, S., Hong, Y. M., Owada, M. K., Kakunaga, T., and Kannagi, R. (1989). *Journal of Biological Chemistry*, **264**, 6948–55.
56. Blackwood, R. A. and Ernst, J. D. (1990). *Biochemical Journal*, **266**, 195–200.
57. Haigler, H. T., Schlaepfer, D. D., and Burgess, W. H. (1987). *Journal of Biological Chemistry*, **262**, 6921–30.

58. Martin, F., Derancourt, J., Capony, J-P., Watrin, A., and Cavadore, J-C. (1988). *Biochemical Journal*, **251**, 777–85.
59. Drust, D. S. and Creutz, C. E. (1988). *Nature*, **331**, 88–91.
60. Hom, Y. K., Sudhof, T. C., Lozano, J. J., Haindl, A. H., and Rocha, V. (1988). *Journal of Cell Physiology*, **135**, 435–42.
61. Soric, J. and Gordon, J. A. (1986). *Journal of Biological Chemistry*, **261**, 14490–5.
62. Rothhut, B., Comera, C., Prieur, B., Errasfa, M., Minassian, G., and Russo-Marie, F. (1987). *FEBS Letters*, **219**, 169–75.
63. Ernst, J. D., Hoye, E., and Blackwood, R. A. (1989). *Biochemical and Biophysical Research Communications*, **161**, 959–64.
64. Tait, J. F., Gibson, D., and Fujikawa, K. (1989). *Journal of Biological Chemistry*, **264**, 7944–9.
65. Woolgar, J. A., Boustead, C. M., and Walker, J. H. (1990). *Journal of Neurochemistry*, **54**, 62–71.
66. Kobayashi, R., Nakayama, R., Ohta, A., Sakai, F., Sakuragi, S., and Tashima, Y. (1990). *Biochemical Journal*, **266**, 505–11.
67. Mollenhauer, J. and von der Mark, K. (1983). *EMBO Journal*, **2**, 45–50.
68. Seaton, B. A., Head, J. F., Kaetzel, M. A., and Dedman, J. R. (1990). *Journal of Biological Chemistry*, **265**, 4567–9.
69. Funakoshi, T., Heimark, R. L., Hendrickson, L. E., McMullen, B. A., and Fujikawa, K. (1987). *Biochemistry*, **26**, 5572–8.
70. Martin, F., Derancourt, J., Capony, J-P., Colote, S., and Cavadore, J-C. (1987). *Biochemical and Biophysical Research Communications*, **145**, 961–8.
71. Mathew, J. K., Krolak, J. M., and Dedman, J. R. (1986). *Journal of Cellular Biochemistry*, **32**, 223–34.
72. Fauvel, J., Vicendo, P., Roques, V., Ragab-Thomas, J., Granier, C., Vilgrain, I., Chambaz, E., Rochat, H., Chap, H., and Douste-Blazy, L. (1987). *FEBS Letters*, **221**, 397–402.
73. Owens, R. J. and Crumpton, M. J. (1984). *Biochemical Journal*, **219**, 309–16.
74. Creutz, C. E., Pazoles, C. J., and Pollard, H. B. (1978). *Journal of Biological Chemistry*, **253**, 2858–66.
75. Creutz, C. E., Snyder, S. L., Husted, L. D., Beggerly, L. K., and Fox, J. W. (1988). *Biochemical and Biophysical Research Communications*, **152**, 1298–303.
76. Glenney, J. R. and Tack, B. F. (1985). *Proceedings of the National Academy of Sciences, USA*, **82**, 7884–8.
77. Osborn, M., Johnsson, N., Wehland, J., and Weber, K. (1988). *Experimental Cell Research*, **175**, 81–96.
78. Findlay, J. B. C. and Evans, W. H. (ed.) (1987). *Biological membranes: A practical approach.* IRL Press, Oxford.
79. Sudhof, T. C. (1984). *Biochemical and Biophysical Research Communications*, **123**, 100–7.
80. Rhoads, A. R., Lulla, M., Moore, P. B., and Jackson, C. E. (1985). *Biochemical Journal*, **229**, 587–93.
81. Gerke, V. (1989). *FEBS Letters*, **258**, 259–62.
82. William, F., Haigler, H. T., and Kraft, A. S. (1989). *Biochemical and Biophyiscal Research Communications*, **160**, 474–9.
83. Glenney, J. R. and Zokas, L. (1988). *Biochemistry*, **27**, 2069–76.

84. Isacke, C. M., Lindberg, R. A., and Hunter, T. (1989). *Molecular and Cellular Biology*, **9**, 232–40.
85. Pepinsky, R. B., Sinclair, L. K., Douglas, I., Liang, C. M., Lawton, P., and Browning, J. L. (1990). *FEBS Letters*, **261**, 247–51.
86. Isacke, C. M., Trowbridge, I. S., and Hunter, T. (1986). *Molecular and Cellular Biology*, **6**, 2745–51.
87. Zokas, L. and Glenney, J. R. (1987). *Journal of Cell Biology*, **105**, 2111–21.
88. Hayashi, H., Owada, M. K., Sonobe, S., Domae, K., Yamanouchi, T., Kakunaga, T., Kitajima, Y., and Yaoita, H. (1990). *Biochemical Journal*, **269**, 709–15.
89. Schlaepfer, D. D. and Haigler, H. T. (1987). *Journal of Biological Chemistry*, **262**, 6931–7.
90. Reeves, J. P. and Dowben, R. M. (1969). *Journal of Cellular Physiology*, **73**, 49–60.
91. Powell, M. A. and Glenney, J. R. (1987). *Biochemical Journal*, **247**, 321–8.
92. Davidson, F. F., Dennis, E. A., Powell, M., and Glenney, J. R. (1987). *Journal of Biological Chemistry*, **262**, 1698–705.
93. von Kieckebusch, A., Fritsche, U., Vogel, V., Witzemann, V., and Whittaker, V. P. (1988). *Biochimica et Biophysica Acta*, **957**, 131–7.
94. Newman, R., Tucker, A., Ferguson, C., Tsernoglou, D., Leonard, K., and Crumpton, M. J. (1989). *Journal of Molecular Biology*, **206**, 213–19.
95. Davidson, F. F. and Dennis, E. A. (1989). *Biochemical Pharmacology*, **38**, 3645–51.
96. van Binsbergen, J., Slotboom, A. J., Aarsman, A. J., and de Haas, G. H. (1989). *FEBS Letters*, **247**, 293–7.
97. Conricode, K. M. and Ochs, R. S. (1989). *Biochimica et Biophysica Acta*, **1003**, 36–43.
98. Rothhut, B., Russo-Marie, F., Wood, J., DiRosa, M., and Flower, R. J. (1983). *Biochemical and Biophysical Research Communications*, **117**, 878–84.
99. Clark, M. A., Littlejohn, D., Conway, T. M., Mong, S., Steiner, S., and Crooke, S. T. (1986). *Journal of Biological Chemistry*, **261**, 10713–18.
100. Spudich, J. A. and Watt, S. (1971). *Journal of Biological Chemistry*, **246**, 4866–71.
101. Gerke, V. and Weber, K. (1985). *Journal of Biological Chemistry*, **260**, 1688–95.
102. Comera, C., Rothhut, B. Cavadore, J. C., Vilgrain, I., Cochet, C., Chambaz, E., and Russo-Marie, F. (1989). *Journal of Cellular Biochemistry*, **40**, 361–70.

11

Quantification of the affinity of drugs acting at the calcium channel

BARRY A. KENNY, ANDREW T. KILPATRICK, and MICHAEL SPEDDING

1. Functional calcium channels

Calcium plays a vital role as an intracellular second messenger and is responsible for a variety of stimulus–response events. Intracellular calcium in resting cells is maintained at very low levels (less than 10^{-7} M) much of it being bound and sequestered at a variety of sites including endoplasmic and sarcoplasmic reticulum and intracellular Ca^{2+} binding proteins. Intracellular Ca^{2+} can rise as a result of increased permeability of the plasma membrane by the selective opening of Ca^{2+} channels allowing Ca^{2+} to pass into the cellular cytoplasm along its electrochemical gradient. The gating of these channels is regulated by changes in membrane potential (voltage sensitive Ca^{2+} channels) and the channels are the site of action of compounds classed as calcium antagonists. Voltage sensitive Ca^{2+} channels (VSCC) have been shown, in electrophysiological experiments, to consist of three types: T (transient channel currents), N (neuronal current, coupled to transmitter release), and L channels (long lasting currents with slow inactivation) (1). The putative structures of the heart and skeletal muscle L channel has been published (2), and there appear to be distinct subunits (α_1, α_2, β, γ, δ). The amino acid composition of the subunits may vary with the tissue allowing the possibility of L channel isoforms. The α_1 subunit has been shown to have high affinity for various classes of calcium antagonists and the quantification of their interactions are the subject of this review and will be limited to studies of L channels in membranes.

2. Ligand-binding experiments

Radioligand-binding studies have demonstrated quite clearly that high-affinity binding sites exist for [^{3}H]-1,4-dihydropyridines (DHPs) in a number of tissues (3). At least three discrete sites have been demonstrated for different classes of calcium antagonist acting at the α_1 subunit on the L channel. *Table 1* lists the ligands available to label these sites. It is also evident from the results of many laboratories that membrane preparations from brain tissue and smooth, cardiac and skeletal muscle differ in their sensitivity towards [^{3}H]-1,4 DHPs in terms of their susceptibility to allosteric regulation by various classes of calcium antagonist, and by other factors such as divalent cations and thermodynamic effects. Careful manipulation of experimental conditions can highlight differences between the action of the various classes of calcium antagonist and can thus probe for more tissue/site selective drugs.

It is therefore critical that analysis of binding data be carried out under different experimental protocols in order to fully describe the mode of interaction of any given compound.

Table 1. Radioligands available to label sites on calcium channel membrane preparations

Class I	**Class II**	**Class III**	
[^{3}H]-nitrendipine	[^{3}H]-verapamil	[^{3}H]-fluspirilene (site iv)	[^{3}H]-SR 33557 (site v)
[^{3}H]-nifedipine	[^{3}H]-D-*cis*-diltiazem		
[^{3}H]-nimodipine	[^{3}H]-(−) D 600		
[^{3}H]-(+) PN 200–110			
[^{3}H]-Bay K 8644			

3. Pharmacological specificity

3.1 Displacement and dissociation of radioligands

Displacement studies carried out in a variety of different membranes using [^{3}H]-1,4 DHPs illustrate selective modes of interaction between the various classes of Ca^{2+} antagonist. Structurally-related DHPs (antagonists and activators) displace [^{3}H]-1,4 DHP ligands maximally with high affinity and in almost all cases with Hill slopes not significantly different from unity, suggesting a competitive type of interaction. Furthermore, saturation studies with the same radioligands indicate high affinity binding to single populations of binding sites. The nature of the interaction of various classes of compound is often described in terms of Scatchard plot transformations of saturation data (*Figure 1a, b*), whereby changes in K_d indicate a competitive type of interaction (at the same receptor site), whereas a compound producing changes in B_{max} acts non-competitively (i.e. the reduction in binding is produced as a result of an interaction at some other site, thereby reducing the total number of binding sites available). Structurally related DHPs (class I agents) behave in a competitive manner in competition experiments (*Figure 2*), saturation experiments (*Figure 1b*) and in dissociation experiments (sometimes referred to as 'chase' experiments).

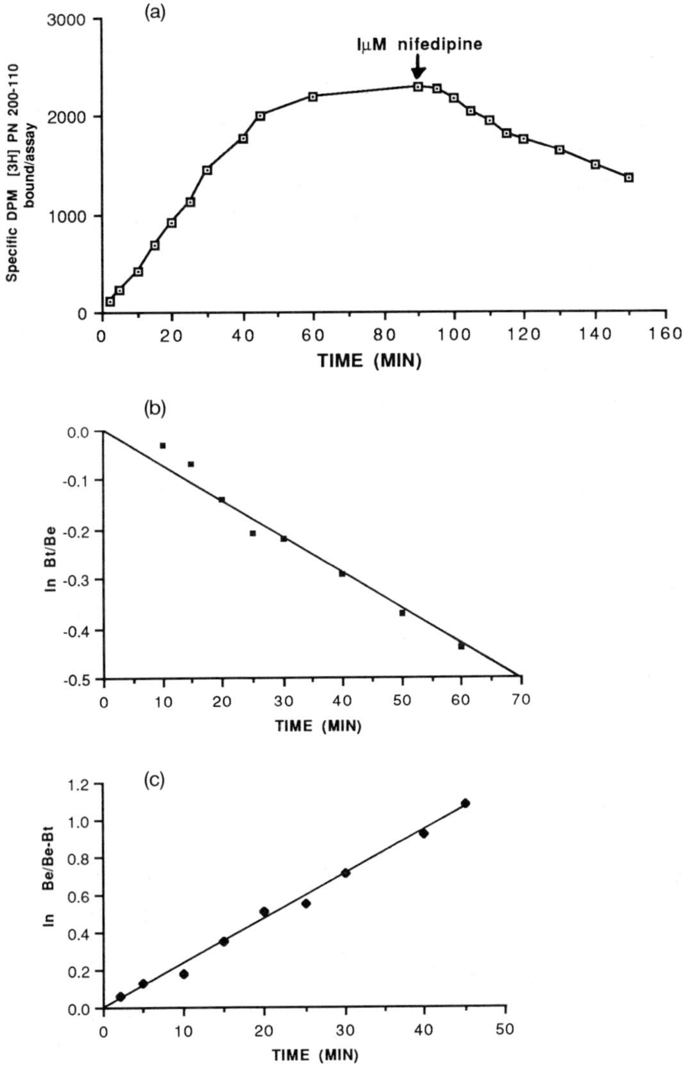

Figure 1. (a) Time-course for the association and dissociation of [^{3}H] (+) PN 200–110 binding in rat cortex at 25 °C. The association rate was determined from the equation:

$$K_1 = \frac{K_{obs} - K_1}{[\text{PN } 200\text{–}110]}$$

where K_{obs} is the psuedo first-order rate-constant for association and is 0.024 min^{-1}, [PN 200–110] is the concentration of [^{3}H] in the assay. Bt and Be represent the amount of specifically bound ligand at time *t* and at time equilibrium. The dissociation constant (K_{-1}) was determined after dissociation was initiated at steady state by 1 μM nifedipine and was found to be 0.0066 min^{-1}. (b) and (c) show the log linear transformation of the data:

$$K_d = \frac{K_{-1}}{K_1} = 0.03 \text{ nM}.$$

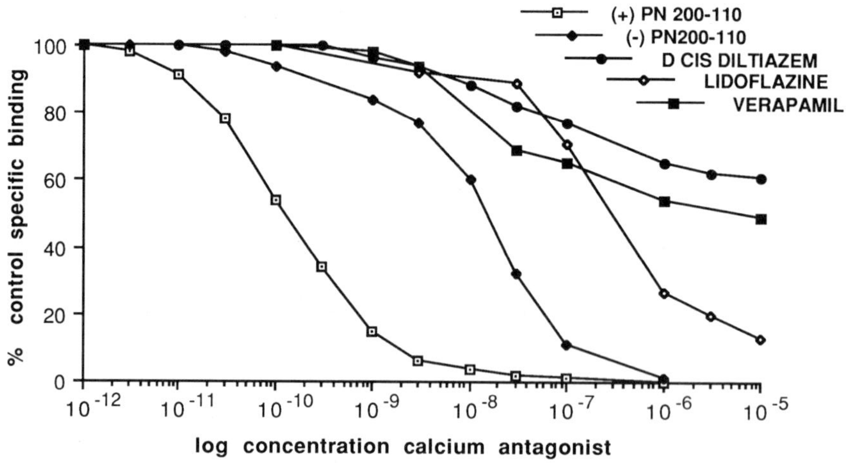

Figure 2. Inhibition of [^{3}H]-PN 200–110 binding to rat cortex membranes by calcium antagonists. Incubations with drugs were carried for 2 h under subdued light at 25 °C in a total volume of 2.0 ml. Specific binding was determined by 1 μM nitrendipine and represented 5% of the total binding. D-*cis* diltiazem and verapamil displaced only 39 and 48% of the binding maximally up to concentrations of 10 μM.

Dissociation experiments are performed to demonstrate allosteric effects of some compounds that are acting in a non-competitive manner. Following the attainment of equilibrium between the binding site and the radioligand, two methods are available to dissociate the radioligand from its site. The first method is to add a large excess of unlabelled DHP (*Figure 3*) effectively producing a one way process of dissociation, association being negligible as the radiolabelled ligand is unlikely to reassociate to the binding site because of overwhelming competition by unlabelled DHP. A second approach in this type of experiment is to dilute the assay mixture thus reducing the likelihood

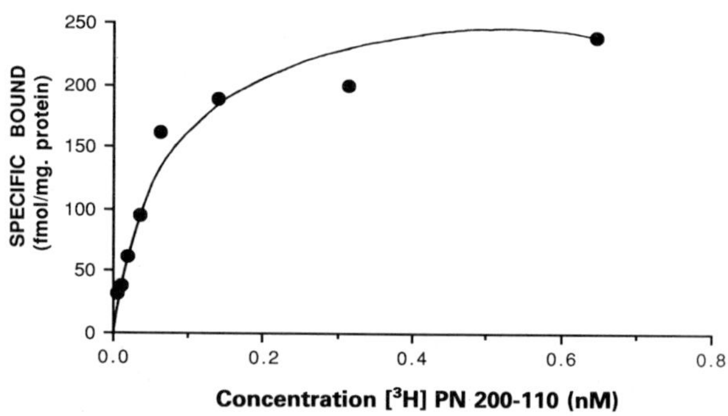

Figure 3 (a). Saturation isotherm of [^{3}H]-PN 200–110 binding to rat cortex at 25 °C.

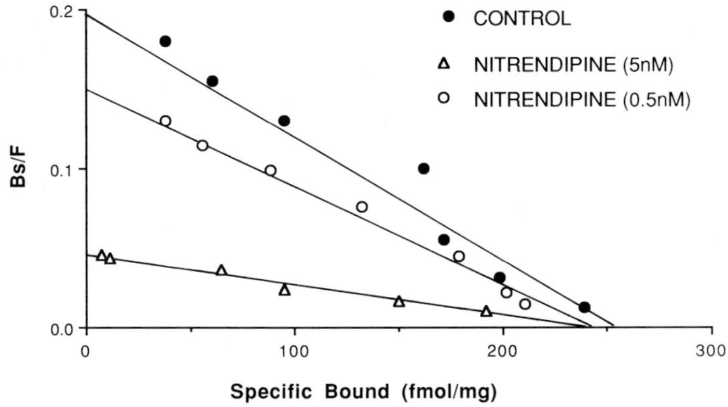

Figure 3 (b). Scatchard transformation of the same data; Bs represents the specifically bound ligand and F the free concentration of ligand. Competitive interaction with nitrendipine is shown. *Control values*: K_d 0.045 nM, B_{max} 265 fmol/mg protein. In the presence of 5.0 nM nitrendipine; K_d 0.24 nM, B_{max} 212 fmol/mg and in the presence of 0.5 nM nitrendipine; K_d 0.052 nM, B_{max} 226 fmol/mg.

of reassociation because of the very low concentration of radioligand resulting from the dilution process. These 'chase' experiments performed in the presence of competitive agents (class I) display no change in the rate of dissociation of the DHP radioligand from the calcium channel complex (*Figure 4*), whereas agents allosterically modulating the DHP site will increase or decrease the rate of dissociation of the radiolabelled DHP.

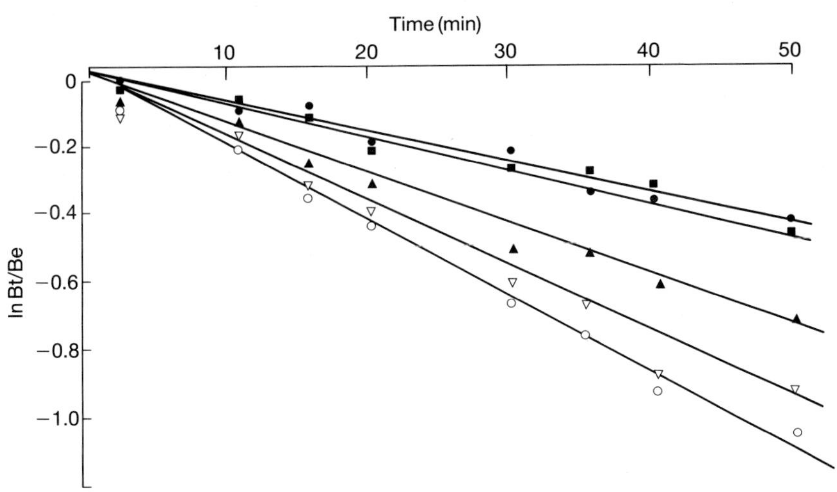

Figure 4. The effect of calcium antagonists on the dissociation rate of the calcium channel–ligand complex in rat cortex at 25 °C initiated at steady by the addition of 1 μM nitrendipine alone (●), or in the presence of 10 nM nifedipine (■), 1 μM fluspirilene (▲), 10 μM verapamil (○), and 10 μM lidoflazine (▽).

The interactions of non-DHP calcium antagonists with the channel are more difficult to define but as a result of extensive biochemical and pharmacological characterization at least two further subgroups of calcium antagonists have been defined:

- Class II: verapamil, diltiazem, diclofurime.
- Class III: lipophilic diphenylalkylamines including cinnarazine, flunarizine, lidoflazine, pimozide, fluspirilene.

Interactions of the class II agents in [^{3}H]-DHP binding studies demonstrate that these agents (typified by verapamil and diltiazem) modulate binding in a negative and positive allosteric manner respectively. The extent of the modulation is dependent upon other factors such as thermodynamics, the radioligand used, and tissue specificity. Type II calcium antagonists displace [^{3}H]-DHP binding incompletely, the extent of maximal displacement being dependent on experimental conditions. Under appropriate conditions D-*cis*-diltiazem induces a concentration-dependent increase in the amount of radioligand bound as a result of its positive allosteric interaction. These allosteric interactions are very susceptible to experimental conditions and can lead to major problems in the interpretation of data unless a full range of experiments are performed to examine the effects of all the experimental variables. As *Figure 5* illustrates, the extent to which [^{3}H]-1,4 DHP binding is modulated by class II antagonists is temperature-, tissue-, and ligand-dependent. In rat cortex the binding of [^{3}H]-nitrendipine is modified by diltiazem to a greater extent than is [^{3}H]-PN 200–110. Similarly, the binding of [^{3}H]-PN 200–110 is more susceptible to modulation in skeletal muscle than in brain homogenates in that the positive heterotropic increase in binding, in the presence of diltiazem, is evident at 25 °C in skeletal muscle membranes but not in brain membranes. It is quite clear that compounds such as D-*cis*-diltiazem that modulate DHP binding would not be easily detected in [^{3}H]-PN 200–110 displacement studies in rat cortex. The failure of standard displacement studies to predict compounds that do not act competitively with DHP binding sites indicates that additional radioligand screens ([^{3}H]-diltiazem and [^{3}H]-verapamil) should be used. However, the respective allosteric interactions of verapamil and D-*cis*-diltiazem at the DHP binding site can be illustrated by examination of the dissociation rates of the ligand–channel complex at equilibrium, in that verapamil causes an increase in the dissociation rate whereas D-*cis*-diltiazem causes a slowing in the rate of dissociation. Transformation of saturation data with [^{3}H]-DHP ligands in various tissues indicates that the allosteric effects of these compounds can result from both an increased or reduced affinity of the channel for the ligand (4) and from more channels being made available for the [^{3}H]-DHP ligand (5) depending on the ligand and tissue in question. It has been clearly established that in skeletal muscle the positive allosteric effect of diltiazem results from an increase in B_{max}, and the extent of this increase is much greater for [^{3}H]-

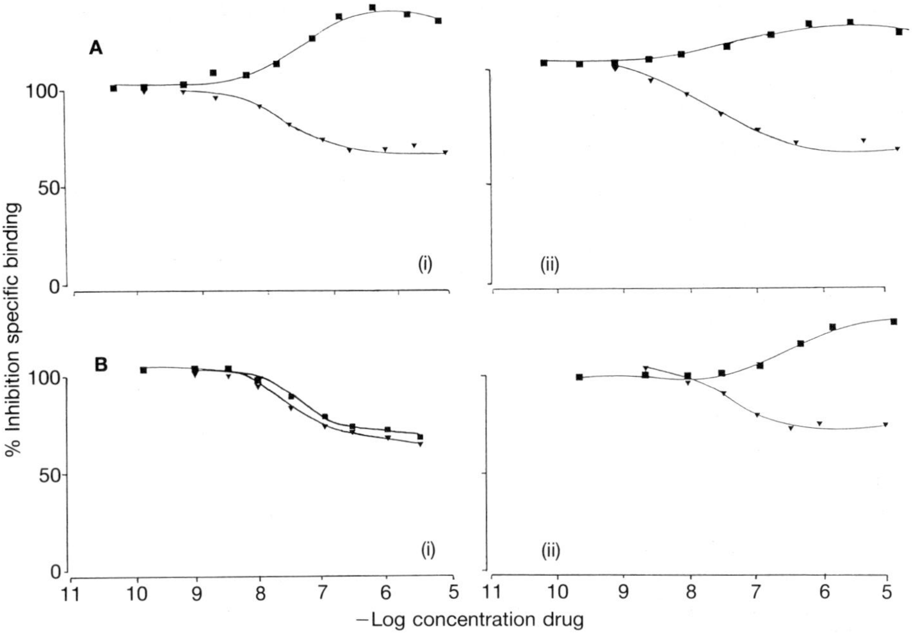

Figure 5. A: Allosteric regulation of [^{3}H]-DHP binding at 37 °C by verapamil (▼) and D-*cis*-diltiazem (■) in rat cerebral cortex with (i) 0.2 nM [^{3}H] nitrendipine and (ii) 0.03 nM [^{3}H]-(+)PN 200–110. **B**: Comparative allosteric properties of verapamil (▼) and D-*cis*-diltiazem (■) in (i) rat cerebral cortex, and (ii) rabbit skeletal muscle membranes, with [^{3}H]-(+)-PN 200–110 at 25 °C.

nimodipine and [^{3}H]-nitrendipine than for [^{3}H]-PN 200–110 (5). However, in brain tissue the same effect is accounted for by a change in K_d for the same ligands (6). The fact that the degree of regulation in any given tissue is highly ligand-(and temperature-)dependent has prompted the suggestion that different DHP ligands have different abilities to hold the channel in certain affinity states or conformations thereby making them more or less susceptible to allosteric control from other sites. This hypothesis is supported by the observation that across the spectrum of [^{3}H]-DHP ligands affinity constants vary from 0.03 nM to 0.20 nM, but each ligand identifies a similar number of sites. Why is it then that a compound can regulate the binding of a given radioligand under the same experimental conditions in differing tissues by different extents (the effect of D-*cis*-diltiazem on saturation transformations in skeletal muscle and brain, for example)? If the conformation of binding sites and their respective affinity states was not uniform then this would certainly account for the differences encountered between tissues, and more importantly, this would explain why high-affinity binding sites for agents like [^{3}H]-fluspirilene can only be demonstrated in skeletal muscle. The fact that this tissue contains a very high density of calcium channels in comparison to other preparations is unlikely to account solely for the binding properties of this type of agent, since studies with [^{3}H]-PN 200–110 in rat cortex indicate that fluspirilene acts in a non-competitive, negatively allosteric manner and thus behaves like other class III agents. Fluspirilene, lidoflazine, and flunarizine all increase the rate of dissociation of the calcium channel–DHP ligand complex, consistent with this type of interaction. Fluspirilene produces a change in K_d and a decrease in B_{max} when saturation transformations are carried out with [^{3}H]-PN 200–110 in rat cortex. As with class II compounds, displacement studies alone do not fully describe their interaction with the [^{3}H]-DHP binding site (*Figure 2*).

When a full range of experiments are carried out, it would appear that the high-affinity binding sites in skeletal muscle labelled by agents such as [^{3}H]-fluspirilene and [^{3}H]-SR 33557 are distinct from the sites of action of the other classes of calcium antagonist (7, 8), and as such, are subject to allosteric control by other classes of calcium antagonist.

4. Influence of cations

Whilst the presence of low contaminating concentrations of Ca^{2+} (that would be present in Tris–HCl washed membranes) is sufficient for high affinity [^{3}H]-DHP binding, pre-treatment of the membranes with chelating agents such as EDTA, or the more Ca^{2+} specific EGTA, clearly demonstrate the divalent cation dependence necessary for high-affinity binding. After depletion of cations with such agents, binding isotherms can be re-established by the re-addition of cations, so called 'refill' curves, with the cations having EC_{50} values in the μM range for reconstitution, the order of potency being $Mn^{2+} > Ca^{2+} > Mg^{2+}$ (3). It has been claimed that the channel, in the inactivated state, in the presence of high-affinity DHP antagonists, binds Ca^{2+} so tightly that it is unable to cross the membrane, the high-affinity binding component being subject to control by voltage and drugs as described above. It has been shown that the increase in binding resulting from re-addition of divalent cations is due to an increase in B_{max} without any change in K_d (9). Restoration of binding of EDTA-treated membranes indicates a divalent cation requirement with a similar charge density to Ca^{2+}. Furthermore, VSCC probably have at least two to four interacting divalent cation sites (29).

The binding of the other classes of [^{3}H]-calcium antagonists is regulated in a different manner to that described for [^{3}H]-DHPs. The binding of both [^{3}H]-verapamil (as studied in skeletal muscle membranes) and [^{3}H]-diltiazem (in brain) are inhibited by divalent cations. Ca^{2+} inhibits the binding of both ligands with an IC_{50} of 0.3 mM. The rank order of potency is $Ca^{2+} = Mn^{2+} > Mg^{2+}$ (10). When the positive and negative allosteric effect of diltiazem and verapamil are studied on [^{3}H]-DHP binding, the effect of Ca^{2+} is to decrease the affinity of the compounds in their interaction without any effect on their maximal potency (11) suggesting a similar interaction of both verapamil and diltiazem sites with the Ca^{2+} recognition site.

5. DHP binding to functional preparations

A variety of binding studies have been described on intact tissue preparations yielding valuable evidence as to the nature of interaction at L channels. These studies are important to assess the physiological relevance of parallel studies on membrane fractions. In the majority of cases these experiments are carried out by incubating [^{3}H]-ligand under similar conditions as described for *in vitro* isolated tissue studies where contractility is measured. The experiment is terminated by washing the tissue to remove excess radioligand and then extracting the radioactivity for scintillation counting. The principles for data analysis remain essentially unchanged. Studies on vascular smooth muscle demonstrate a dependence upon the extent of depolarization present in the tissue in order to attain high-affinity DHP binding. Although an increase in affinity is observed for [^{3}H]-PN 200–110 upon depolarization in portal vein (12), for and in mesenteric artery (13), and the K_i value for PN 200–110 was decreased, nitrendipine was unaffected in the same competition experiments. In other studies, however, the increase in binding resulting from depolarization has been attributed to an increase in the number of channels available, as seen in embryonic chick cardiac cells (14). Binding studies to intact mesenteric artery (13) illustrate how the interaction of other classes of calcium antagonist are modified at the DHP receptor site; this is exemplified by flunarizine, which failed to modify DHP binding in normal physiological medium, but in KCl depolarized solutions it displaced [^{3}H]-PN 200–110 from isolated tissues or from membrane fractions isolated from the media, indicating that the drug only interacts with the DHP receptor site when the membrane is depolarized or disrupted. However, these interactions of flunarizine occurred at concentrations higher than that required to inhibit KCl-induced contractions of the same tissue and this suggests a site of action distinct from the DHP binding site, providing further evidence for multiple binding sites within the channel.

Protocol 1. Preparation of tissue homogenates

A. *Brain tissue (rat cerebral cortex)*

1. Cerebral cortices are dissected free from remaining brain tissue over ice and placed in ice-cold 50 mM Tris–HCl buffer (pH 7.4 at 4 °C). At this stage tissues can be frozen in liquid nitrogen and thawed later, thus enabling membranes to be produced in batches.

Protocol 1. *Continued*

2. Homogenize tissue in 20 vol. (about 18–20 ml per cortex) ice-cold (4 °C) 50 mM Tris–HCl, using a polytron tissue disruptor (setting 8–10) with two bursts of 10 sec duration.
3. Centrifuge homogenate at 48 000 *g* for 15 min in a cooled centrifuge at 4 °C.
4. Carefully decant off the supernatant and resuspend the pellet as step 2, but homogenizing for several seconds only. Repeat step 3.
5. Repeat steps 3 and 4 twice.
6. Resuspend final pellet in 50 mM Tris–HCl at a protein concentration of 2–3 mg/ml. Store in 4–5 ml aliquots frozen under liquid N_2. Using the described protocol, 1 aliquot will be sufficient for 130 assay samples to be prepared, with a protein concentration of about 80 μg per 2 ml assay.

B. *Cardiac tissue*

Rat hearts are cleaned free from connective tissue and fat, finely minced with scissors, and homogenized as in as A, step 2. The homogenate is filtered through a single layer of cheesecloth before proceeding as A, step 3.

C. *Skeletal muscle*

1. Removal of all the hind skeletal muscle from a 1.5–2.5 kg rabbit yielding about 300 g total tissue weight (at this stage it is best to store the tissue frozen and re-thaw as required, as the total amount of tissue exceeds most preparative capacities).
2. Homogenize tissue in ice-cold 0.3 M sucrose/40 mM Mops, pH 7.5, containing 1 μM pepstatin, and 0.1 mM phenylmethylsulfonylfluoride (PMSF) using a Waring blender, with four 20-sec bursts, each separated by 10 sec, using a ratio of 1 g tissue to 5 ml buffer.
3. Centrifuge at 3500 *g* for 30 min in a cooled centrifuge (4 °C).
4. Remove supernatant and check pH. Bring back to >pH 7.0 using 2 M KOH.
5. Centrifuge at 10 000 *g* for 20 min. Decant supernatant, and adjust to 0.5 M KCl using the solid salt.
6. Centrifuge at 100 000 *g* (ultracentrifuge) for 2 h, resuspend pellet and centrifuge at the same speed for 30 min.
7. Resuspend final tissue pellet in 50 mM Tris–HCl and store in liquid N_2 in aliquots of 4.0 ml.

6. Ligand-binding assay procedure

Assays are normally carried out using polypropylene disposable assay tubes (as these show negligible binding properties for most [^{3}H]-radioligands) in a Tris–HCl buffer (pH 7.4 at 25 °C), and in a total volume of 2.0 ml.

Incubations are carried out at the desired temperature (normally 25 °C) for 2 h. It is necessary to have previously measured the time to equilibrium by measuring the specific binding at selected time-points after the addition of membranes to the assay media, using the lowest concentration of ligand to be used, as analysis of binding data at non-equilibrium will yield anomalous results. The amount of membrane protein in each assay will depend on the relative density of binding sites and under the conditions described for membrane preparation each assay would typically contain (in micrograms of protein), cerebral cortex 40–60, cardiac membranes 60–80, and skeletal muscle 8–15. Binding experiments are generally performed under subdued light because many DHPs are light-sensitive. Data points for most assays are determined in duplicate or triplicate to eliminate variation between samples. Non-specific binding is determined in the presence of an unlabelled 1,4 DHP (normally at 1 μM), but different from the [^{3}H]-1,4 DHP in the assay (incorrect definition of non-specific binding can be a source of error in the analysis of binding data, especially in saturation experiments; see below).

Assays using other [^{3}H]-ligands such as [^{3}H]-verapamil or [^{3}H]-diltiazem can be carried out essentially as described with modification to protein, assay volume, incubation times, etc. (see refs 5 and 15); however, filters are pre-soaked with 0.05% polyethylenimine to reduce filter binding of the ligands. The same protocol is used to reduce filter binding properties of [^{3}H]-fluspirilene, which only demonstrates high-affinity binding in skeletal muscle when assayed under the described conditions.

Incubations are terminated by rapid filtration of assay samples over Whatman GF/B glass-fibre filters that have previously been washed with buffer (50 mM Tris–HCl) using a 24-place cell harvester (Brandel) under an applied vacuum of 23 mmHg. The filters are then immediately washed twice with 0.5 ml ice-cold buffer, briefly dried under vacuum and then left to dry further on clean tissue-paper. Bound ligand is quantified, using liquid scintillation counting, and the amount bound per sample expressed as d.p.m. At this stage care must be taken to allow sufficient time for the samples to attain equilibrium with the particular scintillation cocktail, a property which will depend on the physicochemical properties of the ligand, and samples are generally left for at least 12 h prior to counting.

It is worth noting that almost all DHPs are highly lipophilic and accumulate in cell membranes, where they may be in equilibrium with their receptor sites. Due to the lipophilic nature of these compounds, dilutions must be carefully prepared and the free ligand concentration accurately monitored as the ligand may well adhere to test-tubes, tubing, and pipette tips.

7. Data analysis

Equilibrium binding parameters, receptor affinity and number (*K*d and B_{max}) are obtained using an iterative non-linear least-square curve-fitting program as described (16). This and many similar other programs, including Graph PAD (Institute for Scientific Information, Philadelphia), and the Lundan program (Lundan Software, Cleveland), have been described for computer systems (including Apple 11, IBM-PC, as well as mini-computers and main frames). Such programs will provide accurate estimations of ligand binding parameters from saturation data; however, care must be taken to avoid the many experimental artefacts arising from these experiments, such as non-equilibrium, use of impure ligands, improper definition of non-specific binding, poor recovery of unfiltered assay media in measuring amount of [^{3}H] in a particular assay, and displaced non-specific binding (which could be interpreted as heterogeneity in binding sites). An extensive account of these artefacts is described in full, as well as the computerized approach used to resolve these data (17).

Binding isotherms from displacement studies are analysed using a non-linear least-squares parametric curve-fitting program capable of estimating IC_{50} (concentration of drug displacing 50% of specific binding) and iterative curve fitting to a single or two site model (18) using a BBC microcomputer. The inhibitory constant for the competing drug is calculated using the equation (19) where

$$K_i = \frac{IC_{50}}{1 + \frac{[L]}{K_d}}$$

where $[L]$ represents the free concentration of radioligand, and K_d is the equilibrium dissociation constant.

This equation can generally be applied when certain criteria are fulfilled;

(a) The concentration of free ligand is not limiting (i.e. <10% of the total added radioligand is bound).

(b) The action is competitive, within the laws of mass action (Hill slopes not significantly different from unity).

(c) The assay is carried out at a protein concentration where the tissue dependence of binding is linear.

(d) The concentration of radioligand used is not >> K_d (thus amplifying conversion errors).

(e) Unlabelled and labelled compound have the same affinity for the binding site.

Indeed, this equation can be modified so as to calculate K_d and B_{max} from competition data where competitive binding data at a single site is obtained (see ref. 20 for review).

8. Functional definition of affinity for L channels and site of action

In the above discussion it was shown that affinity for L channels may be measured in membrane homogenates or in isolated tissues, using radioligand binding techniques. However, because affinity may be modified readily by experimental conditions, it is important to verify that the compounds have the appropriate affinity in functional experiments. The selectivity of action of a 'calcium-antagonist' may readily be assessed (21), but the affinity of the compound may prove quite problematic unless certain guide-lines are followed. The most important aspect is the role of membrane potential. It is often assumed that the L channel is in the inactivated state in ligand binding experiments and it is this state that has high affinity for dihydropyridines. In electrophysiological experiments (22, 23) it has been shown that nitrendipine has less than nanomolar affinity for the inactivated state, but a 1000-fold weaker affinity for the resting (closed) state. As the preponderance of a given channel state in an untreated functional tissue is almost exclusively dependent on membrane potential, this variable will markedly change the affinity for a calcium-antagonist, with depolarizing conditions increasing affinity. It is estimated (22) that 70% of L channels will be in the high-affinity state in depolarized cardiac tissue.

If a particular dihydropyridine has affinity constants for the inactivated state of the channel (K_{inact}) of 0.5 nM and for the resting state (K_r) of 200 nM, then the apparent affinity constants (K_{app}) in tissues maintained at different membrane potentials and with different proportions of channel states will vary very widely:

$$K_{app} = \frac{1}{1/K_r + (1 - r)/K_{inact}}.$$

[from Bean *et al.* $K_r = 200$ nM, $K_{inact} = 0.5$ nM]

For 70% of channels in the inactivated state:

$$K_{app} = \frac{1}{1/200 + (1 - 0.3)/0.5} = 0.7 \text{ nM}.$$

For 1% of channels in the inactivated state:

$$K_{app} = \frac{1}{1/200 + (1 - 0.99)/0.5} = 14.3 \text{ nM}.$$

The proportion of channels in a given state will depend on the time of a given depolarization: smooth muscle, with a resting membrane potential of −55 to −50 mV will be highly susceptible to calcium-antagonists on the basis that there will be a high proportion of channels in the inactivated state, whereas in heart muscle (resting membrane potential −70 mV, little inactivated state; action potential duration 200 msec, i.e. approx 20% of total time, approx. 50–70% inactivated) affinity will be lower, and in nervous tissue where action potentials are very short, affinity will be very low (although some neurons grown in tissue culture may be depolarized and consequently have high affinity for calcium-antagonists (24)). It is evident that where a tissue is depolarized for only brief periods that not only will the total duration time of the inactivated state be important, but that the kinetics of association and dissociation of the antagonist will be critical.

Thus, we (25) developed smooth muscle preparations which would remain stable for a long period of time under constant depolarizing (40 mM K^+) conditions; the preparation described (the taenia from the guinea-pig caecum) shows high affinity for all the drugs which have been reported to have high affinity for the L channel.

Strips of taenia (1–2 mm diameter, 2–2.5 cm relaxed length) are dissected from the caecum of male guinea-pigs (250–350 g) and set up in isolated organ baths containing K^+-depolarizing tyrode solution maintained at 35 °C and gassed with 95% O_2 and 5% CO_2. The composition of the K^+-tyrode solution is (mmol/litre): NaCl 97; KCl 40; $NaHCO_3$ 11.9; NaH_2PO_4 0.4; glucose 5.5: pH 7.4. The solution is Mg^{2+}-free to increase the Ca^{2+} sensitivity and has low phosphate to prevent precipitation of calcium phosphate salts. Contractile responses are measured under isotonic conditions (1 g load); responses are less reproducible under isometric conditions.

Cumulative concentration–response curves are obtained to $CaCl_2$ (30–3000 μmol/litre) by increasing the Ca^{2+} concentration at 3-min intervals in logarithmic increments. It is important to ensure that the tissues do not remain in Ca^{2+}-free conditions for long periods of time (>30 min) or the preparations deteriorate. Similarly, if the preparations are exposed to very high Ca^{2+} concentrations for long periods, deterioration occurs, presumably due to Ca^{2+} overload. In contrast, exposure of the tissues to submaximal concentrations of Ca^{2+} (0.1–3 mM) for long periods (up to 3 h) does not affect subsequent sensitivity to Ca^{2+} or to calcium-antagonists. A 20-min wash-out period (six changes of bathing fluid) is allowed between concentration–response curves. The 100% response is taken as the maximum contractile response of the tissue during the second concentration–response curve, and all subsequent contractions are calculated as a percentage of this value. Concentration ratios are calculated as the ratio of the concentrations of Ca^{2+} which produce a 50% maximal response (EC_{50}) in the presence and absence of the antagonist. The slopes are calculated from the concentrations producing 80% and 20% maximal effects. Apparent pA_2 values are calculated (26) by plotting log (concentration ratio −1) against negative log molar concentration of antagonist.

Apparent pA_2 values obtained under these conditions are:

nifedipine: 9.8 ± 0.1 (slope 1.0 ± 0.1)
verapamil: 7.8 ± 0.1 (slope 1.1 ± 0.1)
diltiazem: 7.6 ± 0.1 (slope 0.9 ± 0.1).

Tissues other than the taenia may be used under similar conditions. However, an important point must be made in that although the drugs are used to antagonize Ca^{2+}-induced contractions, they are not physically competing with Ca^{2+} for a binding site, rather the ability of Ca^{2+} to contract the tissue to 50% of maximum is taken as a measure of the channels blocked. This can be shown clearly with the interaction of calcium-antagonists with calcium channel activators such as Bay K 8644 (27).

The calcium channel activator, Bay K 8644, can also be used to define the site of action of the different classes of drugs defined as already defined, in that Bay K 8644 will competitively reverse the effects of calcium antagonists acting at the DHP site (27) but non-competitively reverse class II agents (27) and drugs such as SR 33557. In contrast, the effects of class III drugs such as fluspirilene, flunarizine, etc., are hardly reversed by Bay K 8644 (28). These functional interactions should be correlated with the appropriate radioligand binding studies to define the site of action of a given calcium antagonist.

References

1. Nowycky, M. C., Fox, A. P., and Tsien, R. W. (1985). *Nature*, **316**, 339–43.
2. Tanabe, T., Takeshima, A., Mikami, A., Flockerzi, A. V., Takahashi, K., Kangawa, M., Kojima, M., Matsuo, H., Hirose, T., and Numa, S. (1987). *Nature*, **328**, 313–18.
3. Glossmann, H., Ferry, D. R., Lubbecke, F., Mewes, R., and Hofmann, F. (1982). *Trends in Pharmacological Sciences*, **3**, 431–7.
4. Lee, H. R., Roeske, W. R., and Yamamura, H. I. (1984). *Life Sciences*, **35**, 721–32.
5. Glossmann, H. and Ferry D. R. (1985). In *Methods in enzymology*, Vol. 109 (ed. L. Birnbaumer and B. W. O'Malley), pp. 721–32. Academic Press, New York and London. **109**, 513–50.
6. Yamamura, H. I., Schoemaker, H., Boles, R. G., and Roeske, W. R. (1982). *Biochemical and Biophysical Research Communications*, **108**, 640–6.
7. Schmid, A., Romey, G., Barhanin, J., and Lazdunski, M. (1989). *Molecular Pharmacology*, **35**, 766–73.

8. Qar, J., Galizzi, J. P., Fosset, M., and Lazdunski, M. (1987). *European Journal of Pharmacology*, **141**, 261–8.
9. Bolger, G. T. and Skolnick, P. (1986). *British Journal of Pharmacology*, **88**, 857–66.
10. Gallizzi, J., Fosset, M., and Lazdunski, M. (1984). *European Journal of Pharmacology*, **144**, 211–15.
11. Schoemaker, H. and Langer, S. Z. (1989). *Journal of Pharmacology and Experimental Therapeutics*, **248**, 710–15.
12. Dacquet, C., Loirand, G., Rakotoarisoa, L., Mironneau, C., and Mironneau, J. (1989). *British Journal of Pharmacology*, **97**, 256–62.
13. Morel, N. and Godfraind, T. (1988). *British Journal of Pharmacology*, **95**, 252–8.
14. Lee, R. T., Smith, T. W., and Marsh, J. D. (1987). *Circulation Research*, **60**, 683–90.
15. Spedding, M. and Mir, A. (1987). *Journal of Cardiovascular Pharmacology*, **9**, 469–77.
16. Munson, P. J. and Rodbard, D. (1980). *Analytical Biochemistry*, **107**, 220–39.
17. Kermode J. C. (1989). *Biochemical Pharmacology*, **38**, 2053–60.
18. MacKinnon, A. C., Brown, C. M., Spedding, M., and Kilpatrick, A. T. (1989). *British Journal of Pharmacology*, **98**, 1143–50.
19. Cheng, Y. C. and Prusoff, W. H. (1973). *Biochemical Pharmacology*, **22**, 3099–108.
20. DeBlasi, A., O'Reilly, K., and Motulsky, H. J. (1989). *Trends in Pharmacological Sciences*, **10**, 223–5.
21. Spedding M. and Cavero I. (1984). *Life Sciences*, **35**, 575–87.
22. Bean B. P. (1984). *Proceedings of the National Academy of Sciences, USA*, **81**, 6388–92.
23. Sanquinetti, M. C. and Kass R. S. (1984). *Circulation Research*, **55**, 336–48.
24. Spedding, M. and Middlemiss D. N. (1985). *Trends in Pharmacological Sciences*, **6**, 309–10.
25. Spedding, M. (1982). *Naunyn-Schmiedeberg's Archiv für Pharmakologie*, **318**, 234–40.
26. Arunlakshana, A. D. and Schild H. O. (1959). *British Journal of Pharmacology*, **14**, 48–58.
27. Spedding, M. (1985). *Journal de Pharmacologie* (*Paris*), **16**, 319–43.
28. Kenny, B. A., Fraser, S., Kilpatrick, A. T., and Spedding, M. (1990). *British Journal of Pharmacology*, **100**, 211–16.
29. Glossmann, H., Ferry, D. R., Goll, A., and Rombusch, M. (1984). *Journal of Cardiovascular Pharmacology*, **6**, s608–21.

12

Sodium–calcium exchange activity in plasma membrane vesicles

JOHN P. REEVES

1. Introduction

The Na^+–Ca^{2+} exchange system is a carrier-mediated transport process in which the movement of Ca^{2+} ions across the membrane is directly coupled to the movement of Na^+ ions in the opposite direction. It is found in the plasma membrane of many types of cells, particularly excitable cells such as nerve and cardiac muscle. The exchange system is electrogenic with a stoichiometry in most cells of 3 Na^+ per Ca^{2+} (1); in retinal rods, however, K^+ movements are coupled to the exchange process and it can function with a stoichiometry of 4 Na^+ per 1 Ca^{2+} + 1 K^+ (2, 3).

The exchange system can mediate Ca^{2+} fluxes in either direction across the cell membrane, depending upon the prevailing electrochemical gradients for Na^+ and Ca^{2+}. The major physiological role of the exchange system is to pump Ca^{2+} out of the cell, using the inwardly-directed electrochemical gradient for Na^+, maintained by the Na^+, K^+–ATPase, as an energy source. Indeed, in cardiac myocytes and retinal rods, the exchange system is the *primary* mechanism for extruding Ca^{2+} from the cell; in most other cell types, the relative importance of the exchanger vs. the Ca^{2+}–ATPase is less firmly established. During cell activation, complex changes in membrane potential, $[Na^+]_i$ and $[Ca^{2+}]_i$ occur and these can exert a dramatic influence on both the rate and the direction of Ca^{2+} movements via the exchange system. Thus, the exchange system plays an exceedingly complicated and versatile role in regulating cellular Ca^{2+} content, particularly during cellular activation (cf. recent reviews in refs 1 and 4).

Many techniques have been applied to the study of Na^+–Ca^{2+} exchange, including direct Na^+ and Ca^{2+} flux measurements, the use of intracellular and extracellular Ca^{2+} indicator dyes, and measurements of membrane currents due to Na^+–Ca^{2+} exchange. This chapter describes a subcellular approach to the study of Na^+–Ca^{2+} exchange, involving the use of plasma membrane vesicles (5, 6). This technique has allowed the application of biochemical methods to the study of exchange activity and has been a critical development in efforts to identify and purify the exchange carrier. In vesicles, exchange activity is assayed by generating transmembrane concentration gradients for Na^+ using appropriate loading and dilution procedures, and measuring Ca^{2+} movements by either filtration or Ca^{2+}-sensitive dyes such as arsenazo-III. The former approach has been used more frequently than the latter, and will be emphasized in the present chapter. The reader is referred to Caroni *et al.* (7), Kadoma *et al.* (8), or Cook and Kaupp (9) for information on the use of indicator dyes to follow Na^+–Ca^{2+} exchange activity.

2. Membrane preparation

The amount of Na^+–Ca^{2+} exchange activity can vary markedly between different tissue sources. Cardiac sarcolemmal vesicles exhibit 50- to 100-times more activity than most other membrane preparations (10); rod outer segments are also a rich source of Na^+–Ca^{2+} exchange activity. For membrane vesicles with low exchange activities, or with high permeabilities to various ions, certain aspects of exchange activity, such as electrogenicity, may be difficult to demonstrate. In these instances, it may be advantageous to solubilize the native vesicles in detergent and reconstitute exchange activity in proteoliposomes, a procedure which often reduces the passive movements of the various ions. This procedure is described in Section 5 of this chapter.

Most methods for preparation of membrane vesicles involve some form of cell or tissue homogenization, followed by differential and sucrose density gradient centrifugation to isolate a membrane fraction enriched in plasma membranes. Na^+–Ca^{2+} exchange activity does not appear to be particularly fastidious about the conditions used in membrane isolation although the use of low temperatures and a cocktail of protease inhibitors is recommended. Since the measurement of transport activity requires the presence of osmotically sealed vesicles, preparations that yield sheets of membranes, or permeabilized membranes, will obviously not be useful. *Protocol 1* presents the protocol followed in our laboratory for the routine preparation of sarcolemmal vesicles from bovine cardiac tissue. This procedure yields 0.15–0.2 mg of membrane protein per gram of ventricular tissue (wet weight); sarcolemmal marker enzymes (e.g. Na^+, K^+–ATPase) are purified more than 50-fold compared to the crude homogenate and initial rates of Na^+–Ca^{2+} exchange under standard assay conditions (cf. below) are typically 1–5 nmol Ca^{2+} per milligram of protein per second.

Protocol 1. Preparation of bovine cardiac sarcolemmal membranes

1. Obtain heart from abattoir as quickly as possible after killing animal.
2. Immediately cut heart into medium-sized chunks (~ 30 cm^3) and immerse in ice-cold 0.9% NaCl in 10 mM histidine (pH 7.5) for transport to laboratory.

Protocol 1. *Continued*

3. Trim fat and endocardium from tissue and homogenize (Polytron PT-35) at 0.5 g wet wt. per millilitre in ice-cold solution A: 0.25 M mannitol/l mM EDTA in 70 mM Tris (pH adjusted to 7.4 with H_2SO_4).[a]
4. Centrifuge at 14 000 g_{max} for 20 min and discard supernatant.
5. Resuspend pellets in fresh solution A and repeat homogenization; centrifuge at 14 000 g_{max} for 20 min and save supernatant.
6. Repeat step 5 and save second supernatant.
7. Centrifuge supernatants at 190 000 g_{max} and resuspend pellets in solution A; layer over 0.64 M sucrose/1 mM EDTA in 20 mM Imidazole/HCl (pH 7.4)[a] and centrifuge at 160 000 g_{max} for 90 min.
8. Collect sarcolemmal membranes at interface, dilute threefold in solution B: 160 mM NaCl/0.1 mM EDTA in 20 mM Mops, adjusted to pH 7.4 with Tris.[a] Centrifuge at 160 000 g_{max} for 30 min.
9. Suspend pellets in solution B at 2–3 mg protein/millilitre and freeze aliquots in cryotubes in liquid N_2. Store in liquid N_2 or in −20 °C freezer.

[a] These solutions also contain 1 µg/ml each of leupeptin, aprotinin, and pepstatin; add 0.2 mM PMSF from a 0.2 M stock solution in isopropanol just before use. This procedure is based on modifications (11) of the method of Kuwayama and Kanazawa (12).

3. Assaying exchange activity

Assays for Na^+–Ca^{2+} exchange activity in vesicles usually involve measurements of Ca^{2+} movements using either $^{45}Ca^{2+}$ or Ca^{2+}-indicators. $^{22}Na^+$ fluxes can also be measured but because of the large background leak of most vesicles for Na^+ and the relatively low affinity for Na^+ of the exchange carrier, these measurements are much more difficult and imprecise than Ca^{2+} flux measurements. Most assays involve generating concentration gradients of either Na^+ or Ca^{2+} by appropriate loading and dilution steps and measuring movements of the exchange partner against a concentration gradient. Occasionally, however, it is advantageous to measure exchange activity under equilibrium conditions, i.e. in the absence of electrochemical gradients for the transported ions (cf. Section 4).

3.1 Na^+-dependent Ca^{2+} uptake

This is the simplest and most widely used approach to measuring Na^+–Ca^{2+} exchange activity in vesicles. The procedure consists of loading the vesicles internally with a high concentration of Na^+ and then diluting them 20-fold or more into a Na^+-free medium containing $^{45}Ca^{2+}$. The outwardly directed concentration gradient of Na^+ created by the dilution step brings about the accumulation of Ca^{2+} against a concentration gradient within the vesicle. The vesicles are then harvested by filtration at various times after the dilution step. The compositions of the solutions used in routine assays for Na^+–Ca^{2+} exchange activity are indicated below.

- Loading medium: 160 mM NaCl in 20 mM 3-(*N*-morpholino)propanesulfonic acid (Mops), pH adjusted to 7.4 with Tris (Mops–Tris).
- Dilution medium: 160 mM KCl in 20 mM Mops–Tris, pH 7.4.
- Quenching medium: 200 mM KCl/0.1 mM EGTA in 5 mM Mops–Tris, pH 7.4 (ice-cold).

3.1.1 Vesicle loading procedure

Vesicles are loaded with Na^+ by incubating them in the loading medium for 30 min. at 37 °C (or overnight at 4 °C). For cardiac sarcolemmal vesicles, the membranes are used at a protein concentration of 1–2 mg/ml; for vesicles from other sources with lower exchange activity, a higher protein concentration may be necessary. With cardiac sarcolemmal vesicles, the presence of endogeneous Ca^{2+} in the loading medium can have a dramatic effect on exchange activity. This is because intravesicular Ca^{2+} lowers the K_m of the exchanger for external Ca^{2+} and givcs an apparcnt stimulation of activity (13). Thus, variations in the amount of endogenous Ca^{2+} associated with the vesicles will produce corresponding variations in exchange activity when different preparations of vesicles are compared. This can be controlled by including 0.1 mM EGTA in the loading medium; this concentration of EGTA does not markedly affect the assay for Na^+–Ca^{2+} exchange activity. Vesicles from other tissue sources do not exhibit this effect of intravesicular Ca^{2+} (1).

3.1.2 Assay for exchange activity

To measure Na^+–Ca^{2+} exchange activity, a 2-μl aliquot of the Na^+-loaded vesicles is diluted into 100 μl of a Na^+-free dilution medium (usually 160 mM KCl/20 mM Mops–Tris, pH 7.4) containing 10–30 μM $^{45}CaCl_2$. Although the vesicles may be added directly to the dilution medium, timing is more precise if the 2-μl aliquot of vesicles is placed near the bottom of the inside wall of a 12 × 75 mm polystyrene tube containing the dilution medium and the reaction is initiated by vortexing the tube. It is helpful to use a vortexing apparatus such as the Thermolyne Maxi-Mix which can be run continuously and has a vibrating platform for mixing; the large area of the platform facilitates proper tube placement for vortexing and the continuous operation minimizes delay in mixing.

The reaction is terminated after an appropriate interval with 5 ml of ice-cold quenching medium and the vesicles are harvested by filtration on Whatman GF/A glass-fibre filters. The filters are pre-soaked in 0.3% polyethylenimine in water prior to use. The polyethylenimine increases the binding of the vesicles to the filter and improves activity; this is particularly important when measuring activity in reconstituted proteoliposomes (cf.

Section 5). The Whatman GF/A filters appear to work just as well as finer grades of glass-fibre filters (e.g. Whatman GF/C) or membrane filters (e.g. Millipore HA filters) and are considerably cheaper. Once the reaction is terminated, the contents of the assay tube are immediately poured into the filtration apparatus and a second 5-ml aliquot of the quenching medium is added to the tube and poured over the filter just before the meniscus in the chimney from the first rinse reaches the filter. A final 5-ml aliquot of the quenching medium is added directly to the filter, again just before the meniscus reaches the filter. We have found that the data are more reproducable if the filter is not exposed to air between rinses. We use a single filter holder mounted on a 2-litre filtering flask; there does not appear to be any advantage to using a multiport filtration manifold since the filters in a single holder can be changed quite rapidly.

For uptake periods of more than 5 sec, timing can be done with a stopwatch; for shorter intervals, a metronome set to beat at 1-sec intervals is essential.

3.1.3 Blanks

To correct for $^{45}Ca^{2+}$ bound non-specifically to the vesicles and filters, blanks are run in which the Na^+-loaded vesicles are diluted into 160 mM NaCl/20 mM Mops–Tris, pH 7.4 containing the same concentration of $^{45}Ca^{2+}$ as the assay medium. Exchange activity is blocked under these conditions because the high concentration of Na^+ competes with Ca^{2+} for binding sites on the exchange carrier and abolishes the driving force for Ca^{2+} accumulation. Another type of blank can also be run, in which the vesicles are loaded internally with a Na^+-free solution (such as 160 mM LiCl) and assayed in the same dilution medium as that used for the Na^+-loaded vesicles. This type of blank often shows slightly higher levels of Ca^{2+} uptake than the first type because endogenous intravesicular Ca^{2+} brings about low levels of $^{45}Ca^{2+}$ accumulation by Ca^{2+}–Ca^{2+} exchange (cf. Section 3.3). For this reason, and also because of the convenience of using a single batch of vesicles, blanks measured in 160 mM NaCl are recommended.

3.1.4 Assay medium

The composition of the dilution medium is not a critical factor in measuring Na^+–Ca^{2+} exchange activity as long as it is free of Na^+ and is not contaminated by Ca^{2+} or inhibitory cations (e.g. lanthanides); thus, with vesicles one can readily measure Na^+–Ca^{2+} exchange activity in sucrose, LiCl, choline Cl, or *N*-methylglucamine Cl. The nature of the anion is not critical as long as it doesn't chelate Ca^{2+}. Certain salts may be contaminated with Ca^{2+}, which lowers the specific activity of the $^{45}Ca^{2+}$ used in the assay and gives an apparent inhibition of exchange activity. This is the case for reagent grade LiCl from several commercial sources, which contains sufficient Ca^{2+} (20 μM in 160 mM LiCl) to lower the apparent activity and should be treated with Chelex (Bio-Rad) prior to use in exchange assays.

Na^+–Ca^{2+} exchange activity is highly pH-dependent (14) and is markedly inhibited by acidic conditions. It is often not appreciated that the pK_a of most buffer constituents is temperature-dependent; for the dilution media used in our laboratory, increasing the temperature from 22 ° to 37 °C will reduce the pH from 7.4 to 7.1! For this reason, it is important to adjust the pH of the dilution medium at the temperature to be used in the assay.

3.1.5 Electrogenicity

Because the stoichiometry of Na^+–Ca^{2+} exchange is 3 Na^+ per Ca^{2+} in cardiac vesicles, exchange activity will generate a membrane potential that will slow the turnover of the exchanger (15, 16). This effect can be prevented by adding an ionophore that increases the conductivity of the membranes. We routinely use the K^+-conducting ionophore valinomycin when assaying exchange activity in KCl. Valinomycin can be added as a 2 mM solution in dimethyl sulfoxide to either the dilution medium (0.5 μM final concentration) or to the vesicle suspension (1–2 μM final concentration). In the former case, the dilution medium must be used within a few minutes since valinomycin is very hydrophobic and will not remain in solution for long. When added directly to the membrane suspension, the valinomycin associates with the vesicles and remains effective indefinitely. For cardiac sarcolemmal vesicles, the addition of valinomycin stimulates activity more than twofold. In other vesicles, valinomycin may have a much smaller effect, either because exchange activity is not great enough to affect the membrane potential, or because the membranes have a high ionic conductivity. In one study with vesicles from smooth muscle, for example, valinomycin did not stimulate activity in native vesicles, but did stimulate activity when the exchanger was reconstituted into proteoliposomes (17); the latter are presumably much less leaky to ions than the former.

The reader may wonder why valinomycin is necessary to increase the K^+ conductance of the vesicles since the cardiac sarcolemma should be highly conductive to K^+ by itself. Experiments with potential sensitive dyes, however, indicate that the vesicles do not exhibit a high K^+ conductivity in the absence of valinomycin (18). This may be because the K^+ channels of the native sarcolemma have inactivated during the vesicle preparation procedures, or because they are so sparsely distributed that only a small percentage of the vesicles contain K^+ channels.

3.1.6 Ca^{2+} binding

Not all of the Ca^{2+} transported to the vesicle lumen is free; indeed, most is probably bound to internal sites, possibly involving membrane phospholipids (11, 19). Experiments have shown that the amount of $^{45}Ca^{2+}$ associated with cardiac sarcolemmal vesicles at equilibrium is 5- to 20-fold higher than that predicted by the intravesicular volume of the vesicles (approximately 8.5 µl/mg protein) (11). The presence of these internal binding sites acts as a sink for internal Ca^{2+} and amplifies the extent of Ca^{2+} accumulation. In many kinds of experiments, this can create difficulties in interpretation. For example, when comparing vesicles from different sources, it is possible that differences in the rate and extent of Ca^{2+} accumulation could be due, in part, to differences in the number and affinity of intravesicular Ca^{2+} binding sites. As another example, the effects of 'inhibitors' on exchange activity could be due to an effect on intravesicular Ca^{2+} binding rather than an inhibition of exchange activity *per se*. For this reason, conclusions derived from Ca^{2+} uptake studies should always be verified, if possible, with corresponding studies of Na^+-dependent Ca^{2+} efflux.

3.2 Na^+-dependent Ca^{2+} efflux

To measure Ca^{2+} efflux, the vesicles are first loaded with $^{45}Ca^{2+}$ by incubating them for several hours at 37 °C (or overnight on ice) in 160 mM KCl/20 mM Mops–Tris, pH 7.4 containing the desired concentration of $^{45}CaCl_2$ (10–100 µM). The assay method itself is very similar to that described above for measuring Na^+-dependent Ca^{2+} uptake: A 2-µl aliquot of the Ca^{2+}-loaded vesicles is placed on the side of the polystyrene assay tube and mixed with 100 µl of the efflux medium containing the desired concentration of Na^+. At various times after mixing, the vesicles are harvested by filtration as described above.

3.2.1 Assay medium

The assay medium for Ca^{2+} efflux usually contains mixtures of 160 mM NaCl and either 160 mM KCl or 160 mM LiCl to yield the desired concentration of Na^+. The rate of passive Ca^{2+} efflux, not mediated by the exchanger, is determined using 160 mM KCl or 160 mM LiCl alone. All the solutions contain 0.1 mM EGTA; this prevents any stimulation of $^{45}Ca^{2+}$ efflux by contaminating unlabelled Ca^{2+} through the process of Ca^{2+}–Ca^{2+} exchange. Using reagent grade KCl, the level of contaminating Ca^{2+} in a 160 mM solution is approximately 1–2 µM, as determined by the arsenazo-III technique described by Ohnishi (20); as discussed above, contaminating Ca^{2+} in solutions of LiCl is generally much higher.

As in the studies of Ca^{2+} uptake, the use of valinomycin is recommended to preclude a slowing of the exchange reaction due to formation of a membrane potential; if the rates of efflux for different Na^+ concentrations are being compared, it is best to use LiCl to balance the osmolarity in the assay medium instead of KCl so that the results will not be affected by the different K-diffusion potentials generated in the presence of valinomycin.

3.2.2 Ca^{2+} binding

Efflux experiments should be conducted at 25 °C instead of 37 °C. At the higher temperature, the rate of Na^+–Ca^{2+} exchange may be sufficiently rapid that the rate limiting step of the efflux process is not exchange activity itself but the dissociation of Ca^{2+} from internal binding sites. One simple test to verify that the rate of Na^+-dependent Ca^{2+} efflux is not limited by dissociation of Ca^{2+} from internal binding sites is to dilute the Ca^{2+}-loaded vesicles into a KCl medium containing 0.1 mM EGTA and the Ca^{2+}-ionophore A23187. The rate of ionophore-mediated Ca^{2+} efflux should be considerably more rapid than Na^+-dependent Ca^{2+} efflux; if not, the temperature should be lowered or the Na^+ concentration reduced to find conditions where exchange activity is the rate-limiting step in the efflux process.

3.3 Ca^{2+}–Ca^{2+} exchange

In addition to operating as a Na^+–Ca^{2+} antiport system, the exchange system can also catalyse Ca^{2+}–Ca^{2+} and Na^+–Na^+ exchanges. The former process is stimulated by monovalent cations such as Li^+, K^+, Rb^+, and, at low concentrations, Na^+, although the stimulating cation does not appear to be co-transported with Ca^{2+} (11). Ca^{2+}–Ca^{2+} exchange can be measured as either $^{45}Ca^{2+}$ efflux or $^{45}Ca^{2+}$ uptake. The vesicles are pre-loaded with either labelled or unlabelled Ca^{2+} as described above for Ca^{2+} efflux (Section 3.2). To study Ca^{2+}–Ca^{2+} exchange as $^{45}Ca^{2+}$ efflux, the pre-loaded vesicles are diluted into an assay medium as described in Section 3.2 except that the assay contains an appropriate quantity (usually 0.1 mM) of unlabelled $CaCl_2$ instead of EGTA. Again, however, the background rate of passive Ca^{2+} efflux must be measured in a medium containing no added Ca^{2+} and 0.1 mM EGTA.

Measuring Ca^{2+}–Ca^{2+} exchange as $^{45}Ca^{2+}$ uptake is more sensitive and precise than measuring $^{45}Ca^{2+}$ efflux because of the low background in the uptake studies. The vesicles are pre-loaded with unlabelled $CaCl_2$ (0.1–1 mM) in a Na^+-free medium (usually 160 mM KCl); it is important to use a Na^+-free medium for the loading step so that the results will not be complicated by the presence of Na^+–Ca^{2+} exchange activity. The Ca^{2+}-loaded vesicles can be treated just like Na^+-loaded vesicles as described in Section 3.1. The presence of valinomycin is not necessary during assays of Ca^{2+}–Ca^{2+} exchange because this is an electroneutral process which is unaffected by changes in membrane potential. Blanks are obtained by measuring $^{45}Ca^{2+}$ uptake in 160 mM NaCl, as in the Na^+–Ca^{2+} exchange assay. Alternatively, the vesicles may be depleted of endogenous Ca^{2+} by prolonged incubation in the presence of 0.1 mM EGTA (3 h at 37 °C or overnight on ice) and then assayed in the medium used for the Ca^{2+}-loaded vesicles.

3.4 Na^+–Na^+ exchange

The basic procedure underlying the Na^+–Na^+ exchange activity measurements is very similar to that for the other exchange activities (21). The vesicles are loaded with the desired concentration of NaCl (usually 160 mM) and then diluted into a medium containing tracer levels of $^{22}Na^+$ plus 160 mM KCl or LiCl and 0.1 mM EGTA. The external Na^+ concentration is defined by the magnitude of the dilution in these studies (i.e. a 20-fold dilution results in 8 mM external Na^+). The outwardly directed concentration gradient of Na^+ creates an 'overshoot' of $^{22}Na^+$ uptake, i.e. a transient peak of $^{22}Na^+$ accumulation which then decays back down to an equilibrium level as the Na^+ gradient dissipates. Passive $^{22}Na^+$ influx is measured in vesicles pre-loaded in Na^+-free medium (+0.1 mM EGTA to deplete the vesicles of internal Ca^{2+}). Because of the relatively low specific activity of the $^{22}Na^+$ under these conditions, the number of c.p.m. on the filters is 10–20 times less than usually observed for $^{45}Ca^{2+}$ uptake; moreover, the background uptake of $^{22}Na^+$ is a substantial fraction of the total uptake due to the relatively high permeability of the membranes to Na^+. For these reasons, measurements of Na^+–Na^+ exchange are much less precise than for Na^+–Ca^{2+} exchange.

4. Equilibrium measurements

4.1 Isotopic equilibrium exchange

For certain applications, it is advantageous to measure exchange activity in the absence of concentration gradients for Na^+ or Ca^{2+}. This approach has proven useful, for example, in studies of Ca^{2+}–Ca^{2+} exchange (11), in estimating the site density of the exchange carrier in reconstituted proteoliposomes (22), and in defining the stoichiometry of the exchange process (16). The general approach is to equilibrate the vesicles with the desired medium by incubation at 37 °C for 3–4 h in a parafilm-sealed tube. The vesicles are then diluted 4-fold into the same medium containing $^{45}Ca^{2+}$ and harvested by filtration as usual. Because there is no accumulation of $^{45}Ca^{2+}$ against a concentration gradient under these conditions, the number of c.p.m. per point is substantially less than for gradient-driven exchange fluxes. For this reason, 5 μl aliquots of vesicles are used instead of the 2 μl aliquots used in the standard procedure. The vesicles are placed near the bottom of a polystyrene tube containing 15 μl of the dilution medium, taking care to separate the two droplets so they will not mix prior to initiating the reaction; this requires some gentleness in handling the assay tubes. As before, the reaction is initiated by vortexing the tubes and terminated by the normal quenching and filtration procedure.

4.2 Stoichiometry measurements

For measuring the stoichiometry of the exchange process, a somewhat different approach is taken (16). In this case, both the loading and the dilution media contain 30 mM NaCl, 20 mM KCl, 110 mM LiCl, and 0.1 mM $^{45}CaCl_2$, so that the vesicles are initially at isotopic equilibrium and no change in the amount of vesicle-associated $^{45}Ca^{2+}$ occurs after mixing unless the equilibrium is perturbed.

The stoichiometry measurements involve creating a K^+-diffusion potential across the membrane to initiate a Ca^{2+} flux; this is done by treating the membranes with 2 μM valinomycin (cf. Section 3.1.5) and altering the [KCl] in the dilution medium. Thus, if the dilution medium contains no KCl, the dilution step generates a fourfold outwardly directed K^+ concentration gradient which, in the presence of valinomycin, would be expected to produce a transient diffusion potential with an initial value of approximately −36 mV (calculated by substituting the internal K^+ concentrations into the Nernst equation). Under these conditions, a transient efflux of $^{45}Ca^{2+}$ from the vesicles occurs (16). Alternatively, if the [KCl] in the dilution medium is increased to 100 mM (by replacing 80 mM of the LiCl), the dilution step creates a fourfold inwardly directed concentration gradient for K^+, for an initial diffusion potential of +36 mV; a transient accumulation of $^{45}Ca^{2+}$ is observed under these conditions (16). The transient $^{45}Ca^{2+}$ movements do not occur in the absence of either Na^+ or valinomycin, indicating that they are mediated by the Na^+–Ca^{2+} exchange system and require the generation of a membrane potential. The requirement of valinomycin provides strong evidence that the K^+ gradients by themselves are not able to bring about exchange-mediated Ca^{2+} movements, indicating that in cardiac sarcolemma, unlike retinal rods (2, 3), exchange activity is not coupled to K^+ fluxes.

The reasoning behind the stoichiometry determinations is that the membrane potential generated by the dilution step constitutes a driving force for exchange-mediated Ca^{2+} movements and that these movements can be blocked by establishing an equal but opposite driving force in the form of a Na^+ gradient. The balance point at which no Ca^{2+} movements occur can be related to the stoichiometry by the following expression:

$$n/(n\ \text{-}2) = \log\ \{[K]_i/[K]_f\}/\log\ \{[Na]_i/[Na]_f\}$$

where n is the stoichiometry and the subscripts i and f refer to the initial and final concentrations of K^+ and Na^+ in the external medium before and after the dilution step (cf. Appendix of ref. 16). This expression assumes that neither the Na^+ nor the K^+ transmembrane gradients dissipate significantly over the time-course of the measurements.

The stoichiometry measurements themselves consist of mixing the vesicles with various dilution media containing a fixed K^+ concentration (to generate a fixed diffusion potential) and various Na^+ concentrations (adjusting the LiCl concentration to yield a total electrolyte concentration of 160 mM). The reaction is terminated after a short interval after mixing (e.g. 2 sec) to minimize dissipation of the ion gradients. The example shown in *Figure 1*, taken from ref. 16, with $[K^+]_i = 20$ mM and $[Na^+]_i = 30$ mM, indicates that for a calculated diffusion potential of +31 mV ($[K^+]_f$ = 65 mM), Ca^{2+} fluxes

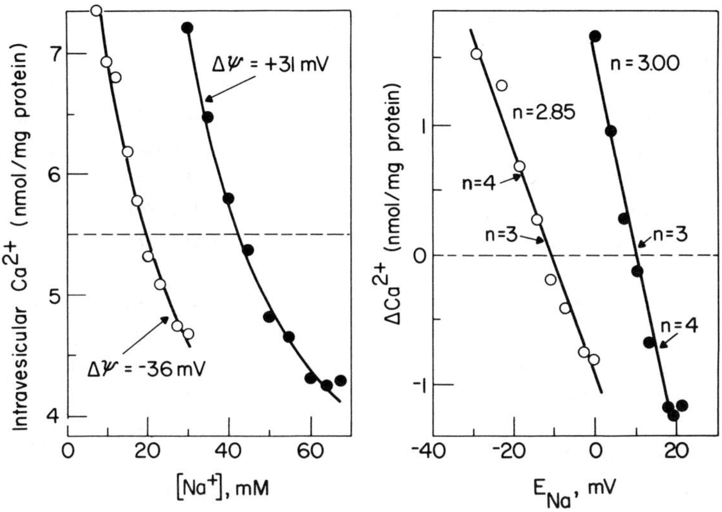

Figure 1. Stoichiometry of Na^+–Ca^{2+} exchange in cardiac sarcolemmal vesicles. *Left*, vesicles were equilibrated in 0.1 mM $^{45}CaCl_2$, 30 mM NaCl, 20 mM KCl, 110 mM LiCl for 3 h at 37 °C in a Parafilm-sealed tube. Valinomycin was added to a final concentration of 2 μM. For each point, a 5 μl aliquot of vesicles was mixed with 15 μl of various solutions containing 0.1 mM $^{45}CaCl_2$, 80 mM KCl ($\Delta\Psi = +31$ mV) and the following concentrations of NaCl (in millimolar): 30, 36.7, 43.3, 50, 56.7, 63.3, 70, 75, and 80. The LiCl concentration was adjusted to give a total electrolyte concentration of 160 mM for each solution. For the points with $\Delta\Psi = -36$ mV, the dilution medium contained no KCl and the following concentrations of NaCl (mM): 0, 3.3, 6.7, 10, 13.3, 16.7, 20.6, 26, and 30. The Ca^{2+} fluxes were terminated 2 sec after mixing. Each point is the mean of triplicate determinations. The *dotted line* represents the Ca^{2+} level of vesicles 2 sec after dilution into a medium with a composition identical to that of the equilibration medium; this also reflects the initial level of Ca^{2+} associated with the vesicles after equilibration. *Right*, the data (*left*) are replotted as the change in intravesicular Ca^{2+} relative to the initial value vs. the equilibrium potential for Na^+ (E_{Na}), calculated from the Nernst equation as E_{Na} (mV) $= 60 \log \{[Na]_i/[Na]_f\}$. The *arrows* indicate the expected points of intersection of each line with the $\Delta Ca^{2+} = 0$ axis for $n = 3$ and $n = 4$.

became zero at $[Na^+]_f = 44.4$ mM; Ca^{2+} influx was observed at lower values of $[Na^+]_f$ and efflux was observed at higher values. The results were consistent with a stoichiometry of 3.0. For a diffusion potential of −36 mV, Ca^{2+} fluxes became zero at $[Na^+]_f = 19.7$ mM, for a calculated stoichiometry of 2.9. The average value of n (± SE) obtained from nine different determinations, six at positive diffusion potentials and three at negative potentials, was 2.97 ± 0.03.

5. Reconstitution of exchange activity

Reconstitution is a process whereby membrane transport proteins are solubilized in a suitable detergent and then re-incorporated in a functional state into phospholipid membranes upon removal of the detergent. The vesicles created by this procedure are called proteoliposomes to distinguish them from native vesicles. Miyamoto and Racker (23) first demonstrated functional reconstitution of Na^+–Ca^{2+} exchange activity from cardiac sarcolemma by solubilizing the native membranes in 2% cholate in the presence of Na^+ and soybean phospholipids (Asolectin), subsequently removing the detergent by dilution and collecting the proteoliposomes by centrifugation. The following paragraphs describe the procedure adopted for reconstitution of Na^+–Ca^{2+} exchange activity in the author's laboratory.

5.1 Phospholipids

Asolectin is a complex mixture of soybean lipids which has proven to be very useful in a wide variety of reconstitution procedures. It can be obtained from various commercial sources, but the source we have used is Associated Concentrates, Woodside, NY 11377, USA. After the lipids are subjected to an acetone precipitation procedure (24), described in detail in *Protocol 2*, they are dissolved in chloroform at 50 mg/ml and stored at −20 °C.

Protocol 2. Acetone precipitation of Asolectin

1. Dissolve 2 g Asolectin in 20 ml 9:1 chloroform:methanol.
2. Put 400 ml of acetone in a glass-stoppered bottle wrapped in aluminium foil and place on magnetic stirrer. Bubble with Ar (or O_2-free N_2) for 5 min; then add 56 µl of *β*-mercaptoethanol and the Asolectin solution.
3. Flush bottle with Ar, seal with stopper and cover with aluminium foil.
4. Stir for 2 h.
5. Weigh two glass centrifuge bottles, fill with the Asolectin suspension and centrifuge for 5 min at low speed.
6. Decant supernatant and discard. Dry pellets under Ar and weigh.
7. Dissolve dried Asolectin in chloroform at 50 mg/ml. Store in foil-wrapped glass-stoppered bottled under Ar at −20 °C.

5.2 Choice of detergent

Our experience indicates that the best detergent for preserving the activity of the cardiac exchanger in the proteoliposomes is 2% cholate. Activity is quickly lost, however, unless exogenous phospholipids (2 mg/ml Asolectin) are included in the solubilization mixture. Cholate is not suitable for certain applications, such as separating proteins on an anion exchange column, and in such cases, non-ionic detergents such as 1% Triton X-100, 30 mM Chaps, or

1% polyoxyethylene 9 lauryl ether ($C_{12}E_9$; polidoconol) can be used. Triton X-100 has the advantage of producing proteoliposomes with a narrow size distribution (K. Philipson, personal communication), unlike the other detergents mentioned, including cholate. The disadvantage of Triton X-100 is that it absorbs strongly at 280 nm and cannot be used in purification procedures that rely on monitoring protein absorbance. Another widely used detergent, octylglucoside, does not yield good exchange activity in proteoliposomes and its use is not recommended.

5.3 Solubilization conditions

For the cardiac exchanger, maintenance of activity in the solubilized state requires the presence of Na^+ in the detergent mixture (22, 23). Studies with 1% Triton X-100 indicated that activity was lost within 30 min on ice if Na^+ (30 mM or more) was not present during solubilization; activity could not be restored by subsequently adding Na^+ to the solubilized membranes nor would Ca^{2+} substitute for Na^+ (22). As mentioned above, the presence of lipid (2 mg/ml Asolectin) is also required to preserve activity when using cholate; for Triton X-100 or $C_{12}E_9$, either lipid or 20% glycerol helps preserve activity, but for short-term solubilizations, neither is absolutely required.

An example of a typical solubilization procedure, a modified version (24) of that described by Miyamoto and Racker (23), is described below. This procedure gives excellent results for cardiac sarcolemmal vesicles, but for activity from other sources, the investigator should attempt to find the conditions optimal for the particular system under study. An aliquot of Asolectin solution in chloroform, prepared as described in *Protocol 2*, is evaporated to dryness under a stream of Ar in a round-bottomed flask or tube; the tube or flask is rotated during evaporation to distribute the drying phospholipid film evenly over the bottom surface. The film is redissolved in a small amount of ethyl ether and dried again under Ar as before. Ar is passed over the dried film until all traces of solvent are gone. The dried phospolipid is then resuspended in a final concentration of 2 mg/ml in 2% (wt/vol.) Na cholate containing 30 mM NaCl and 20 mM Mops–Tris (pH 7.4) and sonicated to clarity in a batch sonicator (Laboratory Supplies Company; specify Racker design when ordering). The final solution is kept on ice until use. In most applications, it is wise to include protease inhibitors in the solubilizing solution, although omitting them has no apparent affect on the activity of the proteoliposomes. We routinely use the cocktail of inhibitors described in *Protocol 1*.

Native sarcolemmal vesicles are collected by centrifugation; the pellets are dissolved in the above solution and allowed to sit for 20 min on ice prior to centrifugation for 30 min. at 160 000 g_{max}. The supernatant, containing the solubilized exchanger, is collected by aspiration and kept on ice. With some membrane preparations, a layer of white material is found on top of the solubilized mixture after centrifugation; this has no apparent effect on activity but can be avoided if necessary by removing the top portion of the mixture before aspirating the rest of the supernatant. There is also a loose layer of incompletely solubilized material at the bottom of the tube, above the tightly packed pellet. Careful aspiration is necessary if you wish to avoid incorporating this material into the solubilized extract; while its presence does no harm to the recovery of activity in the proteoliposomes, the high protein concentration of this material tends to lower the final specific activity.

5.4 Reconstitution procedure

Several different procedures may be used to remove the detergent and reconstitute proteoliposomes, including dilution, dialysis, and the use of adsorbent beads. We have found the following procedure (22), using small columns of adsorbent polystyrene beads (Bio-Beads SM-2; Bio-Rad), to be very versatile and readily adaptable to reconstituting 20–30 separate preparations at once (e.g. fractions from chromatography columns). Moreover, the technique can be applied to detergents such as $C_{12}E_9$, which cannot be removed by dialysis or dilution. Bio-Beads are rinsed several times in 160 mM NaCl/20 mM Mops–Tris pH 7.4 to remove any fines and then placed in small columns formed in Rainin RT 200 disposable microlitre pipette tips to a final bed volume of about 1 ml. The columns are rinsed several times with 160 mM NaCl/20 mM Mops–Tris (pH 7.4) and placed in 10 × 75 mm glass culture tubes so that the column is held by the lip of the tube and the bottom of the column is suspended above the bottom of the tube. The tubes containing the columns are then centrifuged for 1 min at low speed in a table-top centrifuge (e.g. setting No. 1 in an IEC clinical centrifuge). After centrifugation, the columns are placed in fresh 10 × 75 mm glass tubes and used immediately for reconstitution.

A solution of 50 mg/ml Asolectin in 4% cholate containing 500 mM NaCl and 20 mM Mops–Tris (pH 7.4) is prepared following the Ar evaporation procedure described above. A 50 μl aliquot of this mixture is added to 200 μl of the detergent-solubilized vesicles, and the mixture is placed on top of one of the Bio-Beads columns. When the extract contains a low concentration of Na^+, activity is enhanced if 20 μl of 5 M NaCl is added to the extract prior to adding the 50 mg/ml Asolectin. The columns are allowed to sit at room temperature for 30 min. and then centrifuged as before so that the reconstituted mixture is expressed into the 10 × 75 mm glass tubes. The columns are discarded and the reconstituted mixture is assayed for Na^+–Ca^{2+} exchange activity following the procedures adopted for native vesicles. In some cases, an increase in activity is observed if the proteoliposomes are frozen overnight at −20 °C prior to assay; this is not always helpful for cholate- or Triton X-100-solubilized vesicles, but can increase activity nearly twofold for vesicles solubilized in $C_{12}E_9$.

Acknowledgements

I would like to thank the various colleagues who have worked with me over the years in developing the techniques described here, including Diane Ahrens, Carole Bailey, Joo Cheon, John Durkin, Calvin Hale, Pilar de la Peña, Phil Poronnik, Robert Slaughter, and John Sutko. I also wish to thank Ron Kaback for introducing me to the advantages of transport studies in vesicles many years ago and for providing many stimulating discussions and much advice since then.

References

1. Reeves, J. P. (1990). In *Intracellular calcium regulation* (ed. F. Bronner), p. 305. Alan R. Liss, New York.
2. Cervetto, L., Lagnado, L., Perry, R. J., Robinson, D. W., and McNaughton, P. A. (1989). *Nature*, **337**, 740.
3. Schnetkamp, P. P. M., Basu, D. K., and Szerencsei, R. T. (1989). *American Journal of Physiology*, **257**, C153.
4. Allen, T. J. A., Noble, D., and Reuter, H. (ed.) (1989). *Sodium–calcium exchange*. Oxford University Press, Oxford.
5. Reeves, J. P. and Sutko, J. L. (1979). *Proceedings of the National Academy of Sciences, USA*, **76**, 590.
6. Reeves, J. P. (1988). *Methods in enzymology*, Vol. 157 (ed. S. Fleischer and B. Fleischer), p. 505. Academic Press, Orlando, Florida.
7. Caroni, P., Reinlib, L., and Carafoli, E. (1980). *Proceedings of the National Academy of Sciences, USA*, **77**, 6354.
8. Kadoma, M., Froehlich, J., Reeves, J., and Sutko, J. (1982). *Biochemistry*, **21**, 1914.
9. Cook, N. J. and Kaupp, U. B. (1988). *Journal of Biological Chemistry*, **263**, 11382.
10. Reeves, J. P. and Philipson, K. D. (1989). In *Sodium–calcium exchange* (ed. T. J. A. Allen, D. Noble and H. Reuter), pp. 27–53. Oxford University Press, Oxford.
11. Slaughter, R. S., Sutko, J. L., and Reeves, J. P. (1983). *Journal of Biological Chemistry*, **258**, 3183.
12. Kuwayama, H. and Kanazawa, T. (1982). *Journal of Biochemistry*, **91**, 1419.
13. Reeves, J. P. and Poronnik, P. (1987). *American Journal of Physiology*, **252**, C17.
14. Philipson, K. D., Bersohn, M. M., Nishimoto, A. Y. (1982). *Circulation Research*, **50**, 287.
15. Reeves, J. P. and Sutko, J. L. (1980). *Science*, **208**, 1461.
16. Reeves, J. P. and Hale, C. C. (1984). *Journal of Biological Chemistry*, **259**, 7733.
17. Matlib, M. A. and Reeves, J. P. (1987). *Biochimica et Biophysica Acta*, **904**, 145.

18. Schilling, W. P., Schuil, D. W., Bagwell, E. E., and Lindenmayer, G. E. (1984). *Journal of Membrane Biology*, **77**, 101.
19. Philipson, K. D., Bers, D. M., and Nishimoto, A. Y. (1980). *Journal of Molecular and Cellular Cardiology*, **12**, 1159.
20. Ohnishi, S. T. (1979). *Biochimica et Biophysica Acta*, **586**, 217.
21. Reeves, J. P. and Sutko, J. L. (1979). *Federation Proceedings*, **38**, 1199.
22. Cheon, J. and Reeves, J. P. (1988). *Journal of Biological Chemistry*, **263**, 2309.
23. Miyamoto, H. and Racker, E. (1980). *Journal of Biological Chemistry*, **255**, 2656.
24. Hale, C. C., Slaughter, R. S., Ahrens, D. C., and Reeves, J. P. (1984). *Proceedings of the National Academy of Sciences, USA*, **81**, 6569.

13

The Ca^{2+}-ATPase of the plasma membrane

ERNESTO CARAFOLI

1. Introduction

The ejection of Ca^{2+} from cells is a necessity dictated by the existence in the plasma membrane of Ca channels which permit, albeit in a strictly controlled fashion, the penetration of Ca into the cell. Normally, Ca is about 10 000 times less concentrated in the cytosol than in the environment: therefore its ejection is an active process, and is indeed mediated by an ATP-powered pump. The pump was discovered by Schatzmann in 1966 in erythrocytes (1), and was subsequently found in a number of other eukaryotic plasma membranes (for reviews, see refs 2–5). The pump belongs to the P-class of ion-motive ATPases (6–7) and is now well-characterized functionally. More recent work on the purified pump has provided significant information on its structural properties (8). The pump interacts with Ca with high affinity, and can thus be considered as the fine tuner of cytosolic Ca, i.e. it would efficiently eject it from cells even when its concentration is at the low resting level (between 100 and 200 nM) typical of most cytosols. It co-operates with the other plasma membrane Ca exporting system, the Na–Ca exchanger. The latter is a lower affinity, but higher capacity system which is particularly active in excitable cells. Presumably, the exchanger only ejects Ca efficiently when its concentration within the cell has been allowed to rise to levels compatible with its low Ca affinity.

2. General properties of the Ca pump in the plasma membrane

2.1 The reaction cycle

The general mechanism of the pump is that of all other P-type pumps (*Figure 1*). Ca promotes the transfer of the terminal phosphate of ATP to an aspartic acid residue in the pump (9–10): using γ-labelled ATP, the approximate molecular mass of the pump protein in erythrocytes has been estimated at

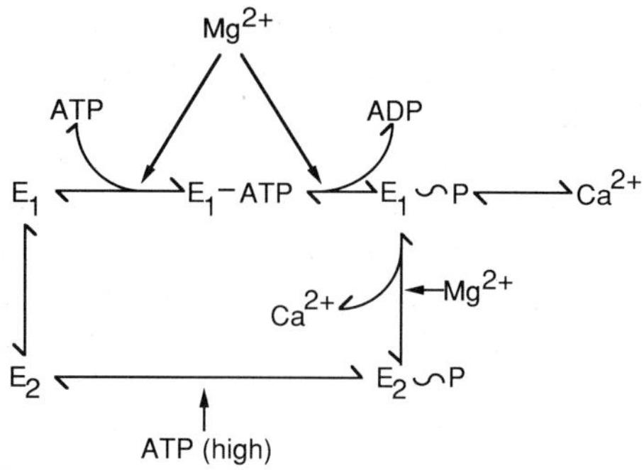

Figure 1. The reaction cycle of the plasma membrane Ca^{2+} pump. The Ca^{2+} translocating step is visualized here in the $E_1 \sim P \rightarrow E_2 \sim P$ transition. As discussed in the text this identification is tentative.

about 140 000 since Ca-dependent incorporation of radioactivity was only observed in a protein corresponding to this mass in electrophoretic gels (10). In the accepted formalism of P-type pumps the enzyme incorporates phosphate in the E_1 conformational state (E_1~-P), the affinity (K_m for Ca^{2+}) in the reaction being as low as 0.2 μM (11). ATP apparently interacts at two sites: one, assumed to be the catalytic site, having a K_m about 1.5 μM (12–13), the other, probably involved in the dephosphorylation of the pump, having a much lower affinity (K_m between 145 and 180 μM (13)). Mg^{2+} promotes a structural transition of the pump, to the E_2~-P conformer (14), which then hydrolyses to E_2 and phosphate in a reaction which is assumed to be accelerated by the binding of ATP to the low-affinity site. In the last step of the reaction cycle, the E_2 conformer returns to the original E_1 state.

The step of the cycle during which the translocation of Ca across the membrane takes place has not been conclusively established, although the phosphorylation step certainly preceeds the translocation, since considerable evidence suggests that the Ca binding site in the phosphorylated state faces the interior of the cell. On the other hand, it appears plausible to relate the translocation process to a conformational change of the protein, i.e., to the step E_1~-P to E_2~-P, as indicated in *Figure 1* (the step $E_2 \rightarrow E_1$ could also be considered, but it appears less logical). At the end of the translocation step, the Ca^{2+} binding site will have to face the external side of the plasma membrane, and have much lower affinity for Ca to facilitate its release. The scheme of *Figure 1* and the discussion in the text are based on the assumption of one Ca equivalent bound and translocated per pump molecule at each run of the cycle. Although some kinetic evidence would suggest that the pump molecule contains two Ca binding sites (15–16), most of the evidence available today, particularly on the purified reconstituted pump (see below), indicates a 1:1 Ca–ATP stoichiometry, i.e. only one Ca would be translocated across the enzyme molecule per reaction cycle (17). However, the possibility of a 2:1 Ca–ATP stoichiometry analogous to that of the Ca pump of sarcoplasmic reticulum (18) is not ruled out (19–20).

As with all other P-type pumps the plasma membrane Ca pump is inhibited by La and by vanadate. One important aspect of the inhibition by La which is specific for the plasma membrane Ca pump is the increased steady-state level of the phosphoenzyme (11, 21), an observation which could be conveniently explained with the inhibition of the hydrolysis of the phosphorylated intermediate by La. The observation is useful, since it increases the sensitivity of the detection of the phosphoenzyme in polyacrylamide gel systems, and permits discrimination between the phosphoenzymes of the plasma membrane Ca pump and that of the sarcoplasmic reticulum pump in mixed preparations (22).

The inhibition by vanadate (13–24) is generally assumed to be due to binding to the aspartyl residue which forms the phosphoenzyme. Its potency depends on the ionic composition of the medium and on the concentration of ATP (25): under optimal conditions, i.e. in the presence of both Mg and K, the K_i may be as low as 2 μM. In analogy with proposals on other P-type pumps it is assumed that vanadate interacts with the E_2 conformer, blocking the $E_2 \rightarrow E_1$ transition and thus stabilizing the E_2 conformation. The vanadate-interacting site is located at the inner side of the plasma membrane (26).

2.2 Activation of the pump

2.2.1 Calmodulin

The observation that calmodulin stimulates the Ca pump in erythrocyte ghosts was preceded by the finding (27) that a soluble cytosolic activator present in erythrocytes activated the pump. Jarrett and Penniston (28) and Gopinath and Vincenzi (29) found that the effects of the activator were mimicked by calmodulin, which thus became established as a natural activator of the pump. The activation by calmodulin persisted in detergent-solubilized erythrocyte membrane preparations (30–31), showing that the effect was due to the direct interaction of the activator with the pump, and consisted of both a K_m and a V_{max} effect. Using erythrocyte membranes made calmodulin-free with chelator washes, the K_m (Ca) of the pump has been found to decrease from 20–30 μM to below 1 μM, and the maximal transport rate to increase up to tenfold (30–32), in line with the idea of an increase of the turnover of the pump. Accordingly, calmodulin stimulates both the rate of phosphorylation and of dephosphorylation of the enzyme (30, 33–34).

2.2.2 Other activating treatments

Phospholipase digestion experiments on erythrocyte membranes showed that the pump is very sensitive to the phospholipid environment (35–36). In the first study reactivation was obtained with all glycerophospholipids tested (35), whereas the second (36) showed a marked preference for acidic phospholipids, and found reactivation also with long-chain polyunsaturated fatty acids. The reason for the experimental discrepancy is not clear, but more recent work with the purified enzyme (see below) has strongly supported the special ability of acidic phospholipids to activate the pump in alternative to calmodulin. The finding that acidic phospholipids (e.g. phosphatidyl-serine) activate the pump may have important physiological implications, owing to their presence in the plasma membrane. In fact, calculations on the isolated reconstituted enzyme to be discussed in more detail below (37) have indicated that the pump in the native environment of the erythrocyte membrane may be permanently activated to about 50% of maximal.

The pump in the plasma membrane can be activated by trypsin and chymotrypsin proteolysis (38) in the absence of calmodulin. The activation by trypsin results in the removal from the cytosolic side of the membrane of a fragment of molecular mass about 30 000 which was suggested to contain the calmodulin-sensitive portion of the pump. (The matter of trypsin proteolysis on the purified ATPase will be discussed in detail below.) (39). More recent work on the Ca-dependent protease calpain has also shown activation, although at a much slower rate than with trypsin (40–42). The activation by calpain reduced the molecular mass of the pump by about 12 000, leaving in the membrane a product of molecular mass about 124 000 which still binds calmodulin during the initial phases of the activation process, but fails to bind it at later times. The finding suggests that calpain attacks the calmodulin binding domain of the pump in a gradual way, resulting eventually in a fragment completely free of the calmodulin binding domain. The matter of calpain is of great interest, since calpain could indeed be active on the pump *in vivo* (at variance with the other proteases mentioned above), and also because calpain, which contains two calmodulin-like sequences in its structure, could be specifically targeted to the calmodulin-binding domain of the pump. This has been supported by recent work on the purified enzyme to be discussed below. It is of interest that the rate of calpain proteolysis, and the activation of the pump, are slowed down by calmodulin (42).

A number of proteins, present in the cytosol or bound to the plasma membrane, have been claimed to activate the pump, but their mechanism of action and physiological significance are obscure (43–46). The activation by a phosphorylation process mediated by the cAMP-dependent protein kinase has been studied in more detail (47–48), and has recently been extended to the purified pump. The activation lowers the K_m (Ca) of the pump in the absence of calmodulin to about 1 μM, and is related to the phosphorylation of the pump protein proper, not to the phosphorylation of an accessory protein that would then activate the ATPase. Recent work on the purified enzyme has shown that only one of the isoforms of the pump present in human tissues is responsive to cAMP (49). Although the cGMP protein kinase has also been claimed to activate the Ca pump in vascular smooth muscle, thus contributing to the relaxing effect of cGMP (50–54), recent work has shown that the pump does not become phosphorylated by the G-kinase (55). Protein-kinase C has also been claimed to activate the pump, in the presence of either diolein or phorbol ester (56): the activation has been claimed to be additive with that produced by calmodulin, and to consist of a five- to sevenfold increase in the V_{max} of both Ca transported and ATP hydrolysis, without effects on the K_m (Ca).

3. The plasma membrane Ca pump in other cell types

Following its discovery in 1966, for a number of years the pump was studied almost exclusively in erythrocytes. Only relatively recently the studies were extended to other cell types, showing that the pump is ubiquitously distributed in eukaryotic cells. Surprisingly, the pump was found also in the plasma membrane of excitable cells like heart (57) of neurons (58), i.e. in cells where Ca extrusion was generally assumed to occur via a Na–Ca exchanger. Although no attempt will be made here to present all the information available on the pump in different cell types (the reader is referred to recent comprehensive reviews, refs 4–5), some particular aspects will be mentioned. The pump in liver plasma membrane appears to differ from all others since it is not stimulated by calmodulin, (59–63) and appears to have a molecular mass slightly higher than that of the erythrocyte enzyme. However, recent work in this laboratory has suggested that the differences probably are less dramatic than generally assumed: the liver enyzme cross-reacts immunologically with that of erythrocytes, and can be made to interact with calmodulin under special conditions (work in preparation). One interesting aspect of the pump is its distribution in different subdomains of the plasma membrane in different cell types. This is true of liver, where the pump seems to be concentrated in the sinusoidal subdomain (work in preparation) and of intestinal mucosa, where the pump activity decreases from the duodenum to the jejunum and ileum (64–65), is inhomogeneously distributed along the villar/crypt axis of the enterocytes, and appears to be located exclusively in the basolateral domains (68–69). In kidney, the pump is only found in distal tubules, and appears to be located exclusively in the basolateral domain of the plasma membrane (70–72). One last point that could be mentioned in this section is the report that the pump of intestinal mucosa is stimulated by the vitamin-D-dependent calcium binding protein calbindin (64–73).

4. The isolated pump

4.1 Purification of the enzyme

The pump was first purified by Niggli *et al.* in 1979 (74), using a calmodulin-affinity column. The column was loaded with a Triton X-100 solubilizate of erythrocyte ghosts which had been made calmodulin-free by extensive washings with EDTA. After washing the column with Ca, elution with EDTA removed a protein of molecular mass about 140 000 which had Ca-dependent ATPase activity (*Figure 2*). Unexpectedly, in the original experiment the activity was not stimulated by calmodulin, at variance with a preparation obtained shortly thereafter essentially by the same methods, but with deoxycholate instead of Triton X-100 as the solubilizing detergent (75). The finding was soon rationalized with the presence in the column and in the elution buffer of Niggli *et al.* (74) of the acidic phospholipid phosphatidyl-serine, which was included to stabilize the enzyme (35), and which evidently produced maximal stimulation of the ATPase (37). Replacement of phosphatidyl-serine with the zwitterionic phospholipid phosphatidyl choline did indeed produce an enzyme that had a lower basal specific activity, but which could be fully stimulated by calmodulin (76).

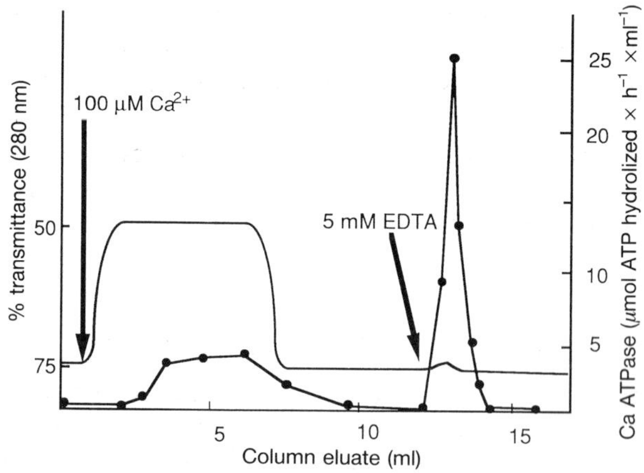

Figure 2. The purification of the Ca^{2+} pump from the erythrocyte membrane using a calmodulin affinity column (ref. 74).

4.2 Functional properties of the purified pump

The isolated pump was reconstituted in liposomes with high transport efficiency, showing a Ca to ATP stoichiometry approaching 1 under conditions where no leakage of Ca occurred (76). Therefore, the experiment provided strong support to the concept of a stoichiometry of 1 between Ca transported and ATP hydrolysed (17). The liposomal system has also permitted to establish that the pump functions as an obligatory Ca: proton exchanger (77). However, whether the exchange is electroneutral or partially electrogenic is still debated: the experiments on the liposomal system have indicated electroneutrality, whereas other experiments on erythrocyte ghosts suggest a 1:1 Ca: proton exchange, i.e. partial electrogenicity.

The purified enzyme has also been used to study in detail the matter of the phospholipid stimulation (37). By reconstituting the enzyme in liposomes of controlled phospholipid composition it has been possible to show that all acidic phospholipids activate [i.e. they lower the K_m (Ca)] in the absence of calmodulin. Whereas optimal activation by phosphatidyl-serine requires very large amounts of the phospholipid (about 5000 molecules per molecule of the ATPase), the molar ratio is dramatically lower in the case of the phosphorylated derivatives of phosphatidyl-inositol (80). Since the latter phospholipids may have rapid turnover rates, and since the products of their receptor-stimulated breakdown (diacylglycerol and inositol-tris-phosphate) have no stimulatory effects on the pump (ref. 81, and unpublished work in this laboratory) the proposal that they modulate the activity of the pump *in vivo* is an attractive possibility (81). As already pointed out above, the liposomal reconstitution experiments (37) have shown that the enzyme in the erythrocyte could be permanently activated to about 50% of the maximal activity by the existing concentration of phosphatidyl-serine. Phospholipids, in fact, activate the enzyme even better than calmodulin, i.e., they lower the K_m (Ca) to about 0.2–0.3 μM, as compared to about 0.5 μM in the case of calmodulin (37, 82–83).

These general properties of the purified erythrocyte pump, proving its functional competence, have also been verified on the pump isolated from other sources, including heart (84), skeletal muscle (85), and vascular smooth muscle (86).

4.3 Proteolysis studies on the purified pump

The most comprehensive proteolytic studies on the isolated pump have employed trypsin (87, 88, 83). They have shown that the protease, at initial times, attacks the enzyme at both the C- and N-termini. The cleavages at the C-terminus remove sequentially from the enzyme hydrophilic portions leaving behind a truncated ATPase: the products formed in succession have molecular masses of 90 000, 85 000, 81 000, 76 000. Their molecular masses are those measured in SDS-polyacrylamide gels, in which portions of the enzyme are separated also from the N-terminus. However, whereas the removal of hydrophilic portions at the C-end of the enzyme is demonstrated conclusively (83), whether after the trypsin cut portions are removed also from the opposite end of the pump is an open question. In fact, the hydrophobicity of the N-terminal portion of the ATPase, and its tendency to associate with the bigger trypsin fragments makes in rather unlikely.

Figure 3 summarizes the results of the experiments on trypsin proteolysis, and the properties of the fragments produced. The fragment of 90 000 still contains the full calmodulin binding domain, that of 81 000 has lost it completely. The activity of the former, measured as Ca-dependent ATP hydrolysis (89) is rather limited, but can be fully stimulated by calmodulin,

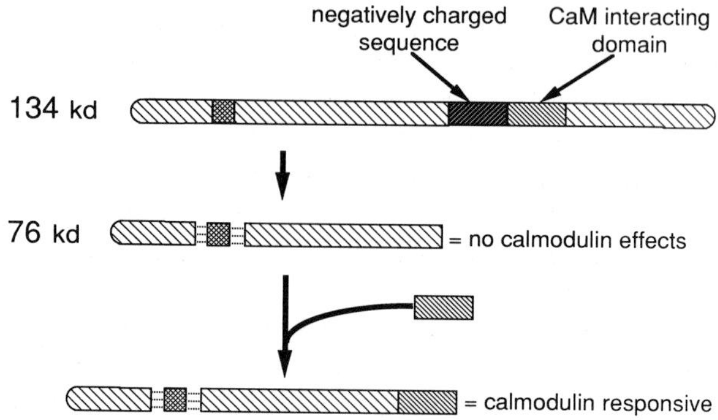

Figure 3. Trypsin proteolysis of the purified Ca pump. The cleavage pattern is based on work quoted in ref. 87–88, and 83. Some of the aspects of the scheme are based on recent unpublished work in this laboratory, e.g. the precise location of the C-and N-terminal cuts, and the location of the putative domain interacting with acidic phospholipids. In the scheme, the provisional assumption has been made that the N-terminal portion of the pump (i.e. a fragment of molecular mass about 33 000) remains bound to the remainder of the pump after the trypsin cut. Since the 76 kd product fails to respond to acidic phospholipids, the putative domain responsive to the latter has been represented as structurally deranged.

whereas the activity of the latter is already fully expressed in the absence of calmodulin. The 85 000 fragment binds calmodulin, but is poorly stimulated by it, suggesting the existence of an inhibitory sequence in the calmodulin binding domain that can only be removed from some ATPase site critical to activity when the C-terminal portion of the calmodulin binding domain is in place (i.e. the 90 000 fragment). All fragments react to acidic phospholipids, but that of molecular mass 76 000 does not. N- and C-terminal sequence of the fragments, combined with information on the primary structure of the pump to be discussed below, have placed the cleavage that produces the 76 000 fragment at the N-terminal end of fragment 81 000 (unpublished). Thus, even if the 33 500 fragment N-terminal to the cut does not become separated from the main body of the enzyme, the domain responsible for phospholipid response is evidently deranged by the trypsin attack leading to the 76 000 fragment. As mentioned above, the calmodulin binding domain of the ATPase is cleaved off also by the Ca-dependent protease calpain, and in this case the attack may have physiological relevance, since calpain is ubiquitously present in cells (90–91). Calpain degrades the calmodulin binding domain in two successive steps, producing two fragments of nearly identical molecular mass, i.e. 124 000. The early product is cleaved about ten residues from the C-terminus of the calmodulin binding domain, and still retains some calmodulin sensitivity (91). The later product has lost all of the calmodulin binding domain, and its activity is fully expressed and insensitive to calmodulin (91). It is of interest that the effect of calpain is inhibited by calmodulin, which evidently competes with calpain for the occupancy of the calmodulin binding domain of the pump.

CNBr has also been applied to the purified pump in a study aimed at the identification, isolation, and sequencing of the calmodulin binding domain (92). The calmodulin binding domain has been labelled prior to CNBr splitting with a cleavable photoactivatable cross-linker bound to calmodulin. After splitting and separation of the fragments, the labelled one was sequenced, yielding a structure that has similarities to the calmodulin binding domain of other calmodulin-modulated proteins. The similarities are not particularly striking in the sequence proper, but they are in the predominance of basically charged amino acids, in the absence of negatively charged amino acids, and in the propensity to form an amphiphilic helix.

5. The primary structure of the pump

The complete primary structure of the pump was deduced in 1988 by Shull and Greeb (93) from complementary DNA isolated from rat brain and by Verma *et al.* (94) from human teratoma cell libraries. In both cases oligonucleotides were constructed from tryptic fragments of the human erythrocyte ATPase (94), or from the conserved sequence of the ATP binding sites of other P-type ion motive ATPases (93). The rat brain has permitted to deduce three isoforms of the pump, having 1176 amino acids (molecular mass 129 500), 1198 amino acids (molecular mass 132 605) and 1159 amino acids (molecular mass 127 300) (95) respectively. The isoform from the human teratoma library contains 1220 amino acids (molecular mass 134 683), and corresponds to the first brain isoform with which it shares 99% of the first 1117 residues. However, it differs significantly in the C-terminal domain, due to differential RNA splicing involving a single exon (96).

Very recently, two isoforms of the pump have been found expressed in human erythrocytes (97), one corresponding to the teratoma isoform, the other, much more abundant, having 1205 amino acids (molecular mass 133 930), and lacking the cAMP-responsive site in the C-terminal region (see below).

5.1 Functional domains in the pump structure

As shown in the secondary structure model of Figure 4, the pump appears to contain ten transmembrane helices, which have been assigned by the usual hydropathy plots. They are grouped at the N-and C-terminal portions of the pump, leaving a large mid-zone of the enzyme free of hydrophobic sequences. The pump protrudes essentially in to the intracellular domain, where about 80% of its mass is located. The domain protruding between putative transmembrane helices 4 and 5 contains the ATP (fluorescein isothiocyanate,

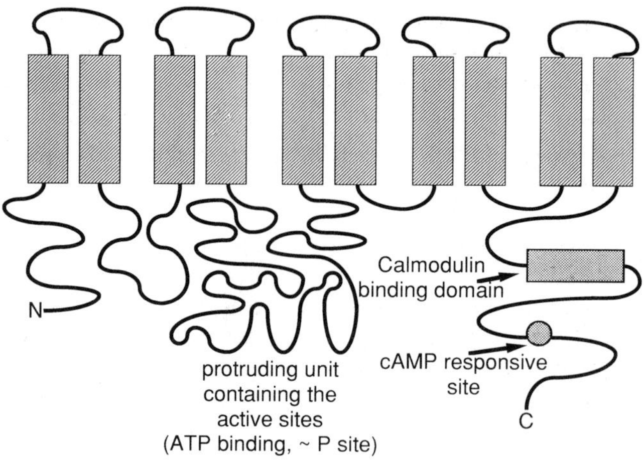

Figure 4. A scheme of the membrane organization of the Ca pump. As indicated, the domain protruding between transmembrane helices four and five is the largest (more than 400 residues). The substrate site for the cAMP-dependent protein kinase has so far been detected only in one of the isoforms of the pump.

FITC) binding site and the site of phosphoenzyme formation (lysine 601, and aspartic acid 475, respectively, in the human teratoma isoform). The domain protruding out of the last putative transmembrane helix contains several interesting domains, i.e. the calmodulin binding domain (residues 1100–1127 in the human teratoma isoform), an acidic stretch immediately N-terminal to it, and a domain having the typical characteristics of the substrate site for the cAMP-dependent protein kinase. This domain, which is present in the human teratoma isoform, but not found in the other, and major, erythrocyte isoform, contains serine 1178, which recent work (98) has shown to become phosphorylated by the cAMP-dependent kinase. The phosphorylation increases the affinity of the ATPase for Ca, to a K_m of about 1.2 μM. As for the acidic sequence N-terminal to the calmodulin binding domain, it would make an attractive candidate as a Ca binding site. However, if involved in Ca binding, this sequence does not play a role in the catalytic cycle proper: as mentioned above, the tryptic fragments of molecular mass 81 000 and 76 000 do not contain this sequence, yet their ATP hydrolysis is stimulated by Ca with normal affinity. Possibly, the sequence plays some sort of regulatory role, i.e. it could regulate the access of Ca to the catalytic binding site, probably located in one or more of the transmembrane helices.

6. Concluding remarks

The Ca pump of the plasma membrane is the high affinity regulator of cell Ca, probably working in co-operation with the lower affinity Na–Ca exchanger.

The pump is one of the targets of calmodulin stimulation, but could also be regulated *in vivo* by the phospholipid ambient. The enzyme is the largest of all P-type pumps: although ubiquitously distributed, it consists of several isoforms, which could possibly be tissue specific. The isoforms could have regulation specificity, e.g. one could by stimulated by c-AMP-directed phosphorylation.

Acknowledgements

The original work described has been made possible by the financial contribution of the Swiss National Science Foundation (Grants No. 3.634.080, 3.189.082, 3.658.084 and 85.159.084). The financial support of the National Institutes of Health is also gratefully acknowledged (Grant No. 28835).

References

1. Schatzmann, H. J. (1966). *Experientia*, **22**, 364.
2. Schatzmann, H. J. (1982). In *Membrane transport of calcium* (ed. E. Carafoli), p. 141. Academic Press, New York and London.
3. Penniston, J. (1983). In *Calcium and cell function*, Vol. IV, p. 100. Academic Press, New York and London.
4. Rega, A. and Garrahan, P. (1975). *Journal of Membrane Biology*, **22**, 313.
5. Carafoli, E. (1990). *Physiological Review*. (In press.)
6. Pedersen, P. L. and Carafoli, E. (1987). *Trends in Biochemical Sciences*, **12**, 146.
7. Pedersen, P. L. and Carafoli, E. (1987). *Trends in Biochemical Sciences*, **12**, 186.
8. Carafoli, E. and Zurini, M. (1982). *BBA Reviews of Bioenergetics*, **683**, 279.
9. Katz, S. and Blostein, R. (1975). *Biochimica et Biophysica Acta*, **389**, 314.
10. Knauf, P. A., Proverbio, F., and Hoffman, J. F. (1974). *Journal of General Physiology*, **63**, 324.
11. Szasz, I., Sarkadi, B., Schubert, A., and Gardos, G. (1978). *Biochimica et Biophysica Acta*, **512**, 331.
12. Muallem, S. and Karlish, S. J. D. (1979). *Nature*, **277**, 238.
13. Richards, D. E., Rega, A. G., and Garrahan, D. J. (1978). *Biochimica et Biophysica Acta*, **511**, 194.
14. Rega, A. F. and Garrahan, P. J. (1975). *Journal of Membrane Biology*, **22**, 313.
15. Ferreira, H. G., and Lew, V. L. (1976). *Nature*, **259**, 47.
16. Lew, V. L., Tsien, R. Y., Miner, C., and Bookchin, R. M. (1976). *Nature*, **298**, 478.
17. Schatzmann, H. J., (1973). *Journal of Physiology, (London)*, **235**, 551.
18. Hasselbach, W. and Wass, W. (1982). *Annals of the New York Academy of Sciences,* **402**, 459.
19. Sarkadi, B. (1980). *Biochimica et Biophysica Acta*, **604**, 159.
20. Sarkadi, B., Szasz, I., Garloczi, A., and Gardos, G. (1977). *Biochimica et Biophysica Acta*, **464**, 93.
21. Schatzmann, H. J. and Bürgin, H. (1978). *Annals of the New York Academy of Sciences*, **307**, 125.

22. Wuytack, F., Raeymaekers, L., DeSchutter, G., and Casteels, R. (1982). *Biochimica et Biophysica Acta*, **693**, 45.
23. Bond, G. H. and Hudgins, P. (1979). *Biochemistry*, **18**, 325.
24. Bond, G. H. and Hudgins, P. (1980). *Biochimica et Biophysica Acta*, **600**, 781.
25. Barrabin, H., Garrahan, P. J., and Rega, A. F. (1980). *Biochimica et Biophysica Acta*, **600**, 796.
26. Rossi, J. P. F. C., Garrahan, P. J., and Rega, A. F. (1981). *Biochimica et Biophysicia Acta*, **648**, 145.
27. Bond, G. H. and Clough, D. L. (1972). *Biochimica et Biophysicia Acta*, **323**, 592.
28. Jarrett, H. M. and Penniston, J. T. (1977). *Biochemical and Biophysical Research Communications*, **77**, 1210.
29. Gopinath, R. M., and Vincenzi, F. F. (1977). *Biochemical and Biophysical Research Communications*, **77**, 1203.
30. Jeffery, D., Roufogalis, B. D., and Katz, S. (1981). *Biochemical Journal*, **194**, 481.
31. Larsen, F. L. and Vincenzi, F. F. (1979). *Journal of Supramolecular Structure*, **9**, 269.
32. Muallem, S. and Karlish, S. J. D. (1981). *Biochimica et Biophysica Acta*, **647**, 73.
33. Muallem, S. and Karlish, S. J. D. (1980). *Biochimica et Biophysica Acta*, **597**, 631.
34. Luthra, M. G., Watts, R. P., Scherer, K. L., and Kim. H. D. (1980). *Biochimica et Biophysica Acta*, **633**, 299.
35. Ronner, P., Gazzotti, P., and Carafoli E. (1977). *Archives of Biochemistry and Biophysics*, **179**, 578.
36. Roelofsen, B., and Schatzmann, H. J. (1977). *Biochimica et Biophysica Acta*, **464**, 17.
37. Niggli, V., Adunyah, E. S., and Carafoli, E. (1981). *Journal of Biological Chemistry*, **256**, 8588.
38. Taverna, R. D., and Hanahan, D. H. (1980). *Biochemical and Biophysical Research Communications*, **94**, 652.
39. Enyedi, A., Sarkadi, B., Szasz, I., Bot, B., and Gardos, G. (1980). *Cell Calcium*, **1**, 299.
40. Au, K. S. (1986). *Biochimica et Biophysica Acta*, **905**, 273.
41. James, P. H., Pruschy, M., Vorherr, T., Penniston, J. T., and Carafoli, E. (1989). *Biochemistry*, **28**, 4253.
42. Wang, K. K., Villabobo, A., and Roufogalis, B. D. (1988). *Archives of Biochemistry and Biophysics*, **260**, 696.
43. Maudlin, D. and Roufogalis, B. D. (1980). *Biochemical Journal*, **187**, 507.
44. Roufogalis, B. D., Elliott, C. T., and Ralston, G. R. (1984). *Cell Calcium*, **5**, 77.
45. Reinlib, L., Clark, A., and Carafoli, (E). (1984). *Journal of Bioenergetics and Biomembranes*, **16**, 611.
46. Loetersztajn, S., and Pecker, F. (1982). *Journal of Biological Chemistry*, **257**, 6638.
47. Caroni, P., and Carafoli, E. (1981). *Journal of Biological Chemistry*, **256**, 3263.
48. Neyses, L., Reinlib, L., and Carafoli, E. (1985). *Journal of Biological Chemistry*, **260**, 10283.
49. Strehler, E. E., James, P. Fischer, R. Heim, R., Vorherr, T., Filoteo, A.,

Penniston, J. T. and Carafoli, E. (1990). *Journal of Biological Chemistry*, **265**, 2835.
50. Furukawa, K.-I., Tawada, Y., and Shigekawa, M. (1988). *Journal of Biological Chemistry*, **263**, 8058.
51. Kobayashi, S., Kanaide, H., and Nakamura, M. (1985). *Science*, **229**, 553.
52. Popescu, L. M., Panoiu, C., Hinescu, M., and Nutu, O. (1985). *European Journal of Pharmacology*, **107**, 393.
53. Sarmiento, J. G. Janis, R. A., and Lincoln, T. M. (1981). *Federation Proceedings*, **40**, 551.
54. Suematsu, E. Hirata, M., and Kuriyama, H. (1984). *Biochimica et Biophysica Acta*, **773**, 83.
55. Baltesnperger, K., Carafoli, E., and Chiesi, M. (1988). *European Journal of Biochemistry*, **172**, 7.
56. Smallwood, J. I., Gügi, B., and Rasmussen, H. (1988). *Journal of Biological Chemistry*, **263**, 2195.
57. Caroni, P. and Carafoli, E. (1980). *Nature*, **283**, 765.
58. Di Polo, R., (1979). *Nature*, **274**, 390.
59. Bachs, O., Famulski, K. S., Mirabelli, F., and Carafoli, E. (1985). *European Journal of Biochemistry*, **147**, 1.
60. Iwasa, Y. Iwasa, T., Higashi, K., Matsui, K., and Miyamoto, E. (1982). *Biochemical and Biophysical Research Communications*, **105**, 448.
61. Kraus-Friedmann, N., Biber, J., Murer, H., and Carafoli, E. (1982). *European Journal of Biochemistry*, **129**, 7.
62. Lin, S. H. and Fain, J. N. (1984). *Journal of Biological Chemistry*, **259**, 3016.
63. Loterztajn, S., Hanoune, J., and Pecker, F. (1981). *Journal of Biological Chemistry*, **257**, 11209.
64. Ghijsen, W. E. J. M. and Van Os, C. H. (1982). *Biochimica et Biophysica Acta*, **689**, 170.
65. Moy, T. C., Walters, J. R. F., and Weiser, M. M. (1986). *Biochemical and Biophysical Research Communications*, **141**, 979.
66. De Jonge, H. R., Ghijsen, W. E. J. M., and Van Os., C. H. (1981). *Biochimica et Biophysica Acta*, **647**, 140.
67. Ghijsen, W. E. J. M., and Van Os, C. H. (1979). *Nature*, **279**, 802.
68. Hildmann, B., Schmidt, A., and Murer, H. (1982). *Journal of Membrane Biology*, **65**, 55.
69. Nellans, H. N., and Popovich, J. E. (1981). *Journal of Biological Chemistry*, **256**, 9932.
70. Borke, J. L., Epstein, F. H. Penniston, J. T., and Kumar R. (1989). *Mount Desert Island Biological Laboratory Bulletin*, **28**, 48.
71. Borke, J. L., Minami, J., Verma, A., Penniston, J. T., and Kumar, R. (1987). *Journal of Clinical Investigations*, **80**, 1225.
72. Kumar, R., Penniston, J. T., and Borke, J. L. (1988). *News in Physiological Sciences*, **3**, 219.
73. Walters, J. R. F., and Weiser, M. M. (1987). *American Journal of Physiology*, **252**, G170.
74. Niggli, V., Penniston, J. T., and Carafoli, E. (1979). *Journal of Biological Chemistry*, **254**, 9955.
75. Gietzen, K., Teicka, M., and Wolf, H. V. (1980). *Biochemical Journal*, **189**, 81.

76. Niggli, V., Adunyah, E. S., and Carafoli, E. (1981). *Journal of Biological Chemistry*, **256**, 8588.
77. Niggli, V., Sigel, E., and Carafoli, E. (1982). *Journal of Biological Chemistry*, **257**, 2350.
78. Waisman, D. M., Gimble, J. M., Goodmann, D. B. O., and Rasmussen, H. (1981). *Journal of Biological Chemistry*, **256**, 415.
79. Rossi, H. J. P. F. C. and Schatzmann, H. J. (1982). *Journal of Physiology*, **327**, 1.
80. Carafoli, E. and Zurini, M. (1982). *BBA Reviews of Bioenergetics*, **683**, 279.
81. Choquette, D., Hakim, G., Filoteo, A. G., Plishker, F. A., Bostwick, G. R., and Penniston, J. T. (1984). *Biochemical and Biophysical Research Communications*, **125**, 908.
82. Niggli, V., Adunyah, E. S., Penniston, J. T., and Carafoli, E. (1981). *Journal of Biological Chemistry*, **256**, 395.
83. Enyedi, A., Flura, M., Sarkadi, B., Fardos, G., and Carafoli, E. (1987). *Journal of Biological Chemistry*, **262**, 6425.
84. Caroni, P. and Carafoli, E. (1981). *Journal of Biological Chemistry*, **256**, 9371.
85. Michalak, M., Famulski, R., and Carafoli, E. (1984). *Journal of Biological Chemistry*, **259**, 15540.
86. Wuytack, F., DeSchutter, G., and Casteels, R. (1981). *Biochemical Journal*, **198**, 265.
87. Zurini, M., Krebs, J., Penniston, J. T., and Carafoli, E. (1984). *Journal of Biological Chemistry*, **259**, 618.
88. Benaim, G., Zurini, M., and Carafoli, E. (1984). *Journal of Biological Chemistry*, **259**, 8471.
89. Benaim, G., Clark, A., and Carafoli, E. (1986). *Cell Calcium*, **7**, 175.
90. Wang, K. K., Villabobo, A., and Roufogalis, B. D. (1988). *Archives of Biochemistry and Biophysics*, **260**, 696.
91. James, P. Vorherr, T., Krebs, J., Morelli, A., Castello, G., McCormick, D. J., Penniston, J. T., De Flora, A., and Carafoli, E. (1989). *Journal of Biological Chemistry*, **264**, 8289.
92. James, P., Maeda, M., Fischer, R., Verma, A. K., Krebs, J., Penniston, J. T., and Carafoli, E. (1988). *Journal of Biological Chemistry*, **263**, 2905.
93. Shull, G. E. and Greeb, J. (1988). *Journal of Biological Chemistry*, **263**, 8646.
94. Verma, A. K., Filoteo, A. G., Stanford, D. R., Wieben, E. D., Penniston, J. T., Strehler, E. E., Fischer, R., Heim, R., Vogel, G., Mathews, S., Strehler-Page, M. A., James, P., Vorherr, T., Krebs, J., and Carafoli, E. (1988). *Journal of Biological Chemistry*, **263**, 14152.
95. Greeb, J. and Shull, G. E. (1989). *Journal of Biological Chemistry*, **264**, 18569.
96. Strehler, E. E., Strehler-Page, M.-A., Vogel, G., and Carafoli, E. (1989). *Proceedings of the National Academy of Sciences, USA*, **86**, 6908.
97. Strehler, E. E., James, P., Fischer, R., Heim, R., Vorherr, T., Filoteo, A., Penniston, J. T., and Carafoli, E. (1990). *Journal of Biological Chemistry*. (In press.)
98. James, P., Pruschy, M., Vorherr, T., Penniston, J. T., and Carafoli, E. (1989). *Biochemistry*, **28**, 4253.

14

Proteins involved in the uptake and release of Ca^{2+} from the sarcoplasmic reticulum

SUSAN L. HAMILTON and CHARLOTTE A. TATE

1. Introduction

Changes in the intracellular concentration of Ca^{2+} in skeletal muscle regulate the state of the muscle contraction and relaxation. A number of different proteins located in the sarcoplasmic reticulum (SR) are involved in the regulation of the Ca^{2+} concentrations; some of the proteins with known roles and others with putative roles are listed in *Table 1*. In this chapter we will detail some of the methods used to study the two proteins most directly involved in controlling Ca^{2+} levels in skeletal muscle: (a) the Ca^{2+} ATPase, the protein responsible for the uptake of Ca^{2+} into the lumen of the SR, and (b) the Ca^{2+}-release channel, the protein responsible for the release of the luminally sequestered Ca^{2+} into the cytoplasm. The signal which triggers the opening of the Ca^{2+} channel in the SR membrane is in some way generated by depolarization of the T-tubule membrane. This interaction between the T-tubule membrane and SR Ca^{2+}-release is termed 'excitation–contraction' (E–C) coupling and may involve a chemical signal such as Ca^{2+} (1, 2) or IP_3 (3, 4) and/or a mechanical coupling between proteins of the two membranes (5). The most likely T-tubule candidate for a role in E–C coupling is the dihydropyridine (DHP) binding protein which may function as both a voltage sensor and a Ca^{2+} channel in the membrane (6, 7). The muscle contracts when the intracellular Ca^{2+} concentration is raised by the efflux of Ca^{2+} from the lumen of the SR and relaxes when the cytoplasmic Ca^{2+} concentration is again lowered by the pumping of Ca^{2+} back into the lumen of the SR by the Ca^{2+} ATPase. The Ca^{2+}-release protein appears to be concentrated in the portion of the SR known as the terminal cisternae or junctional SR while the Ca^{2+} ATPase enzyme is present at high density in the membranes of the longitudinal tubules (non-junctional SR).

Table 1. Sarcoplasmic reticulum proteins

	Apparent MW	Location	Function	Gene cloned?
Ca^{2+} ATPase	100 kd (8, 9)	LT (10, 11)	Ca^{2+} uptake (12)	Y (13, 14, 15)
Sarcalumenin	160 kd (16, 17)	lumen (18, 19)	Binds Ca^{2+} (18, 19)	Y (18)
GP-53K	53 kd (16, 17, 20)	lumen (17, 19)	Coupling of ATP hydrolysis to Ca^{2+} transport (21, 22, 23)	Y (24)
Ca^{2+}-release channel (25)	450 kd (26–29)	JT (28)	Ca^{2+}-release (30)	Y (31)
Calsequestrin (32)	44–67 kd[a] (9, 32)	JT/lumen (10, 20)	Binds Ca^{2+} (9, 33)	Y (34, 35)
165K (170K) (36, 37, 38)	165 kd (36, 37, 38)	JT (38)	Binds Ca^{2+} (38)	Y (39)
106K (40)	106 kd (40)	JT (40)	Ca^{2+} channel 40)	N
Calreticulin (41)	55 kd (41)	Lumen (42)	Binds Ca^{2+} (41, 42)	Y (42)

LT = longitudinal tubules, JT = junctional tubules.
[a] Apparent Mol. wt dependent on gel system.

2. Isolation of non-junctional SR from skeletal and cardiac muscle

There are numerous methods in the literature for isolating SR from striated muscle. For non-junctional SR enriched in the Ca^{2+} ATPase protein from skeletal muscle (*Protocol 1*) we use the method of VanWinkle *et al.* (43), mainly because it is fast and produces a reliable preparation free of contamination by other organelles and the junctional SR. Cardiac SR is isolated by the method of Tate *et al.* (44) and is described in *Protocol 2*. The protein composition of the cardiac preparation includes the Ca^{2+} ATPase, the ryanodine receptor (the Ca^{2+}-release channel), and other proteins associated with the terminal cisternae of the SR, such as calsequestrin (C. A. Tate, unpublished observation). The method of Jones *et al.* (45) can be used to derive a cardiac fraction devoid of the terminal cisternae; however, the yield is very low. The cardiac preparation described in *Protocol 2* is enriched in the Ca^{2+} ATPase protein and is, therefore, suitable for most purposes.

3. Purification of the Ca^{2+} ATPase of sarcoplasmic reticulum

The procedure (43, 46) routinely used in our laboratory to purify the Ca^{2+} ATPase from both skeletal and cardiac muscle SR is outlined in *Protocol 3*.

Protocol 1. Isolation of non-junctional sarcoplasmic reticulum from skeletal muscle (method of VanWinkle *et al.*, ref. 43)

A. *Solutions*

All are ice-cold and made fresh weekly with glass-distilled water.

1. Solution I: 0.9% NaCl; 5 mM Imidazole (pH 7.0).
2. Solution II: 600 mM KCl; 10 mM Tes (pH 7.0).
3. Solution III: 10 mM Tes (pH 7.0).

B. *Procedure*

1. Remove white skeletal muscle from animal, being careful to always use the same muscles. Cut into 1-inch cubes and immerse in Solution I kept on ice.
2. Trim all fascia and connective tissue with scissors on a watch glass on ice (or in cold room). Blot and weigh tissue. Measure out 3 vol. of Solution I.
3. Mince tissue finely in one-half of Solution I. (A small food processor for baby food can be used in this step.) Transfer to homogenization vessel and add remainder of solution.
4. Homogenize with either a Waring-type blender or Omnimixer homogenizer at approximately full speed for 3 × 20 sec each.
5. Pour into tubes and centrifuge at 5000 *g* for 20 min.
6. Filter supernatant through two layers of cheesecloth and centrifuge at 10 000 *g* for 20 min.
7. Decant supernatant and centrifuge in ultracentrifuge at 50 000 *g* for 60 min.
8. Aspirate supernatant and discard. Using one-half original volume, resuspend all pellets (scrape tubes) with Solution II and homogenize in large homogenizer with five passes.
9. Centrifuge at 50 000 *g* for 45 min.
10. Aspirate and discard the supernatant. The pellet will be loose. Add about 1 ml Solution III, swirl, and remove the pellet with a Pasteur pipette. A gelatinous pellet usually remains on the wall of the centrifuge tube; discard it. It contains significant amounts of glycogen.
11. Estimate protein concentration using either the Biuret or Lowry method. Dilute the suspended microsomes with Solution III to 10 mg protein/ml.
12. Centrifuge diluted protein at 7 000 *g* for 30 min. (The pellet will be junctional SR, and the supernatant will be non-junctional SR.)
13. Carefully remove supernatant leaving some of the supernatant on the

Protocol 1. *Continued*

pellet. Centrifuge at 50 000 *g* for 45 min. Wash pellet gently with 1 ml of Solution III, being careful not to disturb the pellet. Aspirate this solution and discard. Resuspend pellet in a small amount of Solution III (10–15 mg protein/ml).

14. **Optional*. To remove any phosphorylase contamination (which can be considerable), layer vesicles at 10 mg protein/ml on 28% sucrose plus 10 mM Tes (pH 7.0) or 20–50% gradient and centrifuge overnight at 50 000 *g*. Resuspend pellet in small amount of Solution III. There may be a small layer near the top of the tube which is most likely T-tubules.

Protocol 2. Isolation of cardiac sarcoplasmic reticulum (method of Tate *et al.*, ref. 44)

A. *Solutions*

All are ice-cold and made fresh weekly with glass-distilled water.

1. Normal saline: 0.9% NaCl.
2. Solution I: 20 mM Tris-maleate (pH 6.8), 0.9% Na-bicarbonate.
3. Solution II: 20 mM Tris-maleate (pH 6.8) and 0.6 M KCl.
4. Solution III: 20 mM Tris-maleate (pH 6.8). *Note*: Osmotic support can be added; use either 100 mM KCl or 0.25 M sucrose.

B. *Procedure*

1. Quickly remove heart from anaesthetized animal and place in ice-cold saline in ice-filled bucket. Drain most of the blood in the first rinse, then put heart in two successive batches of fresh ice-cold saline. There should be little or no blood in the last rinse. Keep the heart in ice-cold saline.
2. In either a cold room, or alternatively, on a watch-glass on ice, trim fat and great vessels from left ventricle. Use only left ventricle and septum. Blot and weigh. Divide tissue into batches of 50–70 g each. Do not homogenize more than 70 g/batch.
3. Measure volume of Solution I equal to 3 ml/g tissue (3 vol.) from each batch (150–210 ml), or 3 ml for 1 g.
4. Chop tissue to fine cubes. For large batches of heart (e.g. 60 g) either chop with scissors in a small amount of the measured Solution I from step 3 (using the rest when you homogenize), or use a food processor with one-half of the measured Solution I. Use three short bursts of the food processor, then transfer the chopped tissue to the homogenization vessel rinsing the sides of the food processor with the remaining Solution I. For smaller batches (1–3 g), first cut the tissue into large pieces and

Protocol 2. *Continued*

then chop finely with scissors in a centrifuge tube. Use the same tube for the first low-speed spin.

5. Homogenize 3 × 15 sec with a Polytron type homogenizer set at approximately 2000 r.p.m. In between each burst (about 60 sec on ice), clean off any connective tissue which may clog the blades. Avoid frothing. There should be no big chunks left although smaller ones may still be present. For homogenization with the Teflon-glass homogenizer and a drill (with drill press), set the speed to 600 r.p.m. and homogenize with ten full passes of the pestle. The first few passes are difficult, since the tissue on the bottom of the homogenizer must be brought up the sides for homogenization. After the first ten passes, put the homogenizer on ice for 30–45 sec, followed by ten additional passes. Pour in centrifuge tubes.
6. Centrifuge at 650 *g* for 5 min, then increase speed to 4000 *g* for an additional 15 min.
7. Filter supernatant through two layers of cheesecloth. Centrifuge the filtered supernatant for 20 min at 8000 *g*.
8. Centrifuge the resultant supernatant at 40 000 *g* for 30 min.
9. Aspirate (do not decant) supernatant. Resuspend pellet in Solution II by gently washing sides of centrifuge tubes using a Pasteur pipette or shaking slowly. Transfer the resuspended pellet to a large size Teflon-glass homogenizer and gently break up large pellet particles with three passes. Only use 40 ml of Solution II for large batches (60–120 g of tissue), or one-half original volume for smaller batches.
10. Centrifuge for 30 min at 40 000 *g*. Aspirate supernatant quickly and resuspend pellet in Solution III about 0.5–1.0 ml. Transfer resuspended pellet to smaller homogenizer and gently break up large pellet particles with five passes.
11. Estimate protein by Biuret method for large batches or the Lowry method for smaller batches. The average yield is 0.3 and 1.2 mg protein/g wet wt. for the dog and rat, respectively. The stability of cardiac SR is improved markedly when the protein concentration of the suspended pellet exceeds 10 mg protein/ml.

Protocol 3. Purification of Ca^{2+} ATPase protein

A. *Solutions*

1. Solution I (solubilizing buffer): 20 mM Tes (pH 8.7); 0.5 M sucrose; 2 M KCl. Add fresh (daily) 2 mM $CaCl_2$, 2 mM $MgCl_2$, 2 mM dithiothreitol (DTT), and 2 mM Na_2ATP.

Protocol 3. *Continued*

2. Solution II (glycerol buffer): 20 mM Tes (pH 7.0); 40% (w/v) glycerol. Add fresh daily 2 mM $CaCl_2$, 2 mM $MgCl_2$, 2 mM DTT, and 2 mM Na_2ATP.
3. Solution III: 10 mM Tes (pH 7.0).
4. Sucrose gradients:
   (a) 20% sucrose, 10 mM Tes (pH 7.0) and 0.5 M KCl. Add fresh (daily) 1 mM $MgCl_2$, 1 mM Na_2ATP, and 1 mM DTT.
   (b) 50% sucrose, 10 mM Tes (pH 7.0) and 0.5 M KCl. Add fresh (daily) 1 mM $MgCl_2$, 1 mM Na_2ATP, and 1 mM DTT.

   Form gradients in SW27 rotor tubes (30–33 ml/tube) and let chill at 4 °C for at least 1 h.

B. *Procedure*

1. Use an initial concentration of SR protein of 20–25 mg protein/ml (>100 mg protein total). If the preparation is too dilute, centrifuge it at 50 000 *g* for 30 min and resuspend in Solution III to 25 mg protein/ml. Save some of the microsomes for comparative purposes.
2. Dilute the protein in an equal volume (1:1) of the solubilizing buffer (Solution I). Stir gently on ice or in a cold room (4 °C) with a stirplate for 30 sec. Purified deoxycholate is added from a 10% (w/v) aqueous solution SLOWLY with slight vortexing to give a final concentration of 0.4 mg of purified deoxycholate/mg SR protein. (The optimal ratio of detergent/protein will have to be determined for your preparation.)
3. Stir the mixture slowly at 4 °C for an additional 5 min. The mixture should be clear.
4. Centrifuge immediately at 200 000 *g* for 20 min. (Have the rotor in the centrifuge ready to go.)
5. Carefully remove the supernatant and measure its volume.
6. Dilute the supernatant with an equal volume (1:1) of glycerol buffer (Solution II). Stir slowly at 4 °C for 20 min. This step removes the detergent from the solubilized preparation, so that the partially purified protein can be pelleted.
7. Centrifuge at 200 000 *g* for 30 min.
8. Resuspend the pellet (partially purified protein) in 3–5 ml of Solution III with three passes of a Teflon-glass homogenizer.
9. Layer the suspended pellet on the 20–50% (w/w) sucrose gradient and spin at 25 000 r.p.m. in a Beckman SW27 swinging bucket rotor for 6 h or overnight.
10. The band around the middle of the gradient is removed and is diluted with an equal volume of Solution III.

Protocol 3. *Continued*

11. Centrifuge suspension at 100 000 *g* for 45 min.
12. Aspirate the supernatant and resuspend the pellet in a small amount of Solution III. Determine protein with the Lowry method.
13. The presence of residual detergent and lipids can be determined by thin-layer chromatography.

The procedure is easy and straightforward with skeletal muscle SR; however, purification of the Ca^{2+} ATPase from cardiac muscle can be more difficult since a higher detergent/protein ratio is frequently required for solubilization.

4. Assays for the calcium transport process of the sarcoplasmic reticulum

The Ca^{2+} ATPase of the SR actively transports Ca^{2+} through the hydrolysis of ATP and other high energy nucleotides. The reaction cycle of the hydrolysis of ATP includes the formation and breakdown of an acylphosphate intermediate (phosphoenzyme) created by the transfer of the terminal phosphate of ATP to aspartate residues on the enzyme. With the formation of the Ca^{2+}-dependent acylphosphate ($Ca{\cdot}E_1P$), Ca^{2+}, bound to high-affinity binding sites, is occluded from the external medium. Subsequently, a series of conformational changes occur in the enzyme with a decrease in the affinity for Ca^{2+} and a release of Ca^{2+} into the intravesicular space. The phosphorylation site of the Ca^{2+}-free intermediate (E_2P) is then exposed to the external medium with the subsequent release of inorganic phosphate from the enzyme. The complete enzyme cycle is repeated until essentially all of the Ca^{2+} in the external medium is sequestered (47).

Although there may be numerous factors controlling the various steps in the hydrolysis of ATP and the translocation of the bound Ca^{2+}, the $Ca{\cdot}E_1\text{-}P \rightarrow E_2\text{-}P$ step is considered rate limiting. This step depends on the release of Ca^{2+} from the enzyme into the intravesicular space. However, when the free Ca^{2+} inside the vesicles reaches a high level, Ca^{2+} cannot be released and the reaction cycle stops. Thus, there is a build-up of $Ca^{2+}{\cdot}E_1\text{-}P$ which in turn inhibits the observed Ca^{2+}-dependent ATP hydrolysis. This inhibition can be relieved by including a Ca^{2+} ionophore (A23187 or ionomycin) or using a Ca^{2+} precipitating agent (oxalate or phosphate) in the reaction medium. Both strategies relieve the feedback inhibition by reducing the free Ca^{2+} inside the vesicle, thereby stimulating Ca^{2+}-dependent hydrolysis of ATP. The magnitude of the stimulation depends on several factors, the most important of which is the integrity of the vesicles. The stimulation will be greater with a preparation of high membrane integrity, or relative absence of 'leakiness'. In our hands, the vesicles isolated from skeletal muscle have a high degree of membrane integrity while the isolated cardiac SR vesicles may consist of 50% 'leaky' vesicles. In fact, the skeletal muscle SR vesicles may be so 'tight' that a Ca^{2+}-dependent hydrolysis of ATP cannot be observed using standard manual techniques unless either an ionophore or a Ca^{2+} precipitating agent is used.

Under optimal conditions the stoichiometric relationship between Ca^{2+} transport and Ca^{2+}-stimulated ATPase activity is two Ca^{2+} ions transported per ATP molecule hydrolysed. In practice, however, this ratio is rarely achieved unless rapid quenching techniques are used. Usually this ratio is achieved during the first few turnovers of the enzyme (or within the first 10–15 msec at 30 °C). Thus, difficulties in achieving the 'optimal' stoichiometric ratio does not mean that the isolated preparation is of poor quality. The preparation must be optimized and characterized (with respect of contaminating organelles, such as sarcolemma) in each laboratory, but extensive attention to obtaining this stoichiometry may be futile.

We present several assays for the activity of the Ca^{2+} transport process of isolated striated SR. In particular we have focused our attention on: (a) ATP-dependent Ca^{2+} accumulation (with and without the Ca^{2+} precipitating agent, oxalate); (b) ATPase activity (hydrolysis of ATP) using two independent methods; and (c) phosphorylation of the Ca^{2+} ATPase by ATP hydrolysis. We chose methods which are simple and require standard equipment found in most biochemical laboratories. References for alternative methods are given along with a few notes about our experiences with them.

4.1 ATP-dependent Ca^{2+} accumulation

The manual filtration assay for determining Ca^{2+} transport by the isolated vesicles is outlined in *Protocol 4*. It is based on the original method of Martenosi and Ferteos (48). Radioisotopic calcium flux is monitored after the addition of ATP to start the reaction. At various times during the reaction, an aliquot of the reaction mixture is filtered through Millipore filters, the filter is washed to remove any trapped radioactivity, and the radioactivity of the filter (trapped in the vesicles which are on the filter) is determined through liquid scintillation techniques. This method is easy and relatively straightforward. It also allows you to monitor Ca^{2+} accumulation with a minor amount of protein. (Only 5 μg protein/ml of skeletal muscle SR is required.) This feature is important when the SR is isloated from a small amount of tissue.

Protocol 4. ATP-dependent calcium accumulation

A. *Solutions*

1. *Reaction medium*: 2 × cocktail (80 mM Tris-maleate (pH 6.8), 200 mM KCl, 4 mM $MgCl_2$, and 20 mM NaN_3); 2 mM $^{45}CaCl_2$ (specific activity = 8000 c.p.m./nmol); 100 mM Na_2ATP; 100 mM potassium oxalate; and 2 mg protein/ml of SR.

Protocol 4. *Continued*

2. *ATP regenerating system*: 100 mM phosphoenolpyruvate (PEP) and 1000 U/ml pyruvate kinase (PK); add 20 μl of PEP and 10 μl of PK when ATP regenerating is needed; adjust water.
3. *Wash solution*: 40 mM Tris-maleate (pH 6.8), 100 mM KCl, and 1 mM EGTA. 10 μM $LaCl_3$ can be added if desired.

B. *Procedures*

1. The assay is run in duplicate; a zero (no ATP) time-point is required.
2. Soak the number of Millipore 0.45 μ (Type HA) filters required in distilled water before starting the reaction.
3. In the reaction test tube, add 500 μl of 2 × cocktail, 20 μl of $^{45}CaCl_2$, 50 μl of diluted SR protein, and 0.42 ml distilled water (final volume = 1 ml). Mix and incubate for 5 min at 30 °C.
4. Mix and filter 100 μl of reaction medium (time zero). Wash one time with 5 ml of wash solution.
5. Start reaction by adding 9 μl of ATP. Mix.
6. Filter 100 μl aliquots at each time-point desired and wash one time with 5 ml of wash solution.
7. *Note*: Cardiac SR has the ryanodine receptor (calcium release channel); therefore, 20 μM ruthenium red can be used which will result in a twofold increase in peak calcium accumulation. We do not routinely use ryanodine because the prolonged pre-incubation results in a markedly attenuated calcium accumulation under control (no ryanodine) conditions. Oxalate-facilitated calcium uptake will not be affected by the presence of ruthenium red.
8. *For oxalate-facilitated calcium uptake*: add 500 μl of 2 × cocktail, 100 μl of potassium oxalate, 12.5 μl of diluted SR protein, 278 μl of distilled water, and 100 μl of $^{45}CaCl_2$. Mix well and incubate at 30 °C for 5 min. *Note*: Always add the $^{45}CaCl_2$ last (after the water). Mix well before and after adding the $^{45}CaCl_2$. This step will avoid calcium oxalate precipitation in the reaction cocktail. Also, the magnesium concentration can be raised to 10 mM, which will keep the calcium oxalate soluble.
9. Follow steps 4–6 above.
10. The recommended times for calcium accumulation are 15, 30, 45 sec, 1 min, 2 min, and 5 min. For oxalate-facilitated calcium uptake, the same time-points are recommended. Other time-points can be added or subtracted as needed.
11. *To determine the specific activity*: take 10 μl of the remaining reaction medium, place on a wet filter which is in the scintillation vial, and add 10 ml of scintillation cocktail.

Protocol 4. *Continued*

12. Remove the filters from the filtration apparatus, place in scintillation vials, and add 10 ml of scintillation cocktail. Wait 30 min, mix well, and count in a liquid scintillation counter for 30 sec/vial. (In our hands this is long enough.)

13. *Calculations*

(a) $$\frac{\text{c.p.m. sample} - \text{c.p.m. zero}}{\text{Factor}} = \text{nmol } ^{45}Ca^{2+}\text{/mg protein}$$

(b) Factor = specific activity × milligrams of protein in filtered aliquot

(c) Specific activity = c.p.m./nmol counted (e.g. c.p.m./0.4 nmol for 40 μM $^{45}Ca^{2+}$ in final reaction).

14. Plot nmol $^{45}Ca^{2+}$/mg protein vs. time. Calculate rate based on linear rate of oxalate-facilitated calcium uptake. You cannot calculate a rate when oxalate is not used.

There are some concerns with the filtration method which must be considered. First, the radioactive Ca^{2+} is only a tracer for Ca^{2+} movements in the vesicles. The $^{45}Ca^{2+}$ exchanges for non-radioactive Ca^{2+} in the vesicles (including that bound to the phospholipids), the filters, and glassware. Ca^{2+} also contaminates distilled and deionized water even under the most rigorous conditions. Since you must know the total calcium content of your reaction medium, we recommend taking the proper precautions, such as:

- using disposable test tubes
- never storing the solutions in glass containers
- determining the Ca^{2+} concentrations of your water and reagents with atomic absorption spectroscopic methods.

Another potential problem is the inability to distinguish between Ca^{2+} influx and Ca^{2+} efflux at any given time point during the reaction. Since the trapped $^{45}Ca^{2+}$ is the *net* accumulation, other approaches must be used. The tracer method outlined by Katz's group (49) is a reasonable approach. On the other hand, when the Ca^{2+} precipitating agent, oxalate, is used, the initial rate of Ca^{2+} accumulation (ATP-dependent, oxalate-facilitated Ca^{2+} uptake) is an excellent approximation of the rate of Ca^{2+} influx into the vesicular lumen. The solubility product of Ca^{2+} oxalate ($K_{sp} = 10^{12}\ M^{-1}$) favours the formation of Ca^{2+} oxalate precipitates once in the intravesicular space. (Ca^{2+} and oxalate are co-transported in a 1:1 ratio; see Tate *et al.*, ref. 50.) Although oxalate is 'non-physiological' in the sense that striated muscle does not contain oxalate, it is the model dicarboxylic, co-transported anion when the net rate of Ca^{2+} influx must be optimized. The initial rapid rate of Ca^{2+} translocation can only be demonstrated with spectrophotometric techniques using high protein concentrations or rapid chemical quenching techniques

(51); this rate is independent of a number of factors which impact on the subsequent rate of oxalate-facilitated Ca^{2+} uptake. The rate of oxalate-facilitated Ca^{2+} uptake, which follows the initial, non-oxalate dependent Ca^{2+} uptake, is what is measured with this technique. See Entman *et al.* (52) and Mermier and Hasselbach (53) for a discussion of the two-component model of ATP-dependent Ca^{2+} translocation.

We routinely use both a dual wavelength spectrophotometric method with chelometric dyes and the Ca^{2+}-sensitive electrode for our studies (44). There are problems, however, with non-specific dye absorption in the spectrophotometric method and a slow response time for the Ca^{2+} electrode method. Both of these latter methods also require a relatively high protein concentration in the reaction medium which can be prohibitive. Although the filtration method requires the use of radioactive Ca^{2+}, we believe that it offers the greatest flexibility in terms of assay conditions, modest requirements for protein in the assay, and, importantly, can be used with the colorimetric assay for ATPase activity in the same reaction medium.

The assay conditions for ATP-dependent Ca^{2+} accumulation and ATP hydrolysis are taken from work from our laboratory. We use Tris-maleate buffer because it is an excellent buffer at pH 6.8. This pH is used because the rate of ATP-dependent, oxalate-facilitate Ca^{2+} uptake is optimal at pH 6.8 (50). Since maleate is a dicarboxylic acid, its inclusion in the assay medium (when oxalate is not used) will stimulate the maximal amount of Ca^{2+} which can be accumulated (54). (With oxalate present, though, maleate will have no apparent effect on the observed rate.) We also use 30 °C, mainly because the rates are slower than at 37 °C. This facilitates the observation of net Ca^{2+} influx under a variety of conditions. Spontaneous Ca^{2+}-release, which happens even in SR vesicles which do *not* have the Ca^{2+}-release channel, has a Q_{10} of 2 (55); thus, the lowering of the temperature below the 'physiological' temperature of 37 °C is advisable.

4.2 ATPase activity

Two methods for determining the rate of ATP hydrolysis are presented in *Protocol 5*. The simplest method is the linked enzyme method, where the oxidation of NADH is followed with a continuous recording spectrophotometer. The principle of the assay is as follows:

(a) The Ca^{2+} ATPase hydrolyses ATP, forming ADP and P_i.

(b) The ADP is converted to ATP via the reaction:

$$\text{ADP} + \text{phosphoenolpyruvate} \xrightarrow{\text{pyruvate kinase}} \text{pyruvate} + \text{ATP}$$

(c) The ATP can then be used by the Ca^{2+} ATPase (ATP regenerating system).

(d) The rest of the reaction is as follows:

$$\text{pyruvate} + \text{NADH} \xrightarrow{\text{lactate dehydrogenase}} \text{lactate} + \text{NAD}^+ + \text{H}^+$$

The reaction is monitored at 340 nm using a conventional recording spectrophotometer. The difference between the total (Ca^{2+} + Mg^{2+}) ATPase activity and the Mg^{2+} (EGTA present) ATPase activity is the Ca^{2+}-stimulated ATPase activity.

Although the linked enzyme method is an excellent one, there are several problems with it. First, oxalate cannot be used, because oxalate inhibits the activity of both pyruvate kinase and lactate dehydrogenase. A colorimetric assay (56) for the determination of inorganic phosphate production is required when oxalate is used. The colorimetric procedure used in our laboratory is also given in *Protocol 5*. Secondly, when different perturbations are used, one must ensure that the linked enzyme system is not affected. This is evaluated by rapidly putting in a known amount of ADP into the reaction cuvette at the conclusion of the assay for the SR and running the chart recorder (or whatever recording device is used) at a fast speed. The rate of phosphorylation of ADP to ATP should be the same across all experimental conditions. When it is not, the colorimetric method must be used. Overall, though, we prefer the linked enzyme method because it allows the rapid evaluation of the ATPase activity and the greatest flexibility in terms of perturbing the assay conditions and receiving immediate results.

The colorimetric method has advantages and drawbacks. The advantage is that it is relatively stable against pertubations which might impact on the linked enzyme system. Another excellent advantage is that one can monitor the production of inorganic phosphate along with Ca^{2+} accumulation using the same reaction. This approach minimizes experimental errors and gives interesting insights into the differential effect of perturbations on different aspects of the Ca^{2+} transport process. The primary drawback is that the method is time-consuming and requires individual test tubes for each time-point. It also requires the use of a blank, like the filtration method for Ca^{2+} uptake, for the spontaneous breakdown of ATP. For this reason we use the sodium form of ATP which is stable at −20 °C. Once you have convinced yourself that a single time-point truly represents the linear rate of the reaction, you can use that time-point by doing the assay in duplicate. Our experience is that the first minute of the reaction represents the reaction rate extremely well (within 98%); the duplicates are within 0.005 absorption units when care is taken. We recommend that the ATP regenerating system, which utilizes pyruvate kinase and phosphoenolpyruvate, be used with the colorimetric method and the filtration method for Ca^{2+} uptake in order to ensure constant ATP and to prevent a build-up of ADP which may alter the reverse kinetics of the enzyme cycle. In our hands, the ability of the system to regenerate ATP is not altered by the presence of oxalate. Creatine phosphate and creatine kinase should not be used. Our experience with this ATP regenerating system (C. A. Tate, unpublished results) demonstrated the presence of an active creatine kinase in the SR isolated from skeletal muscle.

Protocol 5. Ca^{2+} ATPase activity: hydrolysis of ATP

A. *Solutions and reagents*

1. *General reaction media*: 2 × cocktail (80 mM Tris-maleate (pH 6.8), 200 mM KCl, 20 mM NaN_3, 4 mM $MgCl_2$); 2 mM $CaCl_2$, 100 mM EGTA, 100 mM Na_2ATP, and 2 mM calcium ionophore A23187 (dissolved in absolute methanol).
2. *Linked enzyme system*: 100 mM phosphoenolpyruvate (PEP); 4 mM NADH; and pyruvate kinase-lactate dehydrogenase (PK/LDH) mixture (Sigma).
3. *ATP regenerating system*: 100 mM PEP and 100 U/ml of pyruvate kinase (Sigma).
4. *Colorimetric system reagents* [method of Lanzetta *et al.* (56)]:
   (a) *Stock solutions*: 0.45% malachite green HCl (stored in the dark); 4.2% ammonium molybdate in 4 N HCl; Sterox (Coleman Instruments); 34% sodium citrate·2 H_2O (w/v).
   (b) *Working reagent*: Mix 3 parts malachite green stock with 1 part ammonium molybdate stock. Stir 20 min. Filter through Whatman No. 5 filter paper. Add 100 μl of Sterox per 5 ml of 3:1 mixture. Mix well. Store in a dark bottle for 1 day's use, then discard.
   (c) *Precaution*: Malachite green will contaminate just about everything. Be cautious in weighing it out, in rinsing well all glassware which comes in contact with it, and in disposing of the working reagent. Absolute methanol provides an excellent cleaning solution, especially for rapid samplers.

Inorganic phosphate (P_i) standards are made from a commercially available phosphate standard (Sigma) which is diluted with distilled water to make working standards from 25 to 200 nmol/ml.

B. *Procedures*

1. *Linked enzyme system* which monitors the oxidation of NADH at 340 nm.
   (a) Use a continuous recording spectrophotometer set at 340 nm and 30 °C. Set the cuvette volume at 2.0 ml. Two cuvettes are required for each reaction.
   (b) In both cuvettes add 1.0 ml cocktail, 0.2 ml NADH, 40 μl PEP, 25 μl of PK/LDH mix, and 0.675 ml glass distilled water. Add 40 μl $CaCl_2$

Protocol 5. *Continued*

to one cuvette and 40 μl EGTA to the other. Stir with a plastic stirring rod.

(c) Add 20 μl of 2 mg/ml cardiac SR or 1 mg/ml skeletal muscle SR. Mix well. (The addition of protein is critical for reliability. The diluted SR should be mixed before each addition, and the pipette tip should be carefully wiped before adding the protein to the mixture.)

(d) Pre-incubate the reaction medium for 5 min in the spectrophotometer before adding 20 μl ATP to start the reaction.

(e) After a constant rate is obtained, add 2.5 μl of A23187 to the ongoing reaction in order to obtain an estimation of the maximal rate under these conditions and to determine the vesicular integrity.

(f) *Calculations*: The reaction rate is determined by multiplying the change in absorption per minute by 8 for cardiac SR (or 20 μg protein/ml) and 16 for skeletal muscle SR (or 10 μg protein/ml). These factors are based on the formula:

$$\frac{\Delta A}{\text{min}} \times \frac{1}{\varepsilon} \times \frac{1}{\text{mg protein/ml}} = \mu\text{mol/mg protein/min}$$

The oxidation of NADH is monitored at 340 nm, which directly indicates the hydrolysis of ATP by the Ca^{2+} ATPase. The extinction coefficient (ε) for NADH at 340 nm is 6.22 mM^{-1}. The SR isolated from cardiac muscle has an active NADH oxidase; therefore you must monitor the reaction prior to the addition of ATP and subtract any significant change in absorption from your results. To obtain the Ca^{2+}-dependent hydrolysis of ATP, subtract the rate in the presence of EGTA from the rate in the presence of Ca^{2+}.

2. *Colorimetric procedure* which monitors the production of inorganic phosphate at a given time-point:

(a) Use a shaking water bath set at 30 °C and disposable test-tubes. The assay should be done in duplicate. Blanks (without protein) should be included for both total (Ca^{2+} + Mg^{2+}) and (Mg^{2+} + EGTA) ATPase activities.

(b) Into a series of tubes (the number dependent on the total number of reaction tubes), pipette 1.6 ml of 'working reagent'. Leave out at room temp. beside the shaking water bath.

(c) Into the reaction test tubes, add 0.5 ml of 2 × cocktail, 20 μl of either $CaCl_2$ or EGTA, 10 μl of diluted protein (2 mg/ml for cardiac SR or 1 mg/ml for skeletal muscle SR), and 0.46 ml of distilled water. Mix. Other reactants may be included as needed. For example, you may want to use a lower ATP concentration. Since you do not want the

Protocol 5. *Continued*

ATP to be rate-limiting, you can use the 'ATP regenerating system' (20 μl of PEP and 10 μl of PK) and reduce the water to 0.43 ml. With 1 mM ATP, though, this step is usually unnecessary. Other reactants, such as A23187 or oxalate, can be used as necessary. However, do not use phosphate as a calcium precipitating agent.

(d) Incubate at 30 °C for 5 min. Add 10 μl of ATP to start the reaction.
(e) At designated times (i.e 1–5 min) take 100 μl from the reaction medium and add to the 'working reagent' tube. Mix.
(f) Wait 30 sec. Add 200 μl sodium citrate to the tube to stop the reaction.
(g) Wait 15–30 min to read the reaction (the colour is stable up to 60 min).
(h) Read samples at 660 nm.
(i) Treat the standards (20–200 nmol P_i/ml) in exactly the same manner *except*; (*i*) time is not a variable; (*ii*) the temperature is unimportant; and (*iii*) the standards may be diluted into water (no cocktail, etc.) as long as a water blank is included.
(j) Plot the standard curve and use *only* the linear portion of the curve for your calculations. When the standard curve does not go through zero, you may have phosphate contamination.
(k) *Calculations*:

$$\frac{A_{\text{sample}} - A_{\text{blank}}}{\text{slope of standard curve}} \times \frac{1}{\text{mg protein/ml}} = \frac{\mu\text{mol P}_i}{\text{mg protein}}$$

(l) Plot the μmol phosphate/mg protein as a function of time. Determine the slope of the line in the linear portion to obtain the rate.

4.3 Phosphoenzyme content

One manual method for determining the phosphoenzyme content of the SR is presented in *Protocol 6*. This method utilizes the transfer of the terminal phosphate of ATP to the enzyme. It is an 'active' method since the Ca^{2+} ATPase molecules must be competent to hydrolyses the bound ATP and transfer the phosphate to aspartate groups on the enzyme. The method is conducted on ice in order to slow down the reaction, so that the burst of Ca^{2+}-dependent phosphorylation can be observed within the first 10–15 sec. As mentioned above, the Ca^{2+}-dependent phosphorylation of the enzyme is followed by the occlusion of Ca^{2+} with subsequent translocation to the intravesicular space. Thus, knowledge of the Ca^{2+}-dependent phosphoenzyme content of the SR is a reasonable estimate of the number of functional transport units in the SR. However, there is a considerable Ca^{2+}-independent, Mg^{2+}-dependent phosphoenzyme in cardiac SR (57) and, under certain conditions, in skeletal muscle SR (58). Therefore, the content of phosphoenzyme must be determined in the presence and absence of Ca^{2+} or EGTA. Additionally, this method cannot be used to derive the rapid kinetics of phosphate transfer from ATP to the enzyme; like Ca^{2+} translocation and phosphate production (see above) there are two phases: a rapid phase followed by a slower, linear phase. Chemical quenching techniques must be used with an apparatus allowing a 5–15 msec time-frame (59). Also, the manual method cannot distinguish accurately between the E_1P-E_2P states of the enzyme.

Besides giving a reasonable estimate of the number of functional transport units (or competent Ca^{2+} ATPase molecules), the enzyme turnover number can be derived by first determining the V_{max} of the Ca^{2+}-dependent ATPase activity and dividing it by the Ca^{2+}-dependent phosphoenzyme (V_{max}/EP). This analysis will allow the determination of whether the functional transport units have a normal rate of enzyme turnover, for example, in the SR isolated from an experimentally manipulated skeletal muscle [see Tate *et al.* (60) for an example of such an analysis].

Protocol 6. Phosphorylation of the Ca^{2+} ATPase by [γ-^{32}P]-ATP

A. *Solutions*

1. *Reaction mixture*: 2 × cocktail (80 mM Tris-maleate (pH 6.8), 200 mM KCl, 20 mM NaN_3, 3 mM $MgCl_2$); 2 mM $CaCl_2$; 50 mM EGTA (pH 6.8); 1 mg SR/ml; and 1 mM [γ-^{32}P]-ATP (specific activity of 100 000–250 000 c.p.m./nmol).
2. *Quenching solution*: 10% TCA.
3. *Washing solution*: 0.1 M perchloric acid and 0.06 M H_3PO_4.

Note: The ATP regenerating system cannot be used because of problems with determinating the specific activity of the radioactivity.

B. *Procedure*

1. The assay is run on ice in triplicate in individual test tubes. The following tubes are required: blank (no SR), plus $CaCl_2$, and minus $CaCl_2$ (EGTA present) for each time-point.
2. To each tube add 250 μl of 2 × cocktail, 20 μl of either 2 mM $CaCl_2$ or 50 mM EGTA, 25 μl of 1 mg SR/ml, and 155 μl of distilled water. For the zero, leave out the SR and increase the water volume to 205 μl.
3. Mix and incubate on ice for at least 5 min.

Protocol 6. *Continued*

4. Start the reaction by adding 25 µl of 1 mM [γ-^{32}P]-ATP; mix quickly; and put back on ice.
5. At the indicated time (10–60 sec), quench the reaction with 0.5 ml of quenching solution. Mix well and let stand on ice for at least 15 min.
6. Always using the same filtration port or single apparatus (e.g., the single Millipore glass filtering apparatus), mix the quenched reaction, and filter the quenched reaction by pipetting with a Pasteur pipette through a Whatman GF/F filter which has been soaked in the washing solution. Be careful to pipette the quenched reaction in the middle of the filter. Using the same Pasteur pipette, wash the tube with 2 × 8 ml of washing solution.
7. Wash the filter with 80 ml of the washing solution.
8. Place the filter in a liquid scintillation vial, add 15 ml of liquid scintillation cocktail, and count.
9. Wash the chimney and base of the filtration port, paying attention to any 'O' ring which might exist. This step is necessary to wash off any trapped radioactivity which may transfer to the next filter.
10. The specific activity is determined by putting 10 µl of 100 µM [γ-^{32}P]-ATP on a filter in a scintillation vial and adding 15 ml of scintillation cocktail. This should be done in triplicate.

C. *Calculations*

1. $$\frac{\text{c.p.m. sample} - \text{c.p.m. blank}}{\text{Factor}} = \frac{\text{nmol } [^{32}\text{P}_\text{i}]}{\text{mg SR protein}}$$

2. Factor = specific activity × mg SR protein in 0.5 ml reaction
3. Specific activity = c.p.m./nmol

Subtract the Mg^{2+} (EGTA present) phosphoenzyme from total (Ca^{2+} + Mg^{2+}) phosphoenzyme to obtain the calcium-stimulated phosphoenzyme content.

D. *Precautions*

The high radioactivity is necessary to get a good signal (especially with cardiac SR where the total [Ca^{2+} + Mg^{2+}] phosphoenzyme content may be approximately 0.8–1.0 nmol/mg SR protein). Thus, the reaction and washing procedures must be performed behind a shield. The specific activity may be decreased with skeletal muscle SR where the Ca^{2+}-dependent phosphoenzyme content is typically 2–6 nmol/mg SR protein. With a molecular weight of 100 000 kd, the theoretical phosphoenzyme content is 10 if all of the protein is the Ca^{2+} ATPase.

5. Skeletal muscle membranes enriched in junctional SR

To study the SR Ca^{2+}-release mechanisms it is necessary to have membranes enriched in the junctional portion of the SR membranes. A number of procedures have been used to obtain membranes enriched in the Ca^{2+}-release protein. These membranes are also enriched in the protein calsequestrin which is concentrated in the lumenal space of the junctional SR. This Ca^{2+} binding protein is largely responsible for the ability of the SR to sequester large amounts of Ca^{2+} (61). In this section we will describe a procedure for obtaining junctional membranes (*Protocol 7*) which produces active Ca^{2+}-release channels upon reconstitution of membrane fractions into bilayers (62). This procedure does not necessarily produce the most highly purified SR membranes. The procedure described in *Protocol 7* is based on the technique of Rosemblatt *et al.* (63). For those requiring purer fractions we suggest the procedure of Saito *et al.* (11), Chu *et al.* (64), or Meissner (65).

Protocol 7. Fractionation of membranes

A. *Solutions*

All buffers and procedures are at 4 °C.

1. Solution I: 0.3 M sucrose, 20 mM Tris-maleate (pH 7.0).
2. Solution II: 0.3 M sucrose, 0.5 M KCl, 10 mM Mops (pH 7.4).
3. Solution III: 0.3 M sucrose, 0.4 M KCl, 10 mM Mops (pH 7.4).
4. Solution IV: 0.3 M sucrose, 0.1 M KCl, 10 mM Mops (pH 7.4).

B. *Protease inhibitors*

Protease inhibitors are present at all stages of the membrane preparation. The following protease inhibitors are used: PMSF, 100 μM: aminobenzamidine, 100 μM; pepstatin A, 1 μM; leupeptin, 1 μg/ml; and aprotinin, 1 μg/ml.

C. *Fractionation procedure*

1. Remove muscle from the hind legs and back strap of New Zealand white rabbits.
2. Mince 250 g of frozen or fresh tissue with scissors into small (~ 0.5 cm) pieces keeping tissue as cold as possible during the process.
3. Add 250 ml of cold Solution I with protease inhibitors.
4. Homogenize the tissue for 4 × 15 sec at low speed in a Waring blender.
5. Add 500 ml of cold Solution I.
6. Homogenize again for 2 × 30 sec at high speed.

Protocol 7. *Continued*

7. Centrifuge the homogenate at 3000 *g* for 30 min in a Beckman J-21 centrifuge using a JA-14 rotor.
8. Filter the supernatant through cheesecloth.
9. Centrifuge the filtered supernatant at 10 000 *g* for 30 min.
10. Filter this second supernatant again through cheesecloth.
11. Add KCl to the combined filtered supernatants to a final concentration of 0.5 M.
12. Stir for 1 h at 4 °C.
13. Centrifuge for 45 min at 130 000 *g* in a 45 Ti rotor.
14. Resuspend the pellet from this centrifugation in 400 ml of Solution II, using a Ten Broeck homogenizer.
15. Centrifuge for 30 min at 130 000 *g*.
16. Resuspend this microsomal pellet in 10–12 ml of Solution III.
17. Layer on sucrose gradients consisting of 7 ml steps of 20, 25, 30, 35, and 40% (w/w) sucrose in Solution II. (Mark tubes on the outside at positions of the interfaces.)
18. Centrifuge for 14 h at 115 000 *g* on an SW28 rotor.
19. Using a Pasteur pipette, slowly remove, separate, and save material above and at the interfaces.
20. Dilute separated material with 10 mM Mops (pH 7.4) until sucrose is below 10%.
21. Centrifuge at 130 000 *g* for 30 min.
22. Pellets are resuspended in Solution IV with protease inhibitors.
23. T-tubule enriched fractions are obtained from the 10–20% sucrose interface, longitudinal SR enriched fractions are obtained at the 30–35% sucrose interface and junctional SR enriched membranes are obtained from the pellet. The 35–40% interface contains both longitudinal and junctional membranes.

D. *Characterization of membranes*

1. *Assay for T-tubule membranes.* To determine whether a fraction contains T-tubule membranes we assay for the ability to bind [^{3}H]-PN 200–110, a dihydropyridine. Membranes (2–20 μg of protein) are incubated with 2 nM [^{3}H]-PN 200–110 (85 Ci/mmol) in 1 ml of 50 mM Mops (pH 7.4) for 1–2 h at room temp. in the dark. Non-specific binding is determined in the presence of 1 μM unlabelled nitrendipine. Bound radioligand is separated from free by rapid filtration through Whatman GF/F filters. Each filter is immediately washed with 5 × 5 ml of ice-cold

Protocol 7. *Continued*

water. The filters are counted in 10 ml of Beckman Ready Protein scintillant. **Note**: High ionic strength inhibits binding of [^{3}H]-PN 200–110 [Hamilton *et al.* (62)] and should, therefore, be avoided in this assay.

2. Assay for longitudinal SR membranes. The Ca^{2+} ATPase assay *Protocol 5* is used to determined if membranes derived from the longitudinal tubules are present.
3. Assay for junctional tubule membranes. The [^{3}H]-ryanodine binding assay described below *Protocol 8* is used to assay for the presence of membranes derived from the junctional SR.

6. The Ca^{2+}-release channel of the junctional SR

6.1 Properties

6.1.1 Identification of the ryanodine binding protein as the 'foot' protein and the Ca^{2+}-release channel

The T-tubule and SR membranes of skeletal muscle are closely juxtaposed with structures designated 'feet' between them (66, 67). The feet are tetragonal structures composed of four identical subunits and are aligned in rows with a centre to centre distance of 32 nm (66). A protein which is morphologically similar to the feet structures of the triads has been isolated and characterized (27, 28, 29). This protein binds the plant alkaloid, ryanodine, with high affinity, has a subunit molecular weight of 350–450 K and in the native state appears to be a tetramer. Upon reconstitution into bilayers this protein produces Ca^{2+} channel activity similar to that of the Ca^{2+}-release channel reconstituted from native SR membranes (27, 29, 68). The apparent close physical association of the t-tubule and SR membranes has led to the prosposal that the DHPBP and the RBP are coupled and that E–C coupling occurs via a mechanical interaction between the two proteins (69).

6.1.2 Modulation of activity

The activity of the Ca^{2+}-release channel of the SR has been shown to be altered by compounds such as Ca^{2+}, Mg^{2+}, adenine nucleotides, ryanodine, ruthenium red, caffeine (70, 71, 72), and dantrolene (73), and by proteins such as calmodulin (74) and annexin VI (75). Some of these compounds (Ca^{2+}, adenine nucleotides, and caffeine) enhance the activity of the Ca^{2+}-release channel while others (Mg^{2+}, ruthenium red, and calmodulin) decrease its activity.

6.1.3 Radioligand binding

The binding of [^{3}H]-ryanodine to SR membranes is also altered by the agents mentioned above. We have shown that most of these modulators alter [^{3}H]-ryanodine binding by changing the apparent association rate-constant (76). Activators of the channel increase the apparent affinity of the binding site for [^{3}H]-ryanodine by increasing the rate of association while inhibitors of the channel, such as high concentrations of Ca^{2+} and Mg^{2+}, slow the rate of association. In contrast, two inhibitors, ryanodine (at concentrations above 1 μM) and ruthenium red, slow the rate at which bound [^{3}H]-ryanodine dissociates from the high-affinity binding site. These studies suggest that information about the functional state of the channel can be obtained from the kinetics of [^{3}H]-ryanodine binding. *Protocol 8* describes the techniques for both equilibrium and kinetic binding analyses.

Many factors influence [^{3}H]-ryanodine binding. For the protocols described in *Protocol 8* we have chosen conditions to optimize the binding. It should be noted, however, that these conditions may not be the appropriate ones to answer all questions being posed. For example, there is very little difference in the binding of [^{3}H]-ryanodine to membranes from Malignant Hyperthermic pigs compared to controls when the binding is performed in the buffers described in *Protocol 8*; however, significant differences in the K_d are obtained if the binding is performed in the absence of AMP-PCP (Hamilton *et al*, in preparation).

Protocol 8. [^{3}H]-ryanodine binding assays

A. *Solution I*

The solution which we use for high affinity binding of [^{3}H]-ryanodine is: 1 mM AMP-PCP, 300 mM KCl, 100 μM $CaCl_2$, 100 μg/ml BSA, 20 mM Mops (pH 7.0).

B. *Procedures*

1. Equilibrium binding

(a) Use a total assay volume of 250 μl. Add the volume of Solution I which, after other additions, will give 250 μl. 12 × 75 mm glass tubes are convenient for this assay.

(b) For high-affinity binding use concentrations of [^{3}H]-ryanodine of 10 nM, 5 nM, 2.5 nM, 1.25 nM, 0.625 nM, 0.312 nM, 0.156 nM, and 0.078 nM for generation of data for a Scatchard plot. These are prepared by 1:1 serial dilutions of 10 × concentrated stock solution. Add 25 μl aliquots of each 10 × stock solution to six tubes. There will be a total of 48 tubes for each binding assay.

(c) Determine non-specific binding by adding 10 μM unlabelled ryanodine (5 μl of a 0.5 mM solution in ethanol) to three tubes at each

Protocol 8. *Continued*

radioligand concentration from step (b). Add 5 μl of ethanol to the other three tubes at each concentration.

(d) Add the SR membranes. The amount of protein to be added depends on how pure the membranes are. With crude membranes, we suggest about 30 μg while more highly purified membranes may require as little as 5 μg.

(e) Incubate for 8–10 h at room temp. or 3 h at 37 °C. The rate of association of [^{3}H]-ryanodine at room temp. is slow (in Solution I, $k_1 = 0.0010 \pm 0.0004$ nM^{-1} min^{-1}, $n = 9$) and, therefore, long periods of time are required to reach equilibrium. The time to equilibrium should be tested for each new buffer condition as it is slower in some other buffers (75). Extremely long incubation times (>20 h at room temp. or >4 h at 37 °C) result in loss in binding, presumably due to proteolysis. For this reason, it is advisable to include protease inhibitors in the incubation buffers (see below for types and concentrations).

(f) Bound radioligand is separated from free by rapid filtration through Whatman GF/F filters. Immediately after the sample is filtered, wash the filters with 5 × 5 ml of ice-cold 0.3 M KCl. Dissociation is rapid in distilled water and, therefore, water washes should be avoided.

(g) Place filters in 5 ml of Beckman Ready Protein Scintillant and shake for 1 h prior to counting.

2. Association

(a) Equilibrate membranes in the same buffer to be used for association by incubating overnight at 4 °C. SR membranes are used at a protein concentration of 50 μg/ml with a total volume of 4 ml per tube.

(b) To start the association, at time 0 and while vortexing, add [^{3}H]-ryanodine to a final concentration of 5–20 nM. *Note*: The amount of [^{3}H]-ryanodine bound at equilibrium must be less that 10% of the added [^{3}H]-ryanodine to assume pseudo first-order kinetics.

(c) At 10 min intervals, filter a 200 μl aliquot through a Whatman GF/F filter and immediately wash with 5 × 5 ml of cold 0.3 M KCl.

(d) After 1–2 h of 10 min filtrations, filter every 30 or 60 min.

(e) Equilibrium will be reached by about 8–10 h.

(f) Prepare filters for counting as described above.

3. Dissocation

(a) Incubate membranes (2–10 mg/ml) for 4–10 h at room temp. with [^{3}H]-ryanodine (20–40 nM) in the buffer to be used for dissociation.

(b) Initiate dissociation by dilution of 10 μl of the sample into 4 ml of

Protocol 8. *Continued*

buffer. The final [^{3}H]-ryanodine concentration should be less than 0.05 nM to avoid rebinding.

(c) Filter 200 μl aliquots at 1–10 min intervals for the first hour and then at 1 h intervals thereafter. To ensure complete dissociation, filter aliquots at 24 and 36 h. The first aliquot should be filtered at 10–15 sec after dilution.

(d) Prepare filters and count as previously described.

(e) Because high concentrations of ryanodine (greater than 10 μM) slow the dissociation of [^{3}H]-ryanodine from the high affinity site, initiation of dissociation by addition of excess ryanodine is not recommended unless this apparently co-operative phenomenon is being studied.

C. *Data analysis*

The binding data can be fit to a one- or two-site model based on a non-linear least-squares Marquart method (77) to estimate the free parameters, B_{max} and K_d. Parameter uncertainties are provided by χ^2 values. Scatchard analysis of single-site binding data is fit with a least-squares linear regression program. Association kinetics are analysed by linear regression of pseudo first-order plots of ln ($B_{eq}/B_{eq}-B$) vs. time to obtain k_{obs}. The association rate-constants are determined from the slope (when the plot is linear) of a plot of k_{obs} as a function of ligand concentration. Dissociation kinetics are analysed by linear regression of the plot of ln B_t/B_0 vs. time, where B_t is the amount bound at time t and B_0 is the amount bound at time zero. Alternatively, data can be computer-fit as the sum of descending exponentials. Excellent discussions of the theory of radioligand binding are presented by Weiland (78) and Molinoff *et al.* (79) and by Limbird (80) and Chapter 11, this volume.

6.2 Purification

A number of procedures for the purification of the ryanodine binding protein have been published (26–29). The procedure described in *Protocol 9* was devised in our laboratory to obtain large quantities of purified binding protein in a relatively short space of time (81). Because digitonin frequently interferes with the reconstitution of proteins into the bilayer, the final step in this procedure involves the sedimentation of the digitonin solubilized preparation into sucrose gradients containing the detergent Chaps. Using this procedure we routinely obtain purified ryanodine binding protein which retains most of the functional characteristics observed with the membrane bound protein.

Protocol 9. Purification of the [^{3}H]-ryanodine binding protein

A. *Solutions*

1. Solution I: 6 mM Na_2ATP, 520 μM $CaCl_2$, 1M NaCl, 50 mM Mops (pH 7.0).
2. Solution II: 0.1% digitonin, 10 mM Mops (pH 7.4).
3. Solution III:0.4% Chaps, 10 mM Mops (pH 7.4).

B. *Protease inhibitors*

Protease inhibitors are present at all stages of the purification at the following concentrations: 200 μM PMSF, 200 μM aminobenzamidine, 2 mM DTT, 2 μg/ml leupeptin, 2 μg/ml aprotinin, 2 μg/ml pepstatin A, and 10 μg/ml soybean trypsin inhibitor.

C. *Procedure*

All steps are carried out at 4 °C and in the presence of protease inhibitors.

1. Using junctional SR enriched membranes and a protein concentration of 3 mg/ml in buffer I, label part of the ryanodine binding protein by incubating for 1 h at room temp. with 5 nM [^{3}H]-ryanodine.
2. To slow the dissociation of the bound [^{3}H]-ryanodine, add unlabelled ryanodine to 10 μM (Hawkes *et al.* (81)). This step is *not* recommended if the preparation is to be used for reconstitution.
3. Add digitonin to a final concentration of 1%.
4. Incubate on ice for 30 min.
5. Remove insoluble material by centrifuging for 30 min at 120 000 *g* for 30 min.
6. Dilute supernatant with 10 mM Mops (pH 7.4) to a final NaCl concentration of less than 0.1 M.
7. Mix the diluted supernatant with 20 ml of DEAE-trisacryl (which has been extensively washed with solution II).
8. Shake gently for 30 min.
9. Wash resin on a fitted glass filter with 200 ml of 50 mM NaCl in Solution II.
10. Elute with 40 ml of 100 mM NaCl in Solution II. Most of the Ca^{2+} ATPase present will elute at this step.
11. Elute with 40 ml of 200 mM NaCl/Solution II. Most of the DHP binding protein will elute at this step.
12. Elute with 40 ml of 300 mM NaCl/Solution II. Most of the ryanodine binding protein will elute at this step.

Protocol 9. *Continued*

13. Elute with 500 mM NaCl/Solution II. Most of the calsequestrin will elute at this step.

14. Layer the 300 mM NaCl eluate on to two 35 ml 5–20% sucrose gradients (prepared with buffer III).

15. Centrifuge in a 50 VTi rotor for 90 min at 210 000 *g* or a SW28 rotor for 15 h at 115 000 *g*.

16. Collect 2 ml fractions and count an aliquot to determine the position of the ryanodine binding protein (close to the bottom of the gradient).

17. Store purified protein (approx. 3 mg from 230 mg of starting junctional SR enriched membranes) in liquid N_2 in the presence of 2 mM DTT.

6.3 Reconstitution

Both the membrane bound and the purified ryanodine binding protein can be reconstituted into bilayers to produce Ca^{2+} channels with the reported (27, 29, 68) properties of the Ca^{2+}-release channel. The bilayer technique has been widely used to study this channel (82). *Protocol 10* describes this technique.

Protocol 10. Reconstitution of Ca^{2+}-release channel activity into the bilayer

A. *Equipment*

1. *Bilayer apparatus*: The bilayer apparatus used in our laboratory is essentially identical to that described by Hamilton *et al.* (62).

(a) Salt agar bridges equilibrated in the solutions present in each chamber are used to connect the chambers to Ag–AgCl electrodes immersed in 2 M KCl. The Ag–AgCl electrodes are connected to the appropriate inputs of the amplifier.

(b) The small chamber of the bilayer set-up is the chamber to which membranes are added; it is referred to as the *cis* chamber and acts as the voltage control side. The other chamber is the *trans* chamber, and is referred to as the ground.

2. Computer and software for collection and analysis of data: IBM AT286/386 or compatible computer with the following programs: pCLAMP, IPROC, TRANSIT.

B. *Phospholipids*

We have successfully reconstituted the Ca^{2+}-release channel using both natural lipids (brain, heart, or egg) and synthetic lipids. Identical ratios of the different phospholipids were used with both and, therefore, we will describe the procedure using only the synthetic lipids.

Protocol 10. *Continued*

The lipids are obtained in chloroform from Avanti Polar Lipids (Pelham, Alabama, USA). Chloroform is removed by a stream of N_2 and the lipids are resuspended in decane to a total phospholipid concentration of 50 mg/ml and a ratio of 5:3:2 of palmitoyl-oleoyl-phosphatidylethanolamine(POPE): palmitoyl-oleoyl-phosphatidylserine (POPS):palmitoyl-oleoyl–phosphatidyl-choline(POPC).

C. *Formation of bilayer*

1. After cleaning and thoroughly drying the bilayer chamber, place a small drop of the phospholipid solution in decane on the aperture and allow to dry for 5–10 min.
2. Assemble chamber and add agar bridges.
3. Add 25 mM Cs_2SO_4, 10 mM Mops (pH 7.0), 8 μM $CaCl_2$ to each chamber. In our setup, the volumes are 500 μl to the *cis* chamber and 4 ml to the *trans*.
4. Using a micropipette which has been sealed and fire-polished at the end, paint a small amount of lipid across the aperture.
5. Repeat this step until a bilayer is formed.
6. While stirring, establish the ionic gradient by adding Cs_2SO_4 to a final concentration in the *cis* chamber of 125 mM.
7. Remove excess lipid from the aperture by gently moving the glass rod across the opening until the capacitance of the bilayer reaches a value of 150–300 pF.
8. We have successfully used two methods to insert channels from the membranes into the bilayer. The first method is to add a small aliquot of the membranes to the *cis* chamber and wait (stirring frequently) for a fusion event. The second procedure was developed by Dr Enrico Stefani (Baylor College of Medicine) and involves dipping the glass rod into a solution of membranes (at a protein concentration less than 2 mg/ml) and gently painting across the aperture. The latter procedure frequently results in multiple incorporations.
9. The records of the Ca^{2+}-release channel activity are filtered at a frequency of 2.5 kHz.

6.3.1 Analysis of channel activity

Single-channel analysis can be performed with one of several commercially available software programs, including pCLAMP from Axon Instruments Inc., and IPROC from Sachs & Neil. These programs can be used for curve fitting and for analysing the data with respect to dwell-times, amplitudes, latency, and frequency.

A new program TRANSIT will soon be available (contact Dr Tony VanDongen, Department of Molecular Physiology and Biophysics, Baylor College of Medicine, Houston, TX 77030, USA, for availability). TRANSIT is a single-channel analysis environment running on IBM AT286/386 compatible personal computers. It employs a new algorithm for idealization of raw single-channel data that differs in a few important aspects from the more classical algorithms that use amplitude thresholds for the detection of openings. The idealization algorithm used by TRANSIT uses a *slope* threshold to detect transitions between different levels (closed and open conformations). After detecting the transitions, a relative amplitude threshold is used to ensure that all neighbouring levels have a minimal amplitude difference. This second criterion is needed to eliminate spurious transitions which can originate from the noise in the data. To minimize subjectivity, both thresholds are derived from the baseline noise of the data being studied. Therefore, TRANSIT does not require that the user sets a threshold, although both slope and relative amplitude threshold can be modified if the idealization is not satisfactorily. As a result, no a priori knowledge about the channels being analysed is required. Another major advantage of the approach is the ability to reliably idealize single-channel data which contain multiple amplitude levels. Multiple levels can arise when more than one channel is active, or when the channel being studied has multiple (sub)conductance states. There is no limit to the number of different levels that can be analysed.

The idealization converts the raw data into a sequence of time-intervals with a particular amplitude and duration. The idealized data are then used to calculate the single-channel statistics such as the distributions of open times, closed times, burst duration, number of openings per burst, latency until first opening, and single-channel amplitude. In addition, one can calculate diaries, an ensemble average, and various scatter plots. Event filtering can be employed on the idealized data to extract single-channel events with particular properties in the presence of contaminating activity of other channels.

7. Concluding remarks

Our intent in this chapter is to provide relatively straightforward methods to study the proteins involved in the uptake and release of Ca^{2+} from the SR. Certainly there are numerous procedures in the literature which also will provide a satisfactory means of achieving the same goals. The methods described in this chapter are the ones used in our laboratories. For an excellent overview of methods for characterizing other SR processes not covered in this chapter, we suggest the monograph edited by Entman and VanWinkle (83).

References

1. Frank, G. (1982). *Canadian Journal of Physiology and Pharmacology*, **60**, 427–39.
2. Fabiato, A. (1983). *American Journal of Physiology*, **245**, C1–14.
3. Vergara, J., Tsien, R., and Delay, M. (1985). *Proceedings of the National Academy of Sciences, USA*, **82**, 6352–6356.
4. Volpe, P., Salviati, G., DiVirgilio, F., and Pozzan, T. (1985). *Nature*, **316**, 347–9.
5. Schneider, M. and Chandler, W. (1973). *Nature*, **242**, 244–6.
6. Peachy, L. and Franzini-Armstrong, C. (1983). In *Handbook of physiology. Skeletal muscle*, pp. 23–71. American Physiology Society, Bethesda, Maryland.
7. Rios, E. and Brum, G. (1987). *Nature*, **325**, 717–20.
8. MacLennan, D. H. (1970). *Journal of Biological Chemistry*, **245**, 4508–18.
9. Meissner, G., Conner, G., and Fleischer, S. (1973). *Biochimica et Biophysica Acta*, **298**, 246–69.
10. Jorgensen, A. O., Shen, A. C., MacLennan, D. H., and Tokuyasu, K. T. (1982). *Journal of Cell Biology*, **92**, 409–16.
11. Saito, A., Seiler, S., Chu, A. and Fleischer, S. (1984). *Journal of Cell Biology*, **99**, 875–85.
12. Chu, A., Saito, A., and Fleischer, S. (1987). *Archives of Biochemistry and Biophysics*, **258**, 13–23.
13. MacLennan, D. H., Brandl, C. J., Korczak, B., and Green, N. M. (1985). *Nature*, **316**, 696–700.
14. Brandl, C. J., Green, N. M., Korczak, B., and MacLennan, D. H. (1986). *Cell*, **44**, 597–607.
15. Brandl, C. J., deLeon, S., Martin, D. R., and MacLennan, D. H. (1987). *Journal of Biological Chemistry*, **262**, 3768–74.
16. Michalak, M., Campbell, K. P., and MacLennan, D. H. (1980). *Journal of Biological Chemistry*, **255**, 1317–1326.
17. Campbell, K. P. and MacLennan, D. H. (1981). *Journal of Biological Chemistry*, **256**, 4626–4632.
18. Leberer, E., Charuk, J., Green, N. M., and MacLennan, D. H. (1989). *Proceedings of the National Academy of Sciences, USA*, **86**, 6047–51.
19. Leberer, E., Timms, B. G., Campbell, K. P., and MacLennan, D. H. (1990). *Journal of Biological Chemistry*, **265**, 10118–24.
20. Meissner, G. (1975). *Biochimica et Biophysica Acta*, **389**, 51–68.
21. Chiesi, M. and Carafoli, E. (1982). *Journal of Biological Chemistry*, **257**, 984–91.
22. Leonards, K. S. and Kutchai, H. (1985). *Biochemistry*, **24**, 4876–84.
23. Kutchai, H. and Campbell, K. P. (1989). *Biochemistry*, **28**, 4830–9.
24. Leberer, E., Charuk, J. H., Clarke, D. M., Green, N. M., Zubrzycka-Gaarn, E., and MacLennan, D. H. (1989). *Journal of Biological Chemistry*, **264**, 3484–3493.
25. Smith, J. S., Coronado, R., and Meissner, G. (1985). *Nature*, **316**, 446–9.
26. Pessah, I. P., Francini, A. O., Scales, D. J., Waterhouse, A. L., and Casida, J. E. (1986). *Journal of Biological Chemistry*, **261**, 8643–8.
27. Imagawa, T., Smith, J. S., Coronado, R., and Campbell, K. P. (1987). *Journal of Biological Chemistry*, **262**, 16636–43.
28. Inui, M., Saito, A., and Fleischer, S. (1987). *Journal of Biological Chemistry*, **262**, 1740–7.

29. Lai, F. A., Erickson, H. P., Rousseau, E., Liu, Q.-Y., and Meissner, G. (1988). *Nature*, **331**, 315–19.
30. Meissner, G. (1986). *Journal of Biological Chemistry*, **261**, 6300–6.
31. Takeshima, H., Nishimura, S., Matsumoto, T., Ishida, H., Kangawa, K., Minamino, N., Matsuo, H., Ueda, M., Hanaoka, M., Hirose, T., and Numa, S. (1989). *Nature*, **339**, 439–45.
32. MacLennan, D. H. and Wong, P. T. (1971). *Proceedings of the National Academy of Sciences, USA*, **68**, 1231–5.
33. Somlyo, A. V., Gonzalez-Serratos, H., Shuman, H., McClellan, G., and Somlyo, A. P. (1981). *Journal of Cell Biology*, **90**, 577–94.
34. Fliegel, L., Ohnishi, M., Carpenter, M. R., Khanna, V. K., Reithmeier, R. A., and MacLennan, D. H. (1987). *Proceedings of the National Academy of Sciences, USA*, **84**, 1167–71.
35. Zarain-Herzberg, A., Fliegel, L., and MacLennan, D. H. (1988). *Journal of Biological Chemistry*, **263**, 4807–12.
36. Zorzato, F., Margreth, A., and Volpe, P. (1986). *Journal of Biological Chemistry*, **261**, 13252–7.
37. Zorzato, F. and Volpe, P. (1988). *Archives of Biochemistry and Biophysics*, **261**, 324–9.
38. Hofmann, S. L., Brown, M. S., Lee, E., Pathak, R. K., Anderson, R. G., and Goldstein, J. L. (1989). *Journal of Biological Chemistry*, **264**, 8260–70.
39. Hofmann, S. L., Goldstein, J. L., Orth, K., Moomaw, C. R., Slaughter, C. A., and Brown, M. S. (1989). *Journal of Biological Chemistry*, **264**, 18083–90.
40. Zaidi, N., Lagenaur, C. F., Hilkert, R. J., Xiong, H., Abramson, J. J., and Salama, G. (1989). *Journal of Biological Chemistry*, **264**, 21737–47.
41. Ostwald, T. J. and MacLennan, D. H. (1974). *Journal of Biological Chemistry*, **249**, 974–9.
42. Fliegel, L., Burns, K., MacLennan, D. H., Reithmeier, R. A., and Michalak, M. (1989). *Journal of Biological Chemistry*, **264**, 21522–8.
43. VanWinkle, W. B., Bick, R. J., Tucker, D. E., Tate, C. A., and Entman, M. L. (1982). *Journal of Biological Chemistry*, **257**, 11689–95.
44. Tate, C. A., Bick, R. J., Chu, A., VanWinkle, W. B., and Entman, M. L. (1985). *Journal of Biological Chemistry*, **260**, 9618–23.
45. Jones, L. R. and Cala, S. E. (1981). *Journal of Biological Chemistry*, **256**, 11809–18.
46. Chu, A. Bick, R. J., Tate, C. A., VanWinkle, W. B., and Entman, M. L. (1983). *Journal of Biological Chemistry*, **258**, 10543–50.
47. Hasselbach, W. and Oetliker, H. (1983). *Annual Review of Physiology*, **45**, 325–339.
48. Martonosi, A. and Feretos, R. (1964). *Journal of Biological Chemistry*, **239**, 648–58.
49. Takenaka, H., Adler, P. N., and Katz, A. M. (1982). *Journal of Biological Chemistry*, **257**, 12649–56.
50. Tate, C. A., Chu, A., McMillin-Wood, J., VanWinkle, W. B., and Entman, M. L. (1981). *Journal of Biological Chemistry*, **256**, 2934–9.
51. Chiesi, M. and Inesi, G. (1979). *Journal of Biological Chemistry*, **254**, 10370–7.
52. Entman, M. L., Snow, T. R., Freed, D., and Schwartz, A. (1973). *Journal of Biological Chemistry*, **248**, 7762–72.

53. Mermier, P., and Hasselbach, W. (1975). *Zeitschrift für Naturforschung*, **30**, 593.
54. Chu, A., Tate, C. A., Bick, R. J., VanWinkle, W. B., and Entman, M. L. (1983). *Journal of Biological Chemistry*, **258**, 1656–64.
55. Entman, M. L., Bornet, E. P., and Schwartz, A. (1972). *Journal of Molecular and Cellular Cardiology*, **4**, 155–169.
56. Lanzetta, P. A., Alvarez, L. J., Reinach, P. S., and Candia, O. A., (1979). *Analytical Biochemistry*, **100**, 95–7.
57. Tate, C. A., Bick, R. J., Blaylock, S. L., Youker, K. A., Scherer, N. M., and Entman, M. L. (1989). *Journal of Biological Chemistry*, **264**, 7809–13.
58. Carvalho-Alves, P. C. and Scofano, H. M. (1987). *Journal of Biological Chemistry*, **262**, 6610–14.
59. Bick, R. J., VanWinkle, W. B., Tate, C. A., and Entman, M. L. (1983). *Journal of Biological Chemistry*, **258**, 4447–52.
60. Tate, C. A., Bick, R. J., Myers, T. D., Pitts, B. J. R., VanWinkle, W. B., and Entman, M. L. (1983). *Biochemical Journal*, **210**, 339–44.
61. Ikemoto, N., Nagy, B., Bhantnagar, G. M., and Gergely, J. (1974). *Journal of Biological Chemistry*, **249**, 2357–65.
62. Hamilton, S. L., Mejia-Alvarez, R., Fill, M., Hawkes, M. J., Brush, K. L., Schilling, W. P., and Stefani, E. (1989). *Analytical Biochemistry*, **183**, 31–41.
63. Rosemblatt, M., Hidalgo, C., Vergara, C., and Ikemoto, N. (1981). *Journal of Biological Chemistry*, **256**, 8140–8.
64. Chu, A., Dixon, M. C., Saito, A., Seiler, S., and Fleischer, S. (1988). In *Methods in enzymology* , Vol. 157 (ed. S. Fleischer and B. Fleischer), pp. 30–46. Academic Press, Orlando, Florida.
65. Meissner, G. (1984). *Journal of Biological Chemistry*, **259**, 2365–74.
66. Ferguson, D., Schwartz, H., and Franzini-Armstrong, C. (1984). *Journal of Cell Biology*, **99**, 1735–42.
67. Saito, A., Inui, M., Radesmacher, M., Frank, J., and Fleischer, S. (1988). *Journal of Cell Biology*, **107**, 211–19.
68. Hymel, L., Inui, M., Fleischer, S., and Schindler, H. (1988). *Proceedings of the National Academy of Sciences, USA*, **85**, 441–5.
69. Block, B., Imagawa, T., Campbell, K., and Franzini-Armstrong, C. (1988). *Journal of Cell Biology*, **107**, 2587–600.
70. Meissner, G., Darling, E., and Eveleth, J. (1986). *Biochemistry*, **25**, 236–44.
71. Pessah, I. N., Stambuk, R. A., and Casida, J. E. (1987). *Molecular Pharmacology*, **31**, 232–8.
72. Fleischer, S. and Inui, M. (1989). *Annual Review of Biophysics and Biophysical Chemistry*, **18**, 333–364.
73. Danko, S., Kim, D., Streter, F., and Ikemoto, N. (1985). *Biochimica et Biophysica Acta*, **816**, 18–24.
74. Smith, J. S., Rousseau, E., and Meissner, G. (1989). *Circulation Research*, **64**, 352–9.
75. Diaz-Muñoz, M., Kaetzel, M., Harzarika, P., Dedman, J., and Hamilton, S. L. (1990). *Journal of Biological Chemistry*, **265**, 15894–9.
76. Chu, A., Diaz-Muñoz, M., Hawkes, M. J., Brush, K., and Hamilton, S. L. (1990). *Molecular Pharmacology*, **37**, 735–41.
77. Marquardt, D. (1963). *SIAM Journal of Applied Mathematics*, **11**, 430–41.
78. Weiland, G. and Molinoff, P. (1981). *Life Sciences*, **29**, 313–30.

79. Molinoff, R., Wolfe, B., and Weiland, G. (1981). *Life Sciences*, **29**, 427–43.
80. Limbird, L. E. (1986). *Cell surface receptors: A short course on theory and methods.*, Martinus Nijhoff, Boston, Mass.
81. Hawkes, M. J., Diaz-Muñoz, M., and Hamilton, S. L. (1989). *Membrane Biochemistry*, **8**, 133–45.
82. Alvarez, O. (1986). In *Ion channel reconstitution* (ed. C. Miller;, pp. 115–30. Plenum Press, New York.
83. Entman, M. L. and VanWinkle, W. B. (ed.) (1986). *Sarcoplasmic reticulum in muscle physiology*, Vol. I and II CRC Press, Boca Raton, Florida.

15

The role and study of mammalian mitochondrial Ca^{2+}-transport and matrix Ca^{2+}

JAMES G. McCORMACK and MARTIN CROMPTON

1. Introduction

1.1 The mitochondrial Ca^{2+}-transport system and its effectors

The inner membrane of mitochondria of mammalian, and perhaps of all vertebrate, tissues contains Ca^{2+} transport systems that mediate continuous Ca^{2+} cycling across the membrane (*Figure 1*; see refs 1–3 for reviews). The cycle comprises a Ca^{2+} uniporter which allows passive Ca^{2+} entry down the Ca^{2+} electrochemical gradient and one, or possibly two, active egress mechanisms. The principal efflux mechanism involves obligatory exchange between Ca^{2+} and Na^+ with the probable stoichiometry of $2Na^+$: $1Ca^{2+}$; this exchange is linked to the chemiosmotic H^+ circuit via Na^+:H^+ exchange. Mitochondria may also contain a Na^+-independent mechanism for Ca^{2+} efflux, although whether this involves direct Ca^{2+}: H^+ exchange is controversial. Although cycling of Ca^{2+} involves the dissipation of the protonmotive gradient, under normal cellular conditions the calculated overall energy cost is less than 1% of the respiratory capacity of the mitochondria (1, 2).

Ca^{2+} uptake can be inhibited physiologically by Mg^{2+} and activated by spermine, and inhibited artificially by ruthenium red (*Figure 1*). The Na^+-dependent egress mechanism can be inhibited physiologically by increases in extramitochondrial Ca^{2+} and also Mg^{2+}, and artificially by drugs like diltiazem which are more normally used as blockers of the plasma membrane Ca^{2+} channel (*Figure 1*) (1, 2). Overall, the uptake pathway has around tenfold more potential capacity than the egress mechanisms (1, 2).

1.2 The role of mitochondrial Ca^{2+} transport

1.2.1 Under normal cellular conditions

The primary physiological function of this Ca^{2+}-transport system in normal healthy cells is now widely regarded as being in the regulation of matrix Ca^{2+}

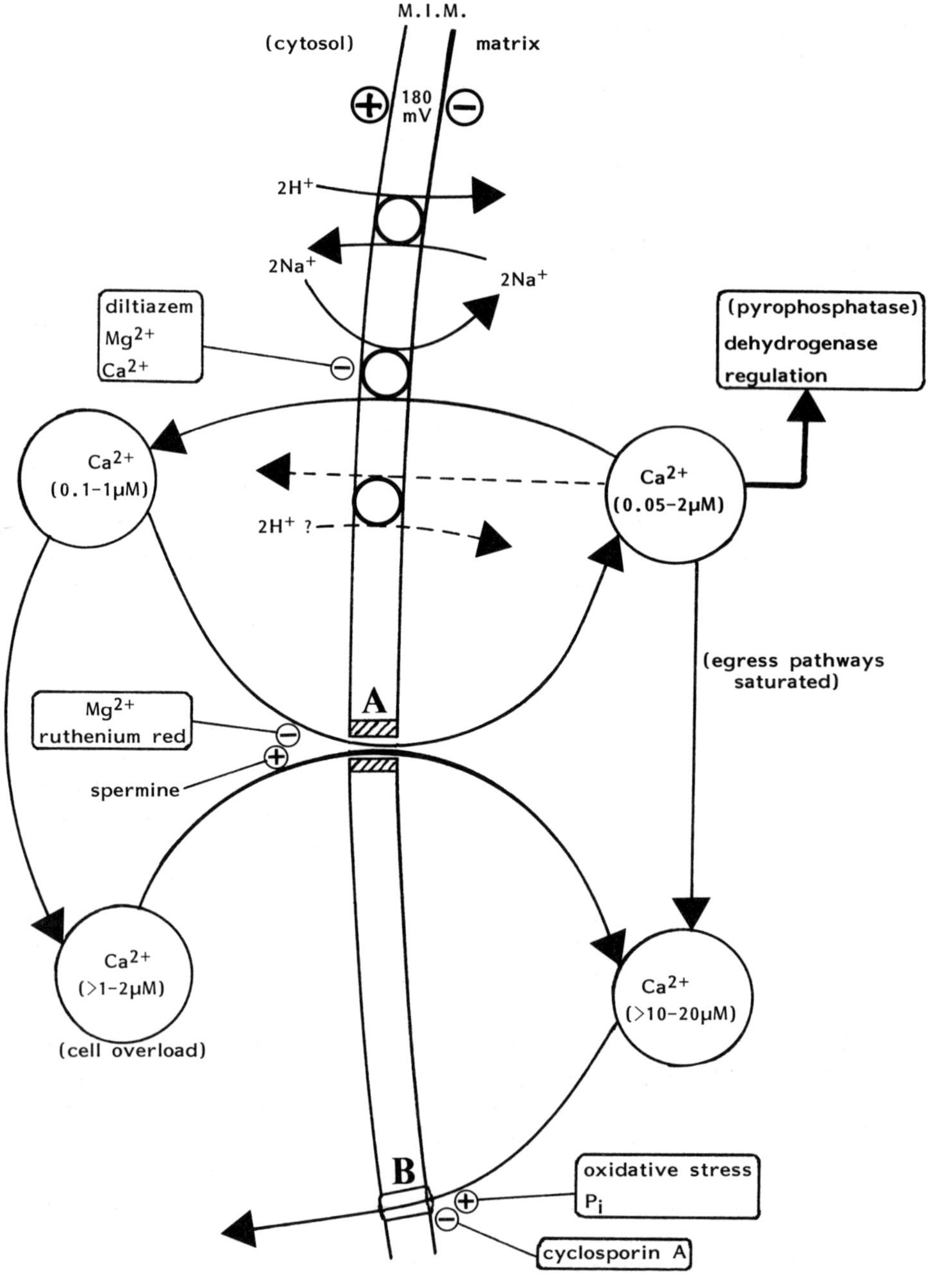

Figure 1. The Ca^{2+}-transport systems of the inner membrane of mammalian mitochondria and their effectors. M.I.M. = mitochondrial inner membrane; **A** = the Ca^{2+}-uniporter; **B** = the Ca^{2+}-operated pore (see text).

concentration within the micromolar range and in relaying changes in cytosolic Ca^{2+} into the matrix (1–3). This is largely because mammalian tissues contain three key oxidative dehydrogenases which can be activated by Ca^{2+} within this range, namely:

- pyruvate dehydrogenase complex (PDH) (EC 1.2.4.1. + 2.3.1.12 + 1.8.1.4)
- 2-oxoglutarate dehydrogenase complex (OGDH) (EC 1.2.4.2. + 2.3.1.61 + 1.8.1.4)
- NAD^+-isocitrate dehydrogenase (NAD-ICDH) (EC 1.1.1.41)

Thus, their activation in this way allows a mechanism for the promotion of ATP production without the need to diminish cellular $NADH/NAD^+$ or ATP/ADP concentration ratios. This therefore allows a means to balance the increases in energy utilization which are brought about by raised cytosolic Ca^{2+} in its activation of processes such as contraction and secretion, and hence also helps to explain the pronounced stability or even increases in these key concentration ratios which are evident under conditions of enhanced oxidative metabolism in stimulated mammalian tissues. There is now much experimental evidence in support of this mechanism occurring (see ref. 4).

1.2.2 Under pathophysiological conditions

It was previously thought that the Ca^{2+}-transport system of mammalian mitochondria allowed them to play a key role in buffering cytosolic Ca^{2+} concentrations (e.g. refs 5 and 6). Certainly, mitochondria can behave as very good buffers of the extramitochondrial Ca^{2+} environment; however, in order to do this the egress pathways have to be saturated for Ca^{2+} (see refs 1 and 3; *Figure 1*), and when this occurs the total calcium content of the mitochondria is well in excess (more than 10 nmol/mg of mitochondrial protein) of the matrix Ca^{2+}-sensitive enzymes' regulatory range (about 0–4 nmol/mg), and also the values of total mitochondrial calcium seen in normal cells (around 1 nmol/mg) *in situ* using X-ray probe microanalysis (e.g. ref. 7).

However, it is likely that this function of the Ca^{2+}-transport system will play a key role in more pathophysiological circumstances, such as ischaemia and reperfusion, where there appears to be abnormal and high cellular Ca^{2+} influx (e.g. ref. 8). Thus, by restricting the rise in cytosolic Ca^{2+} to just over 1–2 μM, this may allow the cell time to recover. However, it is also clear that the resultant overload of mitochondria with calcium can lead to their own damage and hence the derangement of cellular energy production. Recent evidence indicates that this may be due to the operation of an inner membrane pore that opens reversibly under the influence of high matrix Ca^{2+} (*Figure 1*) (9–15). Pore opening is enhanced by P_i and oxidative stress and leads to uncoupling of energy transduction and the egress of low molecular weight (less than around 2 kd) solutes from the matrix and therefore would have severe consequences for cell recovery; opening of the pore can be inhibited by the immunosuppressant cyclosporin A (14) (*Figure 1*).

We now describe the methods associated with the study of these two important roles of the mitochondrial Ca^{2+}-transport system in turn.

2. Mitochondrial Ca^{2+}-transport and matrix Ca^{2+} under normal conditions

2.1 Ca^{2+}-regulation of the matrix dehydrogenases

PDH, NAD-ICDH, and OGDH are only found in the mitochondrial matrix in mammalian tissues. Ca^{2+} activates PDH indirectly by causing increases in the amount of active, non-phosphorylated PDH (PDH_a) through its stimulation of PDH phosphate phosphatase (PDHP-Pase) (16), whereas it activates NAD-ICDH (17) and OGDH (18) more directly by causing marked decreases in their respective K_m values for *threo*-D_s-isocitrate and 2-oxoglutarate.

This chapter will rather focus on the techniques used to study these enzymes in intact mitochondria; however, a few points about the enzymes themselves are now briefly described as these do have a bearing on the interpretation of the results obtained. If required, full details on the extraction of mitochondria or tissues to study the enzymes, or enzyme purification, together with information on assays (all are fairly straightforward), can be found for PDHP-Pase (16, 19), NAD-ICDH (17, 20) and OGDH (18).

Firstly, it should be noted that the regulation by Ca^{2+} is substantial in each case and typically activation of at least fivefold can be achieved. Originally, the effective concentration range for each enzyme was thought to be approximately 0.1–10 μM with half-maximal effects ($K_{0.5}$ values) around 1 μM (16–18). However, more recent estimates indicate that those for NAD-ICDH may be up to an order of magnitude higher than this (20, 21); this may be a device to allow regulation of matrix NADH formation over a wider range of matrix Ca^{2+} concentration. This higher range for NAD-ICDH is especially evident in the presence of high ATP/ADP, and this concentration ratio can also affect the $K_{0.5}$ for Ca^{2+} of OGDH in the same direction (22), but always at lower values than for NAD-ICDH (21). Note that there is an absolute requirement for the presence of AD(T)P in the expression of the Ca^{2+} effects on NAD-ICDH (17), and that NAD-ICDH and PDHP-Pase also require Mg^{2+} (around 1 mM) for activity to be expressed.

Sr^{2+} can mimic Ca^{2+} action on each of these enzymes but at approximately tenfold higher concentrations (16–18); this is also true for Ba^{2+} except that it does not affect PDHP-pase (23). The Ca^{2+}-sensitive properties of these enzymes have been found in extracts of all vertebrate tissues so far studied, but not non-vertebrates (24), which interestingly suggests possible co-evolution with the Ca^{2+}-transport system (25).

PDH, NAD-ICDH, and OGDH all catalyse irreversible oxidative decarboxylations to produce NADH, the principle substrate for oxidative phosphorylation. They are all widely regarded as key regulatory and rate-limiting steps in oxidative metabolism (1, 3), and can also be activated by increases in ADP/ATP and NAD^+/NADH ratios (18); NAD-ICDH and OGDH can also be activated by decreases in pH in the range 7.4–6.6 (these regulations of OGDH and NAD-ICDH are also through changes in substrate K_m), and PDH and OGDH can also be activated by increases in CoA/acetyl CoA and CoA/succinyl CoA respectively (18). Pyruvate also activates PDH through its inhibition of PDH_a kinase (18).

However, regulation of these enzymes by Ca^{2+} is largely distinct and independent from that by these other agents, and this thus allows a means for increasing flux through them without affecting, or even improving (energetically), these key metabolite ratios. In contrast to Ca^{2+}, the regulation by these other metabolites can be observed in extracts from many non-vertebrates (14). However, the potential for regulation by other factors must always be accounted for before any observed effects can be assigned to Ca^{2+} regulation.

The molecular basis for the Ca^{2+} activation of these enzymes is not yet known; however, it would appear that a common subunit is not involved (18). Direct Ca^{2+}-binding to purified preparations of each of the enzymes [PDH/PDHP-Pase (27), NAD-ICDH and OGDH (28)] has already been studied using equilibrium- and flow-dialysis. Interestingly, evidence was found that Ca^{2+} may act as a bridging ligand between PDH-Pase and the E_2 (acetyltransferase) core of PDH (27), and that binding to OGDH was relatively straightforward, whereas binding to NAD-ICDH required the presence of ADP, isocitrate and Mg^{2+}.

There are two other intramitochondrial enzymes which have been reported to be regulated by micromolar Ca^{2+} that should be mentioned, as their Ca^{2+}-regulation is also likely to be important in controlling energy metabolism. However, methods for the study of these are not dealt with here and the appropriate references should be consulted.

Yamada's group have reported that the H^+-ATPase has an inhibitor protein [distinct from the well-characterized Pullman and Monroy inhibitor (29)] which has a molecular weight of about 6.5 kd and which is released from the synthetase by increases in Ca^{2+} in the micromolar range (30–32).

The other enzyme is a pyrophosphatase found in the matrix (*Figure 1*), and Ca^{2+} inhibits this enzyme through calcium pyrophosphate being a powerful competitive inhibitor, with a K_i some 100-fold lower than the K_m for the substrate magnesium pyrophosphate (33–35). Thus, at a probable matrix Mg^{2+} concentration of around 0.3 mM (36, 37), the $K_{0.5}$ value for Ca^{2+} is about 2–4 μM (35). This inhibition leads to increases in matrix pyrophosphate concentration (33, 34), and these in turn are thought to lead to increases in matrix volume by promoting K^+ influx (33–35), perhaps by a mechanism involving the adenine nucleotide translocase (35). Such Ca^{2+}-dependent increases in matrix volume are associated with an activation of the respiratory chain at the level of electron flow into the ubiquinone pool (38), and overall are very similar to those evoked in intact liver cells by Ca^{2+}-mobilizing hormones (34, 35).

The major advances in this field have, however, largely relied on the development of techniques to study Ca^{2+}-regulation of these key dehydrogenases in intact mitochondria. An essential prerequisite has been the development of methods to prepare partially purified mitochondria from rapidly disrupted tissues under conditions which minimize artefactual Ca^{2+} uptake or loss.

2.2 Preparation of mitochondria

The protocol described has been used successfully with several different rat tissues (see refs. 39–43) and should be applicable to most mammalian tissues. To avoid Ca^{2+} redistribution during the preparation of mitochondria (necessary to study effects of, for example, hormone pre-treatment; see Section 2.4.1.*ii* later) it is best to quickly take tissue from anaesthetized animals or from perfusions, and to very rapidly disrupt it (i.e. within 1–2 sec) into at least 6 vol. of ice-cold buffered isotonic media containing EGTA (ethyleneglycol-*bis*-(-*ß*-aminoethyl ether)*N*,*N*,*N'*,*N'*-tetraacetic acid) (*Protocol 1*). Experiments with ^{45}Ca tracer have established that the EGTA prevents any artefactual Ca^{2+} uptake, and that 0–4 °C allows very little Ca^{2+}-egress, even in the presence of Na^{+} (39–43).

Protocol 1. Mitochondrial preparation

Requirements

- Polytron (Kinematica Gmbh, Kriens-Luzern, Switzerland) homogenizer or equivalent, and appropriate (0.5–2 cm diameter) ice-cooled probe.
- Homogenization buffer (buffer A): 250 mM sucrose, 20 mM Tris–HCl, 2 mM EGTA, 1–3% (w/v) essentially defatted albumin[a] (e.g. Sigma A-7906) (pH 7.4, ice-cold).
- Suspension buffer B: as A above, but without the albumin.

1. Rapidly remove required tissue from perfusion or anaesthetized animal, homogenize in at least 6 vol. of buffer A with two or three bursts of 2–3 sec at Mark 4 (longer will disrupt mitochondria as well).
2. Dilute homogenate two- to threefold with buffer A and centrifuge 1000 *g* for 90 sec at 4 °C.
3. Take supernatant or (fatty tissues) infranatant, centrifuge at 10 000 *g* for 5 min at 4 °C.

Protocol 1. *Continued*

4. Suspend pellet (use cold-finger) in 6 ml of buffer B and add 1.4 ml of Percoll (e.g. Sigma P-1644), centrifuge at 10 000 *g* for 10 min at 4 °C[b].
5. Suspend pellet in buffer B and wash once by repeating step 3.
6. Suspend pellet in buffer B to give about 40–80 mg protein/ml.

[a] 3% is only really required for fat tissue [white or brown adipose (19)]; for these tissues 7.5 mM reduced glutathione is also added for further protection.
[b] The Percoll step is not really necessary for heart or fat tissue mitochondria, but with (for example) liver or kidney will result in a 1.5- to 6-fold purification of the mitochondria, and in particular removes (found at the top of the tube) endoplasmic reticulum contaminants, including the IP_3-(*myo*-inositol 1,4,5-trisphosphate)-sensitive pool (44).

The yield of mitochondria using the above is typically 20–40% (42). The Percoll purification step also allows improved detection and confidence in the measurement of mitochondrial parameters, particularly total Ca content (see ref. 44).

2.2.1 The toluene-permeabilization of mitochondria

A useful 'half-way house' for getting more precise information about the regulatory properties of the Ca^{2+}-sensitive matrix enzymes in their natural environment, but at the same time allowing experimental conditions to be carefully controlled, has been to use mitochondria made permeable to all small molecules (less than about 2 kd) by controlled treatment with toluene (21, 45–47) (*Protocol* 2). For instance, it was thus shown that the major effect of Ca^{2+} on PDHP-Pase located within mitochondria is to decrease its K_m for Mg^{2+}, with little or no effects of Ca^{2+} evident at high Mg^{2+}; in contrast, isolated preparations show large activations by Ca^{2+} at all Mg^{2+} concentrations.

Protocol 2. Toluene-permeabilization of mitochondria (adapted from ref. 45).

1. Prepare mitochondria as in *Protocol 1*, except that at step 6 resuspend (to about 20 mg protein/ml) in buffer B which also contains 8.5% (w/v) poly(ethylene glycol) 6000 (buffer C) (at 0–4 °C, pH 7.4), and transfer the required volume (100–500 µl) to a 1.5-ml conical microcentrifuge tube.
2. Add toluene (5 µl/ml) to the suspension and agitate (2–3 cycles/sec) for 2 min in an ice/water bath, gently vortex mixing for 3 sec every 30 sec.
3. Separate mitochondria from medium by centrifugation at 15 000 *g* for 4 min at 4 °C.
4. Remove toluene carefully by aspiration (top layer) and residual supernanant.

Protocol 2. *Continued*

5. Wash once in buffer C and resuspend in C to about 5 mg protein/ml.[a]

[a] Recovery of mitochondrial protein through the permeabilization should be about 50–70% (45); an easy alternative to assess recovery is by measuring recovery of glutamate dehydrogenase activity, e.g. heart or adipose tissue mitochondria have about 0.25 unit of this per milligram of protein (see ref. 45), where a unit is the amount which converts 1 μmol of substrate/min at 30 °C.

2.3 Assay of the Ca^{2+}-sensitive dehydrogenases in intact mitochondria

2.3.1 Pyruvate dehydrogenase (PDH)

i. Incubation of mitochondria

Ca^{2+} affects PDH largely through its effects on PDHP-Pase (Section 2.1). The effects of matrix Ca^{2+} on PDHP-Pase within intact mitochondria are best assayed indirectly by following increases in steady-state PDH_a induced by changes in extramitochondrial Ca^{2+} being relayed into the matrix under conditions (high matrix ATP) where PDH_a kinase activity is likely to be constant. A typical experimental approach is given in *Protocol 3*.

Protocol 3. Demonstration of Ca^{2+} effects on PDH in intact mitochondria

1. Prepare incubation buffer D: 125 mM KCl, 20 mM Tris–HCl, 5 mM KH_2PO_4 (pH 7.3); for work at e.g. pH 7, replace Tris with Mops (4-morpholinepropane sulphonic acid).

2. To buffer D add appropriate respiratory substrates (e.g. 5 mM 2-oxoglutarate with 0.2 mM L-malate)[a] for coupled mitochondria (to provide intramitochondrial ATP), or 1 μM FCCP (carbonyl cyanide *p*-trifluoro-methoxyphenyl hydrazone) plus 2–5 mM Mg·ATP and oligomycin (5 μg/ml) for uncoupled mitochondria (the ionophores A23187 (2 μg/ml) and valinomycin (1 μM) can also be added to ensure complete free Ca^{2+}-equilibrium across the inner membrane, but are not usually necessary; refs 40 and 41).

3. Prepare 1.5-ml conical microcentrifuge tubes each containing, for example, 0.5–1ml of buffer D and appropriate additions, and add appropriate amounts (to give 2–5 mM total ligand) of EGTA and EGTA-Ca stock solutions (of about 200 mM strength) (see ref 42), or other Ca-buffer solutions (e.g. HEDTA) is useful at higher pH (20), but does also bind Mg^{2+} significantly (*N*-(2-hydroxyethyl)-ethylenediamine tri-acetate (see Chapter 7, this volume) to cover a suitable extramitochondrial Ca^{2+} range (see *Figure 4* later).

4. Add mitochondrial suspension to give 0.5–1 mg protein in each tube, mix by hand inversion, and incubate for about 5 min (at 30 °C); this is

Protocol 3. *Continued*

sufficient to achieve new steady-states of PDH_a, but should not lead to de-oxygenation.

5. After incubation, rapidly sediment mitochondria (10 000 *g* for 20 sec), remove supernatant by syringe, and freeze pellet in tube in liquid N_2[b] (samples can be stored at −70 °C for several weeks).

[a] For liver mitochondria (or heart and other mitochondria only if prepared from starved animals), to be able to show Ca^{2+} effects on PDHP-Pase it is usually necessary to add also some pyruvate or its analogue dichloroacetate (e.g. 1 mM) to inhibit PDH_a kinase to some degree as it appears to be too active in these cases (see ref. 43).

[b] Alternatively, for direct extraction and usually directly subsequent assay, incubations can be stopped by using an excess (e.g. 5 vol.) of an ice-cold quenching medium (48); 50 mM Hepes, 3 mM EGTA, 25 mM NaF, 0.1% (v/v) Triton X-100, 1 mM dichloroacetate and 1 mM dithiothreitol, to break the mitochondria under conditions where PDH interconversion is prevented (see next section).

ii. Extraction of PDH

The buffer to do this (*Protocol 4*) contains EDTA to prevent any further interconversion of PDH as both PDHP-Pase and PDH_a kinase require Mg^{2+} (16), together with Triton to disrupt membranes and solubilize PDH, and rat serum as a convenient means of preventing proteolysis of PDH (49).

Protocol 4. Extraction of PDH from mitochondria[a]

1. Prepare extraction buffer E: 100 mM KH_2PO_4, 2 mM EDTA, 0.1% (v/v) Triton X-100, and added on the day, 1 mM dithiothreitol and rat serum (kept stored in freezer) to 50 μl/ml.
2. Add 100–300 μl ice-cold buffer E to frozen pellet and pass up-and-down in a 250–500-μl microsyringe until fully disrupted (about ten strokes).
3. Drop into liquid N_2; store on dry ice or at −70 °C for up to two days.
4. Before assay, thaw, and centrifuge (10 000 *g*, 1 min, 0–4 °C); assay supernatant (below).

[a] Tissue samples can be extracted in the same buffer (the triton is not so necessary here) using the Polytron (20 sec, Mark 5) to fully disrupt the sample; after step 4, assay the infranatant.

iii. Assay of PDH

The preferred method (see ref. 42) is to use a 'following enzyme' which will measure and also remove the acetyl CoA produced in the reaction (and which is a potent end-product inhibitor of PDH_a) as follows:

(a) pyruvate + NAD^+ + CoA $\xrightarrow{PDH_a}$ CO_2 + NADH + acetyl CoA

(b) acetyl CoA + AABS $\xrightarrow{AAT}$ acetylated AABS + CoA

AABS is *p*-(*p*-aminophenylazo)benzene sulfonic acid (Pfaltz & Bauer) which has a mM extinction coefficient of 6.5 at 460 nm where it loses its colour on acetylation, and AAT is arylamine acetyltransferase prepared from pigeon liver acetone powder (*Protocol 5*).

Protocol 5. Preparation of AAT (EC 2.3.1.5) (steps 1–5 based on ref. 50).

1. Homogenize, using an Atomix or a Waring blender or similar, fresh (or thawed but frozen fresh) pigeon livers with 10 vol. of acetone at −10 °C.
2. Filter with suction through Whatman No. 1 paper in a Buchner funnel with a dry-ice trap present.
3. Re-extract 'cake' from filter paper by repeating steps 1 and 2.
4. Add sufficient diethyl ether (pre-treated to remove peroxides by passage through a column of aluminium oxide (Neutral, Grade 1) before use) to cover powder, and filter.
5. Take disc of powder and crumble and dry in fume cupboard for 1–2 h and then leave in evacuated desiccator overnight; store at −70 °C for up to 1 year if well-prepared.
6. When required, homogenize 16 g of powder (Polytron, 2 cm probe, 1 min, Mark 4) in 10 vol. of ice-cold distilled water containing 20 mg of EDTA.
7. Centrifuge for 15 min, 10 000 *g* at 4 °C.
8. Add acetone (freshly opened bottle, analar) at −10 °C drop by drop to the (well-stirred supernatant over 15–30 min to finally give 6.7 ml per 10 ml of original supernatant.
9. Centrifuge 15 min, 10 000 g, −10 °C; take supernatant and repeat step 8 with the same volume of acetone used there, then repeat step 9.
10. Suspend pellet in a minimum volume (about 4 ml) of 10 mM potassium phosphate buffer (pH 7.5) and dialyse for at least 2 h at 0 °C against 2 litres of the same buffer, but at pH 7.
11. Centrifuge at 10 000 *g* for 2 min; take supernatant, dilute to about 4–5 units/ml with pH 7.5 buffer and aliquot, and store at −20 °C for up to 6 months.[a]

[a] Yield should be about 40% and 5–15 units/ml should be obtained; assay in the PDH assay buffer (as below) plus 10 μl of 10 mM acetyl CoA. NB: Commercially obtained powder gives poor yields.

PDH is assayed (usually at 30 °C) after making the following additions to a 1 cm path cuvette:

- 1.5 ml of buffer F; 100 mM Tris–HCl, 0.5 mM EDTA, 1 mM $MgSO_4$ (pH 7.8) containing (added on day of use) 20 µl/ml of a 1 mg/ml AABS solution (Na or Tris salt in water) and 0.2 µl/ml of mercaptoethanol (make up about 50–100 ml of buffer).
- 20 µl of a substrate mix (36 mg thiamine pyrophosphate (cocarboxylase); 23 mg NAD^+; 9 mg pyruvate; 7.5 mg coenzyme A) in 1 ml water; can be stored at −20 °C for up to 2 weeks)
- 20 µl of AAT preparation (60–100 mUnits)
- 10–200 µl of sample

There may be a short lag (1–5 min) before linear rates develop due to the presence of malate and malate dehydrogenase and citrate synthetase competing for the acetyl CoA. Alternative assays include following NADH production in the presence of rotenone (to inhibit NADH oxidase) (48) which has the disadvantage that the acetyl CoA as well as the NADH, which is also a potent end-product inhibitor, will build up, and also the following of $^{14}CO_2$ production from 1–[^{14}C]-pyruvate which suffers from the acetyl CoA build-up as well as the potential problem of considerable non-enzymatic decarboxylation (51).

The sensitivity of the recommended assay can be improved up to tenfold (e.g. if small amounts of sample) by using a fluorimeter and the AAT in the above system to rather acetylate cresyl violet acetate (instead of AABS), as this causes a shift in its fluorescence spectrum from $(ex)_{max}$ 575 nm, $(em)_{max}$ 620 nm to $(ex)_{max}$ 475 nm, $(em)_{max}$ 575 nm (52).

It is usual to express the amount of PDH_a as a percentage of the total activity. The latter is obtained by assaying as above after converting all of the sample PDH phosphate into PDH_a using added exogenous PDHP-Pase (*Protocol 6*).

Protocol 6. Preparation of pig heart PDHP-Pase (based on ref. 53)

1. Obtain 8–10 pig hearts as fresh as possible from slaughterhouse, and keep in ice during transportation.
2. Trim off fat, and blood vessels, and small chamber walls, and cut remaining ventricle walls into 1–2-cm cubes.
3. Homogenize (Waring blender or equivalent) for 1 min with 3–4 litres of ice-cold buffer G (30 mM KH_2PO_4, 1 mM EDTA, pH 7.6, with 5 mM mercaptoethanol added just before use).
4. Check that pH does not fall below 7.0; restore with KOH if necessary.
5. Centrifuge at 5 000 g for 30 min at 4 °C.

Protocol 6. *Continued*

6. Pour supernatant through cheesecloth and re-extract pellets with a further 3–4 litres of buffer G.
7. Adjust combined supernatants to pH 5.4 at 4 °C using 10% (v/v) acetic acid.
8. Centrifuge as in step 5, then resuspend pellets in 2 litres of ice-cold distilled water, and again centrifuge as in step 5.
9. Suspend crude mitochondrial pellets in 0.5 litres of 20 mM KH_2PO_4 (pH 7) containing 5 mM mercaptoethanol (buffer H) and adjust to pH 7 with KOH.[a]
10. Transfer to four round-bottomed (litre) Pyrex flasks and break mitochondria by shell-freezing and thawing three times using liquid N_2 and a water bath (at 30 °C).
11. Centrifuge at 30 000 *g*, 2 h, 4 °C and carefully aspirate off supernatant and incubate it at 30 °C after adding $MgCl_2$ to 10 mM (from a 5 M stock) to convert any PDH phosphate into PDH_a (monitored as above; usually takes 15–30 min and should yield about 400–800 units in total).
12. Cool to 4 °C and slowly (over 30 min) add solid $(NH_4)_2SO_4$ (analar) to give 209 g/litre.
13. Centrifuge, 10 000 *g*, 20 min, 4 °C, and then dissolve pellet in a minimum volume (20–30 ml) of buffer H (check pH 7), and dialyse overnight against 2 litres of the same.
14. Adjust pH (at 4 °C) to exactly 6.1 with 10% acetic acid and centrifuge at 15 000 *g*, 20 min, 4 °C.
15. Restore pH of the supernatant to 7.0 with KOH and freeze in aliquots of 2–5 ml at −70 °C (PDHP-Pase is less stable when separate from PDH).
16. When required, PDHP-Pase is separated from PDH by centrifuging twice for 1 h at 150 000 *g* and 4 °C; the resulting supernatant is checked to be PDH free and can be stored at −20 °C for up to 1 month in aliquots (100–400 μl).

[a] The preparation can be stored at −70 °C at this point if required. The final preparation should be capable of generating about 5 units of PDH_a from added PDH-phosphate/mg protein; this latter can be made by incubating the PDH (after suspension) in the pellets from step 16 with Mg.ATP (5–10 mM; assay decrease in PDH_a until no detectable amounts remain).

Thus conversion of PDH phosphate into PDH_a in samples is achieved by incubating some mitochondrial (or tissue) extract (up to 50 μl) with 20–40 μl of the pig heart PDHP-Pase in the presence of 25 mM $MgCl_2$ and 1 mM $CaCl_2$ (added from a 10 × stock to activate the phosphatase) at 30 °C for 5–15 min (check for completion). Mitochondrial fractions prepared as above would typically contain 30–120 mU of total PDH/mg protein. An alternative approach to estimating total PDH (avoiding making PDHP-Pase) is to incubate a fresh mitochondrial sample in buffer D as in *Protocol 3* but with *only* FCCP present for 5–10 min at 30 °C to deplete them of ATP, and thus all of the PDH should be converted to PDH_a.

2.3.2 2-oxoglutarate dehydrogenase (OGDH)

OGDH is perhaps the most straightforward of these enzymes to assay in intact mitochondria and is often used as the quickest way to check out Ca^{2+} effects, etc.

i. Use of the O_2 electrode

Fortunately, the Ca^{2+}-sensitive properties of OGDH within intact mitochondria from all mammalian tissues so far examined (see refs 19, 22, 39, 54) can readily be demonstrated by simply measuring O_2 uptake (with a Clark-type electrode) by the mitochondria (in buffer D, for example; *Protocol 3*) at a subsaturating concentration (e.g. 0.05–0.5 mM) of 2-oxoglutarate (as the Ca^{2+} effect is on this substrate's K_m), when the restrictions on this imposed by respiratory chain activity are overcome either by using uncoupled mitochondria (1 μM FCCP) or an excess of ADP (e.g. 2 mM), as OGDH appears to become the rate-limiting step for 2-oxoglutarate oxidation under many such circumstances (22, 39, 54). A transport partner for 2-oxoglutarate is usually present (e.g. 0.5 mM), e.g. L-malate, or the non-metabolizable malonate if high blank rates are evident with L-malate (39). With liver mitochondria 37 °C is needed for Ca^{2+} effects to be observed (54), otherwise 25 or 30 °C is sufficient (42). An example of this approach is shown in *Figure 2*.

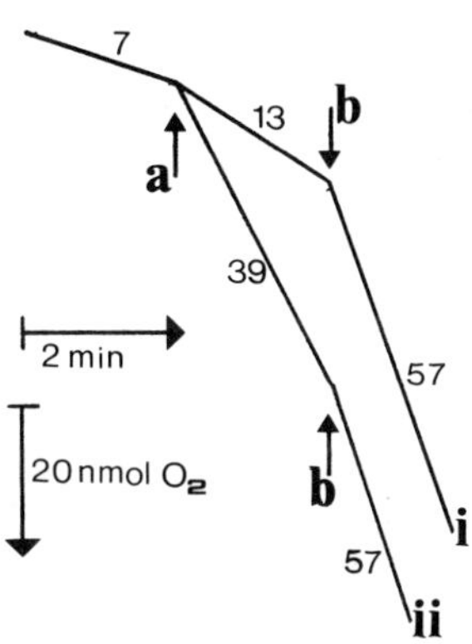

Figure 2. Effects of increasing extramitochondrial Ca^{2+} on O_2 uptake by intact, fully coupled, rat heart mitochondria in the presence of non-saturating and saturating concentrations of 2-oxoglutarate (to show the stimulation of OGDH by increased matrix Ca^{2+}). Mitochondria (0.5 mg protein) were incubated at 30 °C in 1 ml of buffer D (*Protocol 3*) containing 2 mM ADP and 1 mM L-malate, and either (i) 5 mM EGTA (less than 1 nM free Ca^{2+}), or (ii) 5 mM EGTA plus 2.5 mM $CaCl_2$ (as a buffer solution, about 50 nM free Ca^{2+}). Further additions were, at arrow **a**, 0.25 mM 2-oxoglutarate; at **b**, 20 mM 2-oxoglutarate. The adjacent figures indicate nanomoles of O_2 taken up per minute. Data are of the type in ref. 22.

ii. Use of 1-[^{14}C]-2-oxoglutarate

The production of $^{14}CO_2$ from 1-[^{14}C]-2-oxoglutarate by OGDH can also be used to monitor its activity in intact mitochondria under various conditions (see ref. 43), including in the absence or presence of ADP or uncoupler. Its main advantages are simplicity, specificity, and ease of application to many situations (42). A typical experiment is outlined in *Protocol 7* (see ref. 41).

Protocol 7. Assay of OGDH in mitochondria using 1-[^{14}C]-2-oxoglutarate

1. Pre-incubate mitochondria (up to 5 min, 30 °C, 0.5–1 mg protein/ml) in buffer D (*Protocol 3*) (with or without L-malate, EGTA–Ca buffers, etc., as desired).
2. Transfer duplicate 100 µl samples of pre-incubate to small soda-glass test-tubes (30 × 6mm) which are then each placed in 0.5 ml of 2-phenylethylamine (or hydroxylamine) in the base of a standard glass scintillation vial.
3. Seal each vial with a rubber septum and then add (to tubes) the required concentrations of 2-oxoglutarate (e.g. non-saturating, 50 µM; saturating, 3 mM) containing 1-[^{14}C]-2-oxoglutarate at suitable specific radioactivities.
4. After the desired incubation time (e.g. 1–5 min; so that rates of decarboxylation are still linear and usually so that 2–30% of the oxoglutarate is decarboxylated) add 50 µl of 20% (v/v) $HC10_4$ to stop the reaction.
5. Leave samples gently shaking for at least 1 h before removing septum and small test-tube; add 5 ml of appropriate scintillant and count radioactivity in the phenylethylamine (i.e. CO_2 formed).

---

iii. Use of fluorimeter or dual-wavelength spectrophotometer

Perhaps the most flexible and sensitive method for studying the Ca^{2+}-sensitivity of OGDH in intact mitochondria is to use either a fluorimeter (ex, 340 nm; em, 460 nm) or a dual-wavelength spectrophotometer (e.g. 340–370 nm wavelength pair) to monitor mitochondrial NAD(P)H production induced by added, non-saturating, 2-oxoglutarate (see refs 39, 41, 43, 55). The sample is continuously stirred in a thermostatted cuvette unit (e.g. 30 °C) and incubated in buffer D (*Protocol 3*) and appropriate Ca-buffer solutions (and other additions) as for methods *i* and *ii* above. If desired, respiratory chain inhibitors can be added to enhance NAD(P)H build-up, or else ADP or NH_4Cl can be used to dissipate NAD(P)H, depending on where optimal Ca^{2+} effects can be observed (see ref. 43). It is often advantageous to oxidize all endogenous NAD(P)H first of all (43); an experiment of this type is shown in *Figure 3*.

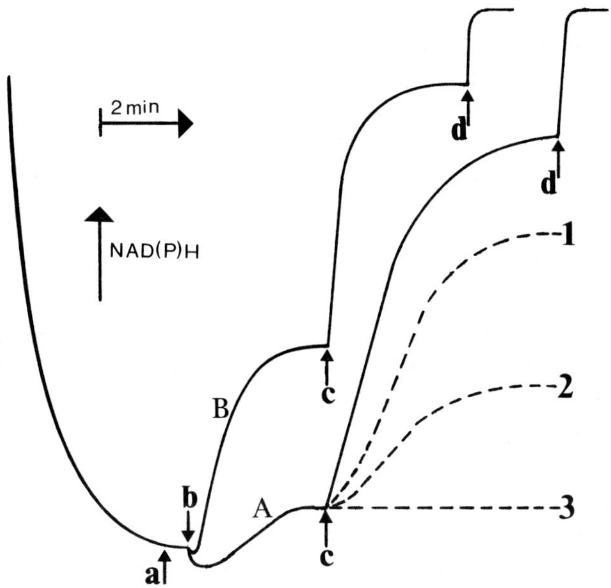

Figure 3. Effects of increasing extramitochondrial Ca^{2+} on the reduction of $NAD(P)^+$ induced by 2-oxoglutarate in intact, fully coupled, rat liver mitochondria (to show effects of matrix Ca^{2+} on OGDH). Mitochondria (about 2 mg protein) were incubated at 30 °C in 2 ml of buffer D (*Protocol 3*) containing initially 0.5 mM EGTA, 0.5 mM malonate, and 0.5 mM NH_4Cl, with either no additions (dashed traces), or else, A, 2 mM EGTA, or B, 2 mM EGTA plus 2 mM $CaCl_2$ (to give about 200 nM free Ca^{2+}) added at arrow **a**. Other additions, at the arrows, were:, **b**, 0.2 mM 2-oxoglutarate; **c**, either (solid lines) 2 mM 2-oxoglutarate, or (dashed lines): **1**, 5 mM EGTA plus 5 mM $CaCl_2$ (to give about 470 nM free Ca^{2+}); **2**, 1.5 mM EGTA plus 1.5 mM $CaCl_2$ (150 nM); **3**, 0.5 mM EGTA alone, or (same trace) as **1**, but plus 1 μg ruthenium red/ml; **d**, 0.5 μg rotenone [to fully reduce $NAD(P)^+$]. NAD(P)H production (arbitrary units) was followed, using a fluorimeter at 340 nm (ex)/460 nm (ex). Data are of the type in ref. 43.

In all the above approaches, non-saturating concentrations of glutamate can replace added 2-oxoglutarate to show Ca^{2+} effects (presumably through its metabolism via glutamate dehydrogenase and subsequent oxidation by OGDH), although as expected higher K_m values (about tenfold) are obtained. *β*-hydroxybutyrate, L-malate or succinate can be used as controls as their oxidation should not be Ca^{2+}-sensitive (see ref. 43).

2.3.3 NAD^+-isocitrate dehydrogenase (NAD-ICDH)

Less work has been done on this enzyme than the other two with intact mitochondria because of a number of difficulties. For instance, no method has as yet been developed to study this enzyme in intact heart mitochondria, which have probably been used most of all for the other two enzymes, as they have a very low activity of the tricarboxylate carrier (56). There is also the major problem with intact mitochondria that the product of NAD-ICDH, but also the non-Ca^{2+}-sensitive $NADP^+$-linked enzyme, is 2-oxoglutarate which is, of course, itself a substrate for the Ca^{2+}-sensitive OGDH, and so far no suitable method has been found to exclusively and unambiguously look at NAD-ICDH in intact mitochondria. Effects of Ca^{2+} on non-saturating isocitrate (or citrate at about tenfold higher concentrations) oxidation by intact liver, kidney, and adipose tissue mitochondria can be observed (see, for example, refs 39, 43, 57), as can similar effects on NAD(P)H production (39, 43), but it is with the above proviso in mind that the effects observed may actually be on OGDH (hydroxymalonate would be used to replace L-malate as a non-metabolizable transport partner if required). Another drawback is that the $^{14}CO_2$-trapping technique also cannot be used satisfactorily in this case as all mammalian mitochondria have high $NADP^+$-linked enzyme activity which may thus catalyse isotopic exchange (43).

2.4 The measurement of matrix Ca^{2+} and mitochondrial Ca^{2+}-transport in intact mitochondria

2.4.1 The use of the Ca^{2+}-sensitive properties of the matrix dehydrogenases

i. Studies on Ca^{2+}-transport across the mitochondrial inner membrane

The above techniques are readily adaptable to studying mitochondrial Ca^{2+}-transport. The major advantages of using this approach, and thus using the enzymes as matrix Ca^{2+}-probes, are that this can allow the monitoring of changes in Ca^{2+} fluxes and steady-state Ca^{2+} distributions across the inner membrane under conditions where matrix Ca^{2+} is carrying out its probable physiological function (see Section 1), and whilst the mitochondria are therefore incubated with buffered and low concentrations of Ca^{2+}, and have low (less than 5 nmol/mg protein) Ca^{2+}-loads, i.e. matching the expected physiological conditions (see refs 1 and 3). This is in contrast to the many earlier studies in this area which generally, because of technique limitations, used high concentrations of Ca^{2+} and thus led to massive Ca^{2+}-loading (see, for example, refs 5, 58–60), and could also hence lead to mitochondrial damage (see, for example, refs. 1–3).

The intramitochondrial concentrations of Ca^{2+} to which the enzymes will be exposed will be determined by the extramitochondrial concentrations of Ca^{2+} and by the relative rates of Ca^{2+} uptake and egress across the inner membrane, and thus also on the presence or absence of the various effectors of the Ca^{2+}-transport systems (see *Figure 1*). An example of this is shown for rat heart mitochondria in *Figure 4* using OGDH activity. It should be noted that in such studies, extramitochondrial Ca^{2+} concentrations again have to be carefully controlled, usually by using EGTA–Ca^{2+} buffer systems (see Chapter 7, this volume) using solutions which have been pre-equilibrated for pH (as EGTA binding to Ca^{2+} releases protons) (see ref. 42). It is also necessary to avoid (except when required) using Mg or Na salts of substrates etc. if their effects on transport are being studied, and to remember that full operation of the cycle (*Figure 1*) requires energization of the mitochondria. The amount of extramitochondrial Ca^{2+} leading to Ca^{2+} accumulation, as opposed to the steady-state relations between extra- and intramitochondrial Ca^{2+} at lower concentrations (as in *Figure 4*), i.e. going higher than the ranges shown in *Figure 4* and leading to egress pathway saturation and potential damage (see Section 3), will also depend on the presence or absence of these effectors.

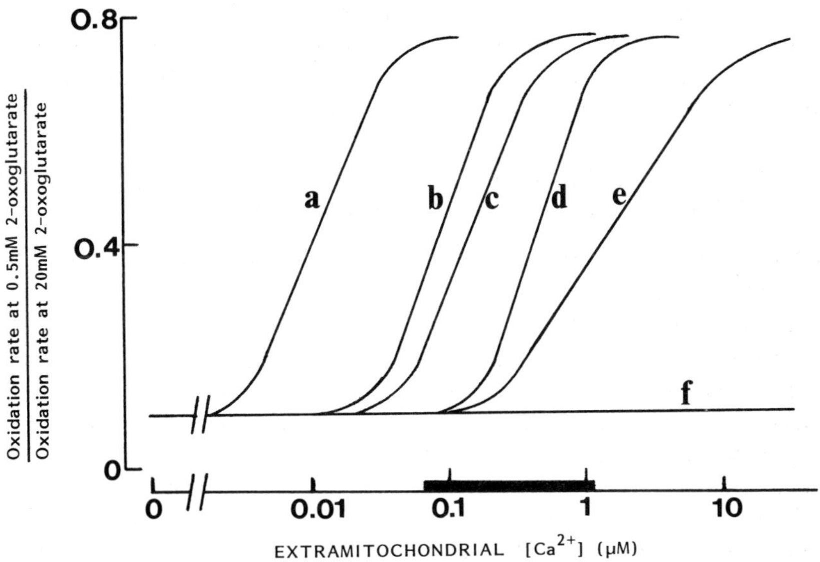

Figure 4. Sensitivity of increases in extramitochondrial Ca^{2+} of OGDH within intact rat heart mitochondria under various conditions. Mitochondria were incubated in experiments of the type shown in *Figure 2* in the presence of 5 mM EGTA and various $CaCl_2$ concentrations (as buffer solutions) to give the free extramitochondrial Ca^{2+} indicated, and together with: **a**, no further additions; **b**, 10 mM NaCl; **c**, 1 mM $MgCl_2$; **d**, Na plus Mg; **e**, 1 μM FCCP; **f**, 1 μM ruthenium red. The bar on the *x*-axis shows the expected physiological range for cytosolic Ca^{2+}. Data are adapted from that in ref. 22.

ii. Studies on the effects of hormone pre-treatment on intramitochondrial Ca^{2+}

The ability to assay for the Ca^{2+}-sensitive properties of PDH and OGDH within intact rat heart (40) and liver (41) mitochondria has been exploited to allow convincing evidence to be obtained in support of the main hypothesis outlined in Section 1, that hormones which increase cytosolic Ca^{2+} also, as a result increase intramitochondrial Ca^{2+} and so cause the increases in PDH_a and oxidative metabolism which are known to be brought about by these same hormones (see refs 1, 3, 40, 41, 44, 61, and 62). The approaches used (40, 41) initially involved the rapid disruption of control and treated tissue and then the preparation of mitochondria under conditions which minimise Ca^{2+} redistribution (see Section 2.2 above). The mitochondria thus prepared were then incubated under conditions where the Ca^{2+}-sensitive enzymes can be assayed as indicators of persistent changes in intramitochondrial Ca^{2+} brought about by the hormone pre-treatments. An example of this from rat liver is shown in *Table 1*. Ca^{2+} can then be egressed from the mitochondria by adding Na^+ (effects of Na^+ blocked by diltiazem; see *Figure 1* and *Table 1*), or else the Ca^{2+}-dependence of the persistent activations due to hormone-pretreatment can also be demonstrated by 'saturating out' the effects by adding sufficient extramitochondrial Ca^{2+} to lead to full Ca^{2+}-dependent activation (*Table 1*). *Table 1* also shows that total mitochondrial Ca measurements [by atomic absorption (44)] parallel those for the enzymes.

Table 1. Effects of control (C) or vasopressin (V) pre-treatment of tissue on activities of PDH and OGDH and total Ca content in subsequently isolated rat liver mitochondria incubated under various conditions[a]

Mitochondrial incubation conditions	PDH_a (as % total PDH)		OGDH (as % V_{max})		Total Ca (nmol/mg protein)	
	C	V	C	V	C	V
Control	17	31[b]	8	13[b]	1.2	2.1[b]
Plus Na^+	14	16	8	8	0.8	1.1
Na^+ plus diltiazem	18	33[b]	9	18[b]	1.1	2.0[b]
At saturating Ca^{2+}	50	51	34	35	4.1	4.3

[a] Data are from ref. 44, or unpublished work of J. G. McCormack; mitochondria were incubated in appropriate media (44), containing 0.5 mM EGTA, for 5 min at 30 °C.
[b] Indicates significant effect of hormone pre-treatment.

The feasibility of the above approach was first of all assessed by loading (control) mitochondria with Ca^{2+} to achieve similar increases in PDH_a as obtained with the hormones acting on the tissues (as described above, e.g. see *Figure 4*), and then 're-preparing' them by the same means as used for the disrupted tissues (see refs 40 and 41). The subsequent measurements of the mitochondrial enzymes' activities and total Ca contents were very similar to those from the hormone pre-treated tissues shown in *Table 1*. Therefore the criteria for establishing whether a hormone is affecting intramitochondrial Ca^{2+} can be summarized as follows; in incubations of mitochondria prepared from hormone-treated tissue (as above), the activities of the matrix Ca^{2+}-sensitive enzymes (with respect to those from untreated controls) should be:

- enhanced in Na^+-free media (containing EGTA) (and in parallel with total Ca content)

- diminished to control values by incubation with Na^+ (e.g. 10 mM)
- enhanced in Na^+-media also containing diltiazem (300 μM)
- maximally stimulated by added Ca^{2+} to the same values as controls can be maximally stimulated

2.4.2 The use of ^{45}Ca pre-loaded mitochondria to study Ca^{2+}-egress

This approach is a more direct way of monitoring the effects of preparation and incubation (as above) on mitochondrial Ca content (40, 41), and is also a very convenient way of looking at Ca^{2+}-egress on its own with Ca-loads within the enzyme regulatory ranges (23, 40, 41). A typical experiment is described in *Protocol 8*.

Protocol 8. Pre-loading of mitochondria with ^{45}Ca and studying Ca^{2+}-egress

1. Incubate (control) mitochondria at 30 °C and about 4 mg protein/ml and with suitable amounts (0.1–0.5 μCi/ml) of ^{45}Ca and unlabelled $CaCl_2$ (20–80 μM) (this is rather than a Ca-buffer to keep specific activity high) to achieve the degree of activation of PDH or OGDH required, in 3–5 ml of buffer D (*Protocol 3*) containing respiratory substrates.
2. After e.g. 3–5 min, add a large excess (at least 6 vol.) of ice-cold buffer B (*Protocol 1*) to stop both uptake and egress, and then re-isolate the mitochondria as in *Protocol 1* (steps 6 and 7).
3. Wash once and re-suspend as in *Protocol 1*.
4. Incubate as required in buffer containing EGTA (see *Table 2*) in appropriate media (e.g. buffer D based, *Protocol 3*).
5. Centrifuge (30 sec, 10 000 *g*); take sample of supernatant for analysis and remove rest, using tissue held in forceps to wipe off any residual supernatant; if required, better separation of media and pellet can be achieved by layering the original suspension over oil (a 50:50 mix of silicone fluids DC 550 and DC 556 gives the correct density) and then centrifuging as above before removing media and oil.[a]
6. Suspend pellet in e.g. 0.5 ml of water [with 0.1% (v/v) Triton X-100 if needed] to disrupt mitochondria.
7. Deproteinize by adding $HClO_4$ to 2% (v/v) (from a 20% stock) and centrifuging for 1 min at 10 000 *g*.
8. Add samples of supernatants to scintillant and count on ^{14}C channel.

[a] Millipore system (and vacuum) or equivalent can also be used to aid separation.

Table 2 shows that little ^{45}Ca is lost at 0 °C (as in mitochondrial isolations), or at 30 °C provided that Na^+ is absent; overall the results are in good agreement with the enzyme and total Ca data (*Table 2* vs. *Table 1*). This system can also be applied to mitochondria prepared from tissue pre-perfused with ^{45}Ca (e.g. 0.2 mCi in a 20 ml re-cycling system for 1 h to equilibrate label), and is also useful for studying the egress mechanism(s) in isolation (usually with ruthenium red present to prevent any further uptake), e.g. looking at the inhibition of egress by divalent cations (23).

Table 2. Effects of various agents and incubation conditions on ^{45}Ca-egress from pre-loaded rat heart mitochondria[a]

Incubation time (min)	**Incubation temp.** (° C)	**^{45}Ca content (as % of initial, zero-time content)** without Na^+	with Na^+ (10 mM)
120	0–4	98	94
2	30	9	36[b]
4	30	80	11[b]
8	30	72	1[b]
2	37	84	11[b]
4, plus diltiazem	30	82	81

[a] Data are taken from ref. 40; incubation media contained respiratory substrates and EGTA.
[b] Indicates significant effect of Na^+.

2.4.3 Use of fura-2-loaded rat heart mitochondria

Recently it has become possible, at least with isolated rat heart mitochondria, to corroborate many of the findings made using the enzymes as matrix Ca^{2+}-probes by a direct chemical means following the entrapment of fluorescent Ca^{2+}-indicators into the mitochondrial matrix (refs 63–68; see also Chapter 1, this volume). A suitable method is described below (based on ref. 68) for fura-2 (*Protocol 9*). This methodology is readily adaptable to quin-2 (62), indo 1 (64), and also the fluorescent pH indicator BCECF (biscarboxyethyl-5(6)-carboxyfluorescein) (65), and also the new Mg^{2+} indicators (63) Mag-indo and Mag-fura; all of these compounds are available from Calbiochem, or Molecular Probes. The methodology is entirely analagous to that used to load cells with such indicators, but obviously involves intramitochondrial rather than cytosolic esterases to cleave the ester forms (69).

Protocol 9. Loading of rat heart mitochondria with fura-2

1. Take suitable volume (e.g. 60 μl for each run on fluorimeter) of original

Protocol 9. *Continued*

mitochondrial suspension (20–40 mg protein/ml in buffer B (*Protocol 1*)) into an Eppendorf tube, and add the acetoxymethyl ester of fura-2 (fura-2/AM) from a stock (at least 10 × required concentration prepared in DMSO) to give 5–10 μM (also do controls with DMSO alone).

2. Vortex briefly (1–2 sec) and incubate for 5 min at 30 °C; this is long enough to get sufficient loading, but not too long to lead to any damage of the mitochondria (68).

3. Transfer approx. 50 μl (1–2 mg protein) to quartz cuvette in fluorimeter (e.g. Perkin Elmer LS-5, with thermostatted (e.g. 30 °C) cuvette unit and magnetic stirrer below the cuvette) containing 2 ml of buffer D (*Protocol 3*) and e.g. 0.5 mM malonate/ 10 mM 2-oxoglutarate (to fully reduce NAD(P)H) and some EGTA initially. Prior to this, a wash step can be incorporated if required.

4. Measure at either 340 nm (ex) to get increases in fluorescence at 500 nm (em) as matrix Ca^{2+} is increased (as the result of adding extramitochondrial Ca^{2+} as EGTA-Ca buffers), or at 380 nm (ex)/500 nm (em) to get decreases, or at 362 nm (ex, the isosbestic point)/500 nm (em) to get no changes with Ca^{2+} and thus to correct for endogenous fluorescence and any non-fura-2 signals [usually very small (68)[a], any extramitochondrial fura-2 (usually less than 2%) can be checked for by doing the same Ca^{2+} additions in the presence of 1 μg ruthenium red/ml (68).

5. At the end of the experiment add either 0.1% (v/v) Triton X-100 to fully permeabilize the mitochondria, or 2 μM ionomycin with 1 μM FCCP to selectively permeabilize them for Ca^{2+} 68)].

6. Add sufficient Ca^{2+} to give F_{max} (amount of Ca^{2+} required calculated from calibrations made earlier using free-acid fura-2), followed by sufficient EGTA to remove all the Ca^{2+} and obtain F_{min} (or alternatively graded EGTA additions to give signals at different Ca^{2+} concentrations to do a calibration), followed by sufficient Mn^{2+} (in excess of amount of EGTA present) to get F_{Mn} and fully quench the signal from fura-2 (69); this ($F_{max} - F_{min}$ or F_{Mn}) (against previous calibrations) also gives a measure of the amount of fura-2 generated in each experiment.[b]

7. The signals are converted to matrix free Ca^{2+} as described in ref. 69, i.e.:

- $[Ca^{2+}]_{matrix} = K_d\,[(F - F_{min})/(F_{max} - F)]$
- ($F_{min} = F_{Mn} + (F_{max} - F_{Mn})/6$; if F_{min} is difficult to obtain (as it often can be))

(The K_d is previously obtained by varying Ca^{2+} using free acid fura-2 under conditions analogous to the experimental conditions; note that the K_d may be pH sensitive over the physiological range (68); this can even be exploited as at

Protocol 9. *Continued*

pH 7 the fura-2-Ca^{2+} saturation will not be reached until higher Ca^{2+} concentration; ref. 68). (See also Chapter 1, this volume).

[a] Fluorimeters are available which have 'choppers' to send light of, for example, 340 and 380 nm (ex) alternatively through the sample, thereby allowing use of a fluorescence ratio method for the signals (69) which maximize changes and makes suitable internal corrections (see Chapter 1, this volume).

[b] Typically, 50–80 pmol of fura-2/mg protein is loaded as above [representing about 25–40% of the added ester and corresponding to about 25–50 μM in the matrix (68)]; this results in suitable signals and complete hydrolysis of matrix fura-2, but not in substantial buffering of matrix Ca^{2+}.

A typical experiment is shown in *Figure 5*. Liver mitochondria have as yet proved much more difficult to load with the indicators (see, for example, refs 66 and 70); this appears to be due to partial hydrolysis of the esters and then the re-export of hydrolysed and partially-hydrolysed forms. Other types of mitochondria have also proved difficult to load, although there is a report (71) on rat brain mitochondria loaded with fura-2; however, the mitochondria were incubated with very high and potentially damaging Ca^{2+} concentrations in this study (millimolar). Fluorescence microscopy of loaded mitochondria attached to coverslips has also been performed (66). It is worth noting that the free acid appears to be very well retained by heart mitochondria, with less than 10% lost after 5 h on ice (65), so that bulk loading and then transfer back on ice for subsequent sampling can be carried out.

The activities of PDH and OGDH can also be measured simultaneously with fura-2 measurements (see Section 2.3 and refs 64, and 68) to get direct measurements of the matrix Ca^{2+} concentrations to which they respond.

Factors which influence the mitochondrial Ca^{2+}-transport system can be readily studied using the fura-2-loaded mitochondria (63–68) (*Figure 5*). Noticeably, in the presence of physiological concentrations of Na^{+} and Mg^{2+}, there is a large degree of positive co-operativity in the transfer of the Ca^{2+} signal into the matrix (68); this is likely to be mostly due to the inhibition of Na^{+}-induced Ca^{2+}-egress by increases in extramitochondrial Ca^{2+} (68) (*Figure 1*). There is also considerable evidence that mitochondria may become loaded with fura-2 and similar indicators when located within intact cells (see, for example, refs 65 and 72), so that in the not too distant future and using fluorescence-imaging (see Chapter 1, this volume), it may become possible to directly measure changes in intramitochondrial Ca^{2+} in intact cells.

2.5 The use of ruthenium red in intact tissues

Following on from the last point above, it is also worth pointing out in this context that, again at least with rat heart, ruthenium red appears to be able to exert its inhibitory effects on mitochondrial Ca^{2+}-uptake in the intact tissue (4, 73–75). Thus it has been shown that the activation of PDH by positive

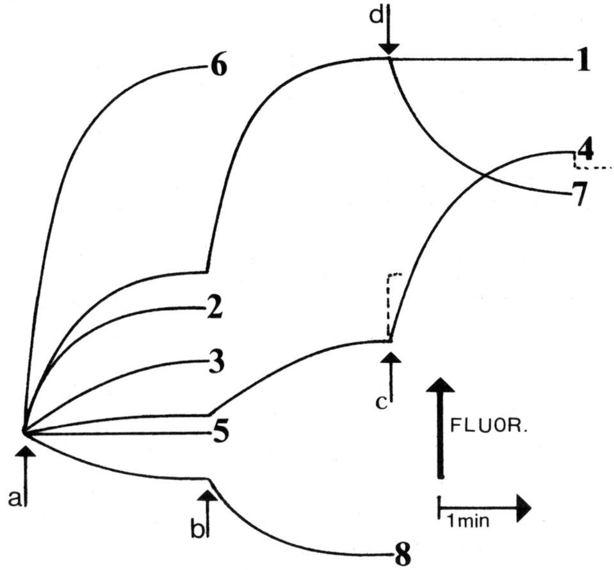

Figure 5. Effects of increases in extramitochondrial Ca^{2+} on matrix Ca^{2+} in rat heart mitochondria incubated under various conditions, assessed by monitoring the fluorescence changes of entrapped fura-2. The fura-2-loaded mitochondria (about 1 mg protein) were incubated at 30 °C in 2 ml of buffer D (*Protocol 3*) containing initially 0.5 mM malonate, 10 mM 2-oxoglutarate [to fully reduce $NAD(P)^+$], and 0.5 mM EGTA, together with: **1**, **7**, and **8**, no further additions; **2**, 10 mM NaCl; **3**, 2 mM $MgCl_2$; **4**, Na plus Mg; **5**, 2 μg ruthenium red; **6**, 0.5 mM spermine. Wavelength settings were 340 nm (ex)/500 nm (ex) or (8 only) 380 nm (ex)/500 nm (em). After a 2-min pre-incubation, additions were made at the arrows as shown: **a**, 0.25 mM EGTA plus 0.25 mM $CaCl_2$ (to give about 25 nM free Ca^{2+}); **b**, 0.75 mM EGTA plus 0.75 mM $CaCl_2$ (105 nM); **c**, (solid line) 4.5 mM EGTA plus 4.5 mM $CaCl_2$ (480 nM); **d**, 10 mM NaCl either (**1**) with or (**7**) without, 300 μM diltiazem. The dashed lines show what happens on adding 1 μM FCCP plus 2 μM ionomycin at the points shown to equilibrate intra- and extramitochondrial Ca^{2+}, indicating that as extramitochondrial Ca^{2+} is raised (plus Na and Mg) the gradient of Ca^{2+} (in:out) across the inner membrane increases. At 362 nm (ex)/500 nm (em) (isosbestic for fura-2–Ca^{2+}) the conditions of **1** would yield a trace similar to that shown for **5**. After the runs the signals can be calibrated and matrix Ca^{2+} calculated after, for example, permeabilization with 0.1% (v/v) Triton X-100 followed by 5 mM HEDTA plus 5 mM $CaCl_2$ (to give 6.3 μM free Ca^{2+}) to get F_{max} and then 10 mM $MnCl_2$ to get F_{Mn} (see *Protocol 9*).

inotropic agents can be prevented by ruthenium red [at about 2.5 μM in heart perfusions (73)], whereas activation by pyruvate (as a PDH_a kinase activator) is unaffected. The ruthenium red was found to have some negative inotropic effects in unstimulated hearts (see refs 4 and 73); however, after stimulation the same degree of inotropy was reached in either its absence or presence, and the Ca^{2+}-dependent activation of cytosolic phosphorylase was unaffected (73). Moreover, in quin-2-loaded myocytes exposed to various interventions designed to elicit increases in cytosolic Ca^{2+} (74), ruthenium red did not affect these increases but again prevented the associated increases in PDH_a. In exploiting this, experiments on perfused hearts using ^{31}P-NMR and ruthenium red have provided strong evidence for a key role of mitochondrial Ca^{2+} in the enhancement of oxidative metabolism under such stimulated conditions (4, 76); this may also explain the pronounced stability of the cellular phosphorylation potential under such conditions (4, 76). In addition, and with relevance to Section 3 below, ruthenium red has been shown to protect hearts and their mitochondria against the excessive Ca^{2+}-accumulation which is associated with conditions such as ischaemia-reperfusion (see, for example, ref. 75). There are also indications that ruthenium red may inhibit mitochondrial Ca^{2+} uptake in other intact mammalian tissue and cell types (see ref. 77).

3. Mitochondrial permeabilization

The mechanism of Ca^{2+}-induced permeabilization has not been resolved, but there are quite firm indications that it involves the reversible opening of an inner mitochondrial membrane pore (9). From the relative permeability to solutes of different size, a pore diameter of about 2 nm is indicated, allowing permeation of small solutes (mol. wt less than 2 kd) at a rate dependent on their size (10). Since H^+ permeates the open pore rapidly, pore opening uncouples respiration and an indication of pore state may be obtained from measurements of inner membrane potential and respiratory control (11). Measurements of this sort, however, provide an indication at best, since interventions of any kind that induce H^+ leakiness or inhibit the respiratory chain will depress membrane potential and decrease respiratory control, so that these parameters do not necessarily reflect non-specific permeabilization. Other indirect measurements may also be ambiguous. For example, pore opening allows leakage of low-molecular-weight matrix constituents, but it may be difficult to exclude the operation of normal transport mechanisms. Pore opening is also associated with light scattering changes consistent with mitochondrial swelling, but Ca^{2+} accumulation may increase matrix volume irrespective of permeabilization (9) (see Section 2.1).

In this laboratory (M. Crompton's) we have developed an assay of pore state in isolated mitochondria which is based on the fundamental property of non-specific permeation. The technique measures the entry and entrapment in the matrix space of solutes that are normally impermeant but which permeate the open pore. In practice, we have found that the measurement of ^{14}C-sucrose permeation is suitable for most purposes. The entrapment technique utilizes the fact that Ca^{2+}-induced pore opening is reversible and, although influenced by a number of factors, has an obligatory requirement for Ca^{2+} (*Figure 1*) so that when Ca^{2+} is chelated, e.g. with EGTA, pore closure invariably occurs. In this way solutes that have entered the matrix owing to Ca^{2+}-induced permeabilization become entrapped in the matrix space, and the degree of entrapment may be used to determine the extent of permeabilization. This basic procedure may be adapted for measurement of rates of permeabilization, rates of resealing on Ca^{2+} removal, and rates of solute permeation via the pore.

3.1 Measurement of mitochondrial permeabilization

This section describes how mitochondria isolated from rat heart and liver may be permeabilized by Ca^{2+} and how permeabilization may be measured using the ^{14}C-sucrose entrapment procedure.

3.1.1 The conditions required for mitochondrial permeabilization

In order to obtain maximal permeabilization mitochondria must accumulate relatively large amounts of Ca^{2+}. Liver mitochondria become maximally permeabilized on accumulation of more than 25 nmol/mg protein, whereas rather higher amounts are required in the case of heart mitochondria (more than 50 nmol/mg). It should be noted, however, that maximal permeabilization does not necessarily imply that all mitochondria are in the fully permeabilized state at any point in time since there is evidence of a steady-state distribution between open and closed states of the pore (10, 11). In this chapter, conditions of maximal permeabilization are taken empirically as those which allow ^{14}C-sucrose to permeate all mitochondria within 5 sec at 25 °C. A further point is that although Ca^{2+} is obligatory, Ca^{2+} accumulation alone does not induce permeabilization, either P_i or oxidative stress also being required (*Figure 1*).

Conditions for Ca^{2+}-induced permeabilization of isolated mitochondria are outlined below [(a)–(b)]; this involves a permeabilization procedure employing Ca^{2+} plus P_i, which are added to mitochondria oxidizing succinate.

(a) Liver or heart mitochondria (suspended in buffer with low (about 10 μM) EGTA) are pre-incubated for 3 min at 25 °C in the following medium (pH 7.2): 120 mM KCl, 5 mM sucrose, 5 mM Hepes (K^+ salt), 3 mM succinate, 1 mg mitochondrial protein/ml, 1 μg rotenone/ml.
(b) Permeabilization is started by addition of 5 mM KH_2PO_4 plus $CaCl_2$ (25 μM for liver mitochondria; more than 50 μM for heart mitochondria). Permeabilization is maximal in 4–8 min.

Under the conditions described, the mitochondria accumulate more than 95% of the added Ca^{2+} within 1 min and become maximally permeabilized in 4–8 min. In the absence of P_i, permeabilization requires oxidative stress (9, 12), i.e. oxidation of the mitochondrial redox couples, NADH/NAD^+, NADPH/$NADP^+$ and GSH/GSSG. Oxidative stress may be imposed by 100 μM *t*-butylhydroperoxide which oxidizes intramitochondrial GSH to GSSG (glutathione peroxidase); this leads in turn to oxidation of NADPH (glutathione reductase) and NADH.

3.1.2 Measurement of permeabilization by ^{14}C-sucrose entry and entrapment in the matrix space

Permeabilization is sufficiently slow that the necessary manipulations may be made manually. A procedure to measure the time course of heart mitochondrial permeabilization at 25 °C (9) is now outlined in *Scheme A* and described in *Protocol 10*.

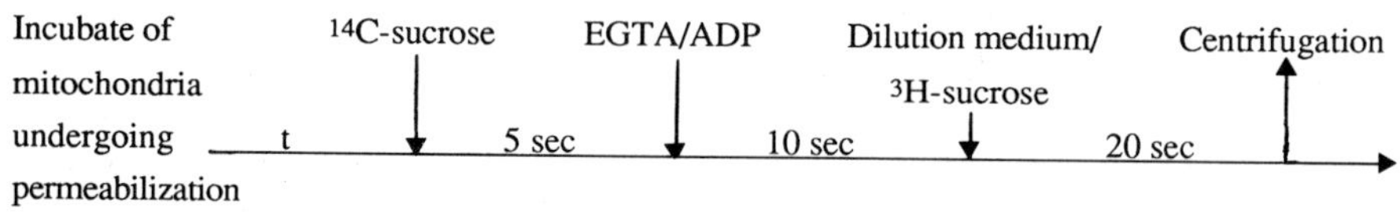

Scheme A

Protocol 10. Mitochondrial permeabilization

1. Commence mitochondrial permeabilization by addition of KH_2PO_4 plus $CaCl_2$ to respiring mitochondria as detailed in (a)–(b) above.
2. At intervals of time thereafter (*t* in *Scheme A*), introduce 0.3-ml portions of the incubate into tubes containing 3 μl of 5 mM [U-^{14}C]-sucrose (approx. 0.2 μCi). Mix immediately. ^{14}C-sucrose is normally impermeant but enters and equilibrates with the matrix space of that fraction of the mitochondria that are permeabilized within 5 sec at 25 °C.
3. Commence mitochondrial resealing by addition of 3 μl of 100 mM EGTA/ 50 mM ADP. The inclusion of ADP ensures rapid resealing; e.g. rat heart mitochondria are resealed completely within 5 sec by EGTA plus ADP, whereas about 1 min is required with EGTA alone (Section 3.2).
4. When resealing is complete (which should be checked independently, Section 3.2) dilute the suspension about fivefold with 120 mM KCl containing 5 mM ^{3}H-sucrose (approx. 0.1 μCi/ml).
5. Sediment the mitochondria by centrifugation in an Eppendorf bench centrifuge (10 000 *g*, 1 min).
6. Determine the ^{14}C and ^{3}H contents of the supernatant and pellet. For this, withdraw a known volume of the supernatant for scintillation counting and as much of the remaining supernatant as possible by aspiration without disturbing the pellet. Dissolve pellet in 2% (v/v) Triton X-100.
7. The supernatant ^{14}C and ^{3}H values are used to correct the pellet ^{14}C values for ^{14}C in the extramitochondrial spaces of the pellets giving ^{14}C entrapped. Sucrose entrapment may be calculated from the following

Protocol 10. *Continued*

relation, where S signifies supernatant, P pellet, and the dilution factor = 5: ^{14}C-sucrose entrapped (nmol/mg protein) =

$$\frac{\{^{14}C_P - [(^{14}C/^3H)_S \times {}^3H_P]\}\ \{[\text{sucrose}]\ \text{of incubate, mM}\}}{(\text{mg pellet protein})\ (^{14}C_S\ \text{per microlitre})\ (\text{dilution factor})}$$

The following points may be noted:

(a) The purpose of dilution is simply to increase the accuracy of the measurements by decreasing the 'background' ^{14}C counts contained in the extramitochondrial space relative to the entrapped ^{14}C values. Dilution is permissible since the rate of efflux of entrapped ^{14}C-sucrose on dilution is very slow (e.g. the half-time is more than 10 min with heart mitochondria at 25 °C (9)). Neverthelesss, it is essential to centrifuge soon after dilution to ensure that losses are negligible.

(b) Maximal permeabilization of heart and liver mitochondria leads to the entry and entrapment of 5–10 nmol of ^{14}C-sucrose/mg of protein when the incubation medium contains 5 mM sucrose. This corresponds to sucrose entry and entrapment in a permeabilized matrix space of 1–2 μl/mg protein. It is therefore quite straightforward to convert entrapment values to values of permeabilized matrix space. An example of the increase in permeabilized matrix space with time is given in *Figure 6*.

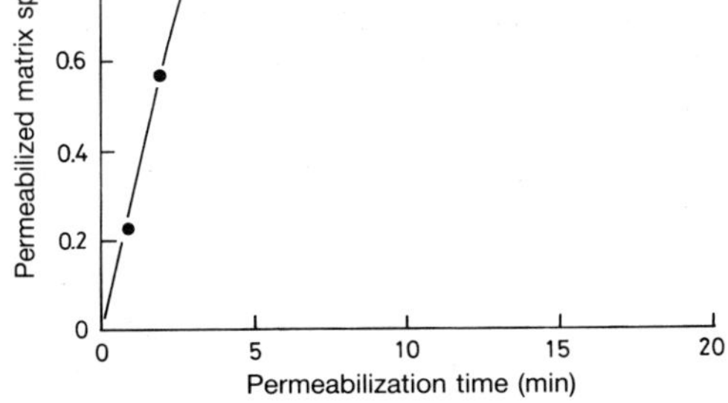

Figure 6. The time-course of rat heart mitochondrial permeabilization in the presence of Ca^{2+} and P_i. Mitochondria were incubated at 25 °C as stated in Section 3.1.1 [(a)–(b)]. Permeabilization was started with 5 mM KH_2PO_4 + 60 μM $CaCl_2$ (zero time in the figure). At intervals thereafter (abscissa) portions of the incubate were transferred to tubes containing ^{14}C-sucrose and entrapment was measured as described in Section 3.1.2. Data are from ref. 9.

(c) The term 'permeabilized matrix space' implies a particular mechanism in which mitochondria become permeabilized relatively slowly and on permeabilization rapidly attain a state of high permeability. This *appears* to be true since the same time-course of ^{14}C-sucrose entrapment is obtained when the time interval between addition of ^{14}C-sucrose and EGTA/ADP is prolonged (9).

3.2 Measurement of mitochondrial resealing

3.2.1 General procedure

The ^{14}C-solute entrapment technique may be adapted to measure the rate at which permeabilized mitochondria reseal on Ca^{2+} chelation with EGTA (9). The procedure is now outlined in *Scheme B* and *Protocol 11*: the composition of the various media added (and also for the pulsed flow experiments later; Section 3.2.2) is given below the scheme.

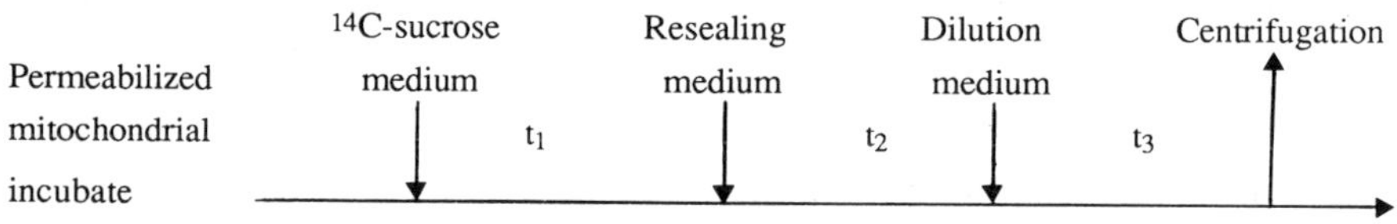

Scheme B

(a) Permeabilized mitochondrial incubate: as in Section 3.1.1

(b) ^{14}C-sucrose medium: 120 mM KCl, 5 mM Hepes, 5 mM KH_2PO_4, 3 mM succinate, ^3H_2O (3 μCi/ml), 5 mM ^{14}C-sucrose (1 μCi/ml), pH 7.2.

(c) Resealing medium: 120 mM KCl, 5 mM Hepes, 5 mM KH_2PO_4, 3 mM succinate, ^3H_2O (1.5 μCi/ml), 5 mM ^{14}C-sucrose (0.5 μCi/ml), 10 mM EGTA, pH 7.2.

(d) Dilution medium: 120 mM KCl, 5 mM Hepes, 5 mM KH_2PO_4, 3 mM succinate, 5 mM EGTA, pH 7.2.

Protocol 11. Mitochondrial resealing

1. Mix maximally permeabilized mitochondria (Section 3.1.1) with an equal volume of medium containing ^{14}C-sucrose [(b) above]. The ^{14}C-sucrose equilibrates with the matrix space in time t_1 (*Scheme B)*; it is essential that the ^{14}C-sucrose equilibrates completely and with permeabilized rat heart or liver mitochondria that is achieved quickly (less than 5 sec at 25 °C).
2. Add an equal volume of resealing medium [(c) above] containing EGTA to the permeabilized mitochondria to commence resealing. It should be noted that the resealing medium contains ^3H_2O and ^{14}C-sucrose to prevent their dilution at this stage. Vary the time interval t_2 (*Scheme B*) so that varying degrees of resealing occur during this period.

Protocol 11. *Continued*

3. Dilute the incubate with a fivefold vol. of dilution medium [(d) above]. Sediment the mitochondria by centrifugation after a time interval t_3 (*Scheme B*). It is important the time period $t_2 + t_3$ is sufficient to allow complete resealing before centrifugation. The rate of EGTA-induced resealing is stimulated by several factors, but the slowest rate of resealing of heart mitochondria (with EGTA alone) normally yields complete resealing within 1 min at 25 °C.
4. After centrifugation, measure the ^{14}C and ^{3}H contents of the supernatant and the mitochondrial pellet, and calculate entrapped ^{14}C-sucrose from the relation given in *Protocol 10* (step 7).

The basic principle of the technique is that on dilution matrix $^{3}H_2O$ would be diluted whereas ^{14}C-sucrose entrapped within resealed mitochondria would not. The degree of ^{14}C-sucrose entrapment depends on the degree of resealing attained in t_2 (*Scheme B*) as well as other factors (Section 3.2.3 below).

3.2.2 The pulsed flow technique

Under many conditions resealing is too fast to be followed by manual procedures, and the rapid mixing, pulsed flow, device of *Figure 7* is used for this purpose (9, 10). The device consists of five syringes (S) each of which is

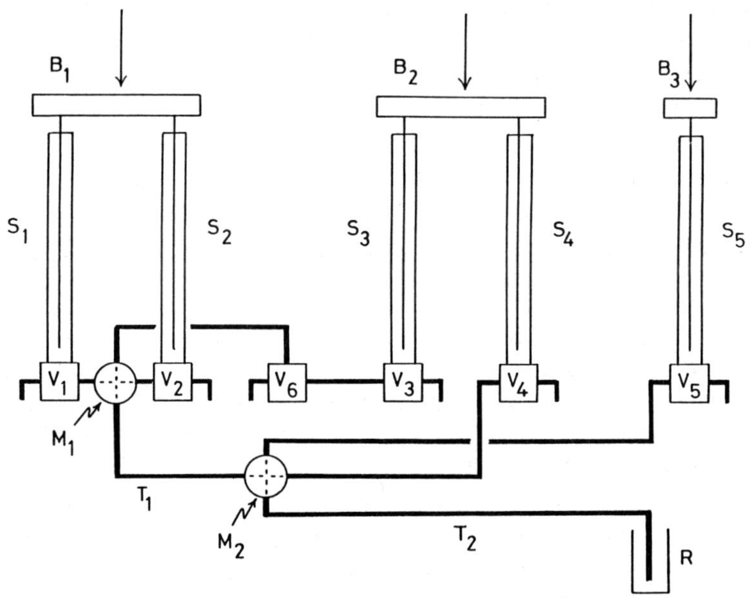

Figure 7. The pulsed flow apparatus. For explanation see Section 3.2.2.

attached to a 3-way valve (V) and connected by tubing to two 4-way mixing chambers (internal diameter 1.6 mm). Syringes S_1 and S_2 are evacuated simultaneously by a compressed air-driven piston acting on bar B_1 and controlled by a solenoid valve. Syringes S_3 and S_4 are evacuated simultaneously in the same way, and S_5 is evacuated independently. The syringes are loaded via the valves V_1–V_5; valve V_6 serves to flush out between runs. The syringes are loaded as follows:

- S_1, permeabilized mitochondrial incubate (200 μl) (Section 3.1.1)
- S_2, ^{14}C-sucrose medium (200 μl) [(b) in Section 3.2.1.]
- S_3, 120 mM KCl (400 μl)
- S_4, resealing medium (400 μl) [(c) in Section 3.2.1]
- S_5, 120 mM KCl (800 μl)

In operation, the contents of S_1 and S_2 are mixed in M_1 by 180 degree opposed flows (>7 m.sec^{-1}) and the mixture collects in tubing T_1 where ^{14}C-sucrose entry takes place. After time delay t_1 (*Scheme B*), evacuation of S_3 and S_4 mixes the contents of T_1 and S_4, and the mixture collects in T_2 where resealing occurs. After a further time delay t_2 (*Scheme B*) evacuation of S_5 drives the contents of T_2 into a fivefold vol. of dilution medium contained in a flat-bottomed receiver R. The diluted suspension is then centrifuged and analysed as described above (Section 3.2.1).

3.2.3 Interpretation of ^{14}C-solute entrapment data

In measurements of the rate of resealing, the relation between ^{14}C-sucrose entrapment (S) and t_2 cannot be interpreted directly in terms of resealing vs. time, since one must take account of the rate at which ^{14}C-sucrose effluxes from the matrix space of unresealed mitochondria on dilution. In effect, this means that one must assume a model for the way in which pore closure takes place. Two such models have been considered (9), one in which pores of the mitochondrial population as a whole close exponentially with time, and one in which pores of any particular mitochondrion close in synchrony (termed resealing) and mitochondria reseal exponentially with time. In practice, entrapment data are empirically consistent with the latter, synchronous pore closure model. Thus solute entrapment (S_t) at any time t_2 may be described by the following relation (9), using the symbols defined below it:

$$S_t = S_\infty \, [1 - e^{-k_r t_2} \, (1 - e^{-k_p/k_r})],$$

S_t = solute entrapment at time t
S_∞ = solute entrapment at time ∞
k_r = apparent rate-constant for mitochondrial resealing
k_p* = apparent rate-constant for solute permeation in the presence of Ca^{2+}

k_p = apparent rate-constant for solute permeation in the absence of Ca^{2+}, i.e. of the unresealed fraction at any point in time during resealing.

Data are best analysed according to the linear form of the equation (10):

$$\ln [1 - S_t/S_\infty] = -k_r t_2 + \ln [1 - e^{-k_p/k_r}] \quad \text{[eqn 1]}$$

Plots of $\ln [1 - S_t/S_\infty]$ against t_2 yield k_r (-slope), the apparent rate-constant for resealing. A typical set of pulsed flow data plotted in this way (*Figure 8*) shows that the rate of EGTA-induced resealing is markedly stimulated by ADP.

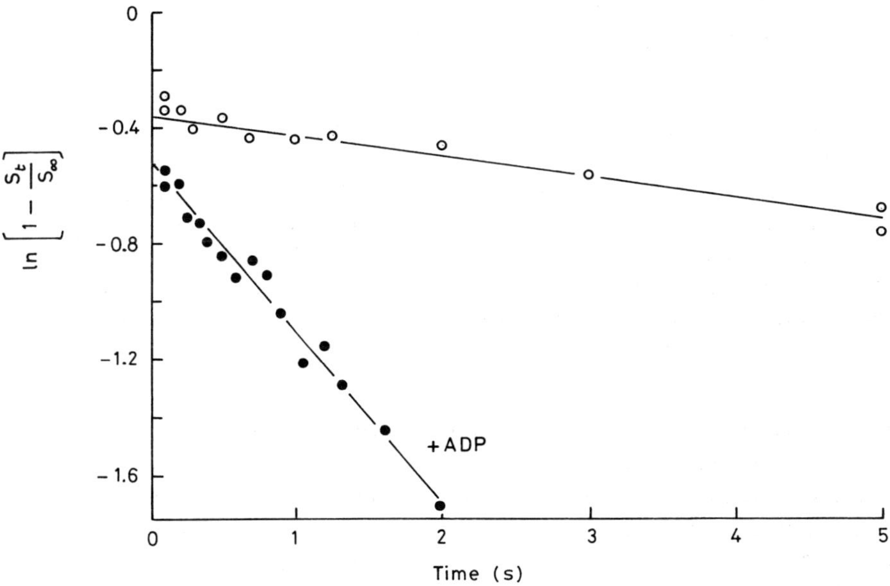

Figure 8. Pulsed flow measurement of mitochondrial resealing. Rat heart mitochondria were maximally permeabilized by incubation for 10 min, as described in Section 3.1.1 [(a)–(b)]. The permeabilized mitochondria were then subjected to pulsed flow at 25 °C according to *Scheme B* with t_1 = 30 sec, t_2 = 0.1–5 sec and t_3 = 60 sec. The resealing medium contained EGTA ± 200 μM ADP. ^{14}C-sucrose entrapment vs. t_2 is analysed according to equation 1. (Reprinted by permission from the *Biochemical Journal*, **266**, 33–9, copyright ©, 1990, The Biochemical Society, London.)

3.3 Measurement of the rate of solute permeation in permeabilized mitochondria

The ^{14}C-solute entrapment technique may be used to measure the rates of ^{14}C-solute permeation into and out of permeabilized mitochondria. One would anticipate that the rate-constants for influx and efflux of any particular solute under the same experimental conditions would be equal, and this is certainly true for ^{14}C-sucrose permeation during resealing (10). Different techniques are used to measure ^{14}C-solute permeation in the presence of Ca^{2+} (k_p*) and during resealing on Ca^{2+} removal (k_p, see above definitions).

3.3.1 Measurement of ^{14}C-sucrose efflux during resealing on Ca^{2+} removal

Since the analysis described in Section 3.2.3 for the determination of the apparent rate-constant for resealing accounts for the rate of ^{14}C-sucrose efflux from non-resealed mitochondria on dilution, the procedure also provides a means of measuring this parameter. When the data are analysed according to equation 1 (as in *Figure 8*) k_p is calculated from the ordinate intercept and the value of k_r. The value of k_p thus determined refers to the permeability of that fraction of the mitochondrial population that is unresealed at any point in time during resealing. The linearity of the plots enables an accurate determination of apparent permeation rate-constants.

3.3.2 Measurement of ^{14}C-sucrose influx in the presence of Ca^{2+} in the absence of net resealing

By a slight modification of *Scheme B* (Section 3.2.1) one may obtain an apparent rate-constant for solute entry into permeabilized mitochondria in the presence of Ca^{2+}. The modifications are:

(a) t_1 is varied (e.g. from 50 msec to 1 sec) so that ^{14}C-sucrose entry is incomplete in this period.

(b) t_2 is prolonged so that complete mitochondrial resealing occurs before dilution. In practice, the time required for resealing is minimized by inclusion of ADP (at least 400 μM) in the resealing medium (*Figure 9*).

With these modifications, ^{14}C-solute entrapment at any time t_1 (S_t) is given by (9, 10):

$$S_t = S_\infty \, [1 - e^{-k_p^* t_1 - (k_p/k_r)}]$$

which rearranges to the linear form:

$$\ln\,[1 - S_t/S_\infty] = -k_p^* t_1 - k_p/k_r \qquad \text{[eqn 2]}$$

Plots of $\ln\,[1 - S_t/S_\infty]$ versus t_1 yield k_p*, the apparent rate-constant for solute permeation in the presence of Ca^{2+}. A typical example of this procedure is given in *Figure 9*.

Clearly, the technique is limited by the rate of resealing (used to stop permeation). If permeation is fast in relation to the rate of resealing, then solute permeation approaches maximal values (i.e. equilibration) irrespective of the interval t_1. We have found that if resealing is induced by EGTA plus ADP the procedure is quite adequate for the determination of permeability

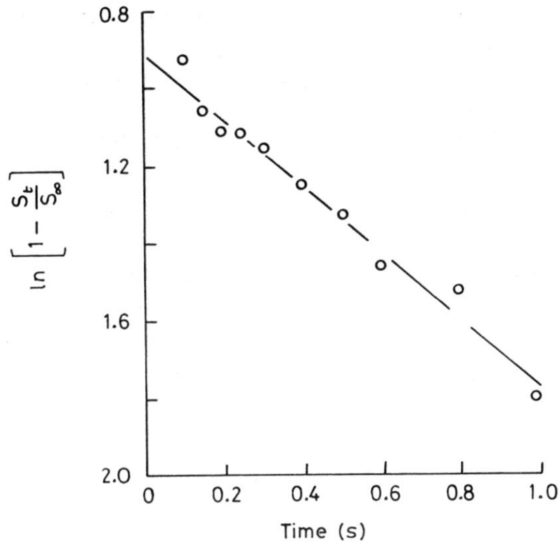

Figure 9. Pulsed flow measurement of ^{14}C-sucrose permeation into permeabilized mitochondria. Rat heart mitochondria were maximally permeabilized as described in Section 3.1.1 [(a)–(b)]. The permeabilized mitochondria were subjected to pulsed flow at 25 °C according to *Scheme B* with $t_1 = 0.1$ sec, $t_2 = 30$ sec, and $t_3 = 30$ sec. The resealing medium contained EGTA + ADP. ^{14}C-sucrose entrapment data versus t_1 are analysed according to equation 2. (Reprinted by permission from the *Biochemical Journal*, **266**, 33–9, copyright ©, 1990. The Biochemical Society, London.)

rate-constants for sucrose and larger solutes in rat heart mitochondria, but flux measurements of smaller, more rapidly permeating solutes would require ways of inducing more rapid resealing (10).

3.4 Practical applications of reversible permeabilization

3.4.1 The manipulation of matrix metabolites

Since permeabilization allows free diffusion of low-molecular-weight solutes across the inner membrane, it may be used to manipulate the low-molecular-weight solute composition of the matrix space. The range of matrix components that may be changed in this way is limited by the size of the pore. One would expect a sharp cut-off in the permeability as the size of the permeating solute approaches the internal dimensions of the pore, and a cut-off is indeed indicated at a solute molecular weight of 1.5 kd (10, 3). It should also be borne in mind that permeating solutes of differing size permeate at different rates. For example, whereas half-time for permeation of sucrose (molecular weight 342) into maximally permeabilized heart mitochondria is less than 1 sec at 25 °C, the half-time for permeation of the indicator arsenazo-III (molecular weight 820) is approximately 30 sec (10). However, heart mitochondria, at least, may be maintained in the maximally permeabilized state for at least 15 min without impairment of resealing capacity (9), which is ample time for depletion of most matrix metabolites. Introduction of selected metabolites to the permeabilized mitochondria prior to resealing allows their incorporation into the matrix space. Thus, liver mitochondria have been depleted and replenished with NAD^+ in this manner (11).

3.4.2 Introduction of Ca^{2+} indicators into the matrix

Reversible Ca^{2+}-induced permeabilization has been used to introduce the Ca^{2+}-indicator arsenazo-III into the matrix component of heart and liver mitochondria (78, 79). In this procedure, high concentrations of arsenazo-III are used as Ca^{2+} chelators to induce resealing. Sufficient quantities of the indicator enter the matrix space and become entrapped to allow spectrophotometric measurements of matrix free [Ca^{2+}]. Of course, the losses of matrix metabolites during permeabilization mean that assays employing entrapped arsenazo-III relate to a metabolite-depleted environment. Nevertheless, the re-introduction of selected metabolites allows specific effects of metabolites on intramitochondrial Ca^{2+}-dependent processes to be investigated. A procedure for loading rat heart mitochondria with arsenazo-III is described in *Protocol 12*.

Protocol 12. Permeabilizing and re-sealing mitochondria to load arsenazo-III

1. Maximally permeabilize mitochondria as described in Section 3.1.1 [(a), (b)]. In order to minimize the amount of arsenazo-III added subsequently (step 2 below), the concentration of mitochondrial protein should be increased to 10 mg/ml and the [Ca^{2+}] to 300 μM. Metabolites that need to be re-introduced into the matrix space should also be added at this stage.
2. After 10 min add 0.1 ml of 70 mM arsenazo-III (K^+ salt, pH 7.0)/ml incubate. This concentration decreases free [Ca^{2+}] below 1 μM and induces resealing.
3. After a further 10 min, sediment mitochondria by centrifugation (40 000 *g* for 1 min).
4. Wash the mitochondrial pellet by resuspension in 210 mM mannitol/70 mM sucrose/10 mM Hepes (pH 7.0) at a concentration of about 1 mg protein/ml, followed by re-centrifugation. Repeat twice. The final mitochondrial pellet is deep purple in colour, characteristic of the Ca^{2+}/arsenazo-III complex, with little or no colour visible in the supernatant.
5. Suspend the mitochondrial pellet in mannitol/sucrose/Tris medium (above) at a concentration of 40–50 mg/ml. Conserve on ice. Rat heart mitochondria loaded in this manner typically contain 2–6 nmol arsenazo-III/mg of protein.

Arsenazo-III-loaded mitochondria accumulate and release Ca^{2+} similar to normal mitochondria. As an example of Ca^{2+} flux measurements, those involving Na^+–Ca^{2+} carrier activity are reported in *Figure 10*. Calibration of intramitochondrial arsenazo-III absorbance versus intramitochondrial free [Ca^{2+}] should be made in incubation medium containing Ca^{2+} buffers (see Chapter 7, this volume) and 1 μM each of the Ca^{2+}-ionophore A23187 and nigericin which serve to abolish gradients of free [Ca^{2+}] and pH respectively across the inner membrane.

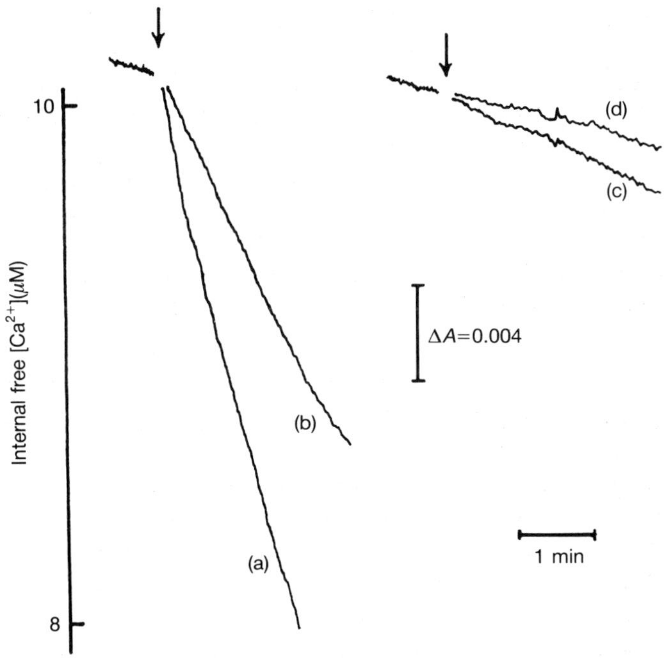

Figure 10. Spectrophotometric measurement of Ca^{2+} fluxes using arsenazo-III-loaded mitochondria. Arsenazo-III-loaded heart mitochondria (3 mg of protein) were suspended in 1.5 ml of incubation medium (pH 6.8) containing 120 mM KCl, 10 mM Hepes, 0.2 mM KH_2PO_4, 4 μM ruthenium red, and 5 mM succinate. Absorbance changes (*A*) were measured in a dual wavelength spectrophotometer operating at 675 minus 685 nm. The following additions were made at the arrows: (a) 10 mM NaCl plus 100 μM EGTA; (b) 10 mM NaCl plus Ca^{2+}/100 μM EGTA (2 μM free Ca^{2+}); (c) 100 μM EGTA; (d) Ca^{2+}/100 μM EGTA (2 μM free Ca^{2+}). Intramitochondrial free [Ca^{2+}] was calibrated in the same incubation medium containing Ca^{2+}/HEDTA (1 mM) buffers, 1 μM A23187 and 1 μM nigericin. (Reprinted by permission from the *Biochemical Journal*, **244**, 533–8, copyright, ©, 1987, The Biochemical Society, London.)

Acknowledgements

Our studies have been supported by grants from the MRC and SERC (UK), the British Heart Foundation, the University of London Central Research

Fund, and the Lister Institute. J.G. McC. was a Lister Institute Research Fellow.

References

1. McCormack, J. G., Halestrap, A. P., and Denton, R. M. (1990). *Physiological Reviews*, **70**, 391–425.
2. Crompton, M. (1985). *Current Topics in Membranes and Transport*, **25**, 231–76.
3. Hansford, R. G. (1985). *Reviews in Physiology, Biochemistry, and Pharmacology*, **102**, 1–72.
4. Unitt, J. F., McCormack, J. G., Reid, D., MacLachlan, L. K., and England, P. J. (1989). *Biochemical Journal*, **262**, 293–301.
5. Fiskum, G. and Lehninger, A. L. (1982). In *Calcium and cell function*, Vol. 2 (ed. W. Y. Cheung), pp. 39–80. Academic Press, New York.
6. Akerman, K. E. O. and Nicholls, D. G. (1983). *Reviews in Physiology, Biochemistry, and Pharmacology*, **95**, 149–201.
7. Wendt-Gallitelli, M. F. (1986). *Basic Research in Cardiology*, **81**, 25–32.
8. Poole-Wilson, P. A., Harding, D. P., Bourdillon, P. D. V., and Tones, M. A. (1984). *Journal of Molecular and Cellular Cardiology*, **16**, 175–85.
9. Crompton, M. and Costi, A. (1988). *European Journal of Biochemistry*, **178**, 489–501.
10. Crompton, M. and Costi, A. (1990). *Biochemical Journal*, **266**, 33–9.
11. Al Nasser, I. and Crompton, M. (1986). *Biochemical Journal*, **239**, 31–40.
12. Crompton, M., Costi, A., and Hayat, L. (1987). *Biochemical Journal*, **245**, 915–18.
13. Haworth, R. A. and Hunter, D. R. (1980). *Journal of Membrane Biology*, **54**, 231–6.
14. Crompton, M., Ellinger, H., and Costi, A. (1988). *Biochemical Journal*, **255**, 357–60.
15. Crompton, M. (1990). In *Calcium and the heart* (ed. G. A. Langer), pp. 167–78. Raven Press, New York.
16. Denton, R. M., and Randle, P. J., and Martin, B. R. (1972). *Biochemical Journal*, **128**, 161–3.
17. Denton, R. M., Richards, D. A., and Chin, J. G. (1978). *Biochemical Journal*, **176**, 899–906.
18. McCormack, J. G. and Denton, R. M. (1979). *Biochemical Journal*, **180**, 533–44.
19. Marshall, S. E., McCormack, J. G., and Denton, R. M. (1984). *Biochemical Journal*, **218**, 249–60.
20. Rutter, G. A. and Denton, R. M. (1989). *Biochemical Journal*, **263**, 445–52.
21. Rutter, G. A. and Denton, R. M. (1988). *Biochemical Journal*, **252**, 181–98.
22. Denton, R. M., McCormack, J. G., and Edgell, N. J. (1980). *Biochemical Journal*, **190**, 107–17.
23. McCormack, J. G. and Osbaldeston, N. J. (1990). *European Journal of Biochemistry*, **192**, 239–44.
24. McCormack, J. G. and Denton, R. M. (1981). *Biochemical Journal*, **196**, 619–24.
25. McCormack, J. G. and Denton, R. M. (1986). *Trends in Biochemical Sciences*, **11**, 258–62.
26. Colman, R. F. (1975). *Advances in Enzyme Regulation*, **13**, 413–33.

27. Teague, W. M., Pettit, F. H., Wu, T. L., Silberman, S. R., and Reed, L. J. (1982). *Biochemistry*, **21**, 5585–92.
28. Rutter, G. A. and Denton, R. M. (1989). *Biochemical Journal*, **263**, 453–62.
29. Pullman, M. E. and Monroy, G. C. (1963). *Journal of Biological Chemistry*, **238**, 3762–9.
30. Yamada, E. W., Shiffman, F. H., and Huzel, N. J. (1980). *Journal of Biological Chemistry*, **255**, 267–273.
31. Yamada, E. W. and Huzel, N. J. (1985). *Cell Calcium*, **6**, 469–79.
32. Yamada, E. W. and Huzel, N. J. (1988). *Journal of Biological Chemistry*, **263**, 11498–503.
33. Davidson, A. M. and Halestrap, A. P., (1987). *Biochemical Journal*, **246**, 715–23.
34. Davidson, A. M. and Halestrap, A. P. (1988). *Biochemical Journal*, **254**, 379–84.
35. Davidson, A. M. and Halestrap, A. P. (1989). *Biochemical Journal*, **258**, 817–21.
36. Corkey, B. E., Duszynski, T. L., Rich, B., Matschinsky, B., and Williamson, J. R. (1986). *Journal of Biological Chemistry*, **261**, 2567–74.
37. Jung, D. W., and Brierley, G. P. (1986). *Journal of Biological Chemistry*, **261**, 6408–15.
38. Halestrap, A. P. (1989). *Biochimica et Biophysica Acta*, **973**, 355–82.
39. McCormack, J. G., Bromidge, E. S., and Dawes, N. J. (1988). *Biochimica et Biophysica Acta*, **934**, 282–92.
40. McCormack, J. G. and Denton, R. M. (1984). *Biochemical Journal*, **218**, 235–47.
41. McCormack, J. G. (1985). *Biochemical Journal*, **231**, 597–608.
42. McCormack, J. G. and Denton, R. M. (1989). In *Methods in enzymology*, Vol. 174 (ed. S. Fleischer and B. Fleischer), pp. 95–118. Academic Press, Orlando, Florida.
43. McCormack, J. G. (1985). *Biochemical Journal*, **231**, 581–95.
44. Assimacopoulos-Jeannet, F. D., McCormack, J. G., and Jeanrenaud, B. (1986). *Journal of Biological Chemistry*, **261**, 8799–804.
45. Thomas, A. P. and Denton, R. M. (1986). *Biochemical Journal*, **238**, 93–101.
46. Midgely, P. J. W., Rutter, G. A., Thomas, A. P., and Denton, R. M. (1987). *Biochemical Journal*, 371–7.
47. Matlib, M. A., Shannon, W. A., and Srere, P. A. (1977). *Archives of Biochemistry and Biophysics*, **178**, 396–407.
48. Hansford, R. G. and Castro, F. (1985). *Biochemical Journal*, **227**, 129–36.
49. Lynen, A., Sedlaczek, E., and Wieland, O. H. (1978). *Biochemical Journal*, **169**, 321–8.
50. Tabor, H., Mehler, A. H., and Stadtman, E. R. (1953). *Journal of Biological Chemistry*, **204**. 127–36.
51. Costantopoulos, G. and Barranger, J. A. (1984). *Analytical Biochemistry*, **139**, 353–8.
52. Solomon, M. and Stansbie, D. (1984). *Analytical Biochemistry*, **141**, 337–43.
53. Severson, D. L., Denton, R. M., Pask, H. T., and Randle, P. J. (1974). *Biochemical Journal*, **140**, 225–37.
54. Johnston, J. D. and Brand, M. D. (1987). *Biochemical Journal*, **245**, 217–22.
55. Hansford, R. G. and Castro, F. (1981). *Biochemical Journal*, **198**, 525–33.
56. Chappell, J. B. and Robinson, B. H. (1968). *Biochemical Society Symposia*, **27**, 123–33.

57. McCormack, J. G. and Denton, R. M. (1980). *Biochemical Journal*, **190**, 95–105.
58. Saris, N. E. and Akerman, K. E. O. (1980). *Current Topics in Bioenergetics*, **10**, 103–79.
59. Nicholls, D. G. and Akerman, K. E.O. (1982). *Biochimica et Biophysica Acta*, **683**, 57–88.
60. Nicholls, D. G. (1987). *Biochemical Journal*, **176**, 463–74.
61. McCormack, J. G. and Denton, R. M. (1989). *Molecular and Cellular Biochemistry*, **89**, 121–5.
62. Denton, R. M. McCormack, J. G., Midgley, P. H. W., and Rutter, G. A. (1989). *Biochemical Society Symposia*, 54, 127–43.
63. Rutter, G. A., Osbaldeston, N. J., McCormack, J. G., and Denton, R. M. (1990). *Biochemical Journal*, **271**, 627–34.
64. Lukacs, G. L., Kapus, A., and Fonyo, A. (1988). *FEBS Letters*, **299**, 219–23.
65. Davis, M. R., Altschuld, R. A., Jung, D. W., and Brierley, G. P. (1987). *Biochemical and Biophysical Research Communications*, **149**, 40–5.
66. Reers, M., Kelly, R. A., and Smith, T. W. (1989). *Biochemical Journal*, **257**, 131–42.
67. Moreno-Sanchez, R. and Hansford, R. G. (1988). *Biochemical Journal*, **256**, 403–12.
68. McCormack, J. G. Browne, H. M., and Dawes, N. J. (1989). *Biochimica et Biophysica Acta*, **973**, 420–7.
69. Cobbold, P. H. and Rink, T. J. (1987). *Biochemical Journal*, **248**, 313–28.
70. Gunter, T. E., Restrepo, D., and Gunter, K. K. (1988). *American Journal of Physiology*, **255**, C304–C310.
71. Komulainen, H. and Bondy, C. H. (1987). *Neurochemistry International*, **10**, 50–64.
72. Steinberg, S. F., Bilezikian, J. P., and Al-Awqati, Q. (1987). *American Journal of Physiology*, **253**, C744–C747.
73. McCormack, J. G. and England, P. J. (1983). *Biochemical Journal*, **214**, 581–5.
74. Hansford, R. G. (1987). *Biochemical Journal*, **241**, 145–51.
75. Ferrari, R., Raddino, R., and Visioli, O. (1982). *Journal of Molecular and Cellular Cardiology*, **14**, 737–740.
76. Katz, L. A., Koretsky, A. P., and Balaban, R. S. (1988). *American Journal of Physiology*, **255**, H185–H188.
77. McCormack, J. G. and Denton, R. M. (1988). *Biochemical Society Transactions*, **16**, 523–7.
78. Al Nasser, I. and Crompton, M. (1986). *Biochemical Journal*, **239**, 19–29.
79. Hayat, L. and Crompton, M. (1987). *Biochemical Journal*, **244**, 533–8.

16

Measurement of free calcium in plant cells

DALE A. CALLAHAM and PETER K. HEPLER

1. Introduction

The measurement of cytoplasmic free calcium concentration ($[Ca^{2+}]$) in single animal cells is widely practised and, although not without problems, it is providing a wealth of new information about the role of this key ion in the process of signal transduction. By contrast, the measurement of $[Ca^{2+}]$ in plant cells is very much more difficult. The rigid plant cell-wall and turgor pressure are difficulties to be overcome in gaining access to the cell interior, and wall-associated enzymes affect the use of some forms of the optical probes. The popular fluorescent dyes, for example, are more difficult to load into plant cells, and once inside the cell they may be subjected to extensive sequestration into non-cytoplasmic compartments, thus rendering them useless as indicators. Chloroplasts and cell-wall autofluorescence interfere with the use of optical methods in many plant cells and other problems arise from conflicting needs to use light as an experimental stimulus as well as a probe. Despite these difficulties, the relationship of $[Ca^{2+}]$ to signal transduction processes in plants is in no way less important than similar studies on animal cells.

This chapter deals with some of the problems encountered in measuring $[Ca^{2+}]$ in plant cells. We focus primarily on the optical probes, including photoproteins, absorbance indicators, and fluorescent dyes, and herein summarize our experience. Calcium-sensitive microelectrodes are discussed for perspective, but their use on plant cells has been limited, and construction and use of these is highly technical so the reader will be referred to more detailed work in this area. We also review the literature pertaining to detection of free intracellular calcium in plants and provide recommendations for the approaches that we think are most favourable.

2. Optical indicators of free intracellular calcium

2.1 Photoproteins

Aequorin is the best known of the several different photoproteins that emit light upon binding calcium. In the range of 200 nM to 100 μM [Ca^{2+}], the light emission increases as approximately [Ca^{2+}]$^{2.5}$. Below about 100 mM [Ca^{2+}], the light output quickly drops toward a level that is Ca-insensitive, and thus the output relationship near the basal [Ca^{2+}] will have a variable exponent and calibration will be difficult at best. Since the aequorin reacting at any moment is a calcium-related fraction of the total aequorin, any reliable quantitation depends on determining an L_{max} value at the end of each experiment (1) and only in retrospect can [Ca^{2+}] values be assigned to the recorded data. If the rate of consumption is high, then the integrated signal for the entire period after loading is required to calculate [Ca^{2+}]. On the positive side, the rate of aequorin consumption is very low if the [Ca^{2+}] remains near basal ($\approx$ 100 nM) levels and thus long-term measurements are possible. Aequorin has been used extensively in certain animal cell systems and thorough reviews on its limitations, use, and mechanism of action are available (2, 3, 4). (Purified native aequorin and relevant data for each batch are available from Mayo Foundation, Attn: Mr. Gary Harrer, Department of Pharmacology, Rochester, MN 55905, USA.) See also Chapter 2, this volume.

Loading of aequorin (mol. wt = 20 000) into cells requires techniques that will admit large molecules while providing a [Ca^{2+}] of <pCa 7.0. Techniques such as scrape loading and ATP-permeabilization that are used to load animal cells (5) are ineffective on plant cells because of the cell wall and we are not aware of their successful use on protoplasts. Electroporation has been used to introduce aequorin into protoplasts (6), although the data given show the technique to be very inefficient: only 800 total detectable counts per cell were initially taken up and this decayed by 80% as the cells resealed. Other studies on animal cells have shown this method to be inefficient for molecules larger than 1 kd (7). Intracellular microinjection using a pressurized micropipette can be used to load individual plant cells with protein solutions. Microinjection and detection of aequorin in plant cells was first accomplished on internode cells of *Nitella* and *Chara* by Williamson and Ashley (8), who centrifuged the cells to create a cytoplasmic plug at one end and then partially plasmolysed the cell to reduce the turgor pressure and allow a small volume of the aequorin solution to be loaded. We have used this method (without centrifugation) to load young internode cells of *Nitella* and stamen hair cells of *Tradescantia*. Specific details for microinjection of plant cells are given in Section 2.4.4.

The reaction with Ca^{2+} discharges a fraction of the aequorin present during each time interval, and at basal [Ca^{2+}] this consumption is on the order of 1×10^{-6}. For even 150×10^{6} molecules of aequorin loaded into a cell at basal

$[Ca^{2+}]$, only 150 photons per second would be emitted and very good detection systems will detect only ≈ 10% of these photons. This very small amount of emitted light requires that the detection system be optimized in every way possible; only the most sensitive and low noise of light detectors are capable of accurately quantitating such low signal levels. Photomultiplier tubes (PMT) that are specially constructed with reduced bialkali photocathode areas and selected for low dark counts offer the best performance for measurements of small cells or aperture-delimited areas of cells. These PMTs offer their best performance when cooled to reduce thermal emission and when operated in the photon counting mode with simple or window pulse height thresholding to discriminate against extraneous electron pulses. Related devices that use a photocathode coupled to one or more microchannel plate electron amplifier stages and a vidicon tube (Photonics Microscopy Inc., 2625 Butterfield Rd, Suite 204–S, Oak Brook, IL 60521, USA) or resistive anode encoder (Surface Science Instruments, 465 National Ave., Montain View, CA 94043, USA) can provide photon-counting two-dimensional imaging of aequorin from single cells (9, 10, 11). The cells used in these studies were quite large and, even so, these experiments push the limits of this technology; measurements of typical small plant cells would be even more difficult. In all of these devices the photocathode has a quantum efficiency of approximatdely 0.2 at 460 nm. A similar factor is reasonable for overall collection and transmission efficiency when microscopes with high numerical aperture lenses are used to collect an image so that the 150 photons in the above example reduce to only 5–10 detected counts per second, a value near the lower limit if any time-resolution is to be obtained.

Of special interest for this chapter is the application of aequorin to studies on plant cells. The first and still the most dramatic demonstration of a $[Ca^{2+}]$ transient associated with stimulus–response coupling in plant cells comes from the work of Williamson and Ashley (8) on the action potentials of Characean algae, where aequorin had been microinjected into the large internode cells. They convincingly showed that the action potential, which causes rapid cessation of cytoplasmic streaming, also generates a corresponding emission of light, from which they were able to estimate that the free $[Ca^{2+}]$ had risen from a resting level of 0.2–0.4 μM to 6–40 μM.

We have recently repeated the experiment of Williamson and Ashley (8) and show in *Figure 1a*, *b*, a photon count recording of Ca-dependent light emission upon elicitation of the action potential from internode cells of *Nitella* microinjected by pressure with aequorin. Plotting the recordings directly at 0.010 sec intervals (*Figure 1b*) reveals the rapidity of $[Ca^{2+}]$ rise and then its much slower decay to resting levels. Williamson and Ashley (8) measured the light from 90% of the cell length, while our system effectively measures the surface along only 400 μm of cell. The use of microscope optics to capture the light from only a small part of the cell enables better resolution of the temporal dynamics of $[Ca^{2+}]$ changes in the cytoplasm than can be obtained

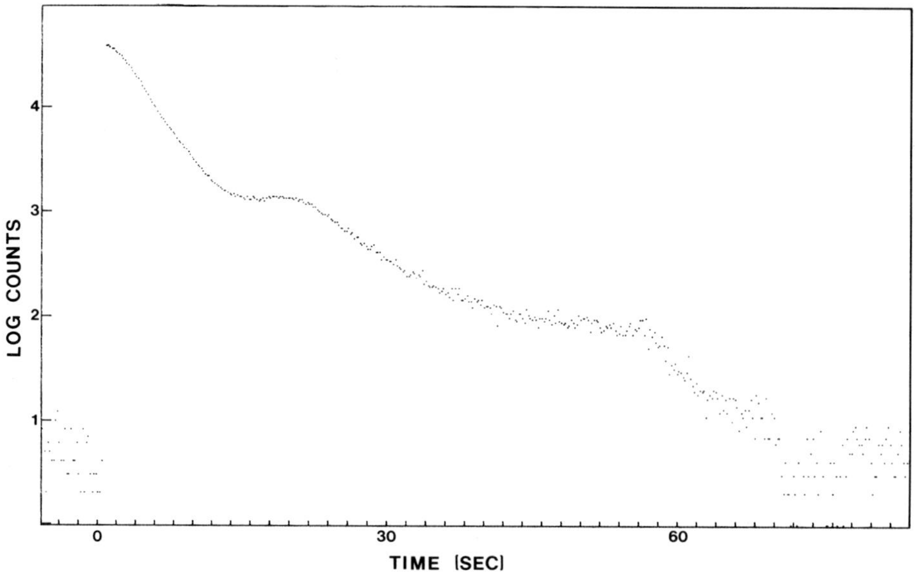

Figure 1 (a). Log plot of photon counts recorded from a 100 × 400 μm area of an aequorin loaded young *Nitella* internode cell during an action potential. Each point is the sum of 20 individual readings of 10 msec duration each. The record shows the abrupt rise in aequorin light emission as calcium enters the cell during the action potential, and the triphasic recovery of $[Ca^{2+}]$ to previous resting levels that is evidenced by the scale expansion of the log plot.

measuring the whole cell, since the action potential and $[Ca^{2+}]$ transient are propagated along the considerable length of these cells. A log plot of the detected counts gives better resolution at low levels and indicates some intermediate plateaux (*Figure 1a*) not reported by Williamson and Ashley (8), suggesting that certain uptake mechanisms are able to sequester Ca quickly to intermediate levels and that the full return to the resting level may take up to 60 sec. Direct microscopic inspection indicates that cytoplasmic streaming restarts after 20–30 sec or before the $[Ca^{2+}]$ has fully reached basal levels. Our studies thus confirm the efficacy of aequorin for $[Ca^{2+}]$ measurements on these large algal cells. In addition, since we have been able to repeat the measurements on cells in which the action potential had been elicited for the second or third time, our studies confirm the fact that the photoprotein is stable and is not fully depleted by the first action potential or through decay over as long as 4 h. We will see later that such long term measurement of $[Ca^{2+}]$ has proven difficult in plant cells using the fluorescent dyes.

Studies using aequorin on protoplasts of carrot and barley show that this indicator can be used on other types of plant cells. Using electroporation, Gilroy *et al.* (6) were able to load a population of cells and to subsequently

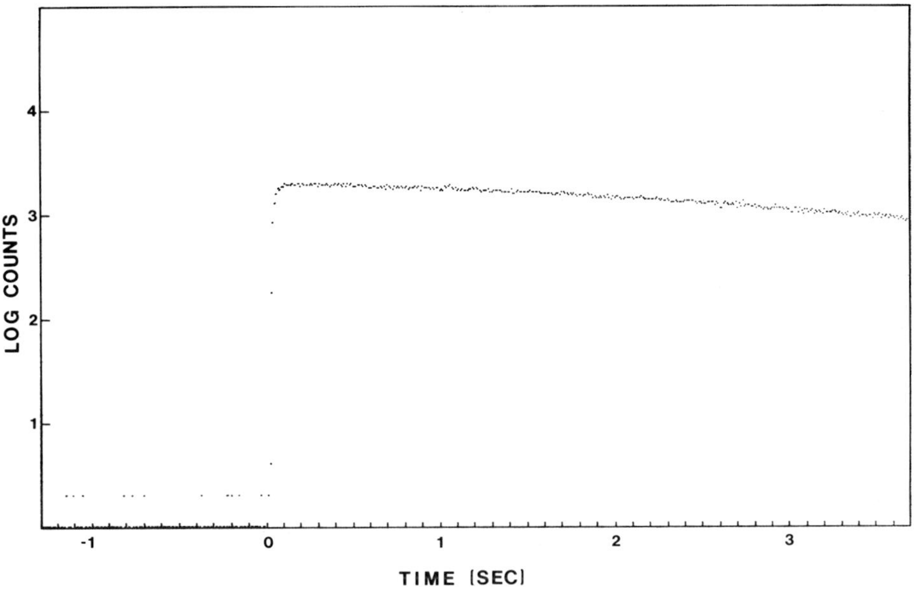

Figure 1 (b). Log plot of the first few seconds of the action potential of *Figure 1a* plotted at 10 msec time-resolution to show the rapid rise in aequorin light emission from [Ca^{2+}] increase.

monitor the signal from this population. The results show, not surprisingly, that the [Ca^{2+}] increases upon treatment with the ionophore ionomycin, and also with azide. The signals they observe benefit from the amplification resulting from the simultaneous measurement of many cells.

We have tried using aequorin on single stamen hair cells of *Tradescantia*, but without much success. The cells we study have a typical volume of only 10–20 pL and detection of the resting [Ca^{2+}] is near the limit of feasibility, using native aequorin as described above. No signal exceding the PMT dark count rate was detected and the total aequorin loaded was difficult to determine since the cuticle of the stamen hair prevents cell lysis by strong detergent solutions and mechanical rupture causes loss of the cytoplasm from the measuring area.

Aequorin presently seems best suited for calcium studies on large cells such as the giant internode cells of the characean algae or eggs of fucoid algae. We do not recommend its use for calcium detection in normal-size plant cells, although we would urge researchers not to ignore the method entirely. A number of recent advances have been made that portend improved prospects for successful use of aequorin in small plant cells. The protein, apoaequorin, has now been produced in quantity in *Escherichia coli* by recombinant DNA techniques and the functional aequorin molecule is readily formed from this product (2); this should improve the purity and availability of aequorin in the future. Shimomura and co-workers (13) have created derivatives of the chromophore and have produced 30 semi-synthetic aequorins with a large range of useful properties, including enhanced light emission at and below 100 nM [Ca^{2+}] and emission spectrum variations proportional to [Ca^{2+}] that allow direct determination of [Ca^{2+}] by emission ratio techniques. One of the aequorin derivatives, hcp-aequorin, has an emission 190 times greater than native aequorin at pCa 7, and this will facilitate detection of [Ca^{2+}] in very small cells, although the higher rate of consumption will limit the length of time that cells can be monitored.

2.2 Absorbance dyes

There are a variety of metallochromic dyes including, murexide, arsenazo-III, antipyrylazo, etc., whose absorbance changes markedly when they bind calcium. We have had considerable experience with arsenazo-III (AA-III) and will therefore restrict our attention to its use. Again, because this dye has been used extensively in studies on nerve and muscle cells, there is much that is known about its properties and characteristics (2, 14, 15). Briefly, the dye exhibits a decrease in absorbance at 530 nm and an increase at 605 and 660 nm when it binds calcium; 580 nm is an isosbestic point, as is 690 nm. The molar extinction coefficient increases from 0.19×10^4/cm/mol to 2.8×10^4/cm/mol at 655 nm as saturating calcium levels are reached. The real variations that occur in the plant cell cytoplasm are probably at the low end of this range but these high values facilitate the use of this dye even with the short path-length of single plant cells and calcium levels so far from the K_d. The stoichiometry of the dye–calcium interaction is not fixed but depends on the concentrations of each in solution (16, 17, 18); at substantial dye concentrations, as are commonly used for *in vivo* determinations, a 2:1 (dye:Ca) ratio may prevail and absorbance changes would increase with the square of the dye concentration, yielding higher sensitivity to [Ca^{2+}] changes. For practical reasons it seems best to use the 660–690 nm wavelength pair to detect differential absorbance since that avoids confusion from pH and magnesium effects (15, 19, 20) and in addition scattering artefacts will be minimized. The differential absorbance change between these two wavelengths then provides a measure of the change in free [Ca^{2+}]. AA-III has a $K_d \approx 50$ μM and thus binds little Ca^{2+} at the basal level (pCa 6.8–7.2). While this means that relatively little buffering of [Ca^{2+}] occurs due to the dye, it also means that only small absorbance changes occur for [Ca^{2+}] changes at or near basal levels, and thus very stable and sensitive differential absorbance detection equipment is required to measure any small changes.

AA-III is negatively charged in distilled water solutions and can easily be microinjected by ionophoresis into living plant cells (Section 2.4.4). The dye concentration in the cell is monitored by following the absorbance change at

530 nm, the isosbestic wavelength, as current is passed. Although the absorbance change can be easily monitored as the dye is loaded, the absorbance at 400 μM dye concentration is small compared with intrinsic absorbance of the cell, the slide, and the culture medium. Also, in contrast with the fluorescent [Ca^{2+}] indicators (Section 2.3), the AA-III concentration cannot be determined accurately later in the experiment since repositioning of organelles and vacuoles or small changes of path length through the medium due to evaporation alter the absolute absorbance of which the dye absorbance is a relatively minor part.

Transmitted light at relatively high intensities (but innocuous wavelengths) is used for this absorbance method, making it possible to use photodiode-based detection systems but we prefer PMT-based systems for the speed and sensitivity where very small cells are measured. We developed a system which generates pulses of narrow bandwidth light using a rotating filter wheel on the illumination side of the specimen, with the transmitted light at each wavelength being synchronously detected by one light detector (22). The benefits of this approach are that the cell is only illuminated with brief pulses of quasi-monochromatic light, minimizing any light-induced damage. Also the use of a single optical path, light detector, and signal processing circuit eliminates any differential drift of detector sensitivity, electronic gain, or electronic offsets that can be associated with separate detection components (14, 23).

Using this equipment coupled to a light microscope we observed significant differential absorbance changes during mitosis in cells loaded with AA-III (22). *In vitro* calibration of the dye in microchamber tubes whose pathlength is similar to that of the cell indicates that with 400 μM AA-III we are able to detect changes in [Ca^{2+}] above 0.3 μM. The method thus is unable to detect the changes near the resting level up to 0.3 μM [Ca^{2+}]. On the positive side, AA-III remains in the cytoplasm and is not compartmentalized like the fluorescent dyes. Moreover, even under continuous irradiation, the dye appears non-toxic to the cells, although its presence causes some of the mitotic events to be slowed, which we attribute to a buffering of the intracellular calcium, possibly through a gradient degrading mechanism (24). Despite our success with this method, we are concerned about its limitations in the measurement of small changes near the resting level of calcium, and for that reason do not provide an enthusiastic endorsement.

2.3 Fluorescent dyes

Through the pioneering efforts of R. Y. Tsien and colleagues several fluorescent dyes, including quin-2, fura-2, indo-1, rhod-2, and fluo-3, have been introduced (25, 26, 27). These dyes have revolutionized calcium detection in living cells, especially those of animal origin. Fluorescence methods are superior to the absorbance methods since the limiting detectivity is over 1000 times greater for fluorescent molecules of high quantum yield (28). Acetoxymethyl ester (AM) derivatives have been developed to enable easy loading of large numbers of cells with the dye, which becomes entrapped in the cytosol as the free anion after cleavage by intracellular esterases (29). As a result of favourable optical properties, fura-2, indo-1, and fluo-3 are able to report on changes of [Ca^{2+}] in single cells at the 30 nM level with only 50–100 μM dye. Finally, the fluorescence signal can be recorded as optical images, thus making it possible to spatially localize changes in [Ca^{2+}]. The five fluorescent [Ca^{2+}] indicators listed above have distinct optical properties and behaviour in the cell; these characteristics and the requirements for detection will be described first. The methods of dye introduction are more general for the whole group, and are described in Section 2.4, as are some specific problems and cautions for the use of these dyes. (These dyes are available from Molecular Probes Inc., PO Box 22010, Eugene, OR 97402, USA.) See also Chapter 1 of this volume.

2.3.1 Quin-2

Quin-2, the first of the practical fluorescent indicators (30), has been used for the detection of [Ca^{2+}] in single plant cells (31) as well as populations (6). It has an excitation spectrum that varies only slightly with [Ca^{2+}], the quantum yield increase (from 0.012 to 0.057) with increasing [Ca^{2+}] being the important parameter (25). The preferred excitation wavelength of 340 nm is poorly transmitted by most microscope optics and excites considerable autofluorescence in most cells. Although the excitation spectrum changes suggest the possibility of using quin-2 in a ratio excitation method (32); in practice, this has not been useful, since the change at 380 nm is small and is sensitive to Mg^{2+}. There are a number of conditions that must be met to obtain meaningful [Ca^{2+}] values from measurements of quin-2 fluorescence when single wavelength excitation is used. The most difficult to satisfy, when used on single cells, is the requirement for calibration with the same dye concentration and optical path length as used for the measurements. Cytoplasmic and organelle rearrangements that change the effective optical path, dye loss from the cell, and dye loss due to photobleaching make the proper use of quin-2 on single cells almost impossible. Even relative [Ca^{2+}] changes are affected by these same factors since there is no compensation afforded by the single wavelength technique. The newer fluorescent indicators that have become available offer improvements over quin-2 in most respects and should completely replace its use.

Quin-2 has been used relatively little for [Ca^{2+}] measurements in plant cells. Keith and co-workers (31) loaded quin-2/AM into wall-less endosperm cells of *Haemanthus* and followed [Ca^{2+}] changes during mitosis. Cork (33) attempted to apply quin-2/AM to a range of plant cells, but with little success. Nobiling and Reiss (34) report success in loading *Lilium* pollen tubes with quin-2/AM and provide some evidence that a tip-to-base declining [Ca^{2+}]

gradient exists, but these studies need to be confirmed with the fluorescent dyes of superior optical properties before they can be accepted.

2.3.2 Fura-2

Fura-2 has been the most popular of the new family of fluorescent indicators, in large part because of the opportunity that it affords for spatial localization of [Ca^{2+}] in single cells (35, 36). The characteristics are similar to quin-2 but the excitation spectrum is at slightly longer wavelengths and the quantum yield is much higher. The ratio of the emission intensities at ≈ 500 nm when excited by 340 nm and 380 nm (360 nm and 380 nm are also used) is proportional to the [Ca^{2+}], providing for ratiometric determinations that are not sensitive to non-limiting variations in the dye concentration or optical path length. Dual wavelength detection together with fluorescence intensity that is 30 times higher than quin-2 has permitted imaging. Fura-2 has a K_d (Ca) ≈ 135 nM and K_d(Mg) ≈ 9.8 mM at pH 7.0, 100 mM KCl, and 20 °C, and so the calcium binding is a little weaker (desirable) and the discrimination against Mg^{2+} considerably better than quin-2 (for details, see ref. 26).

Fura-2 has been quite widely used as a [Ca^{2+}] indicator in plant cells. Brownlee and co-workers (37) loaded fura-2/AM into centric diatoms and subsequently acquired ratio images of the [Ca^{2+}]. Very little detail is apparent in the quantitative images of the two species; the central cytoplasmic region occupied by the nucleus presented the only significant signal level and indicated a [Ca^{2+}] of just above pCa 7.0. Fura-2 anion, loaded by ionophoresis (Section 2.4.4), was also used in imaging applications on root hair cells (38) and *Fucus* rhizoids (39), but the spatial resolution provided only a general sense of [Ca^{2+}] distribution. Both of these reports displayed the ratio image ([Ca^{2+}]) values as a grey scale and thus there was no indication of dye concentration by brightness levels as has been used in pseudocolour representations, and the image regions of poor signal quality are displayed as high [Ca^{2+}]. This fault was overcome when graphical information was plotted from the screen data. Fura-2 has recently been used to demonstrate [Ca^{2+}] changes in guard cells of *Commelina* associated with stomatal closure in response to abscissic acid (40). The dye, introduced by ionophoresis, did not effect the stomatal aperture and moreover it remained in the cytoplasm for 45 min. Very stable [Ca^{2+}] values were obtained from control cells in contrast to the marked [Ca^{2+}] increase seen when separate cells were treated with abscissic acid.

2.3.3 Indo-1

Unlike quin-2 and fura-2, indo-1 exhibits a change in the emission spectrum, rather than the excitation spectrum, when calcium is bound. The emission spectrum shows a peak at 405 nm in the calcium-bound form, and at 490 nm in the calcium-free form; there is an isoemission point at approximately 450 nm when the dye is excited near 355 nm. The K_d(Ca) of indo-1 (250 nM) is nearly the same as that of fura-2. Since emission filters will often distort and displace an image, indo-1 cannot be easily used in quantitative imaging applications, but was recommended for flow cytometry applications (26) because dual wavelength excitation is difficult to achieve in that application.

Indo-1 has been used far less than fura-2 for measurements on single cells although equipment to ratio the emission peaks is no more difficult to construct. In barley aleurone protoplasts subjected to various treatments, indo-1 was loaded from low pH solutions containing the pentapotassium salt of the dye (41, 42) (Section 2.4.2 on acid loading). The indicator was observed only in the cytosol and showed a low loss rate from the cells, enabling measurements for 96 h. Preliminary experiments using ester-loading (Section 2.4.2) had demonstrated that indo-1/AM rapidly accumulatedd in the cytoplasm but was not hydrolysed, and fura-2/AM loaded the cells but rapidly appeared in the vacuole. Lynch and co-workers (43) have used indo-1 for studies of the effects of salinity stress on maize root protoplasts. We have used indo-1 in various experiments with *Tradescantia* stamen hair cells (44) (*Figures 2–4*), introducing the dye by microinjection. With precautions, we have been able to make meaningful measurements for about 30 min of continuous observations and, when indo-1 transport to the vacuole was inhibited with probenecid (see discussion in Section 2.3.5), measurements can be carried out for over 1 h during which time the dye level slowly decreases in the cytosol.

2.3.4 Long-wavelength fluorescent indicators

Fluo-3 and rhod-3 benefit from longer excitation wavelengths than the previous dyes; however, they show intensity modulation of a single emission peak and thus they will suffer from the same difficulties as quin-2 (27). Fluo-3 is the better dye for $[Ca^{2+}]$ detection in most cells since its K_d (400 nM) provides an excellent sensitivity match for changes in $[Ca^{2+}]$ around the basal level, in contrast to rhod-2 whose K_d is substantially higher (1.0 μM). In spite of the full-range 40-fold intensity change that fluo-3 is capable of yielding, the variation over probable plant $[Ca^{2+}]$ levels is less than fourfold. Without the compensation provided by ratiometric detection, small $[Ca^{2+}]$ variations might be undetectable if significant shape changes, streaming, or cytoplasmic rearrangements were to occur. In animal-derived cells that do not show these properties, the low bleaching rate has enabled extended measurements for periods exceeding 1 h (45). A real benefit is that the longer excitation wavelengths allows the use of the growing number of UV-photolabile calcium chelators and caged compounds. As with quin-2, the calibration of the signal from these indicators is difficult (see ref. 27).

2.3.5 Problems with the fluorescent indicators

The remarkable power of these dyes and the enormous impact that they have had cannot be denied but, unfortunately, they have a number of characteristics that can make $[Ca^{2+}]$ measurements very difficult and subject to substantial errors. Most evident is their tendency to compartmentalize, especially in the plant cell vacuole, although cells of animal origin also frequently show sequestration and secretion from the cell (46, 47, 48). It has been our experience with the stamen hair cells of *Tradescantia* that fluo-3 is the most rapidly sequestered (3 min) followed by fura-2 (20 min) and indo-1 (30 min). As has been found with mouse J774 macrophages (47), we find sequestration to be an active process and dye accumulates in the vacuole against a concentration gradient until it is entirely removed from the cytoplasm. In the original publication on fluo-3, Kao *et al.* (45) observed some compartmentalized signal surrounding the nucleus of the fibroblasts, and on lysis with digitonin, a notable amount of fluorescence was not released from the cytoplasm of both ester-loaded and, to a lesser extent, microinjected cells. Where the structure of the cell will enable cytoplasm (only) to be measured, the immediate difficulty will be the loss of dye, and therefore signal quality, as the experiment progresses. However, it is rarely possible to avoid including the vacuolar compartment, and since the proportion of vacuolar volume and its dye concentration may be changing, the measurements will be subject to errors that are not easily corrected. The fluorescent dyes are also secreted from the cytoplasm to the extracellular medium (48, 49) and, as above, the resulting degradation of signal quality can severely affect results during extended measurements. We have found, in some plants, that vacuolar sequestration of the fura-2, indo-1, and other anionic fluorescent dyes can largely be prevented by culturing the cells in 2–5 mM probenecid, an organic anion transport inhibitor (47, 49). Useful levels of indo-1 can be maintained in the cytoplasm for periods exceeding one hour and no obvious effects on cytoplasmic streaming or cell division have been noted. The extent to which compartmentation or vacuolar sequestration will be found cannot be predicted although we have observed these problems in every plant cell tested.

While sequestration and secretion may be the most obvious problems confronting the use of fluorescent dyes, they are not the only ones. Other problems include photobleaching, spectral shifts due to dye–protein binding and viscosity, and the various effects of the dye–light interaction. Photobleaching generally implies the conversion of the dye to a non-fluorescent form that will not alter the spectrum although the effective loss of indicator will lower the signal quality. While this may occur, a more complex phenomenon has been described for fura-2; fluorescent but Ca-insensitive forms are produced by the photoconversion, thus altering the spectral qualities of the resultant mix (and hence the indicated $[Ca^{2+}]$) and reducing the sensitivity to $[Ca^{2+}]$ changes (50). This process violates one of the assumptions on which the ratiometric detection concept is based and, unlike autofluorescence, it produces a changing background that is not easily compensated. Indo-1 was reported to be more sensitive to photobleaching than fura-2 (26) although the characteristics were not detailed.

Viscosity also affects spectral qualities and should be considered when making in vitro calibrations (26, 51). Konishi *et al.* (51) report a blue shift of the fura-2 emission spectrum, which agrees with our observation of higher emission ratios (405 nm/490 nm) from indo-1 when 60% (w/v) sucrose is included in the calibration medium. Cytoplasmic viscosity is probably not uniform from cell to cell nor in various domains within a cell, and further variations with time and physiological state all result in a large degree of uncertainty in any attempt at absolute or even relative [Ca^{2+}] quantitation.

In addition to viscosity-related artefacts there is evidence that as much as 90% of the Ca-free form of fura-2 is bound in amphibian skeletal muscle myoplasm and, *in vitro*, to soluble proteins common to myoplasm (51, 52). This binding results in a 300–400% increase in the K_d and therefore a gross error in determined [Ca^{2+}] can result when suitable compensation is not made. The effects of the altered K_d imply a significant decrease in sensitivity of even relative [Ca^{2+}] changes near the 100 nM [Ca^{2+}] level. Konishi *et al.* (51) suggest that the observed protein binding could be a factor in all cells.

While bleaching, viscosity, and protein binding create problems in the calibration and sensitivity of the dye response, the light–dye interaction is another matter since this can damage the health and performance of the cell. We have recently found that neither indo-1 [50–100 μM] nor the light [70 μW mm^{-2} at 350 nm (10 nm FWHM) incident at the cell] alone blocks mitosis in stamen hair cells of *Tradescantia*, but when coincident the cells often become arrested in metaphase and will fail to enter anaphase. Thus, even with a 100 W tungsten–halogen lamp, which at 350 nm provides much less light intensity than the usual xenon or mercury arc source, there is still sufficient light to cause inhibition in dye-loaded cells. The solution we favour is to reduce the dye concentration, excitation intensity, and time of illumination as much as possible.

Despite the many problems, the fluorescent indicators still offer several marked advantages over the other indicators in regard to sensitivity to [Ca^{2+}] changes at basal levels, ease of use, and ease of detection. After considerable trial and error we have arrived at a procedure that allows us to repeatably measure [Ca^{2+}] changes in single cells during relatively short-term experiments using indo-1. First, we load the cells microionophoretically (see Section 2.4.4) until the dye reaches the 50–100 μM level. With stamen hair cells we take advantage of the process of intercellular transport to move the dye from the injected cell to the target cell on which measurements will be made. Subsequently, the cell is illuminated with 350 nm light, and the fluorescence signal is determined at 405, 450, and 490 nm by a photon-counting PMT with narrow bandwidth interference filters sequentially interposed (*Figure 2*). When a baseline is established, the cell can be challenged either by microinjection or by the addition to the culture medium of an agent that might alter intracellular [Ca^{2+}]. The resulting change in the fluorescence ratio will then provide temporal and quantitative information on the changes in

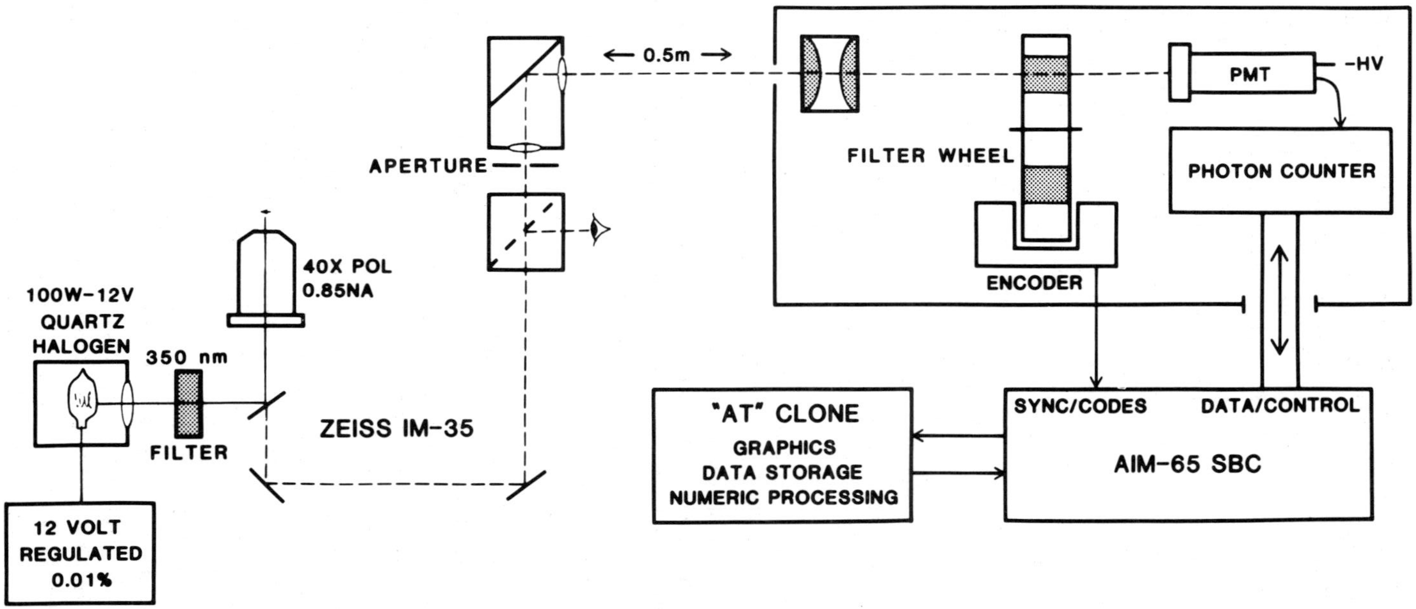

Figure 2. Diagram of the apparatus used for obtaining emission ratio values from indo-1 fluorescence of single cells. Excitation is provided by light from a well regulated tungsten–halogen lamp passed through a 350 nm (10 nm FWHM) interference filter. The emitted fluorescence is collected by the objective which forms a magnified real image at the aperture plane. Light passing the aperture is collimated and diverted across an air space (to avoid vibration of the microscope) to the rotating filter wheel (1800 r.p.m.) that sequentially interposes 405, 450, and 490 nm interference filters. The AIM-65 micro-computer regulates operations, gating and reading the photon counter at appropriate times, and exporting data and time values to the personal computer that provides convenient processing, display, and data storage. (Reproduced from Zhang *et al.*, ref. 44.)

$[Ca^{2+}]$. Sample recordings of indo-1-loaded cells into which Ca^{2+}, the calcium buffer dibromo-BAPTA, and calcium saturated dibromo-BAPTA were injected are shown in *Figures 3* and *4a* and *b*.

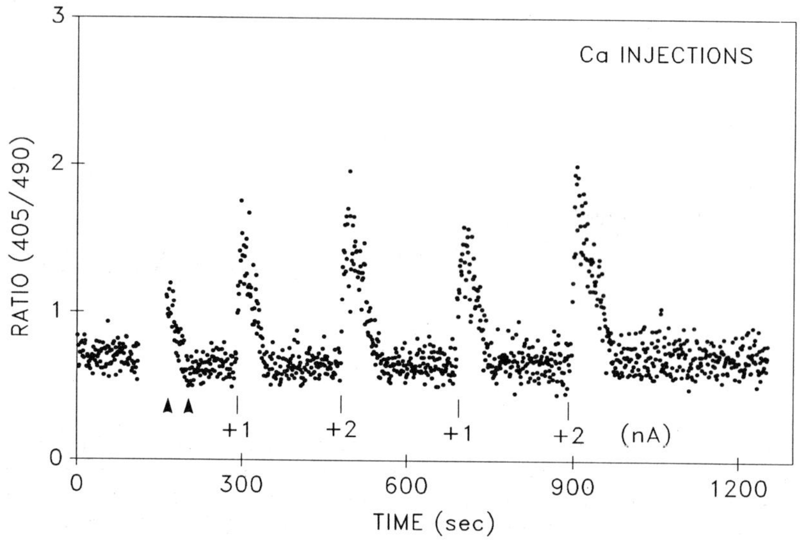

Figure 3. Record of indo-1 emission ratio changes from a single *Tradescantia* stamen hair cell that contained 60 μM indo-1 and was ionophoretically microinjected with Ca^{2+} by passing positive current as indicated. After establishing a baseline, the micropipette was inserted during the blank part of the trace, and the falling ratio (arrows) is recovery from the brief $[Ca^{2+}]$ transient induced by the small wound. Each point is the average of 32 individual ratio determinations but maintains adequate temporal resolution to record the rise in $[Ca^{2+}]$. (Reproduced from Zhang *et al.*, ref. 44.)

We emphasize that the above protocol is only adequate for relatively short-term studies where the measuring times are 20 min or less. To overcome these problems we have recently found that co-culture of the cells with the anion transport inhibitor probenecid, as used by Steinberg *et al.* (49) on macrophages, prevents indo-1 from being sequestered into the vacuole, and allows the time of measurement to be extended for periods exceeding 1 h. Preliminary tests indicate that probenecid does not inhibit streaming or cell division. Although we have not investigated the phenomenon in detail, most of our recordings show a decline in the 405 nm/490 nm ratio beyond 20 min; the total photon counts increase over time with most of the increase at 490 nm and a small decrease at 405 nm. This ratio decline is coupled with a decreased sensitivity to the $[Ca^{2+}]$ increase generated when the cells are treated with 3 mM sodium azide as compared with the same treatment at the beginning of the experiment; however, very large $[Ca^{2+}]$ increases from injection or wounding still give the full expected ratio change. These spectral changes

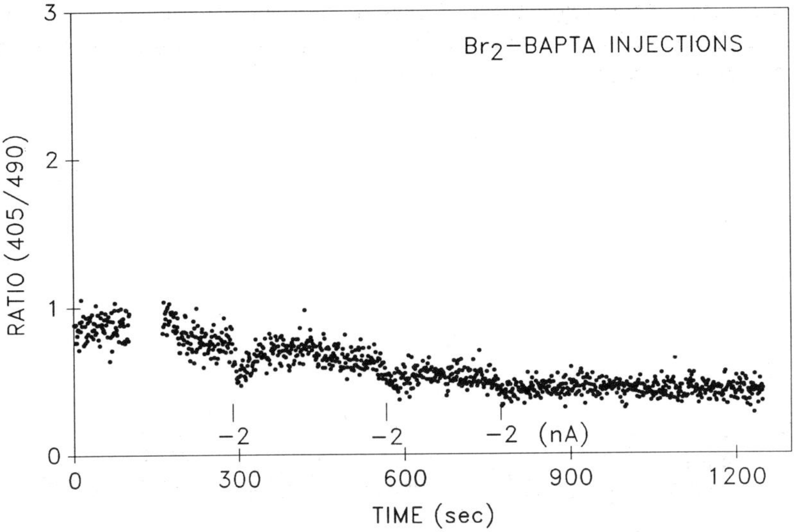

Figure 4 (a). Record of indo-1 emission ratio changes in *Tradescantia* stamen hair cell that was ionophoretically microinjected with calcium-free dibromo-BAPTA, a calcium buffer with ≈ 4 μM K_d(Ca). In spite of the high K_d(Ca), some calcium is bound and the [Ca^{2+}] is transiently depressed, followed by a partial recovery to initial levels.

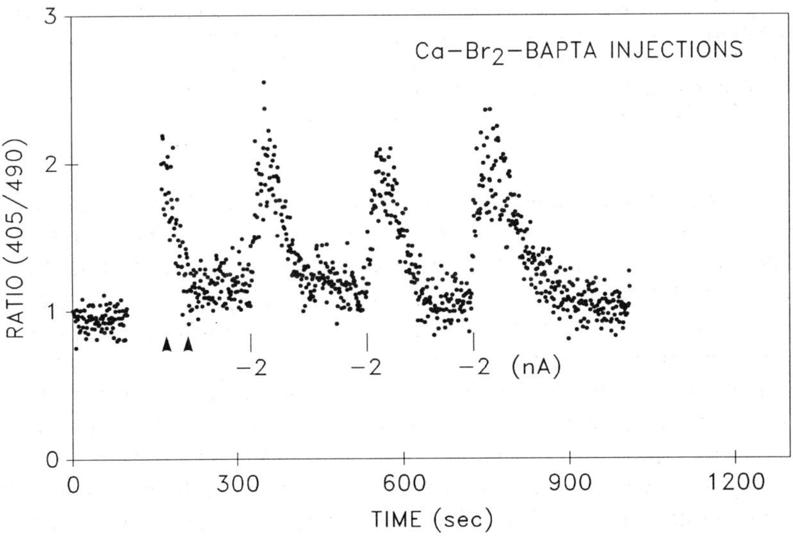

Figure 4 (b). When calcium-saturated dibromo-BAPTA, still carrying a negative charge, is injected by negative current, the lower [Ca^{2+}] of the cell allows dissociation and calcium release, giving transient [Ca^{2+}] increases and buffers the cytosol to slightly higher [Ca^{2+}] levels following each injection. (Reproduced from Zhang *et al.*, ref. 44.)

occur even though the dye remains diffuse in the cytoplasm, and even in the absence of excitation illumination. We do not yet have a full understanding of these observations but they underscore the potential problems that may be encountered.

2.4 Methods of indicator loading

2.4.1 Acetoxymethyl esters (AM)

When using the fluorescent dyes it seems attractive to apply them in their AM form, in which they are supposed to freely pass through lipophilic membranes and, once inside the cell, be cleaved by intracellular esterases rendering the dye localized to the cytoplasmic compartment. While this approach is very attractive, there are a variety of problems, the most serious of which is the presence of cell-wall-associated esterases that cleave the dye before it enters the cell (6, 33). Even if the dye were to enter the cell by the postulated mechanism, there is an extensive endomembrane system just within the plasmalemma (53, 54) allowing the rapid accumulation by the same mechanism into membrane compartments such as the ER and vacuole, which also have extracytoplasmic esterases. We know that many people including ourselves have attempted to load plant cells with the AM form of these dyes but the results have usually been negative. The only reports of success include a study by Nobiling and Reiss (34) on lily pollen and Keith *et al.* (31) on the naturally wall-less cells of *Haemanthus* endosperm. Neither study has been repeated, and in *Haemanthus* endosperm the published images are consistent with the dye being sequestered within endomembranes and not free (diffuse) within the cytoplasm. Ester loading thus does not appear to be a viable method when applied to plant cells.

2.4.2 Acid loading

'Acid loading' refers to a procedure for dye introduction in which the culture medium is made sufficiently acidic to suppress ionization of the 'impermeant', free-acid, form of the dye molecule, allowing it to cross the plasma membrane. Once inside the cell the dye experiences the higher pH of the cytosol and dissociates into its impermeable, anionic state and is trapped in the cell. This method was first described by Gilroy and co-workers (55) who attempted to load plant cells and protoplasts with quin-2. They tried many conditions including modulating the solution pH (pH 2.0–10.0). [$CaCl_2$] (1–50) mM) and [quin-2] (1–50 mM), but achieved no success. This method, however, has been successful when used to load indo-1 into barley aleurone protoplasts (41, 42). Culture medium buffered to pH 4.5 with 10 mM dimethylglutaric acid allowed the most rapid accumulation of the dye, but some loading was observed even at pH 6.5. The presence of added calcium (10 mM) gave a slight suppression of dye accumulation at all pH values. Cytosolic dye concentrations at the end of a 2 h incubation were dependent on the indo-1 concentration in the medium, with 3 μM indo-1 required to obtain threshold loading, and nearly maximal dye accumulation with 15 μM indo-1. Dye concentrations up to 60 μM gave only slight improvement and above 60 μM toxicity was noted.

This technique has also been used successfully to load indo-1 into corn root protoplasts (43). Lynch *et al.* (43) note that any remaining wall material in the preparation became fluorescent and cytosolic dye accumulation was inhibited. Cell wall regeneration more than 30 min after protoplast isolation also inhibited dye loading, but the inclusion of 20 mM galactose in the culture medium suppressed wall formation and allowed dye loading at much later times. The presence of cell wall does not necessarily preclude loading, since it is possible to introduce indo-1 into barley coleoptile epidermal peels floated on pH 4.5 culture medium containing 16 μM indo-1 (Anne Wrona, Cornell University, personal communication). With the exception of the barley coleoptile epidermal cells, we have been unable to load any other walled cells including stamen hair cells of *Tradescantia*, protonemata of *Funaria*, pollen tubes of *Nicotiana* or *Tradescantia*, or guard cells of *Vicia* by this method. A further problem with acid loading is the necessity for 1–2 h incubations at low pH which can require conditions that are far different from experimental conditions used. For instance, in studies with barley aleurone protoplasts, a pH 7.5 treatment was followed by two washes at pH 4.5, a 2 h incubation with indo-1 at pH 4.5, a wash in indo-1-free medium at pH 4.5, and 2 h purification step at pH 5.6 before measuring the indo-1 emission to determine the [Ca^{2+}] which is supposed to be due to the culture conditions which were terminated over 4 h earlier (42). Acid loading cannot yet be regarded as a general solution to indicator introduction.

2.4.3 Electroporation

Electroporation is a general method for cell suspensions that allows one to introduce virtually any molecule into unwalled protoplasts. Not surprisingly, therefore, it has been used to insert [Ca^{2+}] reporters such as sequorin and quin-2 into protoplasts of *Hordeum vulgare*, *Daucus carota*, and *Phaseolus mungo* (6, 55, 56). The conditions that permeabilize each type of cell are different; thus carrot protoplasts were permeabilized by 6–8 kV cm^{-1} (15 μsec duration) for loading quin-2, but 3 kV cm^{-1} (50 μsec) when aequorin was to be introduced (the inefficiency of the aequorin loading was discussed in Section 2.1). Protoplasts of *H. vulgare* were permeabilized to quin-2 by a single 1.1 kV cm^{-1} (100 μsec) pulse while 2 pulses at 3.8 kV cm^{-1} (100 μsec) separated by 30 sec were used to introduce the same dye into protoplasts of *P. mungo*. Although electroporation allows one to introduce any reporter molecule into a large number of cells, the method suffers because it requires one to first make protoplasts, a procedure which in our view is unattractive since the removal of the wall may alter or destroy some crucial aspects of the cell's physiology, including Ca transport and polarity in growth, and non-uniform distribution of cell surface proteins.

2.4.4 Microinjection

While more complicated than some of the methods discussed above, intracellular microinjection allows one to rapidly introduce almost any molecule into the cell. The instrumentation is more elaborate, requiring a high-quality light microscope (preferably inverted), a microelectrode puller, an ionophoretic current source or pressure control apparatus, microelectrode holders, and a stable micromanipulator, in addition to steady hands and much patience. This method is best suited to situations where one or a few cells will be injected and individually monitored, and for that reason, the equipment list will generally include microscope-linked detection equipment. For single-cell studies we think that microinjection offers substantial advantages over other existing methods of dye introduction. See also Chapter 3, this volume.

The act of sticking a fine glass pipette into a living cell may seem disruptive, and, admittedly, we have evidence that there is a brief [Ca^{2+}] spike that attends the insertion process. However, on the other hand, we must emphasize that when microinjection is performed properly, the act of impaling a cell may not perceptibly alter such vital processes as cytoplasmic streaming or anaphase chromosome motion. In *Nitella* the act of impalement triggers an action potential and streaming cessation, but recovery follows a normal time course. Microinjection thus can be relatively non-disruptive and, in addition, it can be applied to a wide variety of cell types and circumstances. Besides stamen hair cells of *Tradescantia*, we have successfully microinjected protonemata of *Funaria*, internode cells of *Nitella*, epidermal and guard mother cells of *Vicia* and *Commelina*, mature guard cells of *Vicia*, insectivorous gland stalk cells of *Drosera*, pollen tubes of *Nicotiana*, epidermal cells of *Hordeum*, and leaf mesophyll and sieve tube members of *Cucurbita*. The last-mentioned example of the sieve tube members supports our working dictum that if we can image the cell in question we can impale it with a microneedle.

It is often assumed, especially in highly vacuolate cells, that impalement of the cell always results in the tip of the needle being located in the central vacuole. For reasons that are not clear, this does not occur, even at times when it is fully expected. The tonoplast is apparently extremely flexible and rather difficult to impale, and even when breached by a micropipette, the cytoplasm usually rearranges placing the micropipette tip back in a cytoplasmic environment. Our experience with *Tradescantia* indicates that, even in highly vacuolate stamen hair cells, the tip of the inserted microneedle nearly always locates to the cytoplasm and not the vacuole, as evidenced by the position of fluorescent tracer dyes.

Microinjection of materials into plant cells uses one of two basic techniques: (i) microionophoresis, for charged, low-molecular-weight materials; and, (ii) pressure injection, for neutral or charged molecules or materials that would not migrate significantly in an electric field. Microionophoresis makes use of the principle that electric currents in solution are carried by ions (e.g. charged dye molecules) rather than the transfer of electrons. Since only ions move, no cell pressure or volume change occurs, making this a very desirable technique for plant cells. Although the ionophoretic current can be controlled quite precisely with fairly simple equipment, the delivery of the chosen substance from the pipette is, unfortunately, not strictly quantitative since there are competing ion movements involved. If care is taken to produce micropipettes of repeatable characteristics and if other conditions are kept the same, the quantity of substance delivered will be very repeatable even if the absolute quantity cannot be directly calculated. In practice, the amount of indicator loaded can be measured using instrumental or visual means. Further information concerning the quantitative and theoretical aspects of microionophoresis can be found in the book by Purves (57).

We use 1 mg ml^{-1} solutions of the indicators for most of our work, and freeze 20-μl aliquots in 100-μl microcentrifuge tubes; a 20-μl aliquot lasts for a week or more of normal work if kept frozen when not used (thawed daily and kept on ice) and is thus very economical. To handle such small samples and place them near the shoulder of the micropipette, we use 1-ml plastic tuberculin syringes, heated to the softening point over a low flame, and drawn out to a fine diameter that will reach inside the micropipette. This manipulation can be mastered with a little practice and the syringes, if kept clean, can be used for long periods.

For indicator loading, micropipettes of 10–20 MΩ resistance (if filled with 3 M KCl) are pulled from 1.0-mm capillary tubing that contains a glass fibre to facilitate loading (WPI Kwik-Fil, World Precision Instruments, New Haven, CT 06513–4445, USA) and, holding the micropipette vertically, a short column (1–2 mm) of the indicator (AA-III, indo-1, fura-2, etc.) is placed inside the micropipette, just at the shoulder taper. After a minute, capillarity assisted by the glass fibre allows the dye to reach the very tip and a small bubble of air will emerge and be released. Although the indicator can be placed at the back end of the micropipette, we have experienced considerable variability of results when this method was used. The stem is then filled with 10 mM KCl, leaving a 0.5 mm air space between the dye and KCl to slow diffusion, and the micropipette is inserted into the KCl-filled holder and secured. The micropipette holders that we use are from WPI and contain a sintered Ag–AgCl pellet to make a connection with the external circuit. A reference electrode to complete the circuit through the solution is required and this is easily formed by connecting a clean, lightly sanded, silver wire to the anode of a 1.5 V battery, a platinum wire to the cathode, and immersing these electrodes in 1 M HCl for a few minutes. This Ag–AgCl wire is then immersed in the bathing solution and connected to the ionophoretic device by ordinary wires. The reference coating should be renewed every few days using this same procedure since the coating is very thin.

An ionophoretic current source can be obtained commercially (World Precision Instruments and other sources), but very simple devices are quite sufficient. A 1.5 V battery in series with a $1 \times 10^9\ \Omega$ resistance produces an adequate current source since the programming resistance is approximately 10^2 greater than the micropipette resistance. This arrangement will deliver about 1 nA (1.5 nA maximum) and the polarity can be reversed by simply reversing the battery. The negative terminal (cathode) should be connected to the micropipette connection for the injection of the anionic calcium indicators. These are very small currents so the wires should be well-insulated (Teflon) and connections kept clean to avoid leakage. The book by Purves (57) contains circuit diagrams for a regulated ionophoretic device, as well as a clear treatment of microelectrode methods.

Impaling plant cells is very delicate work requiring practice and patience. We work with single cells or small pieces of tissue since they present a less confusing optical arrangement. The cells are held in place by a very thin layer of Type VII, low gelling agarose (Sigma, PO Box 14508, St Louis, MO 63178, USA), and then flooding them with a suitable culture medium. The micropipette, carried by the micromanipulator (we use Narashige hydraulic micromanipulators which provide smooth, drift-free movement: available from AZI Corp., PO Box 518, Avon, MA 02322, USA), is brought up to the cell at a 30–45 degree angle, and penetration is most easily made just above the midpoint of the cell where the micropipette will be perpendicular to the cell wall. The micropipette is advanced until the cell is impaled, or if the wall is indented by the micropipette but not penetrated, a light tap on the table-top will often facilitate penetration. An advantage of the more sophisticated ionophoretic devices is the voltage follower (membrane potential measurement) capability that indicates when the cell has been impaled, and can also provide an indication of micropipette conditions, such as high resistance and plugging, during the ionophoretic loading by monitoring the magnitude and noise of the compliance voltage in the micropipette stem.

Once the cell has been impaled, the current can be started and indicator loading followed by continuous or intermittent observation. With indicators below 1000 mol. wt, there may be substantial transport through plasmodesmata into neighbouring cells. In some instances, this phenomenon can be used to advantage by enabling a cell of interest to be loaded without actually impaling it (22). Continuous negative direct current at 0.5–2.0 nA for 20–60 sec per cell provides loading of the anionic indicators. We have not found detrimental effects due to the hyperpolarization that accompanies current of this magnitude, even when sustained for several minutes. Pulsed negative current (39) is often used for ionophoresis, but this appears no better than constant current and may be a function of the equipment that was available. Positive current used to introduce cations like Ca^{2+} has given us erratic results; Ca^{2+} appears to promote rapid sealing of the micropipette tip with wall material in many plant cells, but depolarization of the membrane potential may also be involved. The development of a 'cation plug' due to localized negative charge inside the micropipette tip (57) is reported to favour the flow of cations, rather than anions, contrary to these observations. After loading, the micropipette must be removed very slowly, in stages, allowing the cell to heal.

Pressure injection must be used for uncharged molecules or proteins that will not transport with sufficient rapidity by microionophoresis. The equipment required for forming and positioning electrodes is much the same as described for ionophoresis with the exception that the micropipette puller is usually adjusted to produce more open, lower resistance, pipettes. We have had considerable success using relatively sharp (10 MΩ) pipettes, pulled a little cooler than for ionophoresis, and supplying pressure with a 2.0 ml Gilmont Micrometer Syringe (Thomas Scientific). The syringe is coupled to an Eppendorf Microinjector capillary micropipette holder (No. 471883, with No. 471879 chuck size 0 for 1.0 mm glass) from Carl Zeiss, Inc. (One Zeiss Drive, Thornwood, NY 10594, USA) using Clay Adams PE 20 polyethylene tubing and Clay Adams connectors (Thomas Scientific). The PE 20 tubing is inserted all the way down the shaft of the micropipette holder so that the O-ring that seals the removable chuck also seals the tubing. The syringe and tubing are filled with water, and the micropipette stem is filled with a compatible liquid or else the sample, placed near the shoulder, is isolated from the water by a droplet of silicone oil. The micropipette glass should be cleaned, and the material to be injected should be centrifuged for a few minutes in a table-top ultracentrifuge before loading the micropipette to settle particulates than might block the tip.

Pressure injection of walled plant cells is difficult; the requirement for a sharp needle to penetrate the wall conflicts with the desirability of a large, open tip. Turgor pressure may force materials into the micropipette tip and clog it. The rigid wall limits the volume of material that can be introduced, but in some cases, bringing the cell to incipient plasmolysis may assist injection. Removing the larger micropipettes from the cell is additionally difficult, and several minutes are required for adequate healing. Quantitation of injected volume is relatively crude, primarily using methods such as measuring the diameter of oil droplets that occupied the same column of the micropipette tip as the protein solution. Of the currently available [Ca^{2+}] indicators only aequorin requires pressure injection, although if the fluorescent indicators are successfully coupled to 10–20 000 mol. wt. dextrans to retain them in the cytoplasm, pressure injection may be necessary.

2.5 Optical detection

There is insufficient time or space here to elaborate the details of optical arrangements, detector and detection system electronics, or data acquisition protocols. However, we would like to mention a few general, but important, considerations for the design, selection, and use of custom or commercial instrumentation as applied to plant cells.

2.5.1 Sampling rate considerations

Plant cells are quite variable in shape and distribution of their cytoplasm, requiring careful selection of a proper lens/aperture to minimize optical measuring artefacts. The speed of the Ca^{2+} change, however, demands that attention be given to the performance of the detection system. Fast responses, as in *Nitella*, have been shown to be caused by rapid [Ca^{2+}] changes (8), but slow responses may stem from fast (brief or sustained) or slow [Ca^{2+}] changes, so it is important that preliminary investigations be conducted with a system of sufficient time resolution such that artefacts are not generated in the sampling process. In this regard, the fluorescence ratio measurement techniques deserve special mention since some applications of this method can give results that are subject to considerable artefact. Streaming and cytoplasmic movements in plant cells can present a significantly changing distribution of cytoplasm, organelles, vacuoles, and therefore indicator per unit volume measured. This variable picture requires that paired measures for ratio calculation be obtained with sufficient rapidity that they essentially represent the same cell configuration for noise and artefact to be minimized. When arc lamps are used for excitation, instability can also result in variable fluorescence emission energy. All of these factors will degrade results when manual, or slow automated, wavelength selection methods are used to obtain excitation or emission ratios, and the problems will be worse when measurements are made over a small region of the cell.

Where fura-2 can be used for measurements, systems with characteristics similar to the Spex CM systems (Spex Industries) are favourable since they provide rapid, continuous, alternation of excitation wavelengths and synchronous demultiplexing of the detected signal to produce the ratio values. Indo-1 appears to be a better indicator of [Ca^{2+}] in all plant cells we have examined. Our system for the measurement of [Ca^{2+}] using indo-1 (*Figure 2*) has been designed for measuring emission ratios with the previously mentioned concerns in mind. The filter wheel that provides for emission wavelength selection rotates at 1800 r.p.m. and each wavelength is sampled 8 msec apart. 405 nm and 490 nm are sampled to compute the [Ca^{2+}]-sensitive ratio value, and 450 nm, the isoemission point, is sampled to monitor the dye concentration in the cell. Sampling for 1 msec during the clear aperture of each filter gives a photon count that is equivalent to a snapshot of the instantaneous fluorescence at that wavlength. A selectable number of these readings can be accumulated, trading time-resolution for reduced noise. Examples of data acquired with this sytem from cells that were microinjected with indo-1, and later with various agents, are shown in *Figure 3* and *4a* and *b*. In these presentations 32 paired readings (approximately one second) were averaged to produce each ratio point.

2.5.2 Collection of spatially resolved data

Much of the discussion above has centred around the recording of kinetic data concerning the occurrence of calcium transients, with little attention to the spatial localization of these transients, except that an aperture has perhaps been placed over some region of the cell. Ultimately, one wants to also know where changes occur since they may be highly localized or the ion may be present as in a steep gradient between local sites of release and numerous sites of Ca sequestration (24). Measurements of whole cells or of substantial parts of cells will give results based on the average of the measured volume, and thus signals from very restricted sites of high [Ca^{2+}] would be lost.

Enormous attention is currently being directed to the problem of low-light-level image collection, and substantial progress is being made. Tsien and Poenie (36) provide a detailed discussion of an imaging technique using fura-2 which is basically an extension of the ratio detection method and is thus subject to all of the same problems mentioned in Section 2.5.1 and additional problems related to the acquisition of rapid, spatially resolved, ratiometric measurements. The low emission energy from the sample is now spread over a large number of 'pixels', each of which must achieve an adequate signal-to-noise ratio in order to obtain useful results. This low emission energy, even with high-sensitivity imaging devices, usually requires some degree of frame summing, or integration time, a change of filters, and acquisition of an image at the other wavelength, with a minimum of 1–5 sec required for a complete cycle (58, 59). We have had experience in the laboratory of Prof. David Gross, University of Massachusetts, using a cooled CCD imaging device, with fura-2 excitation provided by a pair of monochromators illuminated by arc lamps. While this apparatus can produce high quality images of weakly fluorescent objects, the time required to collect an image-pair is relatively long, amounting to a few seconds under the conditions used. For reasonable exposure times the excitation light is considerably more intense than is needed for PMT measurements of the fluorescence of whole cells, and pixel-binning (and decreased resolution) was required to obtain meaningful quantitative values. Since increased irradiation may bleach the indicator, alter important physiological functions, or damage the cell, one must weigh these disadvantages with the advantage of obtaining a spatial image of calcium transients. In our own work we have been repeatedly reminded of the difficulty of making [Ca^{2+}] measurements in the first place and we have attempted to keep the analysis as simple but as sensitive as possible. As a result we have much less direct experience with the problems of obtaining spatial information, although we fully recognize that ultimately this form of information will be of enormous value in helping us determine the significance of calcium transients as possible modulators of physiological and developmental processes. See also Chapter 1, this volume.

3. Calcium selective microelectrodes

Calcium selective microelectrodes can now be fabricated in tip sizes less than 1 μm, and these have been used successfully on plant cells. These electrodes, which are larger than electrical recording or microionophoresis electrodes, still have an extremely high resistance due to the ion carrier filling and time-constants (settling to 66% of final value for a step change of $[Ca^{2+}]$ occurs in one time-constant) of the order of 10–12 sec are typical. Their relatively large size will inevitably cause more damage to a cell and, as with any electrode that must be left in a cell, the possibility of drift in the micromanipulator, or sealing over the pipette tip will be potential difficulties. The current state of the art in calcium sensitive microelectrodes makes use of double-barrelled microelectrodes (60, 61) so that the reference electrode is also in the cytoplasm and membrane potential does not affect the recorded potential as it did with earlier techniques. The double-barrelled electrodes, however, are difficult to make and the large, double orifices create problems including possible calcium contamination (61). The useful life of these electrodes is from minutes to one day, especially when used on plant cells, and calibration must be performed before and after each measurement since it may change when the cell is impaled. More than any other technique, the calcium selective microelectrode is subject to reporting from a single point in the cytoplasm and, while this may be more or less the intent, it is quite possible that the tip will be located in a compartment that is not representative of the average cytosolic $[Ca^{2+}]$ in that region as would be indicated by even a small area measurement using optical probes.

In spite of these difficulties, Ca-selective microelectrodes have been used successfully for determinations of $[Ca^{2+}]$ in experiments with plant cells. Brownlee and Wood (62) used separate voltage- and calcium-sensitive electrodes to make measurements near the apex and sub-apical zones of *Fucus* rhizoids and showed a gradient of $[Ca^{2+}]$ across this region, ranging from 2.5 μM at the apex to 0.28 μM in sub-apical zones. They report being able to measure $[Ca^{2+}]$ for up to 30 min. Miller and Sanders (63) also used separate voltage- and $[Ca^{2+}]$-sensitive microelectrodes to follow photosynthetically-induced $[Ca^{2+}]$ reduction in the relatively large cells of *Nitellopsis* in experiments lasting approximately 10 min. Felle (60, 61) has developed and used the newer double-barrelled electrodes to make $[Ca^{2+}]$ measurements during experiments lasting for over 60 min in small cells of *Zea mays* coleoptiles and *Riccia fluitans* rhizoids. See also Chapter 3, this volume.

4. Summary

The measurement of $[Ca^{2+}]$ in most plant cells is still far from routine. Properties such as turgor pressure, cell-wall rigidity and associated enzymes, chlorophyll, prominent vacuoles, and blue-light receptors present special difficulties not usually encountered by those working with animal cells, and demand attention. The most suitable method will depend on the cell or tissue and the experimental protocol required. [Ca^{2+}]-sensitive electrodes, in spite of their technical challenge, are the only method where light is neither used for, nor interferes with, the detection of [Ca^{2+}] and this may dictate their use in some situations. Aequorin is very well suited to the larger cells and it does not enter vacuoles, but measurements of small areas of cells through the microscope push the limits of the technology. With fluorescence methods, excitation or even emitted light can present problems since blue-light is involved in a wide variety of physiological and developmental processes.

Many of the difficulties of [Ca^{2+}] measurement in single cells by optical methods can be circumvented by using cell populations to amplify the faint signals. But this approach has problems since it usually requires that one first prepare protoplasts to achieve dye loading. One then has a system whose biological response may be severely modified due to the removal of the cell wall. Populations of cells, in addition, will give average responses in which the character and time-course of single-cell responses are blurred. Even if populations of cells are loaded, the variable behaviour of the dyes in different cell types makes it necessary to observe the cells microscopically to confirm the distribution of the dye in the cells. We therefore encourage the analysis of single cells.

We feel that the best choice for initial consideration with most cells is optical detection, using one of the fluorescent indicators that permits ratiometric detection. High autofluorescence levels sometimes associated with plant cell-walls or cuticle can mark a system as unsuitable. It is imperative to note critically the behaviour of the dye in the cells, since sequestration to the vacuoles or organelles and dye loss are general but variable parameters. Even where the dye appears to remain in the cytoplasm, dye binding to components of the cell and resultant alterations in the Ca^{2+} sensitivity are hazards that have been noted and mandate *in vivo* calibrations where absolute quantitation is attempted. Careful consideration of the cell and experimental system and careful attention to testing and calibration will help to give an understandable result in a field which currently has many pitfalls, but which holds promise for the future.

Addendum: (Note added in proof)

While this chapter was in press an important development transpired that markedly improves the prospects of measuring and imaging free Ca^{2+} in plant cells. Molecular Probes Inc., Eugene, Oregon have synthesized a dextran-coupled derivative of fura-2. Although pressure injection is required to load the 10 000 mol. wt. dextran/fura-2 complex into cells, once introduced the dye remains almost completely within the free cytoplasmic compartment for several hours (over 4.5 h in stamen hair cells of *Tradescantia*). In addition, the dye appears to have no effect on division in stamen hair cells, or growth and streaming in pollen tubes of *Lilium*, the only systems tested thus far. Remarkably, the dye complex is extremely fluorescent, exceeding that of free fura-2, and thus only relatively small amounts need to be injected into the cell for ratio measurements. For experiments covering periods of 30 minutes or longer we believe that this new product surpasses all others on the market, and that it will permit a host of studies, which heretofore have been difficult or impossible, on the timing, magnitude, and location of free Ca^{2+} transients associated with signal transduction in plant cells.

Acknowledgements

We thank Susan Lancelle and Da-Hong Zhang for help in preparation of this manuscript. This work has been supported by grants from the National Science Foundation (DCB 90–04191, DCB 88–01750) and the United States Department of Agriculture (88–37261–3727).

References

1. Allen, D. G. and Blinks, J. R. (1980). In *Detection and measurement of free Ca^{2+} in cells* (ed. C. C. Ashley and A. K. Campbell), p. 159. Elsevier/North-Holland, Amsterdam.
2. Blinks, J. R., Wier, W. G., Hess, P., and Prendergast, F. G. (1982). *Progress in Biophysics and Molecular Biology*, **40**, 1.
3. Shimomura, O. and Johnson, F. H. (1980). In *Detection and measurement of free Ca^{2+} in cells* (ed. C. C. Ashley and A. K. Campbell), p. 73. Elsevier/North-Holland, Amsterdam.
4. Campbell, A. K., Lea, T. J., and Ashley, C. C. (1980). In *Detection and measurement of free Ca^{2+} in cells* (ed. C. C. Ashley and A. K. Campbell), p. 13. Elsevier/North-Holland, Amsterdam.
5. Cobbold, T. J. and Rink, T. J. (1987). *Biochemical Journal*, **248**, 313.
6. Gilroy, S., Hughes, W. A., and Trewavas, A. J. (1989). *Plant Physiology*, **90**, 482.
7. Baker, P. F., Knight, D. E., and Umbach, J. A. (1985). *Cell Calcium*, **6**, 5.
8. Williamson, R. E. and Ashley, C. C. (1982). *Nature*, **296**, 647.
9. Yoshimoto, Y., Iwamatsu, T., Hirano, K., and Hiramota, Y. (1986). *Development Growth, and Differentiation*, **28**, 583.
10. Jaffe, L. F. (1986). In *Calcium and the cell* (ed. D. Evered and J. Whelan), p. 71. Wiley, Chichester.
11. Whitaker, M. (1985). *Journal of Physiology*, **365**, 5P.
12. Inouye, S., Sakaki, Y., Goto, T., and Tsuji, F. I. (1986). *Biochemistry*, **25**, 8425.

13. Shimomura, O., Musicki, B., and Kishi, Y. (1989). *Biochemical Journal*, **261**, 913.
14. Scarpa, A. (1980). In *Detection and measurement of free Ca^{2+} in cells* (ed. C. C. Ashley and A. K. Campbell), p. 85. Elsevier/North-Holland, Amsterdam.
15. Thomas, M. V. (1982). *Techniques in calcium research*. Academic Press, London.
16. Ahmed, Z., Kragie, L., and Connor, J. A. (1980). *Biophysical Journal*, **32**, 907.
17. Thomas, M. V. (1980). In *Detection and measurement of free Ca^{2+} in cells* (ed. C. C. and Ashley and A. K. Campbell), p. 309. Elsevier/North-Holland, Amsterdam.
18. Palade, P. and Vergara, J. (1982). *Journal of General Physiology*, **79**, 679.
19. Connor, J. A. and Alkon, D. L. (1984). *Journal of Neurophysics*, **51**, 745.
20. Gorman, A. L. F., Levy, S., Nasi, E., and Tillotson, D. (1984). *Journal of Physiology*, **353**, 127.
21. Brown, J. E. and Pinto, L. H. (1980). In *Detection and measurement of free Ca^{2+} in cells* (ed. C. C. Ashley and A. K. Campbell), p. 339. Elsevier/North-Holland, Amsterdam.
22. Hepler, P. K. and Callaham, D. A. (1987). *Journal of Cell Biology*, **105**, 2137.
23. Chance, B., Lagallais, V., Sorge, J., and Graham, N. (1975). *Analytical Biochemistry*, **66**, 498.
24. Speksnijder, J. E., Miller, A. L., Weisenseel, M. H., Chen, T.-H., and Jaffe, L. F. (1989). *Proceedings of the National Academy of Sciences, USA*, **86**, 6607.
26. Tsien, R. Y., Pozzan, T., and Rink. T. J. (1982). *Journal of Cell Biology*, **94**, 325.
26. Grynkiewicz, G., Poenie, M., and Tsien, R. Y. (1985). *Journal of Biological Chemistry*, **260**, 3440.
27. Minta, A., Kao, J. P. Y., and Tsien, R. Y. (1989). *Journal of Biological Chemistry*, **264**, 8171.
28. Clayton, R. K. (1970). *Light and living matter: A guide to the study of photobiology*. McGraw-Hill, New York.
29. Tsien, R. Y. (1981). *Nature*, **290**, 527.
30. Tsien, R. Y., Pozzan, T., and Rink, T. J. (1982). *Nature*, **295**, 68.
31. Keith, C. H., Ratan, R., Maxfield, F. R., Bajer, A., and Shelanski, M. L. (1985). *Nature*, **316**, 848.
32. Kruskal, B. A., Keith, C. H., and Maxfield, F. R. (1984). *Journal of Cell Biology*, **99**, 1167.
33. Cork, R. J. (1986). *Plant, Cell and Environment*, **9**, 157.
34. Nobiling, R. and Reiss, H.-D. (1987). *Protoplasma*, **139**, 20.
35. Williams, D. A., Fogarty, K. E., Tsien, R. Y., and Faye, F. S. (1985). *Nature*, **318**, 558.
36. Tsien, R. Y. and Poenie, M. (1986). *Trends in Biological Sciences*, **11**, 450.
37. Brownlee, C., Wood, J. W., and Briton, D. (1987). *Protoplasma*, **140**, 118.
38. Clarkson, D. T., Brownlee, C., and Ayling, S. M. (1988). *Journal of Cell Science*, **91**, 71.
39. Brownlee, C. and Pulsford, A. L. (1988). *Journal of Cell Science*, **91**, 249.
40. McAinsh, M. R., Brownlee, C., and Hetherington, A. M. (1990). *Nature*, **343**, 186.
41. Bush, D. S. and Jones, R. L. (1987). *Cell Calcium*, **8**, 455.
42. Bush, D. S. and Jones, R. L. (1988). *European Journal of Cell Biology*, **46**, 466.
43. Lynch, J., Polito, V. S., and Lauchli, A. (1989). *Plant Physiology*, **90**, 1271.

44. Zhang, D. H., Callaham, D. A., and Hepler, P. K. (1990). *Journal of Cell Biology*. (In press.)
45. Kao, J. P. Y., Harootunian, A. T., and Tsien, R. Y. (1989). *Journal of Biological Chemistry*, **264**, 8179.
46. DiVirgilio, F., Steinberg, T. H., and Swanson, J. A. (1988). *Journal of Immunology*, **140**, 915.
47. Steinberg, T. H., Swanson, J. A., and Silverstein, S. C. (1988). *Journal of Cell Biology*, **107**, 887.
48. Malgaroli, A., Milani, D., Meldolesi, J., and Pozzan, T. (1987). *Journal of Cell Biology*, **105**, 2145.
49. Steinberg, T. H., Newman, A. S., Swanson, J. A., and Silverstein, S. C. (1987). *Journal of Cell Biology*, **105**, 2695.
50. Becker, P. L. and Fay, F. S. (1987). *American Journal of Physiology*, **253**, C613.
51. Konishi, M., Olson, A., Hollingworth, S., and Baylor, S. M. (1988). *Biophysical Journal*, **54**, 1089.
52. Baylor, S. M. and Hollingworth, S. (1988) *Journal of Physiology (London)*, **403**, 151.
53. Hepler, P. K. (1981). *European Journal of Cell Biology*, **26**, 102.
54. Hepler, P. K., Palevitz, B. A., Lancelle, S., McCauley, M., and Lichtscheidl, I. (1991). *Journal of Cell Science*. (In press.)
55. Gilroy, S., Hughes, W. A., and Trewavas, A. J. (1986). *FEBS Letters*, **199**, 217.
56. Gilroy, S., Hughes, W. A., and Trewavas, A. J. (1987). *FEBS Letters*, **212**, 133.
57. Purves, R. D. (1981). *Microelectrode methods for intracellular recording and ionophoresis*. Academic Press, London.
58. Connor, J. A. (1986). *Proceedings of the National Academy of Sciences, USA*, **83**, 6179.
59. Gross, D. (1990). In *Non-invasive techniques in cell biology* (ed. J. K. Foskett and S. Grinstein), p. 21, Wiley–Liss, New York.
60. Felle, H. (1988). *Planta*, **174**, 495.
61. Felle, H. (1988). *Planta*, **176**, 248.
62. Brownlee, C. and Wood, J. W. (1986). *Nature*, **320**, 624.
63. Miller, A. J. and Sanders, D. (1987). *Nature*, **326**, 397.

A1

Suppliers of specialist equipment

Amersham International, Lincoln Place, Green End, Aylesbury, Buckinghamshire, HP20 2TP, UK.

Analog Devices, PO Box 9106, Norwood, MA 02062–9106, USA, and Station Avenue, Walton-on-Thames, Surrey KT12 1PF, UK. [Amplifiers for ion-selective microelectrodes.]

Axon Instruments Inc., 1101 Chess Drive, Foster City, CA 94404, USA. [Patch and Voltage-clamp systems; acquisition hardware and soft-ware.]

BDH Chemicals Ltd, Broom Road, Parkstone, Poole, Dorset BH12 4NN, UK.

Beckman Instruments Inc., PO Box 10200, Palo Alto, CA 94304, USA.

Bethyl Laboratories Inc., Montgomery, Texas, USA.

Bio-Rad Laboratories, 1414 Harbor Way South, Richmond, CA 94804, USA.

Boehringer-Mannheim GmbH Biochemica, Postfach 310120, D-6800, Mannheim, Germany.

Cairn Research Ltd, Cairn House, Newnham, Sittingbourne, Kent ME9 0LQ, UK.

Calbiochem Corp., PO Box 12087, San Diego, CA 92112–4180, USA.

Cambridge Electronic Design (CED) Ltd, Science Park, Milton Road, Cambridge CB4 4FE, UK. [Patch clamp control and acquisition hardware and software.]

Clark Electromedical Instruments, PO Box 8, Pangbourne, Reading RG8 7HU, UK [Electrode glass, electrode holders, electrode pullers, and Ag–AgCl electrodes.]

Dagan Corporation, 2855 Park Avenue, Minneapolis, Minnesota 55407, USA. [Patch and voltage-clamp systems.]

Edmund Scientific, C916 Edscorp Bldg, Barrington, NJ 08007, USA.

Farnell, Canal Road, Leeds LS12 2TU, UK.

Fluka AG, CH-9470 Buchs, Switzerland. [Ca^{2+} ligand for microelectrodes.]

Micro-Instruments (Oxford) Ltd, 7 Little Clarendon Street, Oxford OX1 2HP, UK. [Micromanipulators.]

Micro Probes Inc., 4849 Pitchford Avenue, Eugene, Oregon 97402, USA.

Millipore, 80 Ashby Road, Bedford, MA 01730, USA.

Narishige USA Inc., 1 Plaza Road, Greenvale, NY 11548, USA. [Micromanipulators.]

Pel-Freeze Biological Inc., Rogers, Alabama, USA.

Perkin-Elmer Cetus, 761 Main Avenue, Norwalk, CT 06859, USA.

Pfaltz & Bauer, Stamford, CT 06902, USA.

Pharmacia-LKB, Bjorkgatan, 30, 751, 82, Uppsala, Sweden.

Promega Corporation, 2800 Woods Hollow Road, Maidson, WI 53711, USA.

Sachs & Neil, SUNY, Buffalo, NY, USA.

Sigma Chemical Co., PO Box 14508, St Louis, MO 63178, USA.

Spex Industries, 3880 Park Avenue, Edison, NJ 08820, USA.

Thomas Scientific, 99 High Hill Road, PO Box 99, Swedesboro, NJ 08085–0099, USA.

Thorn-EMI Electron Tubes Ltd, Bury Street, Ruislip, Middx., UK.

Whatman Ltd, Springfield Mill, Maidstone, Kent ME14 2LE, UK.

World Precision Instruments, 375 Quinipiac Avenue, New Haven, CT 06513, USA, and 15 Dudley Road, Hastings, E. Sussex TN35 5JP, UK. [Electrode glass, electrode holders, micromanipulators, and Ag–AgCl electrodes.]

Index